THEORETICAL STATISTICS

Theoretical Statistics

D.R. COX

D.V. HINKLEY

Imperial College, London

London

CHAPMAN AND HALL

First published 1974
by Chapman and Hall Ltd
11 New Fetter Lane, London EC4P 4EE
© 1974 D.R. Cox and D.V. Hinkley

Typeset by E.W.C. Wilkins Ltd, London
and printed in Great Britain
at the University Printing House, Cambridge

ISBN 0 412 12420 3

Distributed in the U.S.A.
by Halsted Press, a Division
of John Wiley & Sons, Inc., New York

TO

JOYCE AND BETTY

CONTENTS

CONTENTS

CONTENTS

PREFACE

This book is intended for statisticians wanting a fairly systematic development of the theory of statistics, placing main emphasis on general concepts rather than on mathematical rigour or on the detailed properties of particular techniques. The book is meant also to be suitable as a text-book for a second or third course in theoretical statistics. It is assumed that the reader is familiar with the main elementary statistical techniques and further has some appreciation of the way statistical methods are used in practice. Also knowledge of the elements of probability theory and of the standard special distributions is required, and we assume that the reader has studied separately the theory of the linear model. This is used repeatedly as an illustrative example, however.

We have reluctantly decided to exclude numerical examples and to attempt no detailed discussion of particular advanced techniques or of specific applications. To have covered these would have lengthened and changed the character of the book and to some extent would have broken the thread of the argument. However, in the training of students the working of set numerical exercises, the discussion of real applications and, if at all possible, involvement in current applied statistical work are of very great importance, so that this book is certainly not intended to be the basis of a self-contained course in statistics. To be quite specific, the more discursive parts of the book, for example on the usefulness and limitations of significance tests, will probably not be understandable without some experience of applications.

The mathematical level has been kept as elementary as possible, and many of the arguments are quite informal. The "theorem, proof" style of development has been avoided, and examples play a central role in the discussion. For instance, in the account of asymptotic theory in Chapter 9, we have tried to sketch the main results and to set out their usefulness and limitations. A careful account of the general theorems seems, however, more suited for a monograph than to a general book such as this.

Specialized topics such as time series analysis and multivariate analysis are mentioned incidentally but are not covered systematically.

A major challenge in writing a book on theoretical statistics is that of keeping a strong link with applications. This is to some extent in conflict with the essential need for idealization and simplification in presentation and it is too much to hope that a satisfactory compromise has been reached in this book. There is some discussion of the connexion between theory and applications in Chapter 1.

The book deals primarily with the theory of statistical methods for the interpretation of scientific and technological data. There are applications, however, where statistical data are used for more or less mechanical decision making, for example in automatic control mechanisms, industrial acceptance sampling and communication systems. An introduction to statistical decision theory is therefore included as a final chapter.

References in the text have been restricted largely to those giving direct clarification or expansion of the discussion. At the end of each chapter a few general references are given. These are intended partly to indicate further reading, partly to give some brief historical background and partly to give some of the main sources of the material in the text. We felt that a very extensive bibliography would be out of place in a general book such as this.

At the end of every chapter except the first, there is an outline of some of the topics and results that it has not been feasible to include in the text. These serve also as exercises, although as such the level of difficulty is very variable and in some cases much detailed work and reference to original sources will be necessary.

D.R. Cox
D.V. Hinkley
London, 1973

1 INTRODUCTION

1.1 Objectives of statistical analysis and theory

Statistical methods of analysis are intended to aid the interpretation of data that are subject to appreciable haphazard variability. The theory of statistics might then be expected to give a comprehensive basis for the analysis of such data, excluding only considerations specific to particular subject matter fields. In fact, however, the great majority of the theory, at any rate as presented in this book, is concerned with the following narrower matter.

There is chosen a family $\mathcal{F}$ of probability models, often completely specified except for a limited number of unknown parameters. From the data under analysis it is required to answer questions of one or both of the following types:

(a) Are the data consistent with the family $\mathcal{F}$?

(b) Assuming provisionally that the data are derived from one of the models in $\mathcal{F}$, what can be concluded about values of unknown parameters, or less commonly about the values of further observations drawn from the same model?

Problems (a) and (b) are related, but the distinction is a useful one. To a very large extent arithmetical rather than graphical methods of analysis are considered.

To illustrate the discussion consider the standard normal-theory model of simple linear regression. According to this, for data consisting of n pairs $(x_1, y_1), \ldots, (x_n, y_n)$, it is supposed that $y_1, \ldots, y_n$ correspond to random variables $Y_1, \ldots, Y_n$ independently normally distributed with constant variance σ^2 and with expected values

$$E(Y_j) = \gamma + \beta x_j \quad (j = 1, \ldots, n), \tag{1}$$

where γ and β are unknown parameters and $x_1, \ldots, x_n$ are regarded as known constants. This is a family $\mathcal{F}$ of models; a particular model is

1

obtained by specifying values for the parameters γ, β and σ^2. Often, but by no means always, primary interest would be in β.

The problem of type (a) would now be to examine the data for consistency with $\mathcal{F}$, some possibly important types of departure including non-linearity of the dependence of $E(Y_j)$ on x_j, non-constancy of var(Y_j), lack of independence of the different Y_j's and non-normality of the distribution of the Y_j's. For problems of type (b) it would be assumed provisionally that the data are indeed derived from one of the models in $\mathcal{F}$ and questions such as the following would be considered:

Within what limits is it reasonable to suppose the parameter β to lie?

Are the data consistent with a particular theoretical value for, say, the parameter β?

Let $Y^\dagger$ be a further observation assumed to be obtained from the same model and parameter values as the original data, but with

$$E(Y^\dagger) = \gamma + \beta x^\dagger.$$

Within what limits is $Y^\dagger$ expected to lie?

In the theoretical discussion, it will be usual to take a family of models as given. The task is to formulate meaningful and precise questions about such models and then to develop methods of analysis that will use the data to answer these questions in a sensitive way.

In more detail, in order to deal with the first of these questions just mentioned, the first step is to formulate a precise meaning for such limits for β; this is done in the concept of confidence limits (Section 7.2). Then, there usually being many ways of computing confidence limits, at least approximately, the next step is to define criteria giving the best such limits, or failing that, at least reasonably good confidence limits. Finally, general constructive procedures are required for computing good confidence limits for specific problems. This will lead to sensitive procedures for the analysis of data, assuming that the family $\mathcal{F}$ is well chosen. For example, in the special linear regression model (1), we would be led to a procedure for computing from data limits for, say, the parameter β which will in a certain sense give the most precise analysis of the data given the family $\mathcal{F}$.

In this, and corresponding more general theoretical discussions, much emphasis is placed on finding optimal or near optimal procedures within a family of models.

How does all this correspond with the problems of applied statistics? The following points are relevant:

(i) The choice of the family $\mathcal{F}$ of models is central. It serves first to define the primary quantities of interest, for example, possibly the parameter β in (1), and also any secondary aspects of the system necessary to complete the description. Some of the general considerations involved in the choice of $\mathcal{F}$ are considered in Section 1.2.

(ii) Except occasionally in very simple problems the initial choice of $\mathcal{F}$ will be made after preliminary and graphical analysis of the data. Furthermore, it will often be necessary to proceed iteratively. The results of analysis in terms of a model $\mathcal{F}$ may indicate a different family which may either be more realistic or may enable the conclusions to be expressed more incisively.

(iii) When a plausible and apparently well-fitting family is available it is attractive and sometimes important to use techniques of analysis that are optimal or nearly so, particularly when data are rather limited. However, criteria of optimality must always be viewed critically. Some are very convincing, others much less so; see, for example, Sections 5.2(iii) and 8.2. More importantly, it would be poor to use an analysis optimal for a family $\mathcal{F}$ if under an undetectably different family $\mathcal{F}'$ the same analysis is very inefficient. A procedure that is reasonably good for a broad range of assumptions is usually preferable to one that is optimal under very restrictive assumptions and poor otherwise. Thus it is essential to have techniques not only for obtaining optimal methods but also for assessing the performance of these and other methods of analysis under non-standard conditions.

(iv) At the end of an analysis, it is always wise to consider, even if only briefly and qualitatively, how the general conclusions would be affected by departures from the family $\mathcal{F}$. Often this is conveniently done by asking questions such as: how great would a particular type of departure from $\mathcal{F}$ have to be for the major conclusions of the analysis to be altered?

(v) It is important to adopt formulations such that the statistical analysis bears as directly as possible on the scientific or technological meaning of the data, the relation with previous similar work, the connexion with any theories that may be available and with any practical application that may be contemplated. Nevertheless, often the statistical analysis is just a preliminary to the discussion of the underlying meaning of the data.

Some of these points can be illustrated briefly by the regression problem outlined above.

The minimal preliminary analysis is a plot of the points (x_j, y_j) on a scatter diagram. This would indicate, for example, whether a transformation of the variables would be wise before analysing in terms of the model (1) and whether there are isolated very discrepant observations whose inclusion or exclusion needs special consideration. After an analysis in terms of (1), residuals, i.e. differences between observed values and estimated values using the model (1), would be calculated, and analysis of these, either graphically or arithmetically, might then suggest a different family of models. With a more complex situation, an initial family of models may be complicated and one might hope to be able to pass to a simpler family, for example one in which there were appreciably fewer unknown parameters. In the present case, however, any change is likely to be in the direction of a more complex model; for instance it may be appropriate to allow var(Y_j) to be a function of x_j.

It might be argued that by starting with a very complex model, some of the difficulties of (ii)–(iv) could be avoided. Thus one might set up a very general family of models involving additional unknown parameters describing, for example, the transformations of the x- and y-scales that are desirable, the change with x of var(Y), the lack of independence of different observations and non-normality of the distributions. The best fitting model within this very general family could then be estimated.

While this type of approach is sometimes useful, there are two serious objections to it. First, it would make very heavy going of the analysis even of sets of data of simple structure. A more fundamental reason that a single formulation of very complex models is not in general feasible is the following.

In designing an experiment or scheme of observations, it is important to review beforehand the possible types of observations that can occur, so that difficulties of analysis and ambiguities of interpretation can be avoided by appropriate design. Yet experience suggests that with extensive data there will always be important unanticipated features. An approach to the analysis of data that confines us to questions and a model laid down in advance would be seriously inhibiting. Any family of models is to be regarded as an essentially tentative basis for the analysis of the data.

The whole question of the formulation of families of models in the light of the data is difficult and we shall return to it from time to time.

To sum up, the problems we shall discuss are closely related to those of applied statistics but it is very important to ensure that the idealization which is inevitable in formulating theory is not allowed to mislead us in practical work.

1.2   Criteria for the choice of families of models

In the previous section the importance has been explained of choosing a family $\mathcal{F}$ of probabilistic models in terms of which questions are to be formulated and methods of statistical analysis derived. For the more elaborate parts of statistical theory we start from the representation of observations in terms of random variables and the idea that normally the parameters of the underlying probability distributions are the quantities of real interest. Yet this view needs to be treated with proper caution and should not be taken for granted. Where the data are obtained by the random sampling of a physically existing population, the parameters have a reasonably clear meaning as properties of the population. In other cases the probability distributions refer to what would happen if the experiment were repeated a large number of times under identical conditions; this is always to some extent hypothetical and with "unique" data, such as economic time series, repetition under identical conditions is entirely hypothetical. Nevertheless the introduction of probability distributions does seem a fruitful way of trying to separate the meaningful from the accidental features of data. Parameters are to be regarded sometimes as representing underlying properties of a random system and sometimes as giving concise descriptions of observations that would be obtained by repetition under the same conditions.

The methods of most practical value are those that combine simple description of the data with efficient assessment of the information available about unknown parameters.

It is hard to lay down precise rules for the choice of the family of models, but we now list briefly some of the considerations involved. These include:

(a) The family should if possible usually establish a link with any theoretical knowledge about the system and with previous experimental work.

(b) There should be consistency with known limiting behaviour. For example, it may be known or strongly suspected that a curve approaches an asymptote or passes through some fixed point such as the origin. Then, even though this limiting behaviour may be far from the region directly covered by the data, it will often be wise to use a family of models consistent with the limiting behaviour or, at least, to use a family reducing to the limiting behaviour for special parameter values.

(c) So far as possible the models should be put in a form such that the different parameters in the model have individually clear-cut general interpretations.

(d) Further, a description of the data containing few parameters is preferable. This may involve looking at a number of different families to find the most parsimonious representation. There is some chance of conflict with requirements (a) and (b). Indeed in some cases two different analyses may be desirable, one bringing out the link with some relevant theory, the other expressing things in their most economical form.

(e) It is, of course, desirable that the statistical theory associated with the family of models should be as simple as possible.

A fairly recent trend in statistical work places some emphasis on the construction of special models, often by constructing stochastic processes representing the system under investigation. To be at all realistic these often have to be extremely complex. Indeed they may contain so many adjustable parameters that little can be learnt by fitting them to any but extremely extensive data. It is therefore worth stressing that very simple theoretical analyses can be valuable as a basis for statistical analysis. We then choose a family reducing to the theoretical model for special values of certain unknown parameters. In that way we can find how much of the variation present can be accounted for by the simple theory and also characterize the departures from the theory that may be present. It is not at all necessary for a theory to be a perfect fit for it to be a useful tool in analysis.

It will be seen later that the statistical theory of a family is simplified appreciably whenever the family is of the exponential type; see Section 2.2(vi). This provides a fairly flexible set of models and it will often be best to start with a model of this type; for example, nearly all the procedures of multivariate analysis are closely associated with the family of multivariate normal models. The effect of departures can then be considered. It is, however, too much to hope that a reasonably adequate model of exponential type can always be found without violating some other more important requirement.

The requirement of simplicity of statistical analysis may well be in conflict with one or more of the other requirements, for example (a) and (b). Thus the widely used procedure of fitting polynomial curves and surfaces leads to representations whose limiting behaviour is usually unreasonable. This may not matter where the polynomials are used in a limited way for interpolation but seems often likely to be a severe restriction on their usefulness in the interpretation of data; exceptions are when a small quadratic term is used to assess the direction and magnitude of the departure from an expected linear relationship, and when local behaviour in the neighbourhood of a stationary value is under investigation.

1.3   The analysis of complex responses

A widely occurring problem in applied statistics, particularly where automatic recording equipment is used, is the analysis of data in which the response of each individual (subject, animal, batch of material, etc.) is complex and, for example, may consist of a long sequence of observations possibly of several types. Thus in some types of psychological experiment, there will be for each subject a response consisting of a long sequence of successes and failures at some task. In all there may be several such trials for each of many subjects, the whole covering a number of experimental treatments.

It may sometimes be feasible to formulate a single family of models that will embrace all aspects of the data, but more commonly it will be wise to proceed in several steps. First we try to summarize each complex response in a small number of quantities. These may be derived from some formal theoretical analysis, for instance from a simplified mathematical theory of the phenomenon. Or again the formal techniques of time series analysis and multivariate analysis are often guides to the choice of summarizing values. Finally such quantities may be derived from a qualitative inspection and graphical analysis of a sample of individual responses. That is, we construct a few quantities thought to summarize interesting features of the response, such as, for example, average level, trend, amount and nature of local variability. With very complex data it may take much investigation to find the best quantities for further analysis.

Then at the second stage of the analysis the output of the first stage is used as the input data to estimate treatment effects, etc., often by a linear model. There is the following important implication for the discussion in the rest of this book: the observations for analysis and for which we set up a probability model may be the original data or they may be quantities derived from the preliminary analysis of more complex data. This fact gives the relatively simple models that we shall discuss much wider usefulness than would otherwise be the case.

1.4   Plan of the book

Chapter 2 introduces some fundamental ideas about likelihood and sufficiency which are central to much of the subsequent work. The second part of the chapter, which can be omitted on a first reading, compares some of the broad general approaches to statistical inference.

Chapters 3–7, which in some ways are the core of the book, deal with significance testing and with interval estimation. Chapter 8 is concerned with point estimation. Chapter 9 is on asymptotic theory. For large samples this gives approximate solutions to problems for which no simple "exact" solutions are possible by the techniques of Chapters 3–7.

Finally the last two chapters deal with procedures based on the availability of a prior probability distribution for the unknown parameters of the family $\mathcal{F}$, and with decision theory.

To some extent, the chapters can be taken in a different order. For example, some readers may want to take the rather introductory Chapters 10 and 11 relatively early.

It would be very restricting to use throughout a single completely uniform notation and it is virtually impossible to avoid use of the same letter for different things in different contexts. However, as far as is practicable, the following conventions have been followed.

Random variables representing observations or functions calculated from observations are denoted by capital letters such as X, Y and Z. For example, Y denotes the set of random variables $Y_1, \ldots, Y_n$ considered as a column vector. On the whole Y is reserved for the observation under analysis. To distinguish between random variables and the particular realizations of them forming the data under analysis, the observed values will be denoted by corresponding lower case letters, for example y. It is, however, occasionally convenient to be inconsistent and to retain the capital letters for the observations, where there is no danger of confusion.

Random variables that are not observable, for example "errors", are sometimes denoted by the Greek letters ϵ and η. The Greek letter α is almost exclusively reserved for certain probabilities arising in the study of tests and interval estimation.

Fixed constants are denoted by lower case letters $a, b, c, \ldots$

Unknown parameters are usually denoted by the Greek letters θ, $\phi, \ldots$ They may be scalars or vectors depending on the context; when vectors they are to be considered as column vectors. In dealing with particular situations standard symbols are used; for example, the mean and variance of a normal distribution are denoted by μ and σ^2.

For a random variable, say U, the cumulative distribution function is denoted by $F_U(x)$, or by $F_U(x ; \theta)$ if it is required to stress dependence on a parameter θ. That is, $F_U(x) = \text{pr}(U \leqslant x)$, where $\text{pr}(A)$ denotes the probability of the event A. The corresponding probability density function, a term used for both continuous and discrete random variables, is denoted by $f_U(x)$ or $f_U(x ; \theta)$. Thus if U has a Poisson distribution of mean μ, then

$$f_U(x ; \mu) = \frac{e^{-\mu}\mu^x}{x!} \quad (x = 0, 1, 2, \ldots),$$

whereas if V has a normal distribution of mean μ and variance σ^2, written $N(\mu, \sigma^2)$, then

$$f_V(x ; \mu, \sigma) = \frac{1}{\sqrt{(2\pi)}\sigma} \exp\left\{-\frac{(x-\mu)^2}{2\sigma^2}\right\}.$$

The p dimensional multivariate normal distribution of mean the column vector μ and covariance matrix Σ is denoted by $MN_p(\mu, \Sigma)$. A direct extension of the notation for densities is used for joint and conditional densities.

Standard notation is used for expectations, variances and covariances. Where it is required to show the parameter value under which, say, an expectation is calculated, we write, for example, $E(Y ; \theta)$.

Bold fount is restricted to matrices; the transpose of a matrix or vector is denoted by a superscript T.

An estimate of a parameter θ will sometimes be denoted $\tilde{\theta}$; the notation $\hat{\theta}$ will be restricted to maximum likelihood and least squares estimates. All such estimates $\tilde{\theta}$ are functions of the data, i.e. correspond to random variables. Therefore if the general conventions were to be followed a capital letter would be used for the random variable, but this is often inconvenient. Sums of squares and mean squares arising in linear model analyses have been denoted SS and MS, sometimes with an appropriate suffix, it being clear from the context whether random variables or observed values are involved.

Asterisks are very largely reserved for tabulated constants arising in significance tests, etc. By convention we use values corresponding to an upper tail area. For example, k_α^* denotes the upper α point of the standard normal distribution, i.e. $\Phi(k_\alpha^*) = 1 - \alpha$, where $\Phi(.)$ is the standard normal integral.

Abbreviations have been kept to a minimum. The only ones widely used are i.i.d. (independent and identically distributed), p.d.f. (probability density function), referring to both continuous and discrete random variables, and m.l.e. (maximum likelihood estimate).

Bibliographic notes

Discussion of the relation between theory and application is most often found in reports of general lectures. Presidential addresses to the Royal Statistical Society by Fisher (1953), Pearson (1956), Kendall (1961), Bartlett (1967), Yates (1968) and Barnard (1972) all in part bear on this issue; see also Kempthorne (1966). Neyman (1960) emphasizes the role of special stochastic models. Tukey and Wilk (1966) have argued in favour of an increased emphasis on graphical methods and on descriptive statistics generally, i.e. on methods which are not based on explicit probabilistic arguments.

Of the many books on statistical methods, that of Daniel and Wood (1971) particularly well illustrates the interplay between the analysis of data and the choice of model. Of those on theory, the later chapters of Kempthorne and Folks (1971) give an introduction to many of the topics of the present book from a viewpoint similar to that taken here. Silvey (1970) gives a concise mathematical introduction. The comprehensive book of Rao (1973) emphasizes distribution theory and the linear model. The three volumes of Kendall and Stuart (1967—69) are particularly valuable in providing an introduction to a wide range of special topics.

A connected account of the history of statistical inference is not available at the time of writing. A collection of papers on various historical topics is edited by Pearson and Kendall (1970); there are a number of biographical articles in *International Encyclopedia of Social Sciences* (Sills, 1968).

2 SOME GENERAL CONCEPTS

2.1 The likelihood

(i) Definition

Let observations $y = (y_1, \ldots, y_n)$ be realised values of random variables $Y = (Y_1, \ldots, Y_n)$ and suppose that an analysis of y is to be based on the provisional assumption that the probability density function (p.d.f.) of Y belongs to some given family $\mathcal{F}$. It is not known, however, which particular p.d.f. in $\mathcal{F}$ is the true one. Any particular p.d.f. $f(y)$ specifies how the density varies across the sample space of possible y values. Often, it is useful to invert this property, and to examine how the density changes at the particular observed value y as we consider different possible functions in $\mathcal{F}$. This results in a comparison between possible densities based on ability to "explain" y, and to emphasise this we define the *likelihood* of $f(.)$ at a particular y by

$$\text{lik}\{f(.);y\} = f(y). \tag{1}$$

Usually it is convenient to work with the natural logarithm, denoted by $l\{f(.);y\}$ and called the *log likelihood*

$$l\{f(.);y\} = \log f(y). \tag{2}$$

In most applications, we consider families $\mathcal{F}$ in which the functional form of the p.d.f. is specified but in which a finite number of unknown parameters $\theta = (\theta_1, \ldots, \theta_q)$ are unknown. Then the p.d.f. of Y for given θ can be written $f(y ; \theta)$ or, where desirable, as $f_Y(y ; \theta)$. The set of allowable values for θ, denoted by Ω, or sometimes by Ω_θ, is called the *parameter space*. Then the likelihood (1) is a function of θ and we can write, always for $\theta \in \Omega$,

$$\text{lik}(\theta ; y) = f(y ; \theta), \quad l(\theta ; y) = \log f(y ; \theta). \tag{3}$$

If it is required to stress the particular random variable for which the likelihood is calculated we add a suffix, as in $\text{lik}_Y(\theta\,;y)$. It is crucial that in these definitions θ is the argument of a function and can take any value in Ω. A very precise notation would distinguish between this use of θ and the particular value that happens to be true, but fortunately this is unnecessary, at least in the present chapter.

Suppose that Y is a continuous vector random variable and that we consider a one-one transformation to a new vector random variable Z with non-vanishing Jacobian $\partial y/\partial z$. Then to any $f_Y(.)$ in $\mathcal{F}$ there corresponds a p.d.f. for Z given by

$$f_Z(z) \;=\; f_Y(y)\left|\frac{\partial y}{\partial z}\right|,$$

where z is the transformed value of y. Thus, taking the parametric case for simplicity, we see that the likelihood function based on the observed value z of Z is

$$\text{lik}_Z(\theta\,;z) \;=\; \text{lik}_Y(\theta\,;y)\left|\frac{\partial y}{\partial z}\right|. \tag{4}$$

This result suggests that if we are interested in comparing two possible values θ_1 and θ_2 of θ in the light of the data and wish to use the likelihood, it is ratios of likelihood values, rather than, say, differences, that are relevant. For such a comparison cannot reasonably depend on the use of y rather than z.

Very commonly the component random variables $Y_1, \ldots, Y_n$ are mutually independent for all densities in $\mathcal{F}$. Then we can write

$$f_Y(y) \;=\; \prod_{j=1}^{n} f_{Y_j}(y_j) \;=\; \prod_{j=1}^{n} f_j(y_j),$$

say, and in the parametric case we have for the log likelihood

$$l_Y(\theta\,;y) \;=\; \sum_{j=1}^{n} \log f_j(y_j\,;\theta) \;=\; \sum_{j=1}^{n} l_j(\theta\,;y_j). \tag{5}$$

When the densities $f_j(y)$ are identical, we unambiguously write $f(y)$.

(ii) Some examples
The following examples give some instances of the calculation of likelihoods and of particular results that will be required later.

Example 2.1. Bernoulli trials. Consider n independent binary observations, i.e. the jth observation is either a "success", $y_j = 1$, or a "failure", $y_j = 0$, the probability θ of success being the same for all trials ($j = 1, \ldots, n$). The observations $y = (y_1, \ldots, y_n)$ then form a sequence of n ones and zeroes and the probability of any particular sequence is a product of terms θ and $1 - \theta$, there being a θ for every one and a $1 - \theta$ for every zero. Thus

$$\text{lik}_Y(\theta ; y) = \theta^r (1 - \theta)^{n-r}, \tag{6}$$

where $r = \Sigma y_j$ is the number of ones in the observed y. To complete the specification we must give the parameter space which would usually, but not necessarily, be the closed interval $0 \leqslant \theta \leqslant 1$.

Example 2.2. Number of successes in n Bernoulli trials. Suppose that we have exactly the situation of the previous example, except that instead of observing the sequence y we observe only the total number r of successes. We represent this by a random variable R having a binomial distribution. Then

$$\text{lik}_R(\theta ; r) = \binom{n}{r} \theta^r (1 - \theta)^{n-r}. \tag{7}$$

Note that if we are interested in the ratio of likelihoods at say θ_1 and θ_2, then (7) and (6) are equivalent.

Example 2.3. Inverse Bernoulli sampling. Suppose that again we have Bernoulli trials but that new trials continue until a preassigned number r of successes has been obtained and that the total number n of trials necessary to achieve this is observed. Then n is the observed value of a random variable N having a negative binomial distribution and

$$\text{lik}_N(\theta ; n) = \binom{n-1}{r-1} \theta^r (1 - \theta)^{n-r}. \tag{8}$$

Again this is equivalent to (6) and (7) in the sense explained in Example 2.2. Different random systems are, however, involved in the three cases.

Example 2.4. Normal-theory linear model. Suppose that the observations y, considered as an $n \times 1$ column vector, form a realization of the vector random variable Y with $E(Y) = \mathbf{x}\beta$, where $\mathbf{x}$ is a known $n \times q_x$ matrix of rank $q_x \leqslant n$, and β is a $q_x \times 1$ column vector of unknown parameters. Suppose also that $Y_1, \ldots, Y_n$ are independently normally distributed with unknown variance σ^2. Then

$$\text{lik}_Y(\beta, \sigma^2; y) = (2\pi)^{-\frac{1}{2}n}\sigma^{-n}\exp\left\{-\frac{(y - x\beta)^T(y - x\beta)}{2\sigma^2}\right\}. \quad (9)$$

We define the residual sum of squares SS_{res} and least squares estimates $\hat{\beta}$ in the usual way by

$$(x^Tx)\hat{\beta} = x^Ty, SS_{res} = (y - x\hat{\beta})^T(y - x\hat{\beta}).$$

Then it is easily shown that

$$\text{lik}_Y(\beta, \sigma^2; y) = (2\pi)^{-\frac{1}{2}n}\sigma^{-n}\exp\left\{-\frac{SS_{res}}{2\sigma^2} - \frac{(\hat{\beta} - \beta)^Tx^Tx(\hat{\beta} - \beta)}{2\sigma^2}\right\}. \quad (10)$$

Thus the likelihood depends on y only through SS_{res} and $\hat{\beta}$. Note that if $q_x = n$, then $SS_{res} = 0$ and the likelihood is unbounded at $\sigma = 0$.

The result (10) covers in particular the special case when $Y_1, \ldots, Y_n$ are i.i.d. in a normal distribution of unknown mean μ and unknown variance σ^2, $N(\mu, \sigma^2)$, say. Then (10) is easily seen to simplify to

$$\text{lik}_Y(\mu, \sigma^2; y) = (2\pi)^{-\frac{1}{2}n}\sigma^{-n}\exp\left\{-\frac{\Sigma(y_j - \bar{y}.)^2 + n(\bar{y}. - \mu)^2}{2\sigma^2}\right\}, \quad (11)$$

where $\bar{y}. = \Sigma y_j/n$.

Example 2.5. Discrete time stochastic process. We can always write the joint probability density of random variables $Y_1, \ldots, Y_n$ in the form

$$f_Y(y) = f_{Y_1}(y_1)f_{Y_2|Y_1}(y_2|y_1)f_{Y_3|Y_2, Y_1}(y_3|y_2, y_1) \cdots$$
$$f_{Y_n|Y_{n-1}, \ldots, Y_1}(y_n|y_{n-1}, \ldots, y_1).$$

This is particularly useful for processes developing in time and in particular for Markov processes for which

$$f_{Y_r|Y_{r-1}, \ldots, Y_1}(y_r|y_{r-1}, \ldots, y_1) = f_{Y_r|Y_{r-1}}(y_r|y_{r-1}),$$

so that for such processes

$$f_Y(y) = f_{Y_1}(y_1)\prod_{j=2}^{n} f_{Y_j|Y_{j-1}}(y_j|y_{j-1}). \quad (12)$$

In particular consider the two-state Markov chain with transition matrix

$$\begin{bmatrix} \theta_{00} & \theta_{01} \\ \theta_{10} & \theta_{11} \end{bmatrix}.$$

If the state of the system is assumed known at time zero all the terms in (12) are given by one of the θ_{rs}'s, so that

$$\text{lik}(\theta\,;y) = \prod_{r,s} \theta_{rs}^{m_{rs}}, \tag{13}$$

where the elements of the matrix $((m_{rs}))$ give the number of one-step transitions from state r to state s. Note that the result (13) applies directly to all discrete state Markov chains with stationary transition matrices.

Example 2.6. Time-dependent Poisson process. To calculate the likelihood for a stochastic process in continuous time involves in principle a new idea in that we are no longer dealing with a finite number or even with a countable number of random variables. However, if we consider a suitable limiting process a likelihood can usually be calculated. Consider a non-stationary Poisson process of rate $\rho(t)$ observed for the time interval $[0, t_0)$. Let events be observed at times $y_1, \ldots, y_n$. Divide the interval $[0, t_0)$ into a large number m of subintervals each of length h so that $mh = t_0$, and denote these intervals by $[a_j, a_j + h)$ for $j = 1, \ldots, m$. Then by the defining properties of the Poisson process, the subinterval $[a_j, a_j + h)$ contributes a factor $\rho(a_j)h + o(h) = \rho(y_i)h + o(h)$ to the likelihood if $a_j \leqslant y_i < a_j + h$ for some i, whereas if an event did not occur in $[a_j, a_j + h)$ the contribution is $1 - \rho(a_j)h + o(h)$. Probabilities referring to disjoint intervals are independent, so that the likelihood is

$$\prod_{i=1}^{n} \{\rho(y_i)h + o(h)\}\ \prod_{j}^{*} \{1 - \rho(a_j)h + o(h)\}, \tag{14}$$

where $\prod^{*}$ is the product over all j such that $[a_j, a_j + h)$ contains none of $y_1, \ldots, y_n$. As $h \to 0$ this second product tends to

$$\exp\left\{-\int_0^{t_0} \rho(u)du\right\}.$$

If we omit the factor h^n in (14), the omission corresponding to the conversion from a probability to a probability density, it follows that we may take

$$\text{lik}\{\rho(t)\,;y_1, \ldots, y_n\} = \left\{\prod_{j=1}^{n} \rho(y_j)\right\}\ \exp\left\{-\int_0^{t_0} \rho(u)du\right\}. \tag{15}$$

In particular if $\rho(t) = \alpha e^{\beta t}$, the likelihood is

$$\text{lik}(\alpha, \beta \,; y_1, \ldots, y_n) = \exp\left\{ n \log \alpha + \beta \Sigma y_j - \frac{\alpha}{\beta}(e^{\beta t_0} - 1)\right\}, \qquad (16)$$

and in the stationary Poisson case with $\beta = 0$

$$\text{lik}(\alpha \,; y_1, \ldots, y_n) = \alpha^n e^{-\alpha t_0}. \qquad (17)$$

Example 2.7. A likelihood that cannot be specified simply. For most probability models the likelihood can be written down immediately, once the model is properly specified. This is not always the case, however, as the following example shows. Suppose that a series of point events is observed in continuous time and that the model is as follows. Points with integer coordinates are displaced by random amounts that are i.i.d. in $N(0, \sigma^2)$. Only the displaced points are observed; the corresponding order of the originating events is not known. The likelihood can then not be written down in a useful form as a function of σ, especially when σ is large compared with one.

(iii) Mathematical difficulties

There are some mathematical difficulties in a general definition of likelihood for continuous random variables arising from the non-uniqueness of probability density functions. These can be changed on sets of measure zero without changing the probability distribution, and hence likelihood also is in a sense not uniquely defined. However in particular applications there is a "regular" form for the density and it is sensible to define likelihood in terms of this. While a measure-theoretic treatment is possible the fact that all observations are rounded, i.e. essentially discrete, justifies the use of the "regular" version. Indeed in a few cases it will be crucial to remember that continuous distributions are used only as approximations. Similar remarks apply to the likelihood for stochastic processes in continuous time.

(iv) Extended definitions of likelihood

In the definitions (1) and (3) of likelihood, we take the p.d.f. of all the random variables representing the data. Sometimes in dealing with complex problems it is useful to apply a one-one transformation of Y into a new vector random variable which is partitioned into two parts V and W. Provided that the transformation does not depend on the unknown parameters, and this we assume, we can transform the data y into (v, w).

The function of the unknown parameters obtained by considering the p.d.f. of V at the observed value v, i.e.

$$\text{lik}_V(\theta ; v) = f_V(v ; \theta) \tag{18}$$

is called the *marginal likelihood* for the original problem, based on V. It is the likelihood that we would get if v alone were observed. Again, in some situations we may work with the distribution of W conditionally on $V = v$, i.e. define

$$\text{lik}_{W|V}(\theta ; w|v) = f_{W|V}(w|v ; \theta), \tag{19}$$

which we call a *conditional likelihood* for the original problem.

We consider (18) and (19) only in order to obtain functions simpler than $\text{lik}_Y(\theta ; y)$. While we could consider (18) and (19) for any convenient V and W, there would in general be a loss of information relevant to θ in so doing. We would like to use (18) or (19) only when, for the particular purpose intended, all or nearly all of the information is retained. Unfortunately it is difficult to express this precisely and for that reason we shall not make extensive direct use of marginal and conditional likelihoods, although they will be implicit in much of the discussion of Chapter 5. An example will illustrate the possible gain of simplicity.

Example 2.8. One-way analysis of variance. Consider data represented by random variables $(Y_{11}, Y_{12}; Y_{21}, Y_{22}; \ldots; Y_{m1}, Y_{m2})$ that are independently normally distributed with variance σ^2 and $E(Y_{jk}) = \mu_j$ $(j = 1, \ldots, m)$. That is, there are m pairs of observations, each pair having a different mean. The following discussion extends immediately if there are more than two observations in each group. The unknown parameter is $\theta = (\mu_1, \ldots, \mu_m, \sigma^2)$, and the likelihood of the full data is

$$\text{lik}_Y(\theta ; y) = (2\pi\sigma^2)^{-m} \exp\{-\Sigma\Sigma(y_{jk} - \mu_j)^2/(2\sigma^2)\}$$
$$= (2\pi\sigma^2)^{-m} \exp\{-\Sigma(\bar{y}_{j.} - \mu_j)^2/\sigma^2\} \exp\{-\Sigma\Sigma(y_{jk} - \bar{y}_{j.})^2/(2\sigma^2)\}, \tag{20}$$

where $\bar{y}_{j.}$ is the mean of the jth pair; note that $\Sigma\Sigma(y_{jk} - \bar{y}_{j.})^2 = \text{ss}_w$ is the usual within-groups sum of squares. It is tempting to conclude that, although σ occurs throughout (20), the information about σ is largely contained in the final factor. Direct examination of (20) is, however, difficult especially for large m, because it is a function of $m + 1$ parameters.

The situation is clarified somewhat by considering a marginal likelihood. Introduce the orthogonal transformation

$$v_j = (y_{j1} - y_{j2})/\sqrt{2}, \quad w_j = (y_{j1} + y_{j2})/\sqrt{2} \quad (j = 1, \ldots, m).$$

The corresponding random variables are independently normally distributed with variance σ^2 because of the orthogonality of the transformation. Further, $V_1, \ldots, V_m$ are i.i.d. in $N(0, \sigma^2)$ and hence the marginal likelihood based on V is

$$\mathrm{lik}_V(\sigma^2 ; v) = (2\pi\sigma^2)^{-\frac{1}{2}m} \exp\left(-\frac{\Sigma v_j^2}{2\sigma^2}\right) = (2\pi\sigma^2)^{-\frac{1}{2}m} \exp\left(-\frac{\mathrm{SS}_w}{2\sigma^2}\right).$$
(21)

Note that only part of the leading factor in (20) appears in (21). Consideration of the marginal likelihood has replaced a problem with $m + 1$ unknown parameters by a problem with a single parameter. In this particular case, because V and W are independently distributed, (21) can be regarded equally as the conditional likelihood based on V given $W = w$.

If we are concerned solely with σ, and $\mu_1, \ldots, \mu_m$ are unknown and arbitrary, use of (21) is certainly convenient. Is there, however, any loss of information about σ when w is ignored? The distribution of W, a vector of m components, involves all $m + 1$ unknown parameters and it is plausible that, not knowing $\mu_1, \ldots, \mu_m$, we cannot extract any information about σ from w. It is difficult to make this notion precise; the topic is mentioned again in Section 9.2.

2.2   Sufficient statistics

(i) Definition
Suppose that observations $y = (y_1, \ldots, y_n)$ form a realization of a random variable Y and that a family $\mathcal{F}$ of possible distributions is specified. A *statistic* is a function $T = t(Y)$; in general T is a vector. Corresponding to the random variable T is an observed value $t = t(y)$. Note that the distinction between a statistic and an estimate is that the former is not necessarily calculated with the objective of being close to a meaningful parameter.

A statistic S is said to be *sufficient* for the family $\mathcal{F}$ if the conditional density

$$f_{Y|S}(y|s)$$

is the same for all the distributions in $\mathcal{F}$. In a parametric case this means that

$$f_{Y|S}(y|s\,;\theta) \tag{22}$$

does not depend upon θ, $\theta \in \Omega$. For the reason explained in Section 2.1 (iii), possible non-uniqueness of the conditional density for continuous random variables need not worry us.

Note that if S is sufficient, so is any one-one function of S.

Example 2.9. Poisson distribution. Suppose that $\mathcal{F}$ specifies that $Y_1, \ldots, Y_n$ are i.i.d. in a Poisson distribution of mean μ. Then $S = Y_1 + \ldots + Y_n$ is sufficient. To prove this, we calculate the conditional distribution (22). To obtain this, note that

$$f_Y(y\,;\mu) = \prod_{j=1}^{n} \frac{e^{-\mu}\mu^{y_j}}{y_j!} = \frac{e^{-n\mu}\mu^{\Sigma y_j}}{\Pi y_j!},$$

whereas

$$f_S(s\,;\mu) = \frac{e^{-n\mu}(n\mu)^s}{s!}.$$

It follows that

$$f_{Y|S}(y|s\,;\mu) = \begin{cases} \dfrac{(\Sigma y_j)!}{\Pi y_j!}\,\dfrac{1}{n^{\Sigma y_j}} & (\Sigma y_j = s), \\[2mm] 0 & (\Sigma y_j \neq s). \end{cases} \tag{23}$$

Because this does not involve μ, the sufficiency is proved. Note that (23) is the multinomial distribution with $s = \Sigma y_j$ trials and probabilities $(1/n, \ldots, 1/n)$, a result of practical value. The sufficient statistic could equally well be taken as the mean $\Sigma Y_j/n$.

Example 2.10. Uniform distribution. Suppose that $Y_1, \ldots, Y_n$ are i.i.d. in a uniform distribution over $(0, \theta)$, and that $S = \max(Y_j)$. Now

$$f_Y(y\,;\theta) = \begin{cases} 1/\theta^n & (s < \theta), \\ 0 & \text{otherwise,} \end{cases}$$

$$f_S(s;\theta) = \begin{cases} ns^{n-1}/\theta^n & (s<\theta), \\ 0 & \text{otherwise,} \end{cases}$$

it being assumed that all the observations are non-negative. Thus

$$f_{Y|S}(y|s;\theta) = \begin{cases} \dfrac{1}{ns^{n-1}} & (\max(y_j) = s<\theta), \\ 0 & \text{otherwise.} \end{cases} \tag{24}$$

At first sight (24) does depend on θ. However, the restriction $s<\theta$ in (24) is automatically satisfied for all values of θ for which the conditional distribution is defined.

Example 2.11. Order statistics. Suppose that $Y_1, \ldots, Y_n$ are i.i.d. in a continuous distribution and that $\mathcal{F}$ consists of all densities of continuous random variables. Let S be the set of order statistics $(Y_{(1)}, \ldots, Y_{(n)})$, where $Y_{(1)} \leqslant Y_{(2)} \leqslant \ldots \leqslant Y_{(n)}$. The main distributional properties of order statistics are summarized in Appendix 2. Now

$$f_Y(y) = \prod_{j=1}^{n} f(y_j),$$

$$f_S(s) = \begin{cases} n! \prod_{j=1}^{n} f(s_j) & (s_1 \leqslant s_2 \leqslant \ldots \leqslant s_n), \\ 0 & \text{otherwise.} \end{cases}$$

Thus the conditional density of Y given $S = s$ is

$$f_{Y|S}(y|s) = \begin{cases} 1/n! & \text{if } \{y\} \text{ is a permutation of } \{s\}, \\ 0 & \text{otherwise.} \end{cases} \tag{25}$$

Because this does not depend on the density $f(.)$, the sufficiency of S is proved. Note that (25) expresses nothing more than the obvious fact that, given the ordered values, all permutations of them are equally likely under $\mathcal{F}$.

Example 2.12. The likelihood ratio. Suppose that there are just two possible densities for the vector Y, namely $f_0(y)$ and $f_1(y)$. Let $S = f_1(Y)/f_0(Y)$. That is, two points with the same value of S have the same ratio of likelihoods. We prove the sufficiency of S, taking the discrete case for simplicity. Let Σ_s^* denote summation over all y such that $f_1(y)/f_0(y) = s$. Then

$$\text{pr}\left\{Y = y \left| \frac{f_1(Y)}{f_0(Y)} = s; f_0\right.\right\} = \frac{\text{pr}(Y = y \cap S = s; f_0)}{\text{pr}(S = s; f_0)}$$

$$= \frac{f_0(y)}{\Sigma_s^* f_0(y)} = \frac{s f_0(y)}{\Sigma_s^* s f_0(y)}$$

$$= \frac{f_1(y)}{\Sigma_s^* f_1(y)}$$

$$= \text{pr}\left\{Y = y \left| \frac{f_1(Y)}{f_0(Y)} = s; f_1\right.\right\}.$$

A similar argument applies if there are more than two possible densities in $\mathcal{F}$. Thus if there are $q + 1$ possible densities $f_0(y), \ldots, f_q(y)$, the set of likelihood ratios

$$f_1(y)/f_0(y), \ldots, f_q(y)/f_0(y)$$

is sufficient. The choice of $f_0(y)$ as a reference point is arbitrary and there is an obvious modification if $f_0(y) = 0$. In the parametric case the corresponding result is that for a fixed θ_0 the set of all values

$$f(y; \theta)/f(y; \theta_0) = \text{lik}(\theta; y)/\text{lik}(\theta_0; y) \tag{26}$$

for $\theta \in \Omega$ forms a sufficient statistic. That is, if two points y_1 and y_2 have proportional likelihood functions, they have the same value of a sufficient statistic. This result is pursued in part (iv) of this section, in connexion with minimal sufficiency.

(ii) Factorization theorem
The examples of the previous section illustrate the idea of a sufficient statistic but do not show how to find the sufficient statistic in any particular case. Example 2.12 does, however, suggest that it is the structure of the likelihood function that indicates the form of the sufficient statistic. This is expressed formally in the factorization theorem:

A necessary and sufficient condition that S be sufficient for θ in the family $\mathcal{F}$ is that there exist functions $m_1(s, \theta)$ and $m_2(y)$ such that for all $\theta \in \Omega$,

$$\text{lik}(\theta; y) = m_1(s, \theta)m_2(y). \tag{27}$$

' If S is sufficient, then $f_{Y|S}(y|s)$ does not depend on θ and can be written as $m_2(y)$. Thus

$$
\begin{aligned}
\text{lik}(\theta\;;y) = f_Y(y\;;\theta) &= f_{Y|S}(y|s)f_S(s\;;\theta) \\
&= m_1(s,\theta)m_2(y),
\end{aligned}
$$

say. Conversely if (27) holds, we may calculate $f_{Y|S}(y|s\;;\theta)$ as

$$
\frac{f_Y(y\;;\theta)}{f_S(s\;;\theta)} = \begin{cases} f_Y(y\;;\theta)/\sum\limits_{z\,:\,s(z)=s} f_Y(z\;;\theta) & \text{in discrete case,} \\[2mm] f_Y(y\;;\theta)/\int\ldots\int f_Y(z\;;\theta)\,|J|\,dz & \text{in continuous case.} \end{cases}
\tag{28}
$$

In the second formula we have changed to new variables (s, z), introduced a Jacobian, J, and integrated with respect to z. If we substitute (27) into (28), the term $m_1(s,\theta)$ cancels and the conditional distribution thus does not involve θ.

Note that in (27) we can, if we wish, arrange that $m_1(s,\theta)$ is the p.d.f. of S. We call (27) the *factorization theorem*; it will be used repeatedly later. Two examples suffice for now.

Example 2.13. Poisson distribution (ctd). For the situation of Example 2.9, where $Y_1, \ldots, Y_n$ are i.i.d. in a Poisson distribution of mean μ, the joint probability is

$$
\prod_{j=1}^{n} \frac{e^{-\mu}\mu^{y_j}}{y_j!} = \left(e^{-n\mu}\mu^{\Sigma y_j}\right)\cdot\left(\prod \frac{1}{y_j!}\right).
$$

We may take the two factors in this as respectively $m_1(s,\mu)$ and $m_2(y)$, so that $S = \Sigma Y_j$ is sufficient.

Example 2.14. Cauchy distribution. If $Y_1, \ldots, Y_n$ are i.i.d. in a Cauchy distribution of location parameter θ, the joint density is

$$
\frac{1}{\pi^n} \prod_{j=1}^{n} \frac{1}{\{1 + (y_j - \theta)^2\}}
$$

and no factorization involving a function s of fewer than n dimensions is possible; see Example 2.16 for proof. By the general result of Example 2.11 the order statistics of the sample are sufficient.

(iii) Interpretation of sufficiency
Consider two individuals both involved in observations associated

with the family $\mathcal{F}$, as follows:
    Individual I observes y, a value of the random variable Y;
    Individual II proceeds in two stages:
        (a) he observes s, a value of the random variable S having the
            p.d.f. $f_S(s\,;\theta)$,
        (b) he then observes y, a value of a random variable having
            the p.d.f. $f_{Y|S}(y|s)$, not depending on θ.
The following two statements are very plausible.
    (1) Because the final distributions of Y for the two individuals are
identical, the conclusions to be reached from a given y are identical
for the two individuals.
    (2) Because Individual II, in stage (b), is sampling a fixed distri-
bution, i.e. is in effect drawing values from a table of random numbers, only stage (a) is informative, so long as the correctness of $\mathcal{F}$ is
postulated.
    If both (1) and (2) are accepted, it follows that if y is observed
then the conclusions to be drawn about θ depend only on $s = s(y)$,
so long as $\mathcal{F}$ is the basis of the analysis.
    We shall discuss this further later. The argument can be looked at
in two slightly different ways. On the one hand the argument can be
thought convincing enough to make it a basic principle that, so long
as the correctness of $\mathcal{F}$ is accepted, the conclusions to be drawn
should depend only on s. Alternatively one may simply note that all
the optimality criteria that we shall consider later lead to the use of
s and we may regard the above argument as an explanation of that.
    Note that although restriction to the use of s may achieve a big
reduction in dimensionality we still have to decide what to do with s,
or how to interpret it.

(iv) Minimal sufficiency
If in a particular problem S is a sufficient statistic for θ, then so too
is (S, T) for any statistic $T = t(Y)$. Of course, we would rather deal
with S than with (S, T) since our object is to summarize the data
concisely. If no further reduction from S while retaining sufficiency
is possible, then S is said to be *minimal sufficient*; S is necessarily a
function of all other sufficient statistics that can be constructed.

Example 2.15. Binomial distribution. Let $Y_1, \ldots, Y_n$ be independent
Bernoulli random variables with parameter θ and $S = \Sigma Y_j$. Then
$V = g(S)$ is a summary or simplification of S only if $g(r_1) = g(r_2) = v$

for some v and $0 \leqslant r_1 \neq r_2 \leqslant n$. But for $s = r_1, r_2$

$$\mathrm{pr}(S = s \,|\, V = v) \; = \; \frac{\mathrm{pr}(S = s \cap V = v)}{\mathrm{pr}(V = v)}$$

$$= \; \frac{\binom{n}{s} \theta^s (1 - \theta)^{n-s}}{\binom{n}{r_1} \theta^{r_1} (1 - \theta)^{n-r_1} + \binom{n}{r_2} \theta^{r_2} (1 - \theta)^{n-r_2}},$$

which depends on θ. Thus V is not sufficient and S is minimal sufficient.

We want to use minimal sufficient statistics wherever possible. Sometimes the appropriate factorization of the likelihood is obvious on inspection, particularly for a single parameter. In other cases we can use an important close relationship between minimal sufficiency and ratios of likelihoods to derive the minimal sufficient statistic.

Any statistic, and therefore in particular any sufficient statistic S, divides the sample space into equivalence classes, each class containing all possible observations y with a common value of s. The fact that if S is minimal sufficient so too is any one-one function of S indicates that it is the set of equivalence classes that determines the essential nature of the reduction by minimal sufficiency, rather than the particular labelling of the equivalence classes.

Consider the partition created by putting all points with proportional likelihood functions into the same equivalence class, i.e. define the classes

$$\mathcal{D}\,(y) \; = \; \left\{ z \; ; \frac{f_Y(z\,;\theta)}{f_Y(y\,;\theta)} \; = \; h(z, y), \text{ for all } \theta \in \Omega \right\}; \qquad (29)$$

if $z \in \mathcal{D}(y_1)$ and $\mathcal{D}(y_2)$, then $\mathcal{D}(y_1) = \mathcal{D}(y_2)$. This partitioning is minimal sufficient. To see that it is sufficient, note that the conditional distribution of Y within its equivalence class is independent of θ. To show that it is minimal sufficient, consider any other sufficient statistic $V = v(Y)$ which, by the factorization theorem (27), is such that

$$f_Y(y\,;\theta) \; = \; m_1^\dagger \{ v(y), \theta \} m_2^\dagger(y) \; = \; f_Y(z\,;\theta) \, \frac{m_2^\dagger(y)}{m_2^\dagger(z)},$$

if y and z are such that $v(y) = v(z)$. But this implies that y and z are equivalent in the sense of (29). Therefore the partition (29) includes that based on V, proving the minimal sufficiency of (29).

Thus we inspect the likelihood ratio $f_Y(z ; \theta)/f_Y(y ; \theta)$ in order to find which y and z should be assigned the same value of the minimal sufficient statistic.

Example 2.16. Cauchy distribution (ctd). If $Y_1, \ldots, Y_n$ are i.i.d. in the Cauchy distribution of Example 2.14, the likelihood ratio is

$$\frac{f_Y(z ; \theta)}{f_Y(y ; \theta)} = \frac{\Pi\{1 + (y_j - \theta)^2\}}{\Pi\{1 + (z_j - \theta)^2\}}$$

and is thus a rational function of θ. For the ratio to be independent of θ, all powers of θ must have identical coefficients in numerator and denominator. This happens if and only if $(y_1, \ldots, y_n)$ is a permutation of $(z_1, \ldots, z_n)$. Therefore the minimal sufficient statistic is the set of order statistics $(Y_{(1)}, \ldots, Y_{(n)})$.

It was shown in Example 2.11 that the order statistics are sufficient for the full family of continuous distributions; it follows from their minimal property for the Cauchy distribution that *a fortiori* they are minimal for the larger family.

From now on, by sufficient statistic we always mean minimal sufficient statistic.

(v) Examples
We now consider three further examples which serve both to illustrate the factorization theorem and to give some results of intrinsic importance.

Example 2.17. Normal-theory linear model (ctd). The likelihood for the normal-theory linear model was calculated in (10) and involves the observations only through $(\hat{\beta}, SS_{res})$. These are therefore sufficient statistics for the unknown parameters (β, σ^2). If, however, the variance is known and equal to say σ_0^2 we can in (10) separate off the factor

$$\exp\left(-\frac{SS_{res}}{2\sigma_0^2}\right)$$

and treat it as the function $m_2(y)$ in the factorization theorem (27). It then follows that $\hat{\beta}$ is sufficient for β. In particular, when the

random variables are i.i.d., the sample mean and estimate of variance
are sufficient and when the variance is known the mean is sufficient.

Example 2.18. Uniform distribution of zero mean. Suppose that
$Y_1, \ldots, Y_n$ are i.i.d. with uniform density over $(-\theta, \theta)$. Then the
likelihood is

$$
\begin{cases}
\left(\dfrac{1}{2\theta}\right)^n & (\max|y_j| \leqslant \theta), \\
0 & (\max|y_j| > \theta).
\end{cases}
\tag{30}
$$

Hence the sufficient statistic is $\max|Y_j|$, or equivalently
$\max(-Y_{(1)}, Y_{(n)})$, where $Y_{(1)} = \min(Y_j)$, $Y_{(n)} = \max(Y_j)$ are the
extreme order statistics.

This may be compared with the rather simpler result of Example
2.10 that for the uniform distribution over $(0, \theta)$ the largest value
is sufficient.

A simple extension of (30) shows that for the uniform distribution
with both terminals unknown, the smallest and largest values are together sufficient. The same sufficient statistics apply for a uniform
distribution of known range but unknown mean, e.g. the uniform
distribution from $\theta - 1$ to $\theta + 1$.

These results generalize immediately to a known distribution
truncated at unknown points, i.e. to the density

$$
\frac{p(y)}{\int_{\theta_1}^{\theta_2} p(u)du} \quad (\theta_1 \leqslant y \leqslant \theta_2),
\tag{31}
$$

where $p(.)$ is a known non-negative function and one or both of θ_1
and θ_2 are unknown parameters. Again the relevant extreme order
statistics are sufficient.

Example 2.19. Life-testing with an exponential distribution of life.
Suppose that in observations on n individuals, r "die" after times
$y_1, \ldots, y_r$, whereas the remaining $m = n - r$ are still "alive" after
times under test of $y'_1, \ldots, y'_m$; there are a number of situations in
life-testing where data have to be analysed with an appreciable number of lives incomplete. If completed lives are represented by random
variables $Y_1, \ldots, Y_n$ which are i.i.d. with p.d.f. $\rho e^{-\rho y}$ $(y \geqslant 0)$, the
likelihood is

$$\prod_{j=1}^{r} \rho e^{-\rho y_j} \prod_{k=1}^{m} e^{-\rho y_k'}, \tag{32}$$

the terms for the incomplete lives being the probabilities of times to death exceeding $y_1', \ldots, y_m'$. Thus the likelihood is

$$\rho^r e^{-\rho y_.},$$

where $y_. = \Sigma y_j + \Sigma y_k'$ is the total time at risk; the sufficient statistic is $(R, Y_.)$. This result is the continuous time analogue of the result that in any set of Bernoulli trials the number of successes and the total number of trials form the sufficient statistic.

(vi) Exponential family of distributions
Suppose first that there is a single parameter and that $Y_1, \ldots, Y_n$ are mutually independent with

$$f_{Y_j}(y ; \theta) = \exp \{a(\theta)b_j(y) + c_j(\theta) + d_j(y)\}, \tag{33}$$

where $a(.), b_j(.), c_j(.), d_j(.)$ are known functions. Then

$$f_Y(y;\theta) = \exp \{a(\theta) \Sigma b_j(y_j) + \Sigma c_j(\theta) + \Sigma d_j(y_j)\}, \tag{34}$$

so that $\Sigma b_j(Y_j)$ is sufficient. In particular, if the Y_j are identically distributed, $\Sigma b(Y_j)$ is sufficient. Several interesting special distributions have the form

$$\exp \{a(\theta)b(y) + c(\theta) + d(y)\}, \tag{35}$$

among them the normal, gamma, binomial and Poisson distributions. For example, to see that the gamma distribution with known index k_0 belongs to the family we write

$$f_Y(y ; \rho) = \rho(\rho y)^{k_0 - 1} e^{-\rho y}/\Gamma(k_0)$$

$$= \exp \{-\rho y + k_0 \log \rho + (k_0 - 1) \log y - \log \Gamma(k_0)\}.$$

Thus $a(\rho) = -\rho$, $b(y) = y$, $c(\rho) = k_0 \log \rho - \log \Gamma(k_0)$ and $d(y) = (k_0 - 1)\log y$, and the sufficient statistic for ρ from i.i.d. random variables $Y_1, \ldots, Y_n$ is $\Sigma b(Y_j) = \Sigma Y_j$.

One-one transformations of variable or parameter do not affect the general form of (35), i.e. whether or not a distribution belongs to the simple exponential family. Thus we can, provided that $a(.)$ and $b(.)$ are monotonic, transform to a new parameter $\phi = -a(\theta)$ and a new variable $Z = b(Y)$. The p.d.f. for Z has the simple form

$$f_Z(z \; ; \phi) \;=\; \exp\{-z\phi + c^\dagger(\phi) + d^\dagger(z)\}, \tag{36}$$

and the sufficient statistic for ϕ based on i.i.d. random variables
$Z_1, \ldots, Z_n$ is ΣZ_j. The new parameter ϕ is often called the *natural parameter* for the problem, for several technical and practical reasons. One of these is that the ratio of likelihood functions at ϕ_1 and $\phi_2 < \phi_1$ is an increasing function of the sufficient statistic. Also it will turn out that comparisons of different sets of data are most easily achieved in terms of comparisons of the natural parameter values. For example, the natural parameter for the binomial distribution is $\phi = \log\{\theta/(1-\theta)\}$, the so-called log odds ratio. The theory of the comparison of two binomial distributions is simplest not in terms of $\theta_1 - \theta_2$ but in terms of $\phi_1 - \phi_2$. Whether this is really the best parametrization in terms of which to make the comparison depends in addition on circumstances other than mathematical simplicity.

Example 2.20. Exponential family linear model. Suppose that the Y_j's independently have p.d.f.'s $f_{Y_j}(y_j \; ; \theta_j)$ belonging to the same exponential family but with different parameter values θ_j, all of which are themselves functions of a single parameter ψ; that is, $\theta_j = \theta_j(\psi)$. The joint distribution in the simplified form (36) is

$$f_Z(z \; ; \phi_1, \ldots, \phi_n) \;=\; \exp\left\{-\sum_{j=1}^n z_j\phi_j + \sum_{j=1}^n c^\dagger(\phi_j) + \sum_{j=1}^n d^\dagger(z_j)\right\}.$$

Thus if the dependence of θ_j on ψ implies a linear relationship, $\phi_j = a_j\psi$, with a_j constant, then $\Sigma a_j Z_j$ is sufficient for ψ. Linearity on any other scale will not produce a single sufficient statistic for ψ.

The generalization of the exponential family to vector parameters is to consider the density for Y_j given $\theta = (\theta_1, \ldots, \theta_q)$ as

$$f_{Y_j}(y_j \; ; \theta) \;=\; \exp\left\{\sum_{k=1}^m a_k(\theta)b_{jk}(y_j) + c_j(\theta) + d_j(y_j)\right\}. \tag{37}$$

Then the joint p.d.f. for independent variables $Y_1, \ldots, Y_n$ can be written as

$$f_Y(y \; ; \theta) \;=\; \exp\left\{\sum_{k=1}^m a_k(\theta)s_k(y) + c(\theta) + d.(y)\right\} \tag{38}$$

where

$$s_k(y) \;=\; \sum_{j=1}^n b_{jk}(y_j) \quad (k = 1, \ldots, m).$$

The sufficient statistic for θ is therefore $S = (S_1, \ldots, S_m)$. The dimensions m and q of s and θ respectively are not necessarily equal. The case $m < q$ might occur if some non-linear relationship exists between the components of θ, but usually this will not happen. Most common is the case $m = q$, which arises in Example 2.4, the normal-theory linear model of full rank. There the dimensionality of $\theta = (\beta, \sigma^2)$ is $q = q_x + 1$ and it follows directly from (10) that $m = q$, the sufficient statistic being $(\hat{\beta}, \text{SS}_{\text{res}})$.

The case $m > q$, while not very common in applications, can arise in a perfectly natural way. We give one example.

Example 2.21. Normal distribution with known coefficient of variation. Consider a normal distribution in which the ratio γ_0 of standard deviation to mean is known, $\sigma = \gamma_0\mu$, say. The p.d.f. is

$$\frac{1}{\gamma_0\mu\sqrt{(2\pi)}} \exp\left\{-\frac{(y-\mu)^2}{2\gamma_0^2\mu^2}\right\}$$

$$= \exp\left\{-\frac{y^2}{2\gamma_0^2\mu^2} + \frac{y}{\gamma_0^2\mu} - \frac{1}{2\gamma_0^2} - \frac{1}{2}\log(2\pi\gamma_0^2\mu^2)\right\}$$

which is of the form (37) with $m = 2$. If $Y_1, \ldots, Y_n$ are i.i.d. with this distribution, the sufficient statistic is $(\Sigma Y_j, \Sigma Y_j^2)$ or equivalently $(\bar{Y}., \text{MS})$ with $\bar{Y}. = \Sigma Y_j/n$ and $\text{MS} = \Sigma(Y_j - \bar{Y}.)^2/(n-1)$.

Similar examples can be formed from other distributions. A rather different situation is illustrated by Example 2.19, with censored data from an exponential distribution. Here again $q = 1$, $m = 2$, with the sufficient statistic being number of uncensored observations and the total time on test.

When $q = m$, the p.d.f. (37) can, by transformation of variables and parameters, be taken in a simple form somewhat analogous to (36), namely

$$f_Z(z; \phi) = \exp\left\{-\sum_{r=1}^{q} \phi_r b_r^\dagger(z) + c^\dagger(\phi) + d^\dagger(z)\right\}.$$

It is then appealing to assign components of the sufficient statistic

$$S = \left\{\sum_{j=1}^{n} b_1^\dagger(Z_j), \ldots, \sum_{j=1}^{n} b_q^\dagger(Z_j)\right\}$$

to the corresponding components of ϕ. This will be done implicitly in Section 5.2, in connexion with testing hypotheses about components of θ, but for the moment we keep to the original definition where the whole sufficient statistic is a vector associated with the whole parameter as a vector.

In general, the dimension of the sufficient statistic will not be smaller than the sample size unless the distribution is a member of the exponential family. This is illustrated by Example 2.14, the Cauchy distribution, by the Weibull distribution of unknown index, and by Example 2.11, showing that when $\mathcal{F}$ is the family of all continuous distributions, the order statistics are sufficient; it is easily shown in this last example that no further reduction is possible. We do not here attempt to prove the equivalence of sufficient reduction and the exponential family, but for the one-dimensional case ($m = q = 1$) Exercise 2.11 outlines a method of establishing the connexion.

To emphasise that sufficiency is a property of the sampling model, as well as of the distributions being sampled, we give the following example.

Example 2.22. Change-point model. Suppose that $u_1, \ldots, u_n$ are fixed constants, all different, and that for some unknown ξ, Y_j has the distribution $N(0, 1)$ if $u_j < \xi$ and the distribution $N(\mu, 1)$ if $u_j \geqslant \xi$. Then the sufficient statistic for the unknown parameter $\theta = (\mu, \xi)$ is the full set $Y_1, \ldots, Y_n$, and no reduction is possible.

(vii) Completeness

A mathematically important idea is that of completeness. If S is a sufficient statistic for θ in the family of distributions indexed by $\theta \in \Omega$, then S is called *complete* if a necessary condition for

$$E\{h(S); \theta\} = 0 \quad (\theta \in \Omega) \tag{39}$$

is
$$h(S) = 0 \quad (\theta \in \Omega),$$

except possibly on sets of measure zero with respect to all the distributions concerned. A weaker concept is that of *bounded completeness*, for which $h(S)$ must be bounded. The property of completeness guarantees uniqueness of certain statistical procedures based on S; this we discuss in later chapters.

Example 2.23. Normal mean. Let $Y_1, \ldots, Y_n$ be i.i.d. in $N(\mu, 1)$, and let $S = \bar{Y}_. = \Sigma Y_j/n$. Then, because $\bar{Y}_.$ is $N(\mu, 1/n)$, the identity (39) becomes

$$\int_{-\infty}^{\infty} h(s)e^{-\frac{1}{2}s^2} e^{ns\mu} ds = 0 \quad (-\infty < \mu < \infty).$$

But the integral is a bilateral Laplace transform, so that by the appropriate inversion theorem we deduce that $h(s) \exp(-\frac{1}{2}s^2)$ is identically zero except on sets of Lebesgue measure zero. Thus $h(s) \equiv 0$, and S is complete.

Example 2.24. Binomial distribution (ctd). Let $Y_1, \ldots, Y_n$ be independent Bernoulli random variables with parameter θ. Then $S = \Sigma Y_j$ is sufficient for θ. Values of $h(s)$ other than for $s = 0, \ldots, n$ have zero probability and are of no concern; let $h(s) = h_s$. The identity (39) becomes

$$\sum_{s=0}^{n} h_s \binom{n}{s} \theta^s (1-\theta)^{n-s} = 0 \quad (0 \leqslant \theta \leqslant 1). \tag{40}$$

Here the sum is a polynomial of degree n which is identically zero, which implies $h_1 = \ldots = h_n = 0$, so that again S is complete. In fact the vanishing of the h_s's follows if (40) holds for at least $n + 1$ distinct values of θ. That is, S is complete for much smaller parameter spaces than $[0, 1]$.

It can be shown (Lehmann, 1959, Section 4.3) that for random variables i.i.d. in the exponential family density (38), $\dim(S) = \dim(\theta)$, i.e. $m = q$, is necessary and sufficient for S to be complete. Thus in the situation of Example 2.21, the normal distribution with known coefficient of variation, with $m > q$, the sufficient statistic $(\Sigma Y_j, \Sigma Y_j^2)$ is not complete; this is easily verified directly because

$$\frac{n + \gamma_0^2}{1 + \gamma_0^2} \Sigma Y_j^2 - (\Sigma Y_j)^2$$

has expectation zero for all μ. In general if a sufficient statistic is boundedly complete it is minimal sufficient (Lehmann and Scheffé, 1950, 1955); the converse is false.

(viii) Ancillary statistics
Consider the situation where S is the minimal sufficient statistic for θ and $\dim(S) > \dim(\theta)$. Then it sometimes happens that we can

write $S = (T, C)$, where C has a marginal distribution not depending on θ. If this is so, C is called an *ancillary statistic*. Some writers then refer to T as *conditionally sufficient*, because T is used as a sufficient statistic in inference conditionally on $C = c$. The ancillary statistic C is chosen to have maximum dimension.

Example 2.25. Random sample size. Let N be a random variable with a known distribution $p_n = \text{pr}(N = n)$ $(n = 1, 2, \ldots)$, and let $Y_1, \ldots, Y_N$ be i.i.d. with the exponential family density (35). Then the likelihood of the data $(n, y_1, \ldots, y_n)$ is

$$f_{N,Y}(n, y) = p_n \exp\left\{ a(\theta) \sum_{j=1}^{n} b(y_j) + nc(\theta) + \sum_{j=1}^{n} d(y_j) \right\}.$$

Thus
$$\left\{ \sum_{j=1}^{N} b(Y_j), N \right\}$$

is sufficient for θ, and N is an ancillary statistic, whereas $\Sigma b(Y_j)$ is only conditionally sufficient. Any sample size not fixed in advance, but with known distribution independent of θ, is an ancillary statistic.

Example 2.26. Mixture of normal distributions. Suppose that a random variable Y is equally likely to be $N(\mu, \sigma_1^2)$ or $N(\mu, \sigma_2^2)$, where σ_1 and σ_2 are different and known. An indicator random variable C is observed, taking the value 1 or 2 according to whether Y has the first or the second distribution. Thus it is known from which distribution y comes. Then the likelihood of the data (c, y) is

$$f_{C,Y}(c, y) = \tfrac{1}{2}(2\pi\sigma_c^2)^{-\frac{1}{2}} \exp\{-(y - \mu)^2/(2\sigma_c^2)\},$$

so that $S = (C, Y)$ is sufficient for μ with σ_1^2 and σ_2^2 known. Because $\text{pr}(C = 1) = \text{pr}(C = 2) = \tfrac{1}{2}$ independent of μ, C is ancillary.

Example 2.27. Normal-theory linear model (ctd). In a linear regression problem, suppose that the values of the explanatory variable have a known joint p.d.f. $f_X(x)$, and that, conditionally on $X = x$, the $Y_1, \ldots, Y_n$ are independent, Y_j having the distribution $N(\gamma + \beta x_j, \sigma^2)$. Then the full likelihood of the data is

$$f_{X,Y}(x, y) = f_X(x)(2\pi\sigma^2)^{-\frac{1}{2}n} \exp\left\{ -\frac{1}{2\sigma^2} \sum_{j=1}^{n} (y_j - \gamma - \beta x_j)^2 \right\}.$$

The sufficient statistic for $(\gamma, \beta, \sigma^2)$ is

$$S = (\hat{\gamma}, \hat{\beta}, \text{SS}_{\text{res}}, \Sigma X_j, \Sigma X_j^2), \tag{41}$$

the last two components of which form an ancillary statistic. That is, even if the explanatory variable is random, conditioning on the ancillary statistic would lead to treating the explanatory variable as fixed. The argument extends immediately to the general normal-theory linear model.

These simple examples are intended to suggest that inference about θ should be conditional on the ancillary statistic. We can regard the observed value c as describing that part of the total sample space relevant to the problem at hand. For instance, in Example 2.26, the ancillary statistic tells us which normal distribution was in fact applicable. The fact that some other normal distribution might have been used, but actually was not, seems irrelevant to the interpretation of y. However some difficulties are encountered with ancillary statistics. First, there is no general method for constructing C. Secondly, C may not be unique. The following example, given by Basu (1964) in an extended discussion of ancillarity, illustrates both difficulties.

Example 2.28. Special multinomial distribution. Let $Y_1, \ldots, Y_n$ be i.i.d. with the discrete distribution

$$\text{pr}(Y_j = 1) = \tfrac{1}{6}(1 - \theta), \quad \text{pr}(Y_j = 2) = \tfrac{1}{6}(1 + \theta),$$
$$\text{pr}(Y_j = 3) = \tfrac{1}{6}(2 - \theta), \quad \text{pr}(Y_j = 4) = \tfrac{1}{6}(2 + \theta).$$

If n_l is the number of observations equal to $l(l = 1, \ldots, 4)$, then the joint probability of a particular sequence is

$$f_Y(y; \theta) = 6^{-n}(1 - \theta)^{n_1}(1 + \theta)^{n_2}(2 - \theta)^{n_3}(2 + \theta)^{n_4}. \tag{42}$$

The statistic $S = (N_1, N_2, N_3, N_4)$ is minimal sufficient; of course one component can be omitted in view of the identity $\Sigma N_l = n$. The particular structure here leads to two possible ancillary statistics, namely $C_1 = (N_1 + N_2, N_3 + N_4)$ and $C_2 = (N_1 + N_4, N_2 + N_3)$. Which of these to use in inference about θ depends on which one best separates all possible sets of data into equally informative sets; see also Example 2.37 and Exercise 4.11.

While such non-uniqueness arises rather rarely in applications, the possibility is theoretically disturbing.

The next example indicates a very general set of ancillary statistics.

Example 2.29. Location family. Let $Y_1, \ldots, Y_n$ be i.i.d. in the location family with density $h(y - \theta)$. The order statistics $Y_{(1)}, \ldots, Y_{(n)}$ are sufficient by the arguments of Example 2.11. Except when $\log h(y)$ is a polynomial in y of degree less than n, the order statistics are minimal sufficient. The contrasts between these, as determined, for example, by $C_2 = Y_{(2)} - Y_{(1)}$, $C_3 = Y_{(3)} - Y_{(1)}, \ldots,$ $C_n = Y_{(n)} - Y_{(1)}$, are distributed independently of θ and hence form an ancillary statistic. The remaining component of the sufficient statistic can be taken as $T = Y_{(1)}$, or equivalently as any function of T and C, such as $\bar{Y}_.$. The consequences of studying the conditional distribution of T given the ancillary statistic will be taken up in Example 4.15. The statistic $C = (C_2, \ldots, C_n)$ is called the *configuration*.

It would have been possible to have defined ancillary statistics without the preliminary reduction by minimal sufficiency. However, the fact that inference is wherever possible carried out in terms of the minimal sufficient statistic makes the present definition appealing. The alternative would be to define an ancillary statistic as any function of Y with a distribution independent of θ. That this would lead to additional complications is shown by the following example.

Example 2.30. Bivariate normal distribution with unknown correlation coefficient. Suppose that $(Y_1, Z_1), \ldots, (Y_n, Z_n)$ are i.i.d. in a bivariate normal distribution with zero means, unit variances and unknown correlation coefficient ρ. The joint p.d.f. is

$$\frac{1}{(2\pi)^n (1 - \rho^2)^{\frac{1}{2}n}} \exp\left\{ -\frac{\Sigma(y_j^2 + z_j^2)}{2(1 - \rho^2)} + \frac{\rho\Sigma y_j z_j}{1 - \rho^2} \right\}, \qquad (43)$$

so that the minimal sufficient statistic is $S = \{\Sigma Y_j Z_j, \Sigma(Y_j^2 + Z_j^2)\}$. There does not seem to be an ancillary statistic, i.e. a function of S with a p.d.f. independent of ρ, although $C' = \Sigma(Y_j^2 + Z_j^2)$ is in some reasonable sense approximately ancillary, C' having an expectation $2n$ independent of ρ and variance $4n(1 + \rho^2)$ not too strongly dependent on ρ.

If, however, we allow ancillary statistics that are not functions of S, the position is quite different, because both ΣY_j^2 and ΣZ_j^2 separately are ancillary, the corresponding random variables having chi-squared distributions with n degrees of freedom. Clearly there can be no basis for preferring one of ΣY_j^2 and ΣZ_j^2 to the other as an ancillary statistic, so that if the broader definition of ancillarity were adopted, the problem of non-uniqueness would be accentuated.

The definition of ancillary statistics given earlier is restrictive. For instance, in Example 2.25, concerned with random sample size, it is not really crucial that the distribution of sample size should be known. The essential points in that example are that (i) the observed value of sample size by itself should give no information about θ and that (ii) the conditional distribution of the other component given the ancillary statistic depends only on the parameter of interest. The same points arise in connexion with Example 2.27 concerned with random explanatory variables in regression, where the regression parameters γ, β and σ^2 are of primary interest.

To formulate this extended notion of ancillarity, suppose that the unknown parameter θ is partitioned into two parts $\theta = (\psi, \lambda)$, where λ is not of direct interest. We assume that the parameter space is such that any possible value of ψ could arise in conjunction with any possible value of λ, i.e. that $\Omega_\theta = \Omega_\psi \times \Omega_\lambda$, in an obvious notation, the cross denoting Cartesian product. Let S be the minimal sufficient statistic for θ and suppose that $S = (T, C)$, where

(a) the p.d.f. of C depends on λ but not on ψ;

(b) the conditional p.d.f. of T given $C = c$ depends on ψ but not on λ, for all values of c.

Then we call C ancillary for ψ in the extended sense, and T conditionally sufficient for ψ in the presence of the nuisance parameter λ.

With this new definition, we can deal with the situations of Examples 2.25–2.27 when the distributions of sample size, etc. are arbitrary and unknown, or belong to some parametric family, provided that the variation of the ancillary statistic is independent of the parameter of interest in the way just specified.

(ix) Asymptotic sufficiency

In some problems the minimal sufficient statistic may be of dimension n, the number of observations, and yet approximately for large n a statistic of much lower dimension may be "almost sufficient" in a reasonable sense. This is one aspect of the important matter of finding procedures for complex problems that will have desirable properties asymptotically as $n \to \infty$, and which therefore should have good properties for large but finite n. Chapter 9 develops this topic in detail. Here we give two examples to establish the connexion with sufficiency.

Example 2.31. Maximum likelihood estimates. Suppose that $Y_1, \ldots,$ Y_n are i.i.d. with density $f_{Y_j}(y \, ; \theta)$. The asymptotic results of Section 9.2 show that, under certain conditions on $f_Y(y \, ; \theta)$, the value $\hat{\theta}$ which maximizes the likelihood is such that for a suitably defined function $i(\theta)$, the likelihood is given over the range of interest by

$$f_Y(y \, ; \theta) = f_Y(y \, ; \hat{\theta}) \left[\frac{\sqrt{n}}{\sqrt{\{2\pi i(\theta)\}}} \exp \left\{ -\frac{n(\hat{\theta} - \theta)^2}{2i(\theta)} \right\} + r_n(y \, ; \theta) \right],$$

where $r_n(y \, ; \theta)$ is negligible for large n. Comparing this with the factorization criterion (27), we see that $\hat{\theta}$ satisfies this in the limit and hence can reasonably be called asymptotically sufficient.

Example 2.32. Change-point model (ctd). Suppose $Y_1, \ldots, Y_\gamma$ to be i.i.d. in $N(0, 1)$ and $Y_{\gamma+1}, \ldots, Y_n$ to be i.i.d. in $N(\mu, 1)$ with γ and μ both unknown. Suppose also that there is a restriction on the true value γ, namely $1 \leqslant \gamma \leqslant \gamma_0$ with γ_0 fixed and known. Roughly speaking, for large n we know that there is a change in distribution of the random variables Y_j near the start of the sequence. Then if $\hat{\gamma}$ is the value of γ at which the likelihood is maximized the statistic

$$T_n(\hat{\gamma}) = \sum_{j = \hat{\gamma} + 1}^{n} Y_j / (n - \hat{\gamma})$$

is asymptotically sufficient for μ. The heuristic reason is that $T_n(\hat{\gamma})$ and $T_n(\gamma)$ differ by a negligible amount for large n and the latter statistic is sufficient for known γ.

Some explicit details of this problem are given by Hinkley (1972).

2.3 Some general principles of statistical inference

(i) General remarks
In the remainder of this book we develop the theory of a number of types of statistical procedure. One general theme is that the arguments to be used depend both on the type of question of interest

and on the depth to which it is possible to formulate the problem quantitatively. Thus in Chapter 3 we consider situations where only one hypothesis is formulated, whereas in Chapter 11 not only is a full model available for the data, but also there are quantitative specifications of the additional knowledge available and of the consequences of the various possible decisions, one of which is to be chosen in the light of the data. Now it is to be expected on general grounds that once the very much more detailed specification of Chapter 11 is regarded as given, the ideas necessary to develop "optimum" procedures should be relatively straightforward and uncontroversial, whereas when the specification is much weaker there is relatively more need for *ad hoc* arguments and somewhat arbitrary criteria of optimality.

The types of problem that it is worth discussing can be settled only by consideration of applications. We believe that all the levels of specification discussed in the subsequent chapters are useful. Because of this there is no one approach or set of requirements that are universally compelling. The reader may prefer to go straight to the detailed development starting with Chapter 3. On the other hand, there are some general principles that have bearing on the various approaches to be discussed and therefore we now outline these; it is instructive in thinking about particular arguments to consider which of these general principles are obeyed. Some forward reference is inevitable in this and the next section, but has been kept to a minimum.

Throughout, the provisional and approximate character of models has to be borne in mind.

(ii) Sufficiency principle

Suppose that we have a model according to which the observations y correspond to a random variable Y having p.d.f. $f_Y(y ; \theta)$ and that S is minimal sufficient for θ. Then, according to the sufficiency principle, so long as we accept the adequacy of the model, identical conclusions should be drawn from data y_1 and y_2 with the same value of s.

The argument for this has already been given in Section 2.2 (iii). Once the value of s is known the rest of the data can be regarded as if generated by a fixed random mechanism not depending on, and therefore uninformative about, θ, so long as the assumed model is correct.

A subsidiary but still very important aspect of the sufficiency principle is that the adequacy of the model can be tested by seeing whether the data y, given $S = s$, are reasonably in accord with the known conditional distribution.

(iii) Conditionality principle

Suppose that C is an ancillary statistic either in the simple sense first introduced in Section 2.2 (viii), or in the second and extended sense where nuisance parameters are present. Then the conditionality principle is that the conclusion about the parameter of interest is to be drawn as if C were fixed at its observed value c. The arguments for this are best seen from Examples 2.26 and 2.27. Suppose that in Example 2.26 it is known that the observations are obtained from $N(\mu, \sigma_1^2)$. How can it affect the interpretation of these data to know that if the experiment were repeated some other variance might obtain? We may think of c as an indicator of which "experiment" was actually performed to produce the data. The following hypothetical example further illustrates the relevance of the conditionality principle.

Example 2.33. Two measuring instruments. A measurement can be taken from one of two measuring instruments C_1 and C_2, with a view to determining whether a physical parameter θ is equal to θ_1 or θ_2. The possible values of the measurement represented by the random variable Y are one and two, such that

$$\mathrm{pr}(Y = 1 \mid C_1 ; \theta_2) = \mathrm{pr}(Y = 2 \mid C_1 ; \theta_1) = 1,$$
$$\mathrm{pr}(Y = 1 \mid C_2 ; \theta_2) = \mathrm{pr}(Y = 2 \mid C_2 ; \theta_1) = 0.01.$$

The experiment consists of choosing an instrument at random, where $\mathrm{pr}(\text{select } C_1) = 0.9$ and $\mathrm{pr}(\text{select } C_2) = 0.1$, and then taking a measurement y. It is known which instrument is used. Suppose now that $y = 1$, and that C_2 was used. Then we calculate that

$$\mathrm{lik}(\theta_1 ; y = 1) = 0.099, \quad \mathrm{lik}(\theta_2 ; y = 1) = 0.901,$$

whereas, conditioning on the fact that C_2 was used,

$$\mathrm{lik}(\theta_1 ; y = 1 \mid C_2) = 0.99, \quad \mathrm{lik}(\theta_2 ; y = 1 \mid C_2) = 0.01.$$

The conditional likelihood indicates quite clearly that $\theta = \theta_1$. Almost as clearly the unconditional likelihood indicates that $\theta = \theta_2$, but only because of what might have happened if the more likely instrument

$\mathcal{C}_1$ had been used. Thus directly conflicting evidence about θ is given if we do not condition on the information "$\mathcal{C}_2$ was used," which is ancillary.

(iv) Weak likelihood principle

With the same information as in (ii), the weak likelihood principle is that two observations with proportional likelihood functions lead to identical conclusions. That is, if y_1 and y_2 are such that for all θ

$$f_Y(y_1;\theta) = h(y_1, y_2)f_Y(y_2;\theta)$$

then y_1 and y_2 lead to identical conclusions, so long as we accept the adequacy of the model.

It follows from the construction of Section 2.2(iv) that this is identical with the sufficiency principle.

(v) Strong likelihood principle

Suppose now that two different random systems are contemplated, the first giving observations y corresponding to a vector random variable Y, and the second giving observations z on a vector variable Z, the corresponding p.d.f.'s being $f_Y(y;\theta)$ and $f_Z(z;\theta)$ with the same parameter θ and the same parameter space Ω. Then the strong likelihood principle is that if y and z give proportional likelihood functions, the conclusions drawn from y and z should be identical, assuming of course the adequacy of both models. That is, if for all $\theta \in \Omega$

$$f_Y(y;\theta) = h(y,z)f_Z(z;\theta), \tag{44}$$

then identical conclusions about θ should be drawn from y and from z.

Examples 2.1–2.3 concerning Bernoulli trials can be used to illustrate this. The log likelihood function corresponding to r successes in n trials is essentially the same whether (a) only the number of successes in a preassigned number of trials is recorded, or (b) only the number of trials necessary to achieve a preassigned number of successes is recorded, or (c) whether the detailed results of individual trials are recorded, with an arbitrary data-dependent "stopping rule". In all cases the log likelihood is, apart from a constant $k(r, n)$,

$$r \log \theta + (n - r) \log(1 - \theta),$$

and if the strong likelihood principle is accepted, then the conclusions drawn about θ cannot depend on the particular sampling scheme adopted.

These results are very special cases of ones applying whenever we have a "stopping rule" depending in some way on the data currently accumulated but not on further information about the unknown parameter.

Example 2.34. Sequential sampling. Suppose that observations are taken one at a time and that after each observation a decision is taken as to whether to take one more observation. Given $m-1$ observations $y_1, \ldots, y_{m-1}$, there is a probability $p_{m-1}(y_1, \ldots, y_{m-1})$ that one more observation is in fact taken. The conditional p.d.f. of Y_m given $Y_1 = y_1, \ldots, Y_{m-1} = y_{m-1}$ is written in the usual way. Note that this includes very general forms of sequential sampling in which observations may be taken singly or in groups.

Suppose that the data are $(n, y_1, \ldots, y_n)$. Then the likelihood, i.e. the joint probability that observations are taken in the way specified and give the values actually observed, is

$$p_0 f_{Y_1}(y_1 \, ; \theta) p_1(y_1) f_{Y_2 | Y_1}(y_2 | y_1 ; \theta) \ldots p_{n-1}(y_1, \ldots, y_{n-1})$$

$$f_{Y_n | Y_{n-1}, \ldots, Y_1}(y_n | y_{n-1}, \ldots, y_1 ; \theta) \{1 - p_n(y_1, \ldots, y_n)\}.$$

Thus, so long as the probabilities defining the sampling scheme are known they form a constant factor in the likelihood function and the dependence on the parameters is fixed by the observations actually obtained, in fact by the joint p.d.f. of $Y_1, \ldots, Y_n$. Therefore, if the strong likelihood principle were accepted, the conclusion to be drawn about θ would be the same as if n were fixed. Note, however, that N is not in general an ancillary statistic and that conditioning on its value is not a consequence of the conditionality principle as formulated above.

We noted at the end of the previous subsection that the weak likelihood principle and the sufficiency principle are equivalent. The deduction of the strong likelihood principle from the sufficiency principle plus some form of the conditionality principle has been considered by Birnbaum (1962, 1969, 1970), Barnard, Jenkins and Winsten (1962), Durbin (1970), Savage (1970), Kalbfleisch (1974) and Basu (1973). We shall not go into details, but the following seems the essence of the matter.

Suppose that (44) holds. Now pretend that we have the following experiment, which we call the enlarged experiment:

observe Y with probability $\frac{1}{2}$,

or

observe Z with probability $\frac{1}{2}$.

Now imagine this experiment done and consider the following two outcomes:

(a) Y is observed and $Y = y$,

(b) Z is observed and $Z = z$.

Now the likelihood functions of (a) and (b) in the enlarged experiment being $\frac{1}{2}f_Y(y\,;\theta)$ and $\frac{1}{2}f_Z(z\,;\theta)$, they are by (44) proportional for all θ. Hence, if we can apply the weak likelihood principle or sufficiency principle to the enlarged experiment, then the conclusions from (a) and (b) should be identical. Finally in, say, (a) the event "Y is observed" has fixed probability $\frac{1}{2}$ (c.f. Example 2.26) and if the conditionality principle is applied to the enlarged experiment the inference from (a) should be the same as from the simpler component experiment in which $Y = y$. Similarly for (b), and so the strong likelihood principle has been deduced.

There are several reasons why this argument is not compelling. One (Durbin, 1970) is that in (a) the event "Y is observed" will not be part of the minimal sufficient statistic and hence, at least on the definition used here, does not qualify to be an ancillary statistic. A second and more basic reason concerns the propriety of regarding the enlarged experiment as fit for the application of the weak likelihood principle. If we were to insist that calculations are made conditionally on the experiment actually performed, i.e. were to apply some form of conditionality principle before applying sufficiency, the basis for considering the enlarged experiment would collapse.

The Bayesian methods to be considered in Chapter 10 do satisfy the strong likelihood principle; nearly all the other methods do not.

(vi) Invariance principle

The very simplest form of invariance argument holds when there are two values y' and y'' of the vector random variable Y that have the same value of the p.d.f. for all θ, i.e. $f_Y(y'\,;\theta) = f_Y(y''\,;\theta)$ for all $\theta \in \Omega$. It is then, under the model, a pure convention which is called y' and which y'', i.e. we can interchange y' and y'' without material change of the situation. The invariance principle in this case requires

that the conclusions to be drawn from y' and from y'' are the same.
The next example is a quite direct consequence of this argument.

Example 2.35. Invariance of i.i.d. random variables under permutation. Suppose that $Y_1, \ldots, Y_n$ are, under the model, i.i.d. or, more generally, are exchangeable (Feller, 1971, Ch. VII). An arbitrary permutation of the argument leaves the joint p.d.f. $f_Y(y)$ unchanged. Hence any conclusion drawn from data y should likewise be invariant under an arbitrary permutation, so long as the model is accepted.

Another rather similar example is that if, under the model, $Y_1, \ldots, Y_n$ are i.i.d. with a density symmetrical about zero, then the invariance principle requires that the conclusion depends only on $|y_1|, \ldots, |y_n|$.

Now this form of invariance principle is just a special case of the sufficiency or weak likelihood principle. For if y' and y'' have the same value of the p.d.f. for all θ, they generate equal likelihood functions. For instance, Example 2.35 is equivalent to the sufficiency of the order statistics.

An appreciably more elaborate type of invariance argument involves transformations on the sample space and associated transformations on the parameter space. A simple example of the argument concerns location parameters.

Example 2.36. Location family (ctd). Let $Y_1, \ldots, Y_n$ be i.i.d. with component p.d.f. $h(y - \theta)$. Now consider the transformation from Y_j to $Y_j + a$ $(j = 1, \ldots, n)$, where a is an arbitrary real number. We denote the transformation by g_a, i.e. $g_a y = y + a1$, where 1 is a vector of ones. It is clear that the transformed random variables have the same form of distribution as does Y but with a new parameter, namely $g_a^* \theta = \theta + a$. Formally

$$f_{g_a Y}(z\,;\theta) \;=\; f_Y(z\,;g_a^*\theta).$$

We assume that the parameter spaces for θ and for $g_a^*\theta$ are identical, i.e. are the whole real line.

We can now draw a conclusion about θ by
   (i) analysing y directly;
or (ii) analysing $g_a y$ to draw a conclusion about $g_a^*\theta$ and then applying the inverse transformation $(g_a^*)^{-1}$ to convert the conclusion into one about θ.

The invariance principle requires that (i) and (ii) give identical answers for all a.

To be more specific, suppose that point estimation is involved. From y, we compute $t(y)$ as an estimate of θ. From $g_a y$ we compute $t(y + a1)$ as an estimate of $\theta + a$, and hence $t(y + a1) - a$ as an estimate of θ. The invariance principle thus requires that for all a, $t(y + a1) - a = t(y)$, i.e.

$$t(y + a1) = t(y) + a. \tag{45}$$

This, combined with the quite different kind of reduction achieved in Example 2.35, leads to $t(.)$ being a linear combination of order statistics, the weights being a function of the configuration (Example 2.29) and summing to one.

A crucial point is that the argument requires that there are no external reasons for preferring some values of θ to others.

Similar arguments apply to scale changes. These are essentially arguments of dimensional analysis and, as in any such argument, it is necessary that there should be no other "characteristic" values to take account of. For instance, if the parameter space in the last example were $\theta \geqslant 0$, the argument for (45) would fail. More formally the parameter space would not be invariant under the transformations in question.

A general formulation is as follows. Consider the model $f_Y(y ; \theta)$ with $\theta \in \Omega$. Suppose that for all transformations g in a group $\mathcal{G}$ there exists a unique transformed parameter value $\phi = g^* \theta$ such that the distribution of the transformed random variable is given by $f_Y(y ; \phi)$ and $g^* \Omega = \Omega$. Then, to take the point estimation problem to be specific, any estimate $t(y)$ of θ should, according to the invariance principle, satisfy $t(gy) = g^* t(y)$. It is usually convenient to apply the arguments to the minimal sufficient statistic rather than directly to y.

The following example has been discussed by Basu (1964) and by Barnard and Sprott (1971).

Example 2.37. Special multinomial distribution (ctd). Consider an extension of Example 2.28 in which Y can take on the values $1, 2, \ldots, 6$ with probabilities

$$\text{pr}(Y = j) = \begin{cases} (1 - j\theta)/12 & (j = 1, 2, 3), \\ \{1 + (j - 3)\theta\}/12 & (j = 4, 5, 6), \end{cases}$$

the parameter space being the interval $[-1, 1]$. For this very special multinomial distribution there are six possible ancillary statistics, namely $(N_1 + N_4, N_2 + N_5, N_3 + N_6), \ldots, (N_1 + N_6, N_2 + N_5, N_3 + N_4)$, in an obvious notation. If we define the transformation g by $gY = Y + 3 \pmod 6$ the induced parameter transformation is $g^*\theta = -\theta$.

The only ancillary statistic unaffected by the transformation g is $(N_1 + N_4, N_2 + N_5, N_3 + N_6)$ and on the grounds that the inference should be unchanged by this seemingly harmless transformation there is a unique choice of ancillary statistic.

That this argument is rather special can be seen from Example 2.28, where both ancillaries are invariant under the corresponding transformation.

Some transformations that arise in applying invariance arguments are physically natural in the sense that, for example, it is largely conventional what units are used for physical quantities. Any analysis that was not invariant under a change from imperial to metric units would be extremely suspect. Again, in the last example, the relabelling of the cells is just a change in the conventional order in which the cells are recorded; of course, if the cells had been recorded in an order expected to put the larger probabilities first and if it were required to take account of this information in the analysis, then the basis for the invariance argument would collapse.

Sometimes, however, the transformations considered arise purely from the mathematical structure of the problem and have no direct significance. Then the force of the invariance principle is weakened. The following is an important example.

Example 2.38. Normal-theory linear model (ctd). We return once more to the situation of Example 2.4 of independent normal random variables of constant variance and with expectations given by a linear model. Let the group of transformations be the set of all orthogonal transformations $y \rightarrow ly$. The normality, independence and constancy of variance are unaffected by the transformations and, because $E(Y) = x\beta$ implies that $E(lY) = lx\beta = x(x^Tx)^{-1}x^T lx\beta$, it follows that the associated transformation in the parameter space is

$$\beta \rightarrow (x^Tx)^{-1}x^T lx\beta. \tag{46}$$

Now, while such statistics as the residual sum of squares and the F statistic for testing the sum of squares for regression are indeed invariant under these transformations, it is clear that, at least in the great majority of applications, one would not regard $1y$ and y as equally natural for the initial recording of the data. Putting the point another way, there will often be special relations between the component observations or between the component parameters that make the appeal to invariance unconvincing.

A final general point about invariance arguments is that it will turn out in a decision-making context that there is sometimes a clash between the invariance principle and other apparently more compelling requirements; there can be a uniform loss of expected utility from following the invariance principle.

2.4   Some approaches to statistical inference

(i) General remarks
The five principles discussed in the previous section are all rather abstract in the sense that they concern the general way in which the data should affect the conclusions, i.e. what aspects of the data and model are relevant. They do not concern the exact form and interpretation of the conclusions.

We now turn to the question of the interpretation of the conclusions. Here there are again a number of principles that it is useful to bear in mind. We describe these briefly.

(ii) Strong repeated sampling principle
According to the strong repeated sampling principle, statistical procedures are to be assessed by their behaviour in hypothetical repetitions under the same conditions. This has two facets. Measures of uncertainty are to be interpreted as hypothetical frequencies in long run repetitions; criteria of optimality are to be formulated in terms of sensitive behaviour in hypothetical repetitions.

The argument for this is that it ensures a physical meaning for the quantities that we calculate and that it ensures a close relation between the analysis we make and the underlying model which is regarded as representing the "true" state of affairs.

(iii) Weak repeated sampling principle
The weak version of the repeated sampling principle requires that we should not follow procedures which for some possible parameter

values would give, in hypothetical repetitions, misleading conclusions most of the time.

Example 2.39. Sum of squared normal means. Suppose that $Y_1, \ldots,$ Y_n are independently normally distributed with unit variance and expectations $\mu_1, \ldots, \mu_n$. Now according to some approaches to statistical inference, given $Y_j = y_j$, the parameter μ_j can be treated exactly as if it were normally distributed with mean y_j and unit variance. If this is so and statements about mathematically unrelated parameters can be combined by the laws of probability, it would follow that $\delta^2 = \mu_1^2 + \ldots + \mu_n^2$ is distributed with an expectation $y_1^2 + \ldots + y_n^2 + n = d^2 + n$ and a variance easily shown to be $2n + 4d^2$. The final term n in the expectation enters because, in this special treatment, the expectation of μ_j^2 is $y_j^2 + 1$.

This leads us, approximately, to the conclusion that δ^2 is almost certainly within $O(\sqrt{n})$ of $d^2 + n$, provided that δ^2 and d^2 are of order n. On the other hand, if $D^2 = Y_1^2 + \ldots + Y_n^2$, it is easily shown that

$$E(D^2) = \delta^2 + n, \ \mathrm{var}(D^2) = 2n + 4\delta^2. \qquad (47)$$

That is, for any given δ^2 of order n, the quantity d^2 is in hypothetical repetitions nearly always too large as an estimate of δ^2. To obtain an interval that in hypothetical repetitions would nearly always contain δ^2 we must take an interval of width of the order of $\sqrt{n}$ around $d^2 - n$. This shows that the interval given by the argument treating μ_j as a random variable is, for large n, nearly always wrong. Therefore the argument as proposed is inconsistent with the weak repeated sampling principle. This example is due to Stein (1959), who gave a more thorough mathematical discussion; for further comments, see Pinkham (1966).

While the weak repeated sampling principle is too vague and imprecise in its domain of applicability to be very constructive, it is strong enough to put in serious question some proposals that have been made, like the one above.

(iv) Bayesian coherency principle
Chapter 10 outlines the approach to statistical inference via Bayes's theorem. Briefly, in this all uncertainties are described by probabilities, i.e. unknown parameters have probability distributions both

before the data are available and after the data have been obtained. This can be justified by the supposition that

(a) any individual has an attitude to every uncertain event which can be measured by a probability, called a subjective probability;

(b) all such probabilities for any one individual are comparable;

(c) these subjective probabilities can be measured by choice in hypothetical betting games.

The Bayesian coherency principle is that the subjective probabilities should be such as to ensure self-consistent betting behaviour. It can be shown that this implies that subjective probabilities for any one individual should be manipulated by the ordinary laws of probability and in particular by Bayes's theorem. Uncertainty calculations that are not in accord with this are equivalent to inconsistent betting behaviour.

This principle implies that conclusions about unknown parameters in models have implicitly or explicitly to be in the form of probability statements. Now this implies all the principles of Section 2.3 and in particular the strong likelihood principle. Thus by making the conclusions depend on the method of sampling or stopping rule used to obtain the data, Bayesian coherency is not achieved.

(v) Principle of coherent decision making

In problems where an explicit decision is involved, arguments parallel to those used in connexion with Bayesian coherency show that for any individual each decision and true parameter value have an associated utility such that the optimum decision is found by maximizing expected utility. Note that this is a requirement of internal consistency amongst related decisions by the same individual; there is nothing to say that the utility maximized is the "correct" one in a broader sense. This is discussed further in Chapter 11.

(vi) Discussion

The principles outlined above all have appeal. A crucial question is whether one regards Bayesian coherency as totally convincing, in which case the repeated sampling principle must be sacrificed, at least in its strong form.

While there are certainly situations where the consideration of explicit subjective probabilities is valuable, we feel that there are compelling arguments in most statistical problems for giving precedence to the repeated sampling principle and for not regarding

Bayesian coherency as the primary objective. The reasons for this are

(a) all uncertainties are not comparable; it is not usually sensible to regard a probability determined or estimated from a stable physical random system as on the same footing for all purposes as a subjective impression based on no data as to what value an unknown parameter might take. Indeed, two subjective probabilities with equal numerical values may not carry equal conviction;

(b) the betting games contemplated in defining subjective probability are at best interesting models of learning in the face of uncertainty;

(c) most importantly, the Bayesian coherency principle is a requirement for a certain kind of self-consistency. It contains no direct guarantee that the answers we obtain are related to the objective truth about the system under study. On the other hand, the repeated sampling principles are concerned with the relation between the data and a model of the real world. They deal with the question of how close to the "true" answer we are likely to get by repeated use of a particular technique.

Of course this must not be taken as an attempt to dismiss Bayesian arguments. In some applications the relevant probabilities will be available as frequencies and in others it will be desirable to introduce well-based subjective probabilities into the argument. In any case, at a more theoretical level it is always instructive to analyse a problem from several points of view. We do, however, consider that for the majority of applications repeated sampling arguments are the more relevant and for that reason devote the majority of the book to them.

Having discussed these two types of principle, those of the previous section and now several concerned with interpretation, we can set out briefly four broad approaches to statistical inference, namely via sampling theory, via likelihood theory, via Bayesian theory, and via decision theory.

(vii) Sampling theory

In the sampling theory approach to statistical inference, primary emphasis is put on the strong repeated sampling principle, i.e. on ensuring that we follow procedures with an interpretation in terms of frequencies in hypothetical repetitions under the same conditions.

A typical example is the calculation of a confidence interval for the mean μ of a normal distribution, in the simplest situation where the random variables $Y_1, \ldots, Y_n$ are i.i.d. with known variance σ_0^2. By considering the distribution in repetitions of the random variable $\bar{Y} = \Sigma Y_j / n$, we show that

$$\mathrm{pr}(\bar{Y} - k^*_{\frac{1}{2}\alpha} \sigma_0 /\sqrt{n} < \mu < \bar{Y} + k^*_{\frac{1}{2}\alpha} \sigma_0 /\sqrt{n}) = 1 - \alpha, \qquad (48)$$

where $\Phi(k^*_{\frac{1}{2}\alpha}) = 1 - \frac{1}{2}\alpha$, $\Phi(.)$ denoting the standardized normal integral. Hence the interval $(\bar{y} - k^*_{\frac{1}{2}\alpha} \sigma_0 /\sqrt{n}, \bar{y} + k^*_{\frac{1}{2}\alpha} \sigma_0 /\sqrt{n})$ calculated from the data has the property of being found by a procedure that will include the true value $1 - \alpha$ of the time, in the long run. It is called a $1 - \alpha$ confidence interval and is a fairly typical example of a procedure owing its interpretation, and optimality properties, to the repeated sampling idea.

This approach will not satisfy the strong likelihood principle; for example a different form of confidence interval would in general be required for sequential sampling.

The majority of this book is devoted to the sampling theory approach, but two important provisos must be made and will be repeatedly referred to later.

The first is that the repeated sampling interpretation is hypothetical. Thus in the confidence interval example just mentioned it is not suggested that only one value of α should be chosen and that the interval (48) should be regarded as the single assertion to be made from the data. We would typically consider such intervals at least implicitly for various values of α. Any one of them has the physical interpretation outlined above, but no such question as choosing *the* value of α appropriate for the problem arises. If in fact what is required is to give a single interval statement, this is best regarded as a decision problem and the consideration of utility functions becomes important. Decision problems of this particular type seem, however, to arise extremely rarely in applications.

The second, and even more important, point is that we must satisfy ourselves in any application that the hypothetical long-run statements that we consider are relevant to the particular scientific or technological problem under study. This we do by use of the conditionality principle. By considering distributions that are, in a sense, as conditional as possible without sacrificing information, we ensure that the long run of trials considered is like the data.

There are serious difficulties in a sampling theory approach that does not take account of a conditionality principle, although whether the particular formulation of Section 2.3 (iii) is best is, of course, another matter.

(viii) Likelihood theory

A second approach is to use the likelihood function itself directly as a summary of information. In particular, ratios of likelihoods or differences of log likelihood give the relative plausibilities of two preassigned parameter values, say θ_1 and θ_2.

This approach clearly satisfies weak and strong likelihood principles; in addition the conditionality principle is implicitly satisfied in that any factor in the likelihood that does not involve the parameter cancels when likelihood ratios are formed.

Now the likelihood function plays a central role in the great majority of statistical theory and much of the rest of this book is in effect concerned with extracting from the likelihood function answers to specific questions about the model. According to the likelihood theory, all that is necessary is to inspect the function; in particular, consideration of the properties in repeated sampling of quantities derived from the likelihood function is argued to be misleading and unnecessary.

Except in the simplest situations, such as the one mentioned above of comparing two preassigned values of θ, there are two objections to the likelihood approach. The first is the difficulty of dealing with problems with many parameters. Quite apart from the computational and conceptual difficulty of handling a function of many variables, it will usually be required to focus attention on a small number of components, and further rules outside the scope of "pure" likelihood theory are necessary to do this. Sometimes it will be possible to use one of the extended definitions of likelihood in Section 2.1 (iv). A more serious matter is the possibility of conflict with the weak repeated sampling principle. This can arise when the whole of the likelihood function is inspected and a comparison made in the light of the particular function. Three examples will illustrate this.

Example 2.40. Sequential sampling against a fixed value. Suppose that observations are i.i.d. in $N(\mu, \sigma_0^2)$, where σ_0^2 is known. Let observations be taken until the first time that $|\bar{y}_{n.}| > k\sigma_0/\sqrt{n}$, where k is a constant which might for definiteness be taken as 3, and $\bar{y}_{n.}$ denotes the mean of the first n observations. It is easily shown that, with probability one, sampling stops after some finite n, and we make the inessential approximation that when sampling does stop, $|\bar{y}_{n.}| = k$. Now according to any approach that is in accord with the strong likelihood principle, the fact that this particular stopping rule has been used is irrelevant. This applies in particular to the "pure" likelihood approach.

The likelihood function corresponds to the normal distribution of $\bar{Y}_{n.}$ around μ with variance σ_0^2/n and it is easily shown from this that the likelihood at $\mu = 0$ is $e^{-\frac{1}{2}k^2}$ times that at $\mu = \bar{y}_{n.}$ and thus by choice of k the ratio can be made very small.

That is, even if in fact $\mu = 0$, there always appears to be strong evidence against $\mu = 0$, at least if we allow comparison of the likelihood at $\mu = 0$ against any value of μ and hence in particular against the value of μ giving maximum likelihood. This contradicts the weak repeated sampling principle.

Note, however, that if we confine the analysis to the comparison of the likelihood at $\mu = 0$ with that at some fixed $\mu = \mu'$, this difficulty does not arise; it can be shown that the likelihood at $\mu = 0$ is usually greater than that at $\mu = \mu'$, if in fact $\mu = 0$. In practice, however, such limited comparisons will rarely be adequate.

One resolution of the difficulty is to compare the likelihood function achieved with the distribution of other likelihood functions that would be obtained in hypothetical repetitions of the sampling procedure. Another is to work not with likelihood functions but with averages of likelihood functions with respect to suitable fixed weight functions. These correspond, respectively, to the sampling theory and Bayesian approaches.

Of course, the force of this example in no way depends on the practical reasonableness of the particular sampling rule. If, however, the main objective is to find $sgn(\mu)$ the sampling rule is a perfectly reasonable one. The sample size N treated as a random variable has a distribution unaffected by $sgn(\mu)$ but heavily dependent on $|\mu|$.

Example 2.41. Some point hypotheses. A situation very similar to that discussed in Example 2.40 can arise without sequential sampling (Birnbaum, 1969). Let Y be a random variable taking values $1, \ldots,$ 100. One observation is obtained. Suppose that there are 101 possible distributions conveniently indexed by a parameter θ taking values $0, 1, \ldots, 100$. If $\theta = r$ $(r \neq 0)$, Y is certain to equal r. If $\theta = 0$, all values $1, \ldots, 100$ are equally likely.

Now whatever value y is observed the likelihood at $\theta = 0$ is 1/100th of its value at $\theta = y$. That is, even if in fact $\theta = 0$, we are certain to find evidence apparently pointing strongly against $\theta = 0$, if we allow comparisons of likelihoods chosen in the light of the data. Note, however, that if we confine attention to two preselected parameter values $\theta = 0$ and $\theta = r_0$, comparisons of likelihoods are not systematically misleading. If $\theta = 0$, we find with probability 99/100 an infinite likelihood ratio in favour of the true value, and with probability 1/100 a ratio of 100 in favour of the false hypothesis $\theta = r_0$. This is entirely in accord with the weak repeated sampling principle.

Of course, if $\theta = 0$, the taking of a second observation would, with high probability, reveal the true state of affairs in this particular case.

Example 2.42. One observation from a normal distribution. Suppose that one observation is obtained from $N(\mu, \sigma^2)$, both parameters being unknown. The log likelihood function, being

$$-\tfrac{1}{2}\log(2\pi\sigma^2) - (y - \mu)^2/(2\sigma^2),$$

is unbounded for $\mu = y$ and $\sigma \to 0$. More realistically, if we take account of the inevitably discrete nature of observations, the likelihood rises to a finite maximum attained when σ is small compared with the grouping interval used in recording.

To conclude from this that, μ being unknown, the single observation provides strong evidence that σ is very small would, however, be absurd. For a non-trivial example where the likelihood by itself is totally uninformative, see Exercise 2.16.

(ix) Bayesian theory
Suppose now that, in addition to the p.d.f. $f_Y(y\,;\theta)$ assumed to generate the data, we can treat the parameter θ as the value of a random variable Θ, having a known marginal p.d.f. $f_\Theta(\theta)$. That is, we can suppose the parameter value to have a marginal distribution, usually called the *prior distribution*. Then the data y are generated from the conditional p.d.f. $f_Y(y\,;\theta)$, which we can now write $f_{Y|\Theta}(y|\theta)$. Because we are interested in the particular value of Θ realized in the instance under study, we consider the conditional p.d.f. of Θ given $Y = y$, usually called the *posterior distribution*. It is given by Bayes's theorem as

$$f_{\Theta|Y}(\theta\,|\,y) \;=\; \frac{f_{Y|\Theta}(y\,|\,\theta)f_{\Theta}(\theta)}{\int_{\Omega} f_{Y|\Theta}(y\,|\,\theta')f_{\Theta}(\theta')d\theta'}.\tag{49}$$

If we are interested in particular components of Θ, their posterior p.d.f. is formed by integrating out the nuisance components. It is essential that $f_{\Theta}(\theta)$ is known, although sometimes it is natural to regard Θ as having a parametric p.d.f. $f_{\Theta}(\theta\,;\zeta)$, where the parameter ζ is of lower dimension than θ, and itself is the value of a random variable having a known p.d.f.

If the prior distribution arises from a physical random mechanism with known properties, this argument is entirely uncontroversial. The posterior distribution can be regarded as a hypothetical frequency distribution and the first four principles of Section 2.3 are all satisfied. For example, the strong likelihood principle follows directly from (49). A frequency prior is, however, rarely available. To apply the Bayesian approach more generally a wider concept of probability is required and, in particular, we may invoke the Bayesian coherency principle of Section 2.4(iv). Then the prior distribution is taken as measuring the investigator's subjective opinion about the parameter from evidence other than the data under analysis.

The form of the posterior density is often simplified by taking a mathematical form that combines naturally with the data. The following example illustrates this and enables some of the properties of this approach to be examined.

Example 2.43. Bernoulli trials (ctd). Suppose that as in Example 2.1 observations are made on independent binary random variables $Y_1, \ldots, Y_n$ with $\mathrm{pr}(Y_j = 1) = \theta$. Suppose also that the prior p.d.f. for Θ has the beta form

$$\theta^{a-1}(1-\theta)^{b-1}/B(a, b).$$

Then it follows from Bayes's theorem, (49), that

$$f_{\Theta|Y}(\theta\,|\,y) \;=\; \theta^{a+r-1}(1-\theta)^{b+n-r-1}/B(a+r, b+n-r),$$

where $r = \Sigma y_j$ is the number of ones observed. Note that this would be the likelihood if, in addition to observing r ones in n trials, we had previously observed $a-1$ ones in $a+b-2$ trials on the same system.

Note also that the strong likelihood principle is satisfied in that the data enter only through the likelihood, and that any constant factor will cancel.

For any fixed a and b the influence of the prior density on the final conclusion decreases as n increases.

Example 2.44. Two possible parameter values. The simplest example of the Bayesian argument arises when there are just two possible values of the unknown parameter. This case throws light also on the "pure" likelihood approach.

Denote the two possible parameter values by θ_0 and θ_1 and, to simplify the notation, denote the p.d.f.'s of the data by $f_0(y)$ and $f_1(y)$ and the prior probabilities by π_0 and π_1. Then by (49)

$$f_{\Theta \mid Y}(\theta_j \mid Y = y) = \frac{\pi_j f_j(y)}{\pi_0 f_0(y) + \pi_1 f_1(y)}$$

or

$$\frac{\mathrm{pr}(\Theta = 0 \mid Y = y)}{\mathrm{pr}(\Theta = 1 \mid Y = y)} = \frac{\pi_0}{\pi_1} \times \frac{f_0(y)}{f_1(y)} . \tag{50}$$

Thus if we call the left hand side the *posterior odds* for θ_0 versus θ_1, we have that the posterior odds equals the *prior odds* times the likelihood ratio.

This shows very clearly the influence of the data and the prior information on the final conclusion.

In fact (50) does not depend on there being only two parameter values, but follows directly from Bayes's theorem applied to two particular parameter values. This provides one convenient way of interpreting the likelihood ratio for two preassigned parameter values; see Section 2.4(viii). The ratio is such that for any prior distribution it summarizes the contribution of the data to the posterior odds. In particular, it equals the posterior odds for the special case when the two parameter values have equal prior density.

The Bayesian approach satisfies, in addition to the first four principles of Section 2.3, an extended form of the fifth principle, the invariance principle. The extension is that the transformations involved must apply also to the prior distribution. On the whole, this extended invariance principle is not very interesting.

The arguments against regarding Bayesian coherency as a universally compelling requirement in statistical inference have been reviewed in Section 2.4(vi).

(x) Decision theory

The final general approach to be discussed emphasises the action to be taken in the light of the data. If for each parameter value the consequences of each possible action can be measured by a utility function, then we can evaluate the expected utility of the possible methods of action. We can then rule out certain methods of action on the grounds that they lead to uniformly lower expected utility for all parameter values. A unique optimal action will be defined if a prior distribution is available, in which case the expected utility, averaged with respect to the prior distribution, can be maximized over the spectrum of possible actions. The principle of coherent decision making is then applicable.

In fact it is not usually necessary that the utility function be completely specified in order that an optimal action be defined, as we see in the following extension of the last example.

Example 2.45. Two possible parameter values (ctd). Suppose that, in the situation of Example 2.44, we are to decide between θ_0 and θ_1 as to which is the true parameter value, knowing in advance the utilities of the various consequences. That is, the possible actions are d_0, i.e. decide that $\theta = \theta_0$ and d_1, i.e. decide that $\theta = \theta_1$. Let the utility of decision d_j when in fact $\theta = \theta_i$ be u_{ij}, where obviously $u_{00} > u_{01}$ and $u_{11} > u_{10}$.

The action d_0 or d_1 will be taken on the basis of the data y according to some method or rule described formally by $d(y)$. The expected utility of the method averaged over both prior and sampling distributions is

$$\Sigma\Sigma u_{ij} \pi_i \, \mathrm{pr}\{d(Y) = d_j | \theta = \theta_i\}. \tag{51}$$

It is not difficult to see that this expected utility can be maximized by uniformly maximizing the conditional expected utility given $Y = y$, which directly involves the posterior odds (50). Discussion of this is postponed to Chapter 11. However, the main point of the example is that with the addition of a utility function to the formal specification the problem becomes one of simply choosing a function $d(y)$ to maximize (51), there being no ambiguity about how to proceed.

There are two main reasons why this approach, while it has important applications, is not a satisfactory universal basis for statistical analysis. The first is that even when the decision-making aspect is to the fore, the quantitative information for the full analysis may not be available. In Chapter 11, where we develop decision theory, we shall discuss what can be done in such cases. A second and much more important reason is that many if not most statistical analyses have as their immediate objective the summarization of evidence, the answering of the question "What can reasonably be learned from these data?". Even when a clear-cut decision is involved, the summarization of evidence for discussion is very often required, whereas in scientific rather than technological applications the decision-making element, while often present, is usually secondary.

The general qualitative ideas of decision theory, that is the clarification of the purpose of the analysis, the possible decisions available and their consequences, are, however, very important.

Bibliographic notes

The ideas introduced in this Chapter have a long history and an enormous literature. Much modern work may be considered as stemming from a very remarkable paper of Fisher (1922) which gave, among other things, the idea of sufficiency. Fisher's later work on the foundations of statistical inference stressed likelihood and conditionality and his views on the bases are set out in his last book (Fisher, 1973, 1st ed. 1956); see also particularly his papers of 1925 and 1934. The mathematical niceties of sufficiency are given by Halmos and Savage (1949) and Bahadur (1954). The exponential family was mentioned by Fisher (1922), but its central importance was first shown in detail by Pitman (1936), Koopman (1936) and Darmois (1936), after some or all of whom the family is sometimes named. The more mathematical properties of the family are discussed by Linnik (1968) and the connexions with convex analysis by Barndorff-Nielsen (1973a, b).

Fisher's approach to the choice of statistical procedures depended heavily on likelihood and sufficiency and an approach relying more explicitly on properties in repeated sampling was developed in the 1930's in a series of joint papers by Neyman and Pearson; these are now available as a book (Neyman and Pearson, 1967). At first their work was regarded as an approach alternative but essentially equivalent to Fisher's, but in the late 1930's strong disagreements emerged, stemming partly from Fisher's attaching little importance to the strong repeated sampling principle and partly from Neyman's emphasis on decision making. The mathematical aspects of a sampling theory approach are treated in books by Lehmann (1959) and by Zacks (1971).

Fisher at times came close to direct use of the likelihood and this was developed, for carefully restricted categories of problems, by Barnard (1951) and in particular by Barnard, Jenkins and Winsten (1962). The relation between likelihood, conditionality and sufficiency was examined in an important paper by Birnbaum (1962). General use of a "pure" likelihood approach is strongly advocated by Edwards (1972).

Birnbaum (1969) discussed various principles that might apply to statistical inference and the treatment given here draws heavily on his work, in addition to that of Barnard.

The use of conditional and marginal likelihoods was introduced by Bartlett (1936a, b, 1937) and has been extensively developed by Kalbfleisch and Sprott (1970) and by Andersen (1970, 1973).

Methods based on Bayes's theorem were, in particular, much used by Laplace. Non-frequency prior distributions were criticized in the late 19th century and strongly condemned by Fisher. More recently, however, they have attracted increasing attention. Jeffreys (1961, 1st ed. 1939) developed a general theory based on rational degrees of belief and moreover developed a large number of special Bayesian techniques. Most modern work, however, stresses the subjective nature of prior distributions and stems from the work of Ramsey, de Finetti, Good and Savage; for an excellent review, see Lindley (1971a).

Modern statistical decision theory can be regarded as starting with the paper of Neyman and Pearson (1933b), although some of Gauss's work on least squares was essentially in decision theory form. A general non-Bayesian form was developed by Wald (1950); that is, utilities or losses were assumed known but no prior distributions were considered. Despite the historical importance and influence of Wald's work, his approach now seems of limited interest. In particular the fully Bayesian version of decision theory is the natural endpoint of the ideas on subjective probability mentioned above; see particularly Raiffa and Schlaifer (1961).

The introduction of invariance considerations was originally due to Hotelling (1931) and Pitman (1938); extensive further developments by C. Stein have been very influential even though largely unpublished. Hall, Ghosh and Wijsman (1965) cover aspects particularly relevant to sequential sampling.

Barnett (1973) gives an elementary introduction to and comparison of the main approaches to statistical inference and decision making.

Further results and exercises

1. Verify the factorization (10) of the likelihood for the particular normal-theory linear model representing linear regression through the origin. Obtain also from first principles the factorization for the two-way cross-classification, i.e. for random variables $Y_{j_1 j_2}$ $(j_1 = 1, \ldots, m_1; j_2 = 1, \ldots, m_2)$ with

$$E(Y_{j_1 j_2}) = \mu + \alpha_{j_1} + \beta_{j_2}, \quad \Sigma \alpha_{j_1} = \Sigma \beta_{j_2} = 0.$$

Show that if the model is supplemented by a term $\gamma \alpha_{j_1} \beta_{j_2}$ where γ is unknown, no reduction of dimensionality by sufficiency is achieved.

[Sections 2.1, 2.2(ii)]

2. Suppose that random variables follow the first order autoregressive process

$$Y_r = \mu + \rho(Y_{r-1} - \mu) + \epsilon_r,$$

where $\epsilon_1, \ldots, \epsilon_n$ are i.i.d. in $N(0, \sigma_\epsilon^2)$ and $|\rho| < 1$. Write down the likelihood for data $y_1, \ldots, y_n$ in the cases where the initial value y_0 is

(i) a given constant;

(ii) the value of a random variable with the p.d.f. $N\{\mu, \sigma_\epsilon^2/(1 - \rho^2)\}$, independent of $\epsilon_1, \ldots, \epsilon_n$;

(iii) assumed equal to y_n.

Find the minimal sufficient statistic for the parameter $(\mu, \rho, \sigma_\epsilon^2)$ in each case and show in particular that in cases (i) and (ii) the statistic $(\Sigma Y_j, \Sigma Y_{j-1} Y_j, \Sigma Y_j^2)$ has to be supplemented by end corrections.

[Sections 2.1, 2.2(ii); Bartlett, 1966]

3. In a simple queueing system customers arrive in a Poisson process of rate α. There is a single server and the distribution of service-time is exponential with parameter β, i.e. given that there is a customer present the probability of a service completion in the time interval

$(t, t + \Delta t)$ is $\beta \Delta t + o(\Delta t)$ independently of the system's past history. Prove that, provided there is at least one customer present, the time between successive "events", i.e. arrivals or service completions, is exponentially distributed with parameter $\alpha + \beta$, and that the probability that the event is an arrival is $\alpha/(\alpha + \beta)$ independently of the corresponding time intervals. The system is observed for a fixed time t, starting from a given arbitrary state. By building up the likelihood from (a) the intervals between events in the busy state, (b) the types of these events, (c) the intervals spent in the empty state, (d) the duration of any incomplete interval at the end of the period of observation, show that the likelihood is

$$\alpha^{n_a} \beta^{n_b} e^{-\alpha t} e^{-\beta(t - t_0)},$$

where n_a and n_b are the numbers of arrivals and service completions and t_0 is the time for which no customers are present.

[Section 2.1; Cox, 1964]

4. In the one-way random effects model of analysis of variance, random variables Y_{jk} ($j = 1, \ldots, m; k = 1, \ldots, r$) have the form

$$Y_{jk} = \mu + \eta_j + \epsilon_{jk},$$

where the η_j's and the ϵ_{jk}'s are independently normally distributed with zero means and variances respectively σ_b^2 and σ_w^2. The unknown parameter is thus $(\mu, \sigma_b^2, \sigma_w^2)$. Show that the minimal sufficient statistic is $\{\bar{Y}_{..}, \Sigma(\bar{Y}_{j.} - \bar{Y}_{..})^2, \Sigma\Sigma(Y_{jk} - \bar{Y}_{j.})^2\}$, where $\bar{Y}_{j.} = \Sigma Y_{jk}/r$ and $\bar{Y}_{..} = \Sigma \bar{Y}_{j.}/m$. What is the minimal sufficient statistic if it is given that $\mu = 0$? Generalise the results to cover the case where each Y_{jk} is a $p \times 1$ multivariate normal vector.

[Section 2.2]

5. Independent binary random variables $Y_1, \ldots, Y_n$ are such that the probability of the value one depends on an explanatory variable x, which takes corresponding values $x_1, \ldots, x_n$. Show that for the model

$$\rho_j = \log \left\{ \frac{\mathrm{pr}(Y_j = 1)}{\mathrm{pr}(Y_j = 0)} \right\} = \gamma + \beta x_j,$$

the minimal sufficient statistic is $(\Sigma Y_j, \Sigma x_j Y_j)$; the quantity ρ_j is called the logistic transform. Generalise to the case where the $n \times 1$ vector ρ is given by the linear model $\rho = x\beta$ with x a known $n \times q$

matrix of rank q and β a $q \times 1$ vector of unknown parameters.

[Section 2.2; Cox, 1970]

6. The random variables $Y_1, \ldots, Y_n$ are i.i.d. with probability density equal to $2/(3\theta)$ on $0 \leqslant y \leqslant \frac{1}{2}\theta$, and equal to $4/(3\theta)$ on $\frac{1}{2}\theta < y \leqslant \theta$. Prove that the minimal sufficient statistic consists of the number of observations less than $\frac{1}{2}Y_{(n)}$ and those order statistics above $\frac{1}{2}Y_{(n)}$.

[Section 2.2]

7. Suppose that Y_1 and Y_2 are independent random vectors with a distribution depending on the same parameter θ, and that S_1 and S_2 are corresponding minimal sufficient statistics. Show that for the combined random vector $Y = (Y_1, Y_2)$, $S = (S_1, S_2)$ is sufficient, although not necessarily minimal sufficient. Use the Gaussian first-order autoregressive process as an explicit example to show that the result does not hold in general for dependent random variables. Suggest an extended definition of sufficiency which is such that when dependent data sets are merged, the sufficient statistic for the extended set can always be recovered and illustrate the definition on the Gaussian first-order autoregressive process.

[Section 2.2; Fisher, 1925; Bahadur, 1954; Lauritzen, 1975]

8. In a discrete time stochastic process, $\mathcal{F}_1$ and $\mathcal{F}_2$ are two disjoint sets of time points, with Y_1 and Y_2 the corresponding random vectors. The first is to be observed and from this the second is to be predicted. The function S_1 of Y_1 is called *predictive sufficient* for Y_2 if S_1 is sufficient for the parameter θ and if

$$f_{Y_1, Y_2 | S_1}(y_1, y_2 | s_1 ; \theta) = f_{Y_1 | S_1}(y_1 | s_1) f_{Y_2 | S_1}(y_2 | s_1).$$

Again illustrate the result on the first order autoregressive process, this time with known parameter.

[Section 2.2; Lauritzen, 1975]

9. A population consists of an unknown number θ of individuals. An individual is drawn at random, is marked and returned to the population. A second individual is drawn at random. If it is the marked individual, this is noted; if not, the individual is marked. The individual is then returned to the population. At the mth step, an individual is drawn at random. It is noted whether the individual is marked. If not, it is marked. The individual is then returned to the

population. This continues for n steps; the observation consists of a binary sequence indicating, for each step, whether the individual is or is not marked. Show that the likelihood is proportional to $\theta(\theta-1)\dots(\theta-n+r+1)/\theta^{n-1}$, where r is the number of times that the individual drawn is already marked, and hence that R is a sufficient statistic. Prove that the same conclusion holds

(a) if the number n is chosen by an arbitrary data-dependent sequential stopping rule;

(b) if for each marked individual selected, a record is available of the particular previous trials on which it was observed;

(c) if at the mth step, not one but k_m individuals are drawn at random from the population.

List the assumptions made in this analysis and comment on their reasonableness in investigating biological populations.

[Section 2.2; Goodman, 1953; Cormack, 1968; Seber, 1973]

10. The random variables $Y_1, \dots, Y_n$ obey a second-order linear model, i.e. they are uncorrelated and have constant variance and $E(Y) = x\beta$, where x is $n \times q_x$, $q_x < n$ and of full rank. The vector statistic S is called *linearly sufficient* for β if any statistic uncorrelated with S has expectation independent of β. Comment on the relation of this definition with ordinary sufficiency and show that $(x^T x)^{-1} x^T Y$ is linearly sufficient.

[Section 2.2; Barnard, 1963]

11. Suppose that $Y_1, \dots, Y_n$ are i.i.d. with probability density $f(y;\theta)$ and that $T = t(Y)$ is a one-dimensional sufficient statistic for θ for all values of n. If θ_1 and θ_2 are any two fixed values of θ, show, starting with the factorization (27), that for any θ

$$\frac{\partial}{\partial y_j} \log \left\{ \frac{f(y_j;\theta)}{f(y_j;\theta_1)} \right\} \bigg/ \frac{\partial}{\partial y_j} \log \left\{ \frac{f(y_j;\theta_2)}{f(y_j;\theta_1)} \right\}$$

is independent of y_j and hence is a function of θ alone. Since θ_1 and θ_2 are fixed, deduce from this that $f(x;\theta)$ has the exponential family form (35).

[Section 2.2(vi)]

12. Check the form of the natural parameter for the Poisson distribution (log mean), binomial distribution (log odds), exponential distribution (reciprocal of mean), gamma distribution with index

known (reciprocal of mean), negative binomial distribution with index known (log probability), normal distribution with both parameters unknown (mean divided by variance and reciprocal of variance) and multivariate normal distribution of zero mean and unknown covariance matrix (inverse of covariance matrix). Discuss in each case the extent to which it is sensible to use the natural parameter for interpretation, e.g. in the comparison of different sets of data. Relevant considerations include directness of physical meaning and range of possible values.

[Section 2.2(vi)]

13. Prove that when the exponential family density is taken in its natural form (36),

$$f_Z(z \; ; \phi) = \exp\{-z\phi + c^\dagger(\phi) + d^\dagger(z)\},$$

the cumulant generating function of Z is $c^\dagger(t + \phi) - c^\dagger(\phi)$; use, for this, the condition that the total integral of the density is one. Hence express the cumulants of Z for small ϕ in terms of those for $\phi = 0$; compare with a non-degenerate case for which the exact cumulants are known.

[Section 2.2(vi)]

14. The random variables $Y_1, \ldots, Y_n$ are i.i.d. in the density $\tau^{-1}h\{y - \mu)/\tau\}$. Find a transformation $Z_1, \ldots, Z_n$ of the order statistics $Y_{(1)}, \ldots, Y_{(n)}$ such that $Z_3, \ldots, Z_n$ are invariant under location and scale changes. Hence find the ancillary part of the minimal sufficient statistic.

[Section 2.2(vi); Fisher, 1934]

15. Two independent types of observation are obtained on a Poisson process of rate ρ, the first type corresponding to counts of numbers of events in intervals of fixed length t_0, the second to the measurement of intervals between successive events. The corresponding random variables are $Y_1, \ldots, Y_{n_1}$ i.i.d. in a Poisson distribution of mean ρt_0, and $Z_1, \ldots, Z_{n_2}$ i.i.d. in an exponential distribution of parameter ρ. Prove that the minimal sufficient statistic is $(\Sigma Y_j, \Sigma Z_k)$.
   It is not known for certain whether there is an ancillary statistic, although it seems very likely that there is none. Show that while $V = (\Sigma Y_j)(\Sigma Z_k)$ has expectation independent of ρ, its variance does depend on ρ. Obtain the conditional distribution of ΣY_j given $V = v$.

[Section 2.2(viii)]

16. A finite population of m individuals labelled $1, \ldots, m$ is such that for each individual there is a binary property θ. Denote the values $\theta_1, \ldots, \theta_m$; these are not probabilities, but rather unknown constants, each either zero or one. From the finite population a random sample of size n is drawn without replacement; this yields in order individuals with numbers $i_1, \ldots, i_n$ and the corresponding θ's are then observed. This gives observations $y_1, \ldots, y_n$, where $y_i = \theta_{i_j}$. Show that the likelihood is

$$\begin{cases} (m-n)!/m! & (\theta_{i_j} = y_j; j = 1, \ldots, n), \\ 0 & \text{otherwise,} \end{cases}$$

and that therefore this does not depend on the values of the unobserved θ's. Show further that essentially the same likelihood function is obtained for any sampling scheme, provided that it is defined independently of θ.

Discuss the implications of this for the various approaches to statistical inference. Does it mean that (a) it is impossible to learn anything about the unobserved θ's; (b) the interpretation of a sample survey should not depend on the sampling scheme; (c) at least in this context, the likelihood principle is inapplicable?
[Section 2.2, 2.4(viii); Godambe and Thompson, 1971]

17. For the model $f_Y(y; \theta)$, the statistic S is called *Bayesian sufficient* if, whatever the prior distribution of θ, the posterior distribution involves the data only through s. Show that Bayesian sufficiency is equivalent to "ordinary" sufficiency.
[Sections 2.2, 2.4(ix); Raiffa and Schlaifer, 1961]

3 PURE SIGNIFICANCE TESTS

3.1 Null hypotheses

We start the discussion of particular methods of inference by looking at problems with a very simple structure. Suppose that we have data $y = (y_1, \ldots, y_n)$ and a hypothesis H_0 concerning their density $f_Y(y)$. It is required to examine the consistency of the data with H_0. The hypothesis H_0 is called the *null hypothesis* or sometimes the hypothesis under test; it is said to be *simple* if it completely specifies the density $f_Y(y)$ and otherwise *composite*. Often composite hypotheses specify the density except for the values of certain unknown parameters.

One common type of simple null hypothesis involves i.i.d. components with densities of some well-known standard form. As examples, according to H_0, the densities of $Y_1, \ldots, Y_n$ might be standard normal, Poisson of known mean, or uniform over $(0, 2\pi)$. This last example is of particular interest in connexion with data on directions around a circle, the null hypothesis being that the n directions are independently randomly distributed over the circle. An example of a simple null hypothesis involving non-independent random variables is with binary observations, the null hypothesis being that the values are generated by a two-state Markov chain with known transition matrix.

Typical examples of composite null hypotheses are that the observations are independently distributed in a Poisson distribution of some fixed but unknown mean, or are normally distributed with unknown mean and variance. An example with non-independent observations is that with binary data we might wish to test the null hypothesis that the data form a two-state Markov chain with some unknown transition matrix.

In the discussion of this Chapter the null hypothesis is the only distribution of the data that is explicitly formulated. It is, however, necessary to have also some idea of the type of departure from the null hypothesis which it is required to test. All sets of data are uniquely extreme in some respects and without some idea of what are meaningful departures from H_0 the problem of testing consistency with it is meaningless.

Null hypotheses can arise for consideration in a number of different ways, the main ones being as follows:

(a) H_0 may correspond to the prediction of some scientific theory or to some model of the system thought quite likely to be true or nearly so.

(b) H_0 may divide the possible distributions into two qualitatively different types. So long as the data are reasonably consistent with H_0 it is not possible to establish which is the true type. In such a situation there may be no particular reason for thinking that H_0 is exactly or nearly true.

(c) H_0 may represent some simple set of circumstances which, in the absence of evidence to the contrary, we wish to assume holds. For example, the null hypothesis might assert the mutual independence of the observations, or that they arise from a standard type of normal-theory linear model. Of course, both these null hypotheses are highly composite.

(d) H_0 may assert complete absence of structure in some sense. So long as the data are consistent with H_0 it is not justified to claim that the data provide clear evidence in favour of some particular kind of structure. For instance, the null hypothesis may be that a series of events occur in a Poisson process. So long as the data are reasonably consistent with this it is not sensible to claim that the data justify an assertion of the existence of, for example, clustering.

These categories are not exclusive. For example, a null hypothesis of events forming a Poisson process could arise under any or all of these headings. We stress again that the only distribution for the data entering explicitly into the following discussion is that under H_0. In practice, however, especially for (b) and (d), it is necessary to consider how big a departure from H_0 might arise; that is, in practice, the significance tests of the type developed in this Chapter are then only part of the process of analysis. We call a test of consistency with H_0 developed under the very limiting assumptions of this Chapter a *pure test of significance*.

3.2   Test statistics and their null distributions

Let $t = t(y)$ be a function of the observations and let $T = t(Y)$ be the corresponding random variable. We call T a *test statistic* for the testing of H_0 if the following conditions are satisfied:

(a) the distribution of T when H_0 is true is known, at least approximately. In particular, if H_0 is composite the distribution of T must be, at least approximately and preferably exactly, the same for all the simple hypotheses making up H_0;

(b) the larger the value of t the stronger the evidence of departure from H_0 of the type it is required to test.

Note that if there are values of t that are impossible under H_0 but possible if a departure has occurred of the type being tested, then requirement (b) implies that any such values must be in effect transformed to plus infinity.

For given observations y we calculate $t = t_{obs} = t(y)$, say, and the *level of significance* p_{obs} by

$$p_{obs} = pr(T \geqslant t_{obs}; H_0). \qquad (1)$$

Because the distinction between random variables and their observed values is crucial in this part of the discussion and in order to conform with widely used notation we have distinguished observed values by an appropriate suffix. The random variable P for which (1) is an observed value has, for continuous underlying distributions, a uniform distribution over $(0, 1)$ under the null hypothesis H_0. We use p_{obs} as a measure of the consistency of the data with H_0 with the following hypothetical interpretation:

> Suppose that we were to accept the available data as evidence against H_0. Then we would be bound to accept all data with a larger value of t as even stronger evidence. Hence p_{obs} is the probability that we would mistakenly declare there to be evidence against H_0, were we to regard the data under analysis as just decisive against H_0.

In practice, we are usually interested only in the approximate value of p_{obs}. Note also that we are not at this stage involved explicitly in decisions concerning the acceptance or rejection of H_0; these require a much more detailed formulation of the problem for their satisfactory discussion. In fact, in practice, it is rather rare that a pure test of significance is the only analysis to be made.

A final general point is that we are not especially interested in whether p_{obs} exceeds some preassigned value, like 0.05. That is, we do not draw a rigid borderline between data for which $p_{obs} > 0.05$ and data for which $p_{obs} \leqslant 0.05$. Nevertheless, it is very convenient in verbal presentation of conclusions to use a few simple reference levels like 0.05, 0.01, etc. and to refer to data as being not quite significant at the 0.05 level, just significant at the 0.01 level, etc.

Particular examples of this procedure will be familiar to the reader and we therefore give here only a few examples.

Example 3.1. Directions on a circle. Suppose that the observations are directions round a circle, i.e. of angles between 0 and 2π. Consider the simple null hypothesis that the random variables $Y_1, \ldots, Y_n$ are i.i.d. in the uniform density over $(0, 2\pi)$, that is that the angles are completely random. Suppose that we wish to test for symmetrical clustering around the zero direction. Then the observations being $y_1, \ldots, y_n$, one reasonable test statistic is $\Sigma \cos Y_j$. The requirements (a) and (b) are satisfied in that the distribution of T under H_0 is known, at least in principle, and clustering around $y = 0$ will tend to produce large absolute values of the test statistic.

It is thus necessary to solve the distributional problem in which $Y_1, \ldots, Y_n$ are independently uniformly distributed over $(0, 2\pi)$ and the density of $T = \Sigma \cos Y_j$ is required. The moment generating function of T is

$$\{I_0(s)\}^n, \tag{2}$$

where $I_0(.)$ is the zero-order Bessel function defined by

$$I_0(s) = \frac{1}{2\pi} \int_0^{2\pi} e^{s \cos y} \, dy.$$

In principle the density and cumulative distribution function of T are determined by inversion. In practice, for all but very small n, it will be adequate to use a normal approximation, possibly supplemented by a correction using higher moments, and for this it is enough to note from (2) that under H_0

$$E(T) = 0, \quad \text{var}(T) = \tfrac{1}{2}n, \quad \gamma_1(T) = 0, \quad \gamma_2(T) = -\frac{3}{2n}. \tag{3}$$

For the use of higher moments to improve a normal approximation, see Appendix 1.

Example 3.2. Test of goodness of fit; simple H_0. It will serve to illustrate the difficulties of choosing a test statistic in the type of problem
we are considering to take a more general version of the problem of
Example 3.1. Suppose that under H_0 the random variables $Y_1, \ldots, Y_n$
are i.i.d. in the p.d.f. $g(.)$, a known function. We take the continuous
case to avoid complication of detail.

There are, of course, a very large number of types of departure
which might be of interest, including ones in which the random variables are not independent or do not all have the same distribution.
Here, however, we concentrate on the distributional form of the observations. For this we use statistics that do not depend on the order
in which the observations occur, i.e. which involve the data only
through the order statistics $Y_{(1)} \leqslant \ldots \leqslant Y_{(n)}$, or equivalently through
the sample distribution function

$$\tilde{F}_n(y) = \frac{\text{no. of } Y_j \leqslant y}{n} = \frac{1}{n} \Sigma \, \text{hv}(y - Y_j) = \frac{1}{n} \Sigma \, \text{hv}(y - Y_{(j)}), \quad (4)$$

where $\text{hv}(.)$ is the unit Heaviside function

$$\text{hv}(z) = \begin{cases} 1 & (z \geqslant 0), \\ 0 & (z < 0). \end{cases}$$

Now, if Y has the p.d.f. $g(.)$ and cumulative distribution function
$G(.)$, it is easily shown that the random variable $Z = G(Y)$ is
uniformly distributed on $(0, 1)$. That is, we could without loss of
generality take the null hypothesis distribution to be uniform, or
indeed, any other fixed continuous distribution. While this transformation can be convenient for theoretical discussion, in practice
the choice and interpretation of a test statistic is often best done on
a particular scale that happens to be meaningful for the application.

There are broadly two approaches to the choice of a test statistic.
If we have fairly clear ideas of the type of departure from H_0 of
importance, a small number of functions of the observations may be
indicated and the final test statistic will be some combination of
them. For example, if the null density is the standard normal, then
the sample mean or the sample variance or some combination of the
two can be used to see whether the location and scatter are about
what is to be expected under H_0. Example 3.1 is a case where the
rather special nature of the problem suggests a form of test statistic

in which observations y and $2\pi - y$ contribute equally. The second approach is to use some function of the order statistics or of $\tilde{F}_n(.)$ that will be large for a much broader class of departures from H_0.

There is a great variety of such test statistics, some of which are to be discussed more systematically in Chapter 6, and here we mention only a few:

(a) The Kolmogorov statistics

$$D_n^+ = \sup_y \{\tilde{F}_n(y) - G(y)\},$$

$$D_n^- = \sup_y \{G(y) - \tilde{F}_n(y)\}, \tag{5}$$

$$D_n = \sup_y |\tilde{F}_n(y) - G(y)| = \max(D_n^+, D_n^-)$$

measure the maximum discrepancy between sample and theoretical cumulative distribution functions, one of the first two forms being appropriate when the direction of departure from H_0 is known. Often this statistic is insufficiently sensitive to departures in the tails of the distribution.

(b) Weighted Kolmogorov statistics such as the Cramér-von Mises statistic

$$\int_{-\infty}^{\infty} \{\tilde{F}_n(y) - G(y)\}^2 dG(y) \tag{6}$$

put more weight on the tails.

(c) So-called chi-squared goodness of fit statistics can be formed by grouping the data in some way, finding observed and expected frequencies in the cells so formed and taking the test statistic

$$\sum \frac{(\text{observed freq.} - \text{expected freq.})^2}{\text{expected freq.}}. \tag{7}$$

One possibility is to take frequencies corresponding to the intervals

$$(-\infty, a), [a, a + h), [a + h, a + 2h), \ldots, [a + rh, \infty).$$

Another is to define the groups so that under H_0 the expected frequencies are all equal to n/m, where m is the number of groups. Then the statistic (7) is essentially the sum of squares of the observed frequencies. The two forms of grouping are equivalent when the null hypothesis is that the distribution is uniform, so that the second can be regarded as following from a preliminary transformation of the observations from y to $G(y)$.

(d) When the null hypothesis H_0 is true the expected value of the rth largest observation is

$$g_{nr} = \frac{n!}{(r-1)!(n-r)!} \int_{-\infty}^{\infty} yg(y)\{G(y)\}^{r-1}\{1-G(y)\}^{n-r}dy; \quad (8)$$

see Appendix 2. This suggests that if the ordered observations are plotted against the g_{nr} a straight line through the origin of unit slope should be obtained, at least approximately. More precisely, if a large number of sets of data are plotted in this way there will be no average departure from the unit line, if H_0 is indeed true. This method of plotting in fact is a very valuable one, because it indicates the type of departure, if any, involved. For the present purpose, however, its importance is that any test statistic measuring discrepancy between the order statistics and the g_{nr} is a possible test statistic for the problem. In fact, many of the statistics mentioned above can be viewed in this way. The main importance of this approach will, however, emerge in the next example when we consider composite null hypotheses.

To obtain a test, having chosen a test statistic, the density of the statistic has to be found under H_0, i.e. a distributional problem has to be solved. This is an aspect we shall not emphasize here. Often a preliminary transformation to the uniform distribution is helpful; many of the statistics mentioned above have null hypothesis distributions that do not depend on the particular $g(.)$. Typically the exact distribution can be found and tabulated for small values of n and convenient approximations found for large values of n.

Some of the general issues raised by the present example will be taken up in Section 3.4. Two can, however, be mentioned briefly here. One is that there is a clash between tests that pick out a particular type of departure from H_0 and are insensitive for other types of departure and tests that will pick out a wide variety of types of departure but which are relatively insensitive for any particular type. The second point is that the wide variety of plausible test statistics is embarrassing. Chapters 4—6 deal with the development of test statistics that are "optimal"; nevertheless many important tests have been developed by the more informal methods of the present Chapter.

Example 3.3. Test of goodness of fit; composite H_0. We now consider more briefly the problem of Example 3.2 when the density under the null hypothesis contains nuisance parameters; we write the density $g(y; \theta)$. There is now the additional problem that the test statistic is to have a distribution exactly or approximately independent of θ. As an example, the null hypothesis distribution of the data may be normal with unknown mean and variance.

For the normal distribution, and more generally for densities of the form

$$\frac{1}{\tau} g\left(\frac{y - \mu}{\tau}\right), \tag{9}$$

it is often useful to consider the standardized cumulants of the distribution and, in particular,

$$\gamma_1 = \frac{\kappa_3}{\kappa_2^{3/2}} = \frac{\mu_3}{\sigma^3}, \quad \gamma_2 = \frac{\kappa_4}{\kappa_2^2} = \frac{\mu_4}{\sigma^4} - 3, \tag{10}$$

where κ_r is the rth cumulant, σ the standard deviation and μ_r the rth moment about the mean. It is clear on dimensional grounds that γ_1 and γ_2 are independent of the nuisance parameters and that the distribution of sample estimates depends only on the standardized distribution when $\mu = 0$ and $\tau = 1$, and on the sample size. This leads us to define for such problems estimates of the third and fourth cumulants by

$$\tilde{\kappa}_2 = \frac{\Sigma(Y_j - \bar{Y})^2}{n}, \quad \tilde{\kappa}_3 = \frac{\Sigma(Y_j - \bar{Y})^3}{n},$$

$$\tilde{\kappa}_4 = \frac{\Sigma(Y_j - \bar{Y})^4}{n} - 3\tilde{\kappa}_2^2, \tag{11}$$

and then to calculate

$$\tilde{\gamma}_1 = \frac{\tilde{\kappa}_3}{\tilde{\kappa}_2^{3/2}}, \quad \tilde{\gamma}_2 = \frac{\tilde{\kappa}_4}{\tilde{\kappa}_2^2}. \tag{12}$$

In the normal case these should be close to zero. Even if we consider test statistics based solely on $\tilde{\gamma}_1$ and $\tilde{\gamma}_2$ there are a number of possibilities.

If, say, only the symmetry of the distribution is of concern and only departures in one direction are of importance one can use $\tilde{\gamma}_1$ directly as a test statistic; if departures in either direction are of approximately equal importance, $|\tilde{\gamma}_1|$ or equivalently $\tilde{\gamma}_1^2$ can be used. If both statistics $\tilde{\gamma}_1$ and $\tilde{\gamma}_2$ are of interest, then some composite function is needed; the most obvious possibilities are to take a quadratic form in the two statistics, or the larger of the two after some scaling.

More specifically, in testing goodness of fit with the normal distribution the null hypothesis distributions of $\tilde{\gamma}_1$ and $\tilde{\gamma}_2$ have been tabulated for small n (Pearson and Hartley, 1970; D'Agostino, 1970; D'Agostino and Tietjen, 1971) and for larger n are approximately normal with zero means and variances

$$\frac{6}{n} \text{ and } \frac{24}{n},$$

respectively. For a composite test statistic (D'Agostino and Pearson, 1973) we may for example take

$$\tfrac{1}{6}n\tilde{\gamma}_1^2 + \tfrac{1}{24}n\tilde{\gamma}_2^2, \tag{13}$$

which has for large n approximately a chi-squared distribution with two degrees of freedom, or

$$\max\left(\frac{|\tilde{\gamma}_1|\sqrt{n}}{\sqrt{6}}, \frac{|\tilde{\gamma}_2|\sqrt{n}}{\sqrt{24}}\right), \tag{14}$$

which has approximately the distribution of the larger of two independent standardized semi-normal random variables.

For analogous results for testing goodness of fit with densities $\tau^{-1}g(y/\tau)$ $(y \geqslant 0)$, see Exercises 3.9 and 3.10.

The chi-squared test of Example 3.2 and the Kolmogorov-type statistics can be generalized fairly directly. For chi-squared, the expected frequencies are replaced by fitted frequencies, using the density $g(\,.\;; \tilde{\theta})$, where $\tilde{\theta}$ is a "good" estimate of θ. The approximate distribution theory of this statistic is outlined in Section 9.3(iii); for a simple result, the parameter θ must be "efficiently" estimated from the grouped data. The Kolmogorov statistic can be modified similarly, the given cumulative distribution function in (5) being replaced by an estimate $G(\,.\;; \tilde{\theta})$. The associated distribution theory is difficult; see Durbin (1973). In some special cases the distribution has been determined by simulation.

Finally, we mention test statistics associated with the order statistic plotting procedures mentioned in Example 3.2 (d). When the null hypothesis is of the scale and location type (9), it is sensible to plot the ordered observations against expected values calculated from the standardized null distribution corresponding to $\mu = 0, \tau = 1$. The resulting plot is expected to be linear with location and slope depending on the nuisance parameters. Hence a suitable test statistic must be sensitive to non-linearity in the plot and one such is the squared correlation coefficient, small values being evidence of departure from H_0. Unfortunately, while empirical evidence shows this to be a useful test statistic, determination of the distribution under H_0 is likely to require simulation.

3.3   Composite null hypotheses

A crucial problem in dealing with composite null hypotheses is to find test statistics and distributions that are the same for all members of H_0. The calculation of a level of significance requires that the probability distribution of the test statistic can be computed explicitly in a way not involving nuisance parameters.

Two important ways of finding suitable statistics are (a) by consideration of conditional distributions and (b) by appeal to invariance arguments, especially the dimensional arguments mentioned in Example 3.3. We outline the use of conditional distributions first and then give a further example of the use of invariance. Both methods will be discussed in more detail from a rather different viewpoint in Chapter 5.

In fact, let S be a sufficient statistic for the nuisance parameter in the null hypothesis. Then the conditional distribution of Y given $S = s$ is fixed under H_0 and independent of the nuisance parameter. Therefore any test statistic that is not a function of S can be used and, at least in principle, its distribution under H_0 obtained. Note also that it may be possible to simplify the test statistic in the light of the condition $S = s$.

Example 3.4. Poisson distribution (ctd). As a simple example, suppose that under H_0 the random variables $Y_1, \ldots, Y_n$ are i.i.d. in a Poisson distribution of unknown mean μ. Under H_0 the sufficient statistic for μ is the sample total, i.e. $S = \Sigma Y_j$. Also, by Example 2.9, the conditional distribution of Y given $S = s$ is multinomial, corresponding to s trials with equal probabilities $(1/n, \ldots, 1/n)$.

We can thus find the distribution of any suggested test statistic by summing multinomial probabilities.

One commonly used test statistic is the index of dispersion

$$\frac{\Sigma(Y_j - \bar{Y_.})^2}{\bar{Y_.}}. \tag{15}$$

This is sensible when it is suspected that the Y_j's may be too variable; occasionally, with small values significant, it is used when it is suspected that "competition" between different observations has led to a more homogeneous set of data than would be expected under a Poisson distribution.

Conditionally on $S = s$, the test statistic (15) and its null distribution are exactly those considered in Example 3.2 in connexion with testing goodness of fit with a simple null hypothesis. To a close approximation the distribution can be taken as chi-squared with $n - 1$ degrees of freedom.

Note that conditionally on $S = s$ the test statistic can be written as $\Sigma Y_j^2/s - s$ and that therefore the test statistic could equivalently be taken as ΣY_j^2. This is easier for exact calculations.

Of course, (15) is just one possible test statistic. Two other test statistics that may be useful in particular circumstances are the number of zero Y_j's and $\max(Y_j)$.

A parallel exact discussion can be given when the Poisson distribution is replaced by any other member of the one-parameter exponential family. The appropriate conditional distribution is multinomial only for the Poisson distribution and the chi-squared approximation for the index of dispersion will not hold in general.

Conditioning on the sufficient statistic S in effect converts a composite null hypothesis into a simple one. Any sufficient statistic will induce a conditional distribution independent of the nuisance parameter, but when the statistic is boundedly complete in the sense of Section 2.2(vii), this is the only way of achieving the desired property. To see this, let A be any region in the sample space with the same probability, p say, under all densities in H_0 and let $I_A(.)$ be its indicator function. Then

$$E\{I_A(Y); H_0\} = p.$$

But this can be rewritten, in an obvious notation, in the form

$$E_S[E\{I_A(Y) - p|S; H_0\}] = 0.$$

Bounded completeness then implies that, for almost all s,

$$E\{I_A(Y) - p|S = s; H_0\} = 0,$$

i.e. the region A has the required property conditionally.

It is of conceptual interest that in some cases the argument can in a sense be inverted. For example, in connexion with Example 3.4 it can be shown that if $Y_1, \ldots, Y_n$ are i.i.d. and if for all integer s the conditional distribution of the Y_j's given $S = \Sigma Y_j = s$ is multinomial with equal cell probabilities, then the Y_j's have a Poisson distribution; see Menon (1966) for a general discussion of this for the exponential family.

Finally we consider an example where invariance arguments are used to remove arbitrary parameters. The example is derived from a more detailed treatment by Efron (1971) of the question "does a sequence of numbers follow a simple law?"

Example 3.5. Numerological law. Suppose that observational data is available on some physical phenomenon, giving data pairs (y_1, z_1), $\ldots, (y_n, z_n)$, with n quite small. It is noticed that with remarkably little error, possible observational, the numbers approximately satisfy a simple relationship of the form $y_j = u(z_j ; b_1, b_2)$, when the constants b_1 and b_2 are given appropriate values. There is some background knowledge of the physical phenomenon which does not predict such a simple relationship. Therefore the question is posed "is the approximate relationship spurious?" Presumably if some credibility is attached to the relationship connecting y and z, the search for a physical explanation is sensible. Two familiar historical examples are Newton's gravitational law and Bode's planetary distance law, although the latter is now known to contain clear exceptions.

To fix our ideas, take as an illustration the relationship

$$y_j = a + bz_j \quad (j = 1, \ldots, n), \tag{16}$$

which for suitably chosen a and b approximates well to the given data. For simplicity, we assume the z_j's to be equally spaced. The accepted state of knowledge is simply that y must increase with z, and we wish to assess whether or not (16) is plausible as an advance on this. Without being able to make any probabilistic assumptions concerning possible observational errors, we view (16) as indicating the specific direction in which to test the adequacy of the vague state of accepted prior knowledge. To obtain a suitable statistic for measuring closeness to (16), we appeal to the fact that inference should be invariant under location and scale changes in y. This implies that the statistic used should be a function of standardized differences, such as

$$d_j = \frac{y_{j+1} - y_j}{y_n - y_1} \quad (j = 1, \ldots, n-1).$$

In terms of these invariant quantities, (16) becomes

$$d_j = \frac{1}{n-1},$$

since we have supposed $z_{j+1} - z_j$ is constant. The accepted state of knowledge prior to proposal (16) implies only that the d_j are positive and sum to unity. We represent this formally by considering $d_1, \ldots, d_{n-1}$ as successive spacings $d_1 = u_1, d_2 = u_2 - u_1, \ldots, d_{n-1} = 1 - u_{n-2}$, where $u_1, \ldots, u_{n-2}$ are order statistics from the uniform distribution on the interval $(0, 1)$. This is the null hypothesis, formally expressed as

$$H_0 : f_D(d_1, \ldots, d_{n-1}) = (n-2)! \quad (d_j \geqslant 0, \ \Sigma d_j = 1). \quad (17)$$

A suitable test statistic measuring agreement with (16) is

$$T = \sum_{j=1}^{n-1} \left(D_j - \frac{1}{n-1} \right)^2,$$

whose distribution under H_0 is determined by the uniform distribution (17). For this particular case, then, the proposed relationship (16) plays the role of an alternative hypothesis and small values of T indicate departure from H_0 in that direction. Given a very small value of $p_{obs} = \text{pr}(T \leqslant t_{obs} ; H_0)$, a more elaborate analysis with further data would then proceed from a probabilistic formulation of the model (16), constructed with the aid of any further physical insight into the phenomenon.

Equation (16) is not the only possible simple two-parameter law connecting y and z, so that p_{obs} underestimates the chance under H_0 of being as close as the data to some relationship with the given degree of simplicity. The evaluation of this latter chance is clearly a difficult problem, some comments on which are given in Efron's paper and the published discussion.

3.4   Discussion

We now turn to a number of miscellaneous comments on significance tests. While such tests are widely used in practice, it is very necessary to appreciate the limitations on the arguments involved. In this connexion point (vii) is of particular importance.

(i) Invariance
Any monotonic increasing function of the test statistic T leads to the same p_{obs}.

(ii) Dependence on stopping rule
Because the distributions holding when H_0 fails are not explicitly formulated, use of the likelihood principle is not possible and this and other related principles are of little or no value. However, when there is the possibility of different kinds of sampling rule arising, significance levels do usually depend on the particular rule used; see Exercise 3.13. To this limited extent, pure significance tests can be thought of as conflicting with the strong likelihood principle.

(iii) Selection of test in light of data
Quite often, either explicitly or implicitly, a number of significance tests are applied to the same data. If these concern different questions no particular difficulty arises, although if the different test statistics are highly correlated this needs to be taken into account, at least qualitatively. In other cases, however, the tests bear on essentially the same issue and then the fact that several tests have been applied must be recognized if very misleading conclusions are to be avoided.

We assume in the following discussion that k tests are applied and that the most significant of these is taken. There are, of course, other ways in which the results from several tests can be combined but our assumption is meant to represent the situation where there are a number of different types of departure from H_0 and in which the one finally tested is the most striking of these. Of course, the deliberate practice of applying numerous tests in the hope of finding a "significant" result is not recommended. An extreme case of this practice is described in Exercise 3.14.

Let $P_1, \ldots, P_k$ denote the levels of significance from the separate tests; we are, in effect, taking as test statistic

$$Q = \min(P_1, \ldots, P_k), \tag{18}$$

small values of Q being significant. In the continuous case, $P_1, \ldots, P_k$ are under H_0 uniformly distributed. Now

$$\text{pr}(Q > q_{\text{obs}}; H_0) = \text{pr}(P_j > q_{\text{obs}}, j = 1, \dots, k; H_0),$$

and the significance level of the Q test is thus

$$\text{pr}(Q \leqslant q_{\text{obs}}; H_0) = 1 - \text{pr}(P_j > q_{\text{obs}}, j = 1, \dots, k; H_0).$$

Therefore, if the separate tests are independent, it follows that the significance level is

$$\text{pr}(Q \leqslant q_{\text{obs}}; H_0) = 1 - (1 - q_{\text{obs}})^k \simeq k q_{\text{obs}}, \tag{20}$$

the approximation holding if $k q_{\text{obs}}$ is small.

For example, if the most significant of 5 independent tests is just significant at the 0.05 level, then the true level, allowing for selection, is

$$1 - (1 - 0.05)^5 = 0.226.$$

If the test statistics are not independent, and if calculated from the same data dependence will normally arise, it is necessary in principle to evaluate (19) from the joint distribution of $P_1, \dots, P_k$. This will not often be possible, unless k is very small, but we can obtain a bound as follows.

It is easily shown that for arbitrary events $A_1, \dots, A_k$,

$$\text{pr}(A_1 \cup \dots \cup A_k) \leqslant \Sigma \text{pr}(A_j).$$

But for arbitrary events $B_1, \dots, B_k$, we have that

$$B_1 \cap \dots \cap B_k$$

is the complement of

$$\bar{B}_1 \cup \dots \cup \bar{B}_k,$$

so that

$$\text{pr}(B_1 \cap \dots \cap B_k) = 1 - \text{pr}(\bar{B}_1 \cup \dots \cup \bar{B}_k).$$

Therefore

$$1 - \text{pr}(B_1 \cap \dots \cap B_k) = \text{pr}(\bar{B}_1 \cup \dots \cup \bar{B}_k) \leqslant \Sigma \text{pr}(\bar{B}_j).$$

Thus finally

$$\text{pr}(Q \leqslant q_{\text{obs}}; H_0) \leqslant \Sigma \text{pr}(P_j \leqslant q_{\text{obs}}) = k q_{\text{obs}}. \tag{21}$$

That is, $k q_{\text{obs}}$ is an upper bound for the significance level in the procedure with selection. Clearly, if the tests are highly dependent, the effect of selection will be over-corrected. One difficulty in applying these formulae is that, when the aspect of the null hypothesis tested is selected partly in the light of the data, it may not be clear over precisely what set of statistics selection has occurred.

(iv) Two-sided tests

Often both large and small values of the test statistic are to be re-garded as indicating departures from H_0. Sometimes t can be defined so that t and $-t$ represent essentially equivalent departures from H_0. Then we can take t^2 or $|t|$ as a new test statistic, only large values indicating evidence against H_0. That is, we can define the level of significance as

$$p_{obs} = \text{pr}\{|T| \geqslant |t_{obs}|; H_0\}.$$

More commonly, however, large and small values of t indicate quite different kinds of departure and, further, there may be no very natural way of specifying what are equally important departures in the two directions. Then it is best to regard the tests in the two di-rections as two different tests, both of which are being used, and to apply the arguments of (iii). That is, we define

$$p_{obs}^+ = \text{pr}(T \geqslant t_{obs}; H_0), \quad p_{obs}^- = \text{pr}(T \leqslant t_{obs}; H_0). \qquad (22)$$

Note that $p_{obs}^+ + p_{obs}^- = 1 + \text{pr}(T = t_{obs})$ and is 1 if T is continuously distributed.

This leads us to define

$$Q = \min(P^+, P^-) \qquad (23)$$

as the test statistic and the corresponding level of significance as

$$\text{pr}(Q \leqslant q_{obs}; H_0).$$

In the continuous case this is $2q_{obs}$; in a discrete problem it is q_{obs} plus the achievable one-sided p value from the other tail of the dis-tribution, nearest to but not exceeding q_{obs}.

(v) Tests for several types of departure

The discussion of (iii) is very relevant also to the common problem of testing a null hypothesis for several different kinds of departure. This has arisen, for example, in the discussion of Example 3.3, test-ing for goodness of fit. In some ways the mathematically simplest approach is to set up separate statistics for the different kinds of departure and then to combine them, for example by a positive definite quadratic form, into a function sensitive to departures in some or all of the directions. This is particularly suitable if it is thought likely that there may be relatively small departures from H_0 in several of the respects simultaneously.

On the other hand, this method is open to the following criticism. If a significant departure from H_0 arises we are likely to want to explore in more detail the nature of the departure. It is a serious criticism of the whole formulation of significance testing that it gives us no explicit treatment of this. If, however, we take as test statistic the most significant of the separate statistics for testing the individual kinds of departure, at least the departure giving the most significant effect will be available and this is some guidance as to the way in which H_0 is inadequate.

(vi) Combination of tests from several sets of data

Quite often we apply the same form of test to a number of independent sets of data and then need to combine the answers in some way. The following points arise:

(a) The fact that a test statistic is highly significant in one set of data and not in another is not evidence that the two sets of data themselves differ significantly. The practice of sorting the sets into two, those with a significant departure from H_0 and those not, with the intention of analysing the two groups separately, is almost always bad.

(b) With m sets of data we obtain test statistics $t_1, \ldots, t_m$. If the data are of equal size and structure, these values should be independent and identically distributed with a known distribution, and the situation of Example 3.2 holds. That is, we can treat the test statistics as observations for a second stage of analysis. If the sets of data are of unequal size the statistics can be transformed to have the same distribution under H_0, although then the question of attaching more weight to the larger sets of data arises in any refined analysis.

(c) In (b), the analysis to be applied if all data sets are expected to show any deviation from H_0 in the same direction is different from that if non-homogeneous departures are suspected. Often graphical methods will be useful.

(d) One approach to combining tests for continuous data that relates to points (b) and (c) is to note that, under H_0, P has a uniform distribution on $(0, 1)$. If homogeneous departure from H_0 is anticipated, the successive significance probabilities $P_1, \ldots, P_m$ might be combined to form the summary statistic $T^\dagger = -2 \Sigma \log P_j$. Large values of $T^\dagger$ are evidence against H_0, according to which the random variable $T^\dagger$ has exactly the chi-squared distribution with $2m$ degrees of freedom. A similar approach was mentioned in Section 3.4(iii) for the problem of combining several tests on one set of data. Different methods of combination of significance probabilities could be considered for non-homogeneous situations; for further details see Exercise 3.16.

(e) If departures from H_0 do arise it will usually be required to consider such questions as: are the data homogeneous? is the magnitude of any departure related to explanatory variables describing the sets of data? These questions can only be satisfactorily dealt with by developing a more detailed model in which further parameters are incorporated and, if the data are not homogeneous, related to explanatory variables. Note, in particular, that if the sets are of unequal sizes the magnitude of the test statistic does not by itself indicate the value of the parameter describing "distance" from H_0. Thus, in particular, plotting values of the test statistics against possible explanatory variables, regardless of sample size, can be misleading.

(vii) Limitations of significance tests
The preceding discussion has raised implicitly the main limitations of the pure significance test.

The most obvious one is the difficulty of choosing a suitable test statistic. While the next few Chapters give a formal theory for choosing and assessing test statistics, it is important to realize that this more elaborate discussion involves a more detailed specification of the problem and that choosing such a specification is largely equivalent to, and not necessarily easier than, choosing a test statistic.

A second and more serious limitation is the absence of a direct indication of the importance of any departure from H_0. The significance test merely compares the data with other sets of data that might have arisen and shows how relatively extreme the actual data are. With small amounts of data the value of p_{obs} may be quite large and yet the data may be consistent also with departures from H_0 of great practical importance; a second possibility, probably rarer in applications, is that p_{obs} may be very small but the amount of departure from H_0 of little practical interest. Further, if strong evidence against H_0 is obtained, the significance test gives no more than a guide to the type and magnitude of the departure.

These are, of course, arguments in favour of a much fuller specification of the problem and the close linking of the significance test with a problem of estimation. It is essentially consideration of intellectual economy that makes the pure significance test of interest.

For example, in examining goodness of fit with a normal-theory linear model it would not be sensible to set out a formal model containing parameters for all the main types of departure thought to be of possible interest. Yet it is certainly of value to be able to test goodness of fit of the model and to hope that the test will give some guide as to the types of departure, expected and unexpected, that may be present.

Bibliographic notes

The idea of a pure test of significance goes back at least 200 years, although explicit formulation of the ideas involved is much more recent. In fact, nearly all recent work has concentrated on the situation to be discussed in Chapters 4 and 5 where explicit alternative distributions are formulated. The account of the present Chapter is close to those of Barnard (1962) and Kempthorne (1966); a more formal discussion is due to Stone (1969), who arrived at a slightly different definition of the significance level for the discrete case. Fisher's (1973, 1st ed. 1956, Chapter 4) account puts less emphasis on the test statistic as ranking the possible data in order of discrepancy with the null hypothesis and more emphasis on a so-called logical disjunction "either the null hypothesis is false, or a rare event has occurred", applying when a significant departure from the null hypothesis is observed. Although Fisher was very critical of the idea of treating problems of scientific inference in terms of "decision rules", his own use of significance tests hinged more on the use of conventional levels (5% and 1%) than does the account given here; for a defence of the use of conventional levels, see Bross (1971). Anscombe (1956) discusses the types of situation in which significance tests are used.

There is a large literature on special tests. Mardia (1972) has given a systematic account of the analysis of directional data in two and three dimensions (see Example 3.1) and Lancaster (1969) has given a detailed discussion of chi-squared. David (1970) reviews properties of order statistics. Darling (1957) reviews work on the Kolmogorov distance statistic. The important results on the dispersion test for the Poisson distribution are due to Fisher (1950) and Rao and Chakravarti (1956). A discussion of some aspects of simultaneous significance tests is given by Miller (1966).

Further results and exercises

1. Interpret the test statistic $\Sigma \cos Y_j$ of Example 3.1 in terms of the abscissa of a random walk in the plane with steps of unit length. Suggest test statistics when clustering may be around 0 and π, and also when clustering may be around an unknown direction. For both cases, obtain a normal approximation to the distribution under the null hypothesis of a uniform distribution of angles around the circle.

[Sections 3.2, 3.3]

2. Suggest a test statistic for the null hypothesis that binary data come from a two-state Markov chain with given transition matrix, the departure of interest being that transitions occur more often than the given matrix would suggest.

[Section 3.2]

3. An *absolute test of significance* of a simple null hypothesis is a pure significance test in which the test statistic at an observational point y is a strictly decreasing function of the p.d.f. at y under the null hypothesis. That is, the significance level is the probability under the null hypothesis of obtaining data with a probability density as small as, or smaller than, that of the data actually obtained. Discuss critically whether this is a useful idea; in particular show that for continuous random variables the test is affected by non-linear transformations of the data and that for discrete data the grouping together of points in the sample space could have a major effect on the level of significance.

[Section 3.2]

4. Suppose that there are m possible outcomes to each trial and the data consist of n independent trials under the same conditions. Show that if the numbers of trials of the various types are given by random variables $(Y_1, \ldots, Y_m)$ then

$$\text{pr}(Y_1 = y_1, \ldots, Y_m = y_m) = n! \, \Pi(p_j^{y_j}/y_j!),$$

where $(p_1, \ldots, p_m)$ are the probabilities of the individual outcomes. Show that for large n the log of this probability is, in the relevant range $y_j - np_j = O(\sqrt{n})$, a linear function of

$$\Sigma(y_j - np_j)^2/(np_j).$$

Hence show that for large n the absolute test of significance of the null hypothesis which specifies the p_j's completely is equivalent to the test based on the chi-squared goodness of fit statistic.

[Section 3.2]

5. Show that for continuous distributions the null hypothesis distribution of the Kolmogorov and Cramér-von Mises statistics does not depend on the particular null distribution tested. Prove that the latter statistic can be written in the form

$$\frac{1}{12n^2} + \frac{1}{n} \sum \left\{ G(Y_{(j)}) - \frac{2j-1}{2n} \right\}^2.$$

6. For binary data the null hypothesis is that the data are generated by a two-state Markov chain, the transition matrix being unknown. Suggest some types of departure that it might be of interest to test, indicating an appropriate test statistic in each case.

[Section 3.3]

7. Prove by geometrical considerations that if $Y_1, \ldots, Y_n$ are independently and identically normally distributed the random variables corresponding to the standardized skewness and kurtosis, (12), have distributions that are independent of the estimated mean and variance and which do not depend on the mean and variance of the distribution.

[Section 3.3]

8. List and classify some test statistics for the null hypothesis of normality, based on i.i.d. random variables. Examine for each test statistic the possibility of a generalization to test multivariate normality.

Comment for the multivariate tests on

(a) the ease with which the distribution under the null hypothesis can be found or approximated;

(b) the ease with which the statistic itself can be computed even in a large number of dimensions;

(c) the extent to which the statistic is sensitive to departures from multivariate normality that leave the marginal distributions exactly or nearly normal;

(d) the types of departure from the null hypothesis for which each test is particularly sensitive.

[Section 3.3; Andrews, Gnanadesikan and Warner, 1971;
Shapiro, Wilk and Chen, 1968]

9. Suggest a plot of ordered values useful for testing consistency with
the density $\tau^{-1}g(y/\tau)$ ($y \geqslant 0$), for unknown τ. What would be the
interpretation of a straight line not through the origin? Suggest a test
statistic for such an effect and show how to find its exact null hypothesis distribution when $G(.)$ is exponential.

[Section 3.3]

10. Show that for a gamma distribution of unknown mean and index,
the first three cumulants are such that $\kappa_3\kappa_1 = 2\kappa_2^2$. Hence suggest a
test statistic for examining consistency of data with the gamma
family. Assess methods that might be used to determine the distribution of the statistic under the null hypothesis.

[Section 3.3]

11. Show how exact index of dispersion tests analogous to that of
Example 3.4 can be developed for the binomial and geometric distributions, and for the general one-parameter exponential family distribution. Explain why, under suitable circumstances to be stated, the
chi-squared distribution will be a good approximation to the null
hypothesis distribution in the first case, not in the second and not in
general in the third.

[Section 3.3]

12. The following desirable features have been suggested for graphical
methods of analysis:
(a) conformity with a simple model should be shown by a straight
line plot, or, where the random part of the variability is of direct
interest, by a completely random scatter;
(b) points plotted should be subject to independent errors;
(c) points should be of equal precision;
(d) nonlinear transformations should be such as to emphasize
ranges of values of particular interest.
Discuss various graphical methods for examining agreement with a
normal distribution in the light of these requirements.

[Section 3.4]

13. In five trials, the first four give zero and the last a one. The null hypothesis is that these are Bernoulli trials with the probability of a one equal to $\frac{1}{2}$ and it is required to test for a preponderance of zeroes. Prove that if the number of trials is fixed, so that the number of ones has a binomial distribution, the significance level is 3/32, whereas if trials continued until the first one, so that the number of trials has a geometric distribution, the significance level is 1/16. Discuss this in the light of the strong likelihood principle of Section 2.3(v), noting that in the formulation in which only the null hypothesis is explicitly formulated there is no opportunity for contradiction to arise.

[Section 3.4(ii)]

14. It is proposed that for testing a hypothesis H_0, when no explicit alternatives are available, one can decide between various statistics as to which is most appropriate solely on the basis of their sampling distributions when H_0 is true. This general proposition is then reduced to the limited proposition that if two statistics have the same sampling distribution under H_0, then they are equally effective for testing H_0. To be specific, consider the hypothesis H_0 that given observations $y_1, \ldots, y_n$ are represented by i.i.d. $N(0, \sigma^2)$ random variables $Y_1, \ldots, Y_n$, for which the Student t statistic $\sqrt{n} Y_./\{\Sigma(\bar{Y}_j - \bar{Y}_.)^2\}^{\frac{1}{2}}$ is thought suitable. Refute the above proposal by considering the statistic

$$T' = \frac{\sqrt{n} \Sigma a_j Y_j}{\{\Sigma(Y_j - \Sigma a_j Y_j)^2\}^{\frac{1}{2}}}$$

where $a_j = y_j/\{\Sigma y_k^2\}^{\frac{1}{2}}$ $(j = 1, \ldots, n)$.

[Section 3.4(ii); Neyman, 1952]

15. Suppose that in selecting the most significant of a number of dependent test statistics the joint distributions of pairs of statistics are known. Show how the inclusion-exclusion formula can be used to complement the upper bound (21) by a lower bound for the significance level, adjusted for selection.

[Section 3.4(iii)]

16. Suppose that m independent sets of data of size $n_1, \ldots, n_m$ are tested for compatibility with H_0 with a particular type of alternative

in mind, and that the results of the significance tests are summarized by the significance probabilities $P_1, \ldots, P_m$. Investigate the relative merits of the summary statistics

$$T_1 = -2 \Sigma \log P_j, \quad T_2 = \max(P_j),$$
$$T_3 = \min(P_j), \quad T_4 = \Sigma n_j P_j / \Sigma n_j,$$

bearing in mind the following possibilities: (i) H_0 is expected to be incorrect for at most one set of data, (ii) H_0 is expected to be correct for all or none of the data sets, and any others potentially appropriate.

[Section 3.4(iii)]

4 SIGNIFICANCE TESTS:
SIMPLE NULL HYPOTHESES

4.1 General

In the previous chapter we discussed some key ideas connected with tests of significance. A major step in the procedure as discussed there is the choice of a test statistic. This is done, in effect, by an ordering of the possible sample points in terms of decreasing consistency with the null hypothesis, H_0, bearing in mind the type of departure from H_0 that it is desired to detect. It is natural to look for some more positive guidance in the choice of test statistic and this we now give.

The mathematical theory to be discussed was developed in a remarkable series of joint papers by J. Neyman and E.S. Pearson. Their interpretation of the theory was more directly linked to decision-making than is the account given here; we return to the decision-theoretical aspects in Section 11.6.

In Chapter 3, we supposed that only the null hypothesis is explicitly formulated. We now assume that, in addition to H_0, we have one or more alternative hypotheses representing the directions of the departures from H_0 which it is required to detect. In the present Chapter we consider only the case where H_0 is a simple hypothesis, the corresponding p.d.f. of the vector random variable Y being $f_0(y)$. Let H_A be a particular simple alternative hypothesis, the corresponding p.d.f. being $f_A(y)$. Denote the set of all relevant alternative hypotheses by Ω_A.

There are two rather different cases that can arise. In the first, H_0 is a special case of some family $\mathcal{F}$ of models, assumed provisionally to be adequate for the problem. Then we would normally take Ω_A to be all models in $\mathcal{F}$ except for H_0. On the other hand, if we start with only H_0, we must consider what are the appropriate alternatives. So far as the significance test is concerned, however, the null and alternative hypotheses are not on an equal footing; H_0 is clearly

specified and of intrinsic interest, whereas the alternatives serve only
to indicate the direction of interesting departures. At this stage we
are not concerned directly with which of H_0 or H_A is more likely to
be true or with estimating the magnitude of possible departures from
H_0. The question is solely: is there evidence of inconsistency with H_0?
Of course in applications the extended questions are very likely to
arise, especially when H_0 is embedded in a comprehensive family, but
to cover these the emphasis of the discussion needs to be changed.
Fortunately it turns out that essentially the same mathematical
theory will do. Indeed, Chapters 4–6 could be regarded primarily as
a necessary mathematical preliminary to the discussion of interval
estimation in Chapter 7.

Example 4.1. Directions on a circle (ctd). Suppose that observations
are angles representing, for example, the directions of unit vectors in
two dimensions. We take the null hypothesis, H_0, that $Y_1, \ldots, Y_n$ are
i.i.d. in the uniform distribution over $(0, 2\pi)$. That is,

$$f_0(y) = (2\pi)^{-n} \quad (0 \leqslant y_j \leqslant 2\pi; j = 1, \ldots, n). \tag{1}$$

If we want alternatives that represent a tendency to cluster around
the axis of the polar coordinates, about the simplest possibility is to
take the random variables to be i.i.d. with the density

$$\frac{e^{\theta \cos y}}{2\pi I_0(\theta)}, \tag{2}$$

where $\theta > 0$ is a parameter. In (2) the Bessel function $I_0(.)$ is a normalizing constant defined by

$$I_0(\theta) = \frac{1}{2\pi} \int_0^{2\pi} e^{\theta \cos y} dy. \tag{3}$$

The inclusion in (2) of a cosine term is natural when the observations represent angles and symmetrical clustering is to be represented. A particular alternative hypothesis, H_A, is represented by
(2) with a particular $\theta = \theta_A$; the family, Ω_A, would normally be the
set of all densities (2) with $\theta > 0$. Note that H_0 is recovered by taking
$\theta = 0$. Under H_A the joint density of the data is

$$f_A(y) = \exp(\theta_A \Sigma \cos y_j)\{2\pi I_0(\theta_A)\}^{-n}. \tag{4}$$

If we want to represent the possibility that any clustering around
$\theta = 0$ is balanced by equal clustering around $\theta = \pi$, we merely work
with $2Y_j \bmod (2\pi)$.

Example 4.2. Uniform distribution on unit interval. If we have a null hypothesis representing independent uniform random variables in which the observations are not angles, it will be more convenient to take the range, if known, to be $(0, 1)$. Alternatives that are symmetrical in $(0, 1)$ include the density

$$\frac{y^\theta (1-y)^\theta}{B(\theta + 1, \theta + 1)} \quad (0 \leqslant y \leqslant 1), \tag{5}$$

where the normalizing constant is a beta function. If $\theta > 0$, then the alternative represents clustering around $y = \frac{1}{2}$, whereas $-1 < \theta < 0$ represents clustering near $y = 0$ and 1.

Another special kind of alternative representing an unsymmetrical distribution over $(0, 1)$ is given by the density

$$\frac{y^\theta (1-y)^{-\theta}}{B(\theta + 1, -\theta + 1)}. \tag{6}$$

For some purposes it may be sensible to combine (5) and (6) into the two parameter family

$$\frac{y^{\theta_1}(1-y)^{\theta_2}}{B(\theta_1 + 1, \theta_2 + 1)}. \tag{7}$$

In all cases particular alternatives are represented by taking particular parameter values.

Of course, many other alternatives are possible. Note that the distributions (2), (5), (6) and (7) are all in the exponential family, a fact that will turn out to be mathematically convenient.

4.2   Formulation

In Chapter 3 we defined for every sample point a significance level, the probability of obtaining under H_0 a value of the test statistic as or more extreme than that observed. In the present chapter it is convenient to take a slightly different, although essentially equivalent, approach to the definition of a test. For each α, $0 < \alpha < 1$, let w_α be the set of points in the sample space which will be regarded as significantly different from H_0 at level α. We call w_α a *critical region of size* α. A significance test is defined by a set of such regions satisfying the following essential requirements. First,

$$w_{\alpha_1} \subset w_{\alpha_2} \quad \text{if} \quad \alpha_1 < \alpha_2; \tag{8}$$

this is to avoid such nonsense as saying that data depart significantly from H_0 at the 1% level but not at the 5% level. Next, to ensure that the significance level has the hypothetical physical significance discussed in Chapter 3, we require that, for all α,

$$\text{pr}(Y \in w_\alpha; H_0) = \alpha. \tag{9}$$

To complete the parallel with Chapter 3, we define the significance level p_{obs} of data y by

$$p(y) \equiv p_{\text{obs}} = \inf(\alpha; y \in w_\alpha). \tag{10}$$

Any test can be defined in the above way. We now have to define the best test in the light of the alternative hypotheses. It is convenient to do this for the most part in terms of a fixed but arbitrary α, but we must in practice check that the nesting condition (8) is obeyed by any solution obtained.

4.3   Simple null hypothesis and simple alternative hypothesis

It is best to begin with an artificially easy case in which there is just one simple alternative hypothesis. That is, there are two completely specified densities under consideration, namely

$$H_0: f_0(y), \ H_A: f_A(y).$$

Recall, however, that we are going to treat H_0 and H_A quite unsymmetrically and that we are not discussing the problem of which fits the data better.

Let w_α and w'_α be two critical regions of size α. By (9)

$$\text{pr}(Y \in w_\alpha; H_0) = \text{pr}(Y \in w'_\alpha; H_0). \tag{11}$$

We regard w_α as preferable to w'_α for the alternative H_A if

$$\text{pr}(Y \in w_\alpha; H_A) > \text{pr}(Y \in w'_\alpha; H_A). \tag{12}$$

The region w_α is called the *best critical region* of size α if (12) is satisfied for all other w'_α satisfying the size condition (11). We aim to define a significance test by choosing for all α the corresponding best critical region. The resulting test is such that the significance level P, thought of as a random variable, is, under H_A, stochastically smaller than the level associated with any other system of critical regions. We call $\text{pr}(Y \in w_\alpha; H_A)$ the *size α power* of the test against alternative H_A.

By Example 2.12, the likelihood ratio $f_A(y)/f_0(y) = \text{lr}_{A0}(y)$, say, is a sufficient statistic when just those two distributions are under consideration. This makes it very plausible that the best critical region should depend on the data only through the likelihood ratio. Further, at least in some rough sense, the larger the value of $\text{lr}_{A0}(y)$, the worse the fit to H_0.

For simplicity, we suppose that $\text{lr}_{A0}(Y)$ is, under H_0, a continuous random variable such that for all α, $0 < \alpha < 1$, there exists a unique c_α such that

$$\text{pr}\{\text{lr}_{A0}(Y) \geqslant c_\alpha ; H_0\} = \alpha. \tag{13}$$

That is, if we form critical regions from all sufficiently large values of the likelihood ratio, we can find a unique one of any desired size. We call the region defined by

$$\text{lr}_{A0}(y) \geqslant c_\alpha$$

the size α *likelihood ratio critical region.* The nesting condition (8) is satisfied.

A central result, called the Neyman-Pearson lemma, is that *for any size α, the likelihood ratio critical region is the best critical region.*

To prove this, let w_α be the likelihood ratio critical region and w'_α be any other critical region, both being of size α. Then

$$\alpha = \int_{w_\alpha} f_0(y)dy = \int_{w'_\alpha} f_0(y)dy,$$

so that

$$\int_{w_\alpha - w'_\alpha} f_0(y)dy = \int_{w'_\alpha - w_\alpha} f_0(y)dy. \tag{14}$$

But in $w_\alpha - w'_\alpha$, which is inside w_α, $f_A(y) \geqslant c_\alpha f_0(y)$, whereas in $w'_\alpha - w_\alpha$, which is outside w_α, $f_A(y) < c_\alpha f_0(y)$. Thus, on multiplying (14) by c_α and introducing these inequalities, we have that

$$\int_{w_\alpha - w'_\alpha} f_A(y)dy \geqslant \int_{w'_\alpha - w_\alpha} f_A(y)dy,$$

with strict inequality unless the regions are essentially equivalent. Thus, if we add to both sides the integral over $w_\alpha \cap w'_\alpha$, we have that

$$\int_{w_\alpha} f_A(y)dy \geqslant \int_{w'_\alpha} f_A(y)dy,$$

and this is the required result.

Note that if w_α' had been of size less than α the final inequality would still have followed; that is, it is not possible to improve on the power of the size α likelihood ratio critical region by using a different type of region of size smaller than α.

4.4   Some examples

The determination of a likelihood ratio critical region of size α requires two steps. The first and entirely straightforward one is the calculation of the likelihood ratio $\mathrm{lr}_{A0}(y)$; the second and more difficult part is the determination of the distribution of the random variable $\mathrm{lr}_{A0}(Y)$ under H_0. We now give some illustrative examples.

Example 4.3. Normal mean with known variance. Let $Y_1, \ldots, Y_n$ be i.i.d. in $N(\mu, \sigma_0^2)$, where σ_0^2 is known, and let null and alternative hypotheses specify the mean as follows:

$$H_0: \mu = \mu_0, \quad H_A: \mu = \mu_A.$$

We suppose to begin with that $\mu_A > \mu_0$; this has a crucial effect on the form of the answer. Note that throughout the following argument σ_0^2, μ_0 and μ_A are known constants.

A simple calculation shows that

$$\mathrm{lr}_{A0}(y) = \frac{\dfrac{1}{(2\pi)^{\frac{1}{2}n}\sigma_0^n} \exp\left\{-\dfrac{\Sigma(y_j - \mu_A)^2}{2\sigma_0^2}\right\}}{\dfrac{1}{(2\pi)^{\frac{1}{2}n}\sigma_0^n} \exp\left\{-\dfrac{\Sigma(y_j - \mu_0)^2}{2\sigma_0^2}\right\}}$$

$$= \exp\left\{\frac{n\bar{y}.(\mu_A - \mu_0) - \frac{1}{2}\mu_A^2 + \frac{1}{2}\mu_0^2}{\sigma_0^2}\right\}.$$

Because all the quantities in this, except for $\bar{y}.$, are fixed constants and because $\mu_A - \mu_0 > 0$, a critical region of the form $\mathrm{lr}_{A0}(y) \geqslant c_\alpha$ is equivalent to one of the form $\bar{y}. \geqslant d_\alpha$, where d_α is a constant to be chosen. This completes the first step.

In the second step, we choose d_α to satisfy the size condition (13). But $\bar{Y}.$ is, under H_0, distributed in $N(\mu_0, \sigma_0^2/n)$, so that

$$d_\alpha = \mu_0 + k_\alpha^* \sigma_0/\sqrt{n},$$

where $\Phi(-k_\alpha^*) = \alpha$. Thus the best critical region has the form

$$\frac{\bar{y}. - \mu_0}{\sigma_0/\sqrt{n}} \geqslant k_\alpha^*, \tag{15}$$

and the significance level defined by (10) is exactly that obtained by applying the procedure of Chapter 3 to the test statistic $\bar{Y}.$. Note that (15) does not depend on the particular alternative μ_A, so long as $\mu_A > \mu_0$. If, however, the alternative is such that $\mu_A < \mu_0$, the above argument shows that the critical region should be formed from small values of $\bar{y}.$, as is clear on general grounds. That is, the best critical region is then

$$\frac{\bar{y}. - \mu_0}{\sigma_0/\sqrt{n}} \leqslant -k_\alpha^*. \tag{16}$$

Example 4.4. Exponential family. Suppose that $Y_1, \ldots, Y_n$ are i.i.d. in the single parameter exponential family density (2.35), namely

$$\exp\{a(\theta)b(y) + c(\theta) + d(y)\},$$

and that the hypotheses are $H_0: \theta = \theta_0$ and $H_A: \theta = \theta_A$. Then the likelihood ratio involves the data only through the sufficient statistic $S = \Sigma b(Y_j)$ and the best critical region has the form

$$\{a(\theta_A) - a(\theta_0)\}s \geqslant e_\alpha,$$

for a suitable constant e_α. If $a(\theta_A) - a(\theta_0) > 0$, this is equivalent to $s \geqslant e_\alpha'$, the critical region being the same for all such θ_A.

To find the value of e_α', we need the distribution of S under H_0. If the natural parameterization has been taken, so that $a(\theta) = \theta$, it is easily shown that the moment generating function of S is

$$E(e^{tS}; \theta) = \exp[-n\{c(\theta + t) - c(\theta)\}].$$

Where the exact distribution cannot be obtained in useful form, a normal approximation, possibly with refinements, can be used.

Special cases of this result cover

(a) the problem of Example 4.3 about the mean of a normal distribution;

(b) the analogous problem for the variance of a normal distribution, the mean being known;

(c) the problems of Examples 4.1 and 4.2 concerning the uniform distribution.

For instance, with Example 4.1, the test statistic is $S = \Sigma \cos Y_j$.

Under the null hypothesis

$$E(S;H_0) = \frac{n}{2\pi} \int_0^{2\pi} \cos y \, dy = 0,$$

$$\text{(17)}$$

$$\text{var}(S;H_0) = \frac{n}{2\pi} \int_0^{2\pi} \cos^2 y \, dy = \tfrac{1}{2}n,$$

and for most purposes $S/\sqrt{(\tfrac{1}{2}n)}$ can be treated as a standardized normal variable under H_0. If the alternative in (2) has $\theta_A > 0$, large values of s form the best critical region.

Example 4.5. Interpretation of alternatives. It helps to appreciate the nature and role of the alternative hypothesis to consider an extreme case of Example 4.3. Suppose that we have one observation from $N(\mu, 1)$ and that the hypotheses are H_0: $\mu = 0$ and H_A: $\mu = \mu_A = 10$. The likelihood ratio critical region is formed from large values of y and if, say, $y = 3$, there is a significant departure from H_0 nearly at the 0.1% level. Yet if we had to choose one of H_0 and H_A on the basis of the data, then there is no doubt that, other things being more or less equal, we would prefer H_0. All the significance test tells us is that the data depart from H_0 in the direction of H_A and that the departure is of such a magnitude that can reasonably be accounted as evidence against H_0; by so doing, we are following a procedure that would rarely mislead us when H_0 is true.

The probabilities referred to in these discussions are normally calculated by direct reference to the distribution of the random variable Y. For the reasons discussed in Section 2.3(iii), if there is an ancillary statistic associated with the problem, then probability calculations are, in principle, to be made conditionally on its observed value. It might be thought that this can be done without conflict with the requirement of maximizing power, but unfortunately this is not always so, as the following example shows.

Example 4.6. Observation with two possible precisions. We take for definiteness a simple special case of the situation of Example 2.26. We obtain one observation Z which is normally distributed with mean μ. With probability $\tfrac{1}{2}$, case I, the variance is 1 and with probability $\tfrac{1}{2}$, case II, the variance is 10^6, and it is known which case holds.

Consider the two hypotheses $H_0: \mu = 0$ and $H_A: \mu = \mu_A$. Now the indicator variable c for the type of variance that holds is an ancillary statistic. The whole data are specified by $y = (z, c)$. If we work conditionally on the observed c, the size α critical region is

$$w_\alpha = \begin{cases} z > k_\alpha^* & (c = 1), \\ z > 10^3 k_\alpha^* & (c = 2). \end{cases} \tag{18}$$

That is, we require that

$$\text{pr}(Z \in w_\alpha | C = c \,; H_0) = \alpha, \tag{19}$$

and, subject to this, we require maximum power.

On the other hand, if we do not impose the conditional size condition (19), we apply the Neyman-Pearson lemma directly and calculate the likelihood ratio as

$$\frac{\dfrac{1}{2} \cdot \dfrac{1}{\sqrt{(2\pi)}\sigma_c} \exp\left\{ -\dfrac{(z - \mu_A)^2}{2\sigma_c^2} \right\}}{\dfrac{1}{2} \cdot \dfrac{1}{\sqrt{(2\pi)}\sigma_c} \exp\left(-\dfrac{z^2}{2\sigma_c^2} \right)},$$

where $\sigma_c = 1$ or 10^3 depending on whether $c = 1$ or 2. Thus the best critical region has the form $z/\sigma_c^2 \geqslant d_\alpha$, i.e. $z \geqslant d_\alpha$ if $c = 1$ and $z \geqslant 10^6 d_\alpha$ if $c = 2$. A simple calculation shows that, very nearly, $d_\alpha = k_{2\alpha}^*$. The best critical region is thus, very nearly,

$$z \geqslant k_{2\alpha}^* \quad (c = 1), \quad z \geqslant 10^6 k_{2\alpha}^* \quad (c = 2). \tag{20}$$

This is quite different from the conditional region (18). General theory shows that the average power is greater for (20) than for (18) and detailed calculation confirms this.

This simple example illustrates some important points. If we were concerned with any kind of decision procedure for choosing between H_0 and H_A, there would be no reason for expecting that the error rate under H_0 should be controlled at the same level when $\sigma^2 = 1$ as when it is 10^6, i.e. there would be no reason for imposing (19). Here, however, the object of the significance test is to assess the evidence in the data as regards consistency with H_0. For this purpose, it seems unacceptable that the conclusion reached if in fact it is known that the variance is one should be influenced by the information that the variance might have been 10^6. The point of the example is that the requirement of using a conditional distribution cannot be deduced from that of maximum power and that indeed the two requirements may conflict. In future discussions we suppose that probability calculations are made conditionally on any ancillary statistics.

Note, however, that generally

$$\frac{f_{Y|C}(y\,|c\,;\theta_A)}{f_{Y|C}(y\,|c\,;\theta_0)} = \frac{f_Y(y\,;\theta_A)/f_C(c)}{f_Y(y\,;\theta_0)/f_C(c)} = \frac{f_Y(y\,;\theta_A)}{f_Y(y\,;\theta_0)},$$

so that the likelihood ratio itself is the same calculated conditionally and unconditionally; it is the probability calculation of the distribution under H_0 that is different.

Example 4.7. Uniform distribution of known range. An instructive, if artificial, example of setting up a test is provided by the uniform distribution over $(\theta - \frac{1}{2}, \theta + \frac{1}{2})$. The likelihood corresponding to observations with order statistics $Y_{(1)}, \ldots, Y_{(n)}$ is

$$\begin{cases} 0 & (y_{(1)} < \theta - \frac{1}{2} \text{ or } y_{(n)} > \theta + \frac{1}{2}), \\ 1 & \text{otherwise.} \end{cases} \tag{21}$$

Thus the sufficient statistic is the pair $(Y_{(1)}, Y_{(n)})$ and $Y_{(n)} - Y_{(1)}$ is clearly ancillary.

Now if the null hypothesis is $\theta = \theta_0$, the likelihood ratio for $\theta = \theta_A$ versus $\theta = \theta_0$, $\text{lr}_{A0}(y)$, can take the following values:

(a) infinity, if the data are consistent with θ_A but not with θ_0;

(b) one, if the data are consistent with both hypotheses;

(c) zero, if the data are consistent with θ_0 but not with θ_A.

There is also the possibility that the likelihood ratio is undefined, y being inconsistent with both θ_0 and θ_A; this would lead to the "rejection" of H_0.

Now the level of significance for an observed y is defined formally by

$$p_{\text{obs}} = \text{pr}\{\text{lr}_{A0}(Y) \geqslant \text{lr}_{A0}(y) | Y_{(n)} - Y_{(1)} = c\,;H_0\}$$

and this takes values zero, one and one respectively for cases (a), (b) and (c).

Thus the test has just two achievable significance levels; zero corresponds to complete inconsistency with H_0 and one to complete consistency with H_0. Any other levels can be obtained only by some artificial device. From a common-sense point of view the conclusion that only two levels are achievable seems entirely satisfactory.

Example 4.8. Normal distribution with known coefficient of variation (ctd). In Example 2.21 we discussed, as a simple example of a distribution for which the dimension of the minimal sufficient statistic exceeds that of the parameter, the normal distribution with known coefficient of variation. That is, $Y_1, ..., Y_n$ are i.i.d. in $N(\mu, \gamma_0^2 \mu^2)$. We take the same null and alternative hypotheses as in Example 4.3, namely $\mu = \mu_0$ and $\mu = \mu_A$, with $\mu_A > \mu_0$.

On simplifying the likelihood ratio, the best critical region is found to have the form

$$\left(\frac{1}{\mu_0} + \frac{1}{\mu_A}\right) \Sigma y_j^2 - 2\Sigma y_j \geq d_\alpha. \tag{22}$$

This can, of course, be expressed in terms of the sample mean and variance; qualitatively the complication of the answer as compared with that in Example 4.3 arises because both sample mean and variance contribute information about μ, in the light of the assumed proportionality of standard deviation and mean.

The calculation of d_α requires finding the distribution under H_0 of the right-hand side of (22) and this is possible, at least in principle. Of course the existence of an ancillary statistic would lead to a modification of the distributional calculation but, so far as we know, there is no such statistic.

Quite generally, finding the critical region of given size involves a new distributional calculation for each problem. When the observations are i.i.d. under both H_0 and H_A, we have

$$\log lr_{A0}(Y) = \sum_{j=1}^{n} \log \frac{f_A(Y_j)}{f_0(Y_j)}.$$

Thus, under H_0,

$$E\{\log lr_{A0}(Y)\} = n \int_{-\infty}^{\infty} \log\left\{\frac{f_A(y)}{f_0(y)}\right\} f_0(y)dy \tag{23}$$

and the higher moments and cumulants can be found quite directly, leading to simple approximate forms for the size α critical region. For systematic discussion of quantities like (23), see Kullback (1968).

4.5   Discrete problems

So far in the general discussion we have assumed $\text{lr}_{A0}(Y)$ to have a continuous distribution under H_0 and this has been the case in all the examples except Example 4.7. There we saw in rather extreme form the limitations on the conclusions that arise when the statistic $\text{lr}_{A0}(Y)$ takes on a very restricted set of values.

This phenomenon will be encountered whenever the observed random variable Y has a discrete distribution and we now discuss this case in more detail. The situation is best explained by an example.

Example 4.9. Poisson distribution (ctd). Let Y have a Poisson distribution of mean μ, the two hypotheses being $H_0: \mu = 1$, and $H_A: \mu = \mu_A > 1$. Then

$$\text{lr}_{A0}(y) = \frac{e^{-\mu_A} \mu_A^y}{y!} \bigg/ \frac{e^{-1}}{y!}$$

and the likelihood ratio critical regions have the form $y \geq d_\alpha$.

However, because the random variable Y is discrete, the only possible critical regions are of the form $y \geq r$, where r is an integer, and the corresponding values of α are shown in Table 4.1.

TABLE 4.1

Possible likelihood ratio critical regions for a Poisson problem

r	$\text{pr}(Y \geq r; H_0)$	r	$\text{pr}(Y \geq r; H_0)$
0	1	4	0.0189
1	0.632	5	0.0037
2	0.264	6	0.0006
3	0.080	⋮	⋮

If α is one of the values in Table 4.1, a likelihood ratio region of the required size does exist. As pointed out after the proof of the Neyman-Pearson lemma, no critical region of probability α or less under H_0 can have larger power than the likelihood ratio region.

The levels of significance achievable by the data are thus restricted to the values in Table 4.1; the hypothetical physical interpretation of a significance level remains the same.

By a mathematical artifice, it is, however, possible to achieve like-lihood ratio critical regions with other values of α. This is of no direct practical importance, but is occasionally theoretically useful in comparing alternative procedures. The idea is to use a supplementary randomizing device to determine the conclusion when the observation falls on a boundary. To be explicit, suppose that, in our example, we wanted $\alpha = 0.05$. The region $y \geqslant 4$ is too small, whereas the region $y \geqslant 3$ is too large. Therefore we suppose that all values $y \geqslant 4$ are certainly put in the critical region, whereas if $y = 3$ we regard the data as in the critical region with probability π_3, to be determined. To achieve exactly the required size, we need

$$\text{pr}(Y \geqslant 4; H_0) + \pi_3 \text{pr}(Y = 3; H_0) = 0.05,$$

leading to $\pi_3 = 0.51$.

We call this a *randomized critical region* of size 0.05. In the same way we can achieve any desired α, supplementary randomization being needed except for the special values in Table 4.1.

In a general formulation, it is useful to introduce a size α test function $\pi_\alpha(y)$ by

$$\pi_\alpha(y) = \text{pr}(y, \text{ if observed, is included in the size } \alpha \text{ region}).$$

The condition for exact size α is then

$$E\{\pi_\alpha(Y); H_0\} = \alpha.$$

The "sensible" critical regions all have $\pi_\alpha(.)$ taking values 0 and 1, i.e. with $\pi_\alpha(.)$ as the indicator function of the critical region.

An extension of the Neyman-Pearson lemma (Lehmann, 1959, p. 65) states that, when randomized regions are allowed, the test function $\pi^*(.)$ with

$$E\{\pi^*(Y); H_0\} \leqslant \alpha$$

and with maximum

$$E\{\pi^*(Y); H_A\}$$

is obtained from large values of $\text{lr}_{A0}(y)$, randomizing for points on the boundary in exactly the way illustrated in Example 4.9.

The reason that these randomized critical regions are of no direct statistical importance is that the achievement of some preassigned α, e.g. $\alpha = 0.05$, is irrelevant, at least in the approach adopted here. If we consider instead the calculation of the level of significance

appropriate to given observations, then in Example 4.9 the randomized definition of p_{obs} corresponding to $Y = y$ is

$$\text{pr}(Y > y\,;H_0) + V\,\text{pr}(Y = y\,;H_0), \tag{24}$$

where V is uniformly distributed on $(0, 1)$ independently of Y. The
random variable corresponding to (24) is, under H_0, uniformly distributed on $(0, 1)$. Whereas this may be thought a useful property of
significance levels, it seems entirely secondary to the securing of a
relevant physical interpretation of p_{obs} and this is achieved by the
original definition of Section 3.2, namely

$$p_{\text{obs}} = \text{pr}(Y \geqslant y\,;H_0). \tag{25}$$

An extreme example of the difficulties that can arise with discrete
distributions and randomized tests is provided by the testing of
hypotheses about a binomial parameter using the results of one or
two trials (Cohen, 1958); see Exercise 4.6.

4.6   Composite alternatives

In the discussion so far, it has been assumed that both null and
alternative hypotheses are simple, i.e. completely specify the distribution of the observations. Suppose now that the null hypothesis is
simple but that there is a collection Ω_A of alternative hypotheses;
we denote a simple member of Ω_A by H_A. Two cases now arise. In
the first, we get the same size α best critical region for all $H_A \in \Omega_A$;
in the second, the best critical region, and hence the level of significance, depend on the particular H_A.

In the first case, the theory of the previous sections is immediately
applicable and leads to a significance test which does not require us
to specify which particular $H_A \in \Omega_A$ is of concern. In the second
case, some often rather arbitrary further consideration is needed to
choose a test.

While the first case is very special, it is sufficiently important to
justify some special terminology. If the most powerful size α critical
region against alternative $H_A \in \Omega_A$, then we say that the region is a
uniformly most powerful size α region, for this particular Ω_A. If the
system of critical regions has this property for each α, then the test
itself is called uniformly most powerful. Several examples of uniformly most powerful tests are implicit in earlier examples. Thus in

Example 4.3, concerning the mean of a normal distribution, the critical region (15) does not depend on μ_A, provided that $\mu_A > \mu_0$. More generally, for the one parameter exponential family density of Example 4.4, namely

$$\exp\{a(\theta)b(y) + c(\theta) + d(y)\},$$

the test is uniformly most powerful if either

$$a(\theta_A) > a(\theta_0) \quad (\theta_A \in \Omega_A) \quad \text{or} \quad a(\theta_A) < a(\theta_0) \quad (\theta_A \in \Omega_A).$$

Further, if $a(.)$ is strictly monotonic over some set Ω' of values, then for any $\theta_0 \in \Omega'$, the likelihood ratio test of $\theta = \theta_0$ for alternatives $\theta > \theta_0, \theta \in \Omega'$ or for alternatives $\theta < \theta_0, \theta \in \Omega'$ is uniformly most powerful. This covers many problems about special one parameter distributions and emphasizes the simplifying role of the natural parameter, $\phi = a(\theta)$.

Suppose now that there is no uniformly most powerful test, i.e. that the size α likelihood ratio critical region depends on the particular alternative. Example 4.8 illustrates this. There are now a number of possible approaches including, when the alternatives are indexed by a single parameter,

(a) to pick, somewhat arbitrarily, a "typical" θ_A and to use the test most powerful for that particular alternative;

(b) to take θ_A very close to θ_0, i.e. to maximize the power locally near the null hypothesis;

(c) to maximize some weighted average of power, e.g. when the alternatives are $\theta > \theta_0$, to maximize

$$\int_{\theta_0}^{\infty} \text{pr}(Y \in w_\alpha ; \theta)dk(\theta) \tag{26}$$

for suitable $k(.)$.

Possibility (b), while not very appealing statistically, leads to simple results of general importance and will be developed in detail in Section 4.8. When the alternatives are described by more than one parameter or when they stretch on both sides of the null hypothesis fresh considerations arise; see Sections 4.9 and 4.7.

It is useful to introduce the power function of a test for which the alternatives are defined parametrically, say by θ. This is, at level α,

$$\text{pow}(\theta ; \alpha) = \text{pr}(Y \in w_\alpha ; \theta) = \text{pr}\{p_w(Y) \leq \alpha ; \theta\}, \tag{27}$$

where $p_w(y)$ denotes the level of significance achieved by $Y = y$ using the system of critical regions w. In particular, if the null hypothesis is $\theta = \theta_0$, $\text{pow}(\theta_0; \alpha) = \alpha$.

Usually we consider $\text{pow}(\theta; \alpha)$ as a curve or surface which is a function of θ, one such curve for each α. Alternatively, and somewhat more naturally, we can regard it as giving the cumulative distribution function of P, one such distribution for each θ.

In particular cases it is possible to calculate the power function explicitly. For instance, for the problem of Example 4.3 concerning the mean of a normal distribution,

$$\text{pow}(\mu; \alpha) = \text{pr}\left(\bar{Y}. > \mu_0 + k_\alpha^* \frac{\sigma_0}{\sqrt{n}}; \mu\right)$$

$$= \Phi\left(\frac{\mu - \mu_0}{\sigma_0/\sqrt{n}} - k_\alpha^*\right). \tag{28}$$

Thus

$$\Phi^{-1}\{\text{pow}(\mu; \alpha)\} = \frac{\mu - \mu_0}{\sigma_0/\sqrt{n}} - k_\alpha^*, \tag{29}$$

a linear function of μ with a slope not depending on α. That is, plots of $\Phi^{-1}(\text{power})$ versus μ will be straight lines. Many tests have power functions approximately of the form (28) and the transformation (29) is therefore quite often useful in numerical work with power functions.

There are two broad uses of power functions. The first, which we have already employed, is in the comparison of alternative tests of significance. In the simplest form of the argument, we are just looking for the test with largest power, or making a sensible compromise between tests none of which is uniformly most powerful. In other situations, we may wish to assess the loss of sensitivity arising from the use of a test which, while not of maximum power, has compensating features, such as simplicity or insensitivity to departures from assumptions or to the presence of outliers.

Secondly, we may use power considerations to guide the choice of sample size. For example, (28) defines for the normal mean problem a family of power functions, one for each n. As n increases, the power functions increase more rapidly. Suppose that before the data are obtained, it is required to choose an appropriate number, n, of observations. Let δ be such that $\mu = \mu_0 + \delta$ is an alternative thought to represent a departure from the null hypothesis, $\mu = \mu_0$, of sufficient practical importance that we want to be reasonably confident of finding evidence against H_0, if in fact this particular alternative holds. More formally, we require that

$$\text{pow}(\mu_0 + \delta \, ; \alpha) \; = \; 1 - \beta. \tag{30}$$

That is, β is the probability of failing to find evidence at least at level α, when the alternative holds. By (28) this implies that

$$\Phi\left(\frac{\delta}{\sigma_0/\sqrt{n}} - k_\alpha^*\right) \; = \; 1 - \beta \; = \; \Phi(k_\beta^*),$$

i.e.
$$n \; = \; \frac{(k_\alpha^* + k_\beta^*)^2 \, \sigma_0^2}{\delta^2}. \tag{31}$$

Of course, in general, the right-hand side of (31) will not be an integer, i.e. we cannot find a power curve passing exactly through an arbitrary preassigned point. This is hardly important, however, and can be resolved formally, for instance by taking the smallest n for which the α level power at $\mu_0 + \delta$ is at least $1 - \beta$.

Thus having specified δ/σ_0, α and β, we can find a suitable n; alternatively, for given n, we can see whether the power achieved at $\mu_0 + \delta$ is reasonably satisfactory.

The argument has been given here for a very special problem but can clearly be applied generally, in fact whenever the power function can be calculated numerically, exactly or approximately.

Preliminary rough calculation of sample size is often desirable. The method described here involves rather arbitrary choices of δ and of the probability levels α and β. When more explicit information about costs is available, a decision theoretical discussion of the choice of sample size becomes possible; see Example 11.13.

An alternative version of the present argument can be phrased in terms of widths of confidence intervals; see Chapter 7. A useful, rather rough, form of the argument in terms of tests is to choose a sample size that will make the upper α point of the distribution of the test statistic T under the null hypothesis agree with the expected value under the special alternative of interest, i.e. to require in general that

$$\text{pr}\{T > E(T \, ; H_A) \, ; H_0\} \; = \; \alpha.$$

In the special case discussed above, this is equivalent to taking $\beta = \frac{1}{2}$, i.e. to choosing δ to be the difference from H_0 at which 50% power is required.

A final general comment on the idea of power is that it is never used directly in the analysis of data. When we are interested in the relation between data y and values of the parameter other than the null value, an interval or other form of estimate will be needed; the power function is a general property of a test and does not give us information specific to particular sets of data.

4.7   Two-sided tests

So far we have discussed problems in which, in at least some rough sense, the alternatives lie on one side of the null hypothesis. If, to be specific, we take Examples 4.3 and 4.4, we see a drastic conflict between testing against alternatives $\theta > \theta_0$ and against alternatives $\theta < \theta_0$; as is obvious on general grounds, the opposite tails of the distribution under H_0 are used to form critical regions.

Yet in applications it will often happen that alternatives in more than one direction are of interest. Then no uniformly most powerful test can exist and the situation is indeed more difficult than that arising when we are faced with a choice between two tests both with roughly the right behaviour. We shall suppose for the moment that Ω_A is indexed by a single parameter.

There are two broad approaches to the discussion of two-sided tests. The first, which has been widely used in the literature, is mathematically attractive but statistically normally unsatisfactory. Therefore we deal with it only briefly. Let Ω_A denote the set of alternative hypotheses with H_A a typical member. A critical region w_α of size α is called *unbiased* if

$$\mathrm{pr}(Y \in w_\alpha; H_A) \geqslant \mathrm{pr}(Y \in w_\alpha; H_0) \text{ for all } H_A \in \Omega_A. \qquad (32)$$

We then restrict attention to unbiased regions and among these look for the one with maximum power, if possible for all H_A. If (32) is not satisfied the test is said to be *biased*. The general idea is that we want the probability of falling in the critical region to be smaller under H_0 than under an alternative H_A and (32) requires this to happen for every H_A.

The limitations of this are first that a biased test may exist that is preferable to the best unbiased test for most alternatives. A second and more important point is that $\mathrm{pr}(Y \in w_\alpha)$ is the probability of obtaining a departure significant at the α level regardless of the direction of the observed departure. Now, in the overwhelming majority of applications, it is essential to consider the direction of departure in interpreting the result of the significance test; that is, to conclude that there is evidence that $\theta > \theta_0$ when in fact $\theta < \theta_0$ would normally be very misleading. Yet the composite power function (32) treats such a conclusion in the same way as the correct conclusion.

This argument shows that we are really involved with two tests, one to examine the possibility $\theta > \theta_0$, the other for $\theta < \theta_0$. Each test has its power function of the type considered in Section 4.6 and showing the power for detecting departures in the "correct" direction only. We thus follow the procedure of Section 3.4(iv), i.e. in effect look at both

$$p_{obs}^+ = \text{pr}(T \geqslant t_{obs}; H_0), \quad p_{obs}^- = \text{pr}(T \leqslant t_{obs}; H_0)$$

and select the more significant of the two, i.e. the smaller. To preserve the physical meaning of the significance level, we make a correction for selection. That is, if

$$m_{obs} = \min(p_{obs}^+, p_{obs}^-), \tag{33}$$

we write

$$p_{obs} = \text{pr}(M \leqslant m_{obs}; H_0). \tag{34}$$

In the continuous case $p_{obs} = 2m_{obs}$, but in the discontinuous case

$$p_{obs} = \begin{cases} p_{obs}^+ + \text{pr}(P^- \leqslant p_{obs}^+; H_0) & (p_{obs}^+ \leqslant p_{obs}^-), \\ p_{obs}^- + \text{pr}(P^+ \leqslant p_{obs}^-; H_0) & (p_{obs}^+ \geqslant p_{obs}^-). \end{cases} \tag{35}$$

This corresponds very closely to the commonly used procedure of forming a critical region of size α by combining two regions, each of size $\frac{1}{2}\alpha$, one in each direction. It is different in many cases from the unbiased region.

4.8   Local power

(i) General

While uniformly most powerful tests are available for simple problems about the exponential family, these are in some ways very special situations and we turn now to methods that will give answers more generally. Some approaches were outlined in Section 4.6, and here we examine in detail the second, (b), based on alternatives close to the null hypothesis. As in the previous section, we suppose θ to be one-dimensional.

Denote the p.d.f. of the vector Y by $f_Y(y; \theta)$ and consider the null hypothesis $\theta = \theta_0$ with, for definiteness, alternatives $\theta > \theta_0$. If we take the particular alternative $\theta_A = \theta_0 + \delta$, where δ is small, we have

$$\log \mathrm{lr}_{A0}(y) = \log\{f_Y(y; \theta_0 + \delta)/f_Y(y; \theta_0)\}$$

$$= \delta \frac{\partial \log f_Y(y; \theta_0)}{\partial \theta_0} + o(\delta), \tag{36}$$

under some regularity conditions. Thus, for sufficiently small positive δ, we obtain the likelihood ratio critical region from large values of the random variable

$$U_. = U_.(\theta_0) = u_.(Y; \theta_0) = \frac{\partial \log f_Y(Y; \theta_0)}{\partial \theta_0}. \tag{37}$$

We call $U_.$ the *efficient score* for Y, the argument θ_0 being obvious from the context. In many applications the component random variables $Y_1, \ldots, Y_n$ are independent. Then the log likelihood is the sum of contributions and we can write $U_. = \Sigma U_j$, where

$$U_j = u_j(Y_j; \theta_0) = \frac{\partial \log f_{Y_j}(Y_j; \theta_0)}{\partial \theta_0} \tag{38}$$

is the efficient score for the jth component. Further, $U_.$ is then the sum of independent random variables and this much simplifies discussion of its distributional properties. We now digress to study these, the results being important in a number of contexts.

(ii) Efficient scores and information
In the following discussion we repeatedly differentiate with respect to θ integrals over the sample space and we assume that the order of integration and differentiation can be interchanged. Problems for which this can be done are called *regular*. The precise analytical conditions depend on the context, but are roughly that there is no discontinuity in the underlying distribution with position depending on the parameter θ; see Example 4.12.

We first show that for a regular problem

$$E\{u_.(Y; \theta_0); \theta_0\} = 0; \tag{39}$$

note that it is crucial that the expectation is taken at the same value of θ as is used in evaluating $U_.$. Because $f_Y(y; \theta)$ is a p.d.f., we have:

that

$$\int f_Y(y\,;\theta)dy \;=\; 1.$$

Here, and throughout the following discussion, integration is modified to summation for discrete distributions. On differentiating with respect to θ and putting $\theta = \theta_0$, we have that

$$0 \;=\; \int \frac{\partial f_Y(y\,;\theta_0)}{\partial \theta_0}\,dy \;=\; \int \frac{\partial \log f_Y(y\,;\theta_0)}{\partial \theta_0}\,f_Y(y\,;\theta_0)dy \quad (40)$$

$$= \; E\{U_.(\theta_0);\theta_0\},$$

as required.

It follows that

$$\mathrm{var}\{U_.(\theta_0);\theta_0\} \;=\; E(U_.^2\,;\theta_0) \;=\; i_.(\theta_0), \tag{41}$$

say. The function $i_.(\theta)$ is of general importance and is called the *Fisher information* about θ contained in Y. The dot suffix is a useful reminder that the total data vector is involved.

An important identity for $i_.(.)$ is obtained by differentiating (40) again with respect to θ_0, thereby showing that

$$0 = \int \frac{\partial^2 \log f_Y(y\,;\theta_0)}{\partial \theta_0^2}\,f_Y(y\,;\theta_0)dy + \int \frac{\partial \log f_Y(y\,;\theta_0)}{\partial \theta_0}\,\frac{\partial f_Y(y\,;\theta_0)}{\partial \theta_0}\,dy,$$

i.e.

$$i_.(\theta_0) \;=\; E(U_.^2\,;\theta_0) \;=\; E\left\{-\frac{\partial^2 \log f_Y\,(Y\,;\theta_0)}{\partial \theta_0^2};\theta_0\right\}. \tag{42}$$

We can in principle find the higher moments of $U_.$ by similar arguments.

In the important special case where the Y_j are mutually independent, we have immediately that $i_.(\theta_0) = \Sigma i_j(\theta_0)$, where

$$i_j(\theta_0) \;=\; \mathrm{var}\{U_j(\theta_0);\theta_0\} \;=\; E\left\{-\frac{\partial^2 \log f_{Y_j}(Y_j\,;\theta_0)}{\partial \theta_0^2}\,;\theta_0\right\}$$

is the information from Y_j.

Now suppose that we need the expectation of $U_.(\theta_0)$ evaluated over a distribution with a value of θ near to but not equal to θ_0. We have that

$E\{U_.(\theta_0); \theta_0 + \delta\}$

$$= \int \frac{\partial \log f_Y(y; \theta_0)}{\partial \theta_0} \left\{ f_Y(y; \theta_0) + \delta \frac{\partial f_Y(y; \theta_0)}{\partial \theta_0} + o(\delta) \right\} dy \quad (44)$$

$$= \delta \int \left\{ \frac{\partial \log f_Y(y; \theta_0)}{\partial \theta_0} \right\}^2 f_Y(y; \theta_0) dy + o(\delta)$$

$$= \delta i_.(\theta_0) + o(\delta), \quad (45)$$

the leading term in (44) vanishing because $E\{U_.(\theta_0); \theta_0\} = 0$.
Similarly

$$\mathrm{var}\{U_.(\theta_0); \theta_0 + \delta\} = i_.(\theta_0) + o(\delta), \quad (46)$$

under some regularity conditions. Further terms in the expansions
can be obtained.

If a one-one transformation of parameter is made from θ to
$\phi = \phi(\theta)$, and the efficient score and information evaluated in terms
of ϕ, $U_.$ is multiplied by $\phi'(\theta_0)$ and $i_.(\theta_0)$ by $\{\phi'(\theta_0)\}^2$.

Detailed discussion of the distribution of $U_.$ is simplified when the
Y_j are independent and $U_.$ thus the sum of independent contributions.
A strong central limit effect will often hold and a normal approximation, possibly improved by consideration of third and fourth
cumulants, will usually be adequate; in particular cases it may be
feasible to find the exact distribution of $U_.$, either analytically or
numerically.

When an ancillary statistic C is available we apply the above arguments to the conditional density of Y given $C = c$. It follows from
the remarks at the end of Example 4.6, and can be checked directly,
that the value of the efficient score is unaffected by conditioning; for

$$\log f_{Y|C}(y|c; \theta) = \log f_Y(y; \theta) - \log f_C(c),$$

and the second term is independent of θ because C is an ancillary
statistic.

Further, it follows directly from (40) applied to the conditional
p.d.f. of Y given $C = c$ that

$$E\{U_.(\theta_0)|C = c; \theta_0\} = 0. \quad (47)$$

The conditional variance of $U_.$ given $C = c$, which we can call the
conditional information, $i_.(\theta_0|c)$, is

$$E\left\{-\left.\frac{\partial^2 \log f_Y(y\,;\theta_0)}{\partial\theta_0^2}\right|C=c\,;\theta_0\right\}. \tag{48}$$

It follows from a basic property of conditional expectations that

$$E_C\{i_{.}(\theta_0|C)\} = i_{.}(\theta_0). \tag{49}$$

A measure of the extent to which the ancillary statistic C divides the possible data points into relatively informative and relatively uninformative ones is provided by

$$\mathrm{var}_C\{i_{.}(\theta_0|C)\}. \tag{50}$$

If this is zero, to take an extreme case, then, at least as far as procedures based on the statistic $U_{.}$ and its first two moments are concerned, there would be no point in conditioning on $C = c$. In general, the larger is (50) the more effective is C in separating possible sets of data according to their informativeness about θ.

Finally, the above results generalize immediately when θ is a vector with components $(\theta_1, \ldots, \theta_q)$. The operator $\partial/\partial\theta$ is interpreted as a column vector, so that the efficient score

$$U_{.} = U_{.}(\theta_0) = \frac{\partial \log f_Y(Y;\theta_0)}{\partial\theta_0}$$

is a $q \times 1$ vector with components

$$U_{.r} = \frac{\partial \log f_Y(Y;\theta_0)}{\partial\theta_{r0}} \quad (r = 1, \ldots, q). \tag{51}$$

Exactly as before we have that

$$E\{U_{.}(\theta_0);\theta_0\} = 0$$

and the $q \times q$ covariance matrix of $U_{.}$ has as its (r,s)th element

$$i_{rs}(\theta_0) = E\{U_{.r}(\theta_0)U_{.s}(\theta_0);\theta_0\} = E\left\{-\frac{\partial^2 \log f_Y(Y;\theta_0)}{\partial\theta_{r0}\,\partial\theta_{s0}};\theta_0\right\}. \tag{52}$$

It is convenient to write this in the concise form

$$i_{.}(\theta_0) = E\left\{-\frac{\partial^2 \log f_Y(Y;\theta_0)}{\partial\theta_0^2};\theta_0\right\}, \tag{53}$$

where $\partial^2/\partial\theta^2$ is to be interpreted as $(\partial/\partial\theta)(\partial/\partial\theta)^\mathrm{T}$.

Most of the subsequent formulae involving $U_{\cdot}$ and $i_{\cdot}(\theta_0)$ apply both to scalar and to vector parameters. In accordance with our general conventions we show the information matrix $\mathbf{i}(\theta_0)$ in bold fount whenever we want to emphasize that a vector parameter is involved.

When the components of Y are independent the efficient score and its covariance matrix are sums of contributions,

$$U_{\cdot}(\theta_0) = \Sigma U_j(\theta_0), \tag{54}$$

$$i_{\cdot}(\theta_0) = \Sigma i_j(\theta_0). \tag{55}$$

Before returning to the discussion of locally most powerful tests, it is convenient to give simple illustrations of the formulae involving efficient scores.

Example 4.10. Normal distribution with known variance (ctd).
Suppose first that we have a single random variable Y_1 distributed in $N(\mu, \sigma_0^2)$, where σ_0^2 is known. Then

$$\log f_{Y_1}(y\,;\mu) = -\tfrac{1}{2}\log(2\pi\sigma_0^2) - \tfrac{1}{2}(y-\mu)^2/\sigma_0^2,$$

so that the efficient score for that observation is

$$U_1(\mu_0) = \frac{Y_1 - \mu_0}{\sigma_0^2}.$$

It is obvious directly that $E\{U_1(\mu_0)\,;\mu_0\} = 0$. Further

$$\left(\frac{\partial \log f_{Y_1}(Y_1\,;\mu_0)}{\partial \mu_0}\right)^2 = \frac{(Y_1 - \mu_0)^2}{\sigma_0^4}, \qquad -\frac{\partial^2 \log f_{Y_1}(Y_1\,;\mu_0)}{\partial \mu_0^2} = \frac{1}{\sigma_0^2}.$$

Thus the equality of the alternative versions (42) for the information is verified; in fact

$$i_1(\mu_0) = 1/\sigma_0^2. \tag{56}$$

Note also that $E\{U_1(\mu_0)\,;\mu_0 + \delta\} = \delta/\sigma_0^2$, so that the approximate formula (45) is, in this case, exact.

Suppose now that $Y_1, \ldots, Y_n$ are i.i.d. in $N(\mu, \sigma_0^2)$. Then for the total set of random variables Y, the efficient score is

$$U_{\cdot}(\mu_0) = \Sigma(Y_j - \mu_0)/\sigma_0^2,$$

with

$$i_{\cdot}(\mu_0) = n/\sigma_0^2.$$

Example 4.11. Exponential family (ctd). Suppose that Y_1 has the p.d.f.

$$f_{Y_1}(y \, ; \theta) = \exp\{a(\theta)b(y) + c(\theta) + d(y)\}.$$

Then

$$\frac{\partial \log f_{Y_1}(y \, ; \theta_0)}{\partial \theta_0} = a'(\theta_0)b(y) + c'(\theta_0),$$

$$-\frac{\partial^2 \log f_{Y_1}(y \, ; \theta_0)}{\partial \theta_0^2} = -a''(\theta_0)b(y) - c''(\theta_0). \tag{57}$$

It follows that for this single observation

$$U_1(\theta_0) = a'(\theta_0)b(Y_1) + c'(\theta_0).$$

Therefore $E\{b(Y_1); \theta_0\} = -c'(\theta_0)/a'(\theta_0)$ and hence the expected value of (57) simplifies to give

$$i_1(\theta_0) = a''(\theta_0)c'(\theta_0)/a'(\theta_0) - c''(\theta_0)$$

$$= -a'(\theta_0)\frac{d}{d\theta_0}\left\{\frac{c'(\theta_0)}{a'(\theta_0)}\right\}. \tag{58}$$

In particular, if we take the natural parameterization in which $a(\theta) = \theta$, we have that

$$U_1(\theta_0) = b(Y_1) + c'(\theta_0), \quad i_1(\theta_0) = \text{var}\{b(Y_1); \theta_0\} = -c''(\theta_0). \tag{59}$$

Of course, Example 4.10 is a special case.

If $Y_1, \ldots, Y_n$ are i.i.d. with the above distribution then, using the natural parameterization for convenience, the total efficient score and information are

$$U_.(\theta_0) = \Sigma b(Y_j) + nc'(\theta_0), \quad i_.(\theta_0) = -nc''(\theta_0). \tag{60}$$

Example 4.12. Non-regular problems. The simplest example of a non-regular problem arises with a single random variable Y_1 uniformly distributed over $(0, \theta)$. The start of the argument leading to (40) breaks down because we cannot conclude from

$$\int_0^\theta \frac{1}{\theta}dy = 1$$

that

$$\int_0^\theta \frac{\partial}{\partial \theta}\left(\frac{1}{\theta}\right)dy = 0;$$

there is a second term arising from differentiating the upper limit of the integral.

A less drastic example is provided by the density

$$\tfrac{1}{2}\exp(-|y-\theta|) \quad (-\infty < y < \infty).$$

Here

$$U_1(\theta_0) = \begin{cases} 1 & (Y_1 > \theta_0), \\ -1 & (Y_1 < \theta_0), \end{cases}$$

but is not defined at $Y = \theta_0$. The steps leading to the proof that U_1 has zero expectation are in order, but the second differentiation leading to

$$i(\theta_0) = E\left\{-\frac{\partial^2 \log f_{Y_1}(Y_1 ; \theta_0)}{\partial \theta_0^2}\right\}$$

cannot be done; in fact the second derivative involved is zero almost everywhere, although

$$i_1(\theta_0) = \text{var}\{U_1(\theta_0); \theta_0\} = 1.$$

Another type of irregularity that may occur is the vanishing of $i(\theta)$ on account of the score being identically zero. This can often be avoided by suitable reparameterization, as in Example 4.16.

(iii) Locally most powerful tests

We now apply the results of (ii) to the discussion of locally most powerful tests. According to (37), if we have a null hypothesis $\theta = \theta_0$ and we require maximum power for local alternatives $\theta = \theta_0 + \delta$, then the appropriate test statistic is the total efficient score $U_. = U_.(\theta_0)$, large values being significant for alternatives $\delta > 0$, small values for alternatives $\delta < 0$. The resulting test is locally most powerful in the sense that it maximizes the slope of the power function at $\theta = \theta_0$; see (iv) and Exercise 4.2.

To complete the derivation of a test we need the distribution of $U_.$ under the null hypothesis. This has zero mean and variance $i_.(\theta_0) = n\bar{i}_.(\theta_0)$, introducing for convenience $\bar{i}_.$ as the mean information per observation. For many purposes a normal approximation will be adequate, although this can always be improved by finding higher moments. That is, as a first approximation the upper α point of the distribution of $U_.$ can be taken as

$$k_\alpha^*\{i_.(\theta_0)\}^{\frac{1}{2}} = k_\alpha^* n^{\frac{1}{2}}\{\bar{i}_.(\theta_0)\}^{\frac{1}{2}}. \tag{61}$$

If there is a uniformly most powerful test it must, in particular, have maximum local power, provided that the family of alternatives includes local values, and hence must be identical to the test based on $U_.$.

Example 4.13. Location parameter of Cauchy distribution; unconditional test. Let $Y_1, \ldots, Y_n$ be i.i.d. in the Cauchy distribution with p.d.f.

$$\frac{1}{\pi\{1 + (y - \theta)^2\}} .$$

For the null hypothesis $\theta = \theta_0$ the efficient score from a single random variable Y_1 is

$$U_1(\theta_0) = -\frac{\partial}{\partial\theta_0} \log\{1 + (Y_1 - \theta_0)^2\} = \frac{2(Y_1 - \theta_0)}{1 + (Y_1 - \theta_0)^2} .$$

The test statistic is thus

$$U_.(\theta_0) = 2 \sum_{j=1}^{n} \frac{(Y_j - \theta_0)}{1 + (Y_j - \theta_0)^2} . \tag{62}$$

Now the information from a single observation is

$$i_1(\theta_0) = E\left[\left\{\frac{\partial \log f_{Y_1}(Y_1 ; \theta_0)}{\partial\theta_0}\right\}^2 ; \theta_0\right] = \frac{1}{2} .$$

That is, under the null hypothesis, $U_.$ has zero mean and variance $\frac{1}{2}n$. Further, it is clear that the distribution is symmetrical. Unless n is very small it will for most purposes be enough to use a normal approximation. To investigate this further, however, consider the fourth moment of U_1, namely

$$E\left[\left\{\frac{\partial \log f_{Y_1}(Y_1 ; \theta_0)}{\partial\theta_0}\right\}^4 ; \theta_0\right] = \frac{16}{\pi} \int_{-\infty}^{\infty} \frac{z^4 dz}{(1 + z^2)^5} = \frac{3}{8} .$$

The fourth cumulant of U_1 is thus $-3/8$ and by the additive property of cumulants that for $U_.$ is $-3n/8$; the standard measure of kurtosis is therefore $-3/(2n)$. Thus, unless n is very small, the effect of non-normality on the distribution of $U_.$ is likely to be very slight.

Example 4.14. Location parameter; unconditional test. An immediate generalization of the previous example is obtained by considering $Y_1, \ldots, Y_n$ i.i.d. with density $h(y - \theta)$. The efficient score is

$$\Sigma h'(Y_j - \theta_0)/h(Y_j - \theta_0)$$

and the total information

$$n \int_{-\infty}^{\infty} \frac{\{h'(y)\}^2}{h(y)} dy; \tag{63}$$

the integral occurring here is sometimes called the *intrinsic accuracy* of the density $h(.)$. For a normal density, it is by (56) the reciprocal of the variance.

Example 4.15. Location parameter; conditional test. When $h(.)$ in Example 4.14 is not a member of the exponential family, no reduction by sufficiency is possible, except to the set of order statistics. As noted in Example 2.29, the configuration $C_2 = Y_{(2)} - Y_{(1)}, \ldots,$ $C_n = Y_{(n)} - Y_{(1)}$ forms an ancillary statistic. We should therefore examine the distribution of any test statistic conditionally on $C = c$; note however that, in accordance with the discussion of (ii), U. remains the basis of the locally most powerful test. When we argue conditionally, U. is to be regarded as a function of the single random variable $Y_{(1)}$; it is easily seen that any other order statistic or combination of order statistics, such as $\bar{Y}.$, could be used instead of $Y_{(1)}$.

We proceed therefore to calculate the conditional p.d.f. of $Y_{(1)}$ given $C = c$. The joint density of the order statistics is

$$n! h(y_{(1)} - \theta) \ldots h(y_{(n)} - \theta).$$

The transformation to new variables $(Y_{(1)}, C)$ has Jacobian one, so that

$$f_{Y_{(1)}, C}(y_{(1)}, c) = n! h(y_{(1)} - \theta) h(y_{(1)} + c_2 - \theta) \ldots h(y_{(1)} + c_n - \theta).$$

The p.d.f. of C is formed by integrating with respect to $y_{(1)}$ and therefore the conditional p.d.f. of $Y_{(1)}$ given $C = c$ is

$$f_{Y_{(1)}|C}(y_{(1)}|c) = \frac{h(y_{(1)} - \theta) \ldots h(y_{(1)} + c_n - \theta)}{\int h(y'_{(1)} - \theta) \ldots h(y'_{(1)} + c_n - \theta) dy'_{(1)}}. \tag{64}$$

Expression (64) has an alternative interpretation; the likelihood of the observations is

$$h(y_1 - \theta) \ldots h(y_n - \theta)$$

and (64) is simply that function normalized to have total integral one, and then regarded as a function of $y_{(1)}$ for fixed θ and c.

This function is fairly easily computed numerically in any particular case.

There is, however, a further difficulty before this can be made the basis of a test of a null hypothesis $\theta = \theta_0$. This is that there is no uniformly most powerful test, except for the normal distribution, and if we use the locally most powerful test statistic, the efficient score, this has to be considered as a function of $Y_{(1)}$, for given $C = c$. Now if $U_.$ is a monotonic function of $Y_{(1)}$ for the particular c involved, we have only to examine the tail area in the conditional distribution of $Y_{(1)}$ under the null hypothesis. But in general the relation between $Y_{(1)}$ and $U_.$ will not be monotonic and thus a more complex calculation is needed to convert the distribution of $Y_{(1)}$ into a significance level for $U_.$.

It is tempting to use $Y_{(1)}$ itself as the test statistic rather than $U_.$; however, if the likelihood function and density of $Y_{(1)}$ are appreciably multimodal, this may be rather bad.

While in principle the conditional test described here seems the appropriate one to use, little seems to be known about the relative merits of it and the simpler-to-use unconditional test of Examples 4.13 and 4.14. In a general way, it is clear that there will be little difference between them except for small values of n, because the configuration C will become a stable representation of the known density $h(.)$ for large n.

Unfortunately the requirement of maximum local power is convincing only in so far as it leads to near-to-optimum power in regions of higher power. Thus the situation in which the locally most powerful test has poor power for alternatives where the power is, say, 0.5 or more, is not good. To investigate this further we can argue approximately as follows.

By (45) and (29), the alternative δ_β for which the power is approximately $1 - \beta$ is given by

$$E\{U_.(\theta_0); \theta_0 + \delta_\beta\} = (k_\alpha^* + k_\beta^*)[\mathrm{var}\{U_.(\theta_0); \theta_0\}]^{\frac{1}{2}},$$

i.e.

$$\delta_\beta \simeq n^{-\frac{1}{2}} \{\bar{i}_.(\theta_0)\}^{\frac{1}{2}}(k_\alpha^* + k_\beta^*). \tag{65}$$

Note that for a fixed average amount of information δ_β tends to zero as n increases for any fixed α and β. That is, for sufficiently large n, the local test statistic is appropriate for a region of any pre-assigned power. More specifically, if we want maximum power near

the alternative (65), we can use a higher order expansion of the likelihood ratio, retaining in (36) terms of order δ_β^2 and choosing δ_β
according to (65). This leads to the test statistic

$$U_.(\theta_0) + \frac{(k_\alpha^* + k_\beta^*)}{2\{i_.(\theta_0)\}^{\frac{1}{2}}} \left\{ \frac{\partial U_.(\theta_0)}{\partial \theta_0} + i_.(\theta_0) \right\} , \tag{66}$$

where the constant $i_.(\theta_0)$ has been included in the second term to
make the exact expectation zero.

The moments of this statistic under the null hypothesis can be
evaluated by the techniques outlined in (ii).

A further general point concerning the locally most powerful test
is that in its derivation it is assumed that the log likelihood is differentiable with respect to the parameter at the null value; in one-sided
problems it is, however, enough that one-sided differentiation is
possible. Also it is necessary that the first derivative does not vanish
identically; if this is found to happen reparameterization is normally
indicated. These points are illustrated by the following example; see
also Example 9.12.

Example 4.16. Simplified components of variance problem. Let
$w_1, \ldots, w_n$ be given positive constants and let $Y_1, \ldots, Y_n$ be independently normally distributed with zero mean and with

$$\text{var}(Y_j) = E(Y_j^2) = 1 + w_j\theta. \tag{67}$$

It is required to test the null hypothesis $\theta = 0$ against alternatives
$\theta > 0$.

This is formally a well-defined problem for all $\theta > -1/\max(w_j)$,
but with the particular practical interpretation of most interest it
only makes sense for $\theta \geqslant 0$; for that reason, properties of the model
and derivatives for $\theta < 0$ are not really of interest. In this particular
interpretation, the model is regarded as a simplified version of an
unbalanced one-way components of variance problem in which the
single between-group component of variance is the only unknown
parameter. In fact, if the overall mean is zero, if the within-group
component of variance is one and if there are w_j observations in the
jth group, (67) applies with Y_j equal to $\sqrt{w_j}$ times the sample mean
in the jth group and θ the component of variance between groups.

Now if and only if the w_j are all equal, there is a single sufficient
statistic, namely ΣY_j^2, and hence a uniformly most powerful test.

Suppose then that the w_j are not necessarily all equal and that we look for a locally most powerful test. For this we start from the log density of Y_j, namely

$$- \tfrac{1}{2}\log\{2\pi(1 + w_j\theta)\} - Y_j^2/\{2(1 + w_j\theta)\}. \tag{68}$$

Thus

$$U_j(\theta) = -\frac{w_j}{2(1 + w_j\theta)} + \frac{w_j Y_j^2}{2(1 + w_j\theta)^2}, \tag{69}$$

and, minus the second derivative of (68) being

$$-\frac{w_j^2}{2(1 + w_j\theta)^2} + \frac{w_j^2 Y_j^2}{(1 + w_j\theta)^3},$$

we have, on taking expectations, that

$$i_j(\theta) = \frac{w_j^2}{2(1 + w_j\theta)^2}. \tag{70}$$

At $\theta = 0$, we have that

$$U_j = U_j(0) = \tfrac{1}{2}w_j(Y_j^2 - 1), \tag{71}$$

$$U. = \Sigma U_j = \tfrac{1}{2}\Sigma w_j(Y_j^2 - 1), \tag{72}$$

$$i. = \Sigma i_j(0) = \tfrac{1}{2}\Sigma w_j^2. \tag{73}$$

Note that if, as is very natural, we had written $\theta = \lambda^2$ and had tried to find the efficient score for λ at $\lambda = 0$, a degenerate answer would have resulted because, at the point in question, $d\theta/d\lambda = 0$. This would have suggested reparameterization, although the same answer could be obtained by taking higher terms in the expansion (36) of the likelihood ratio.

Thus (72) and (73) give the form of the locally most powerful test. Large values of $U.$ are significant, and, under the null hypothesis, $E(U.) = 0$ and $\text{var}(U.) = i.$, as is easily verified directly. For more precise work the moment generating function of $U.$, or more simply $\Sigma w_j Y_j^2$, is readily obtained, and there are various ways of extracting from this the exact distribution or more refined approximations.

In the component of variance interpretation $w_j Y_j^2$ is a squared mean weighted by the *square* of the sample size. As such it is different from the usual analysis of variance sum of squares, in which the weight is the sample size. However, Davies (1969) has shown by direct calculation of power functions that the locally most powerful test statistic sometimes has relatively poor power in the region where the power is 0.50–0.99. This region would often be of more practical importance than that very close to the origin.

The test statistic (66) is an approximation to that giving maximum level α power near to power level β and becomes, in this case,

$$\tfrac{1}{2}\Sigma w_j(Y_j^2 - 1) - \frac{(k_\alpha^* + k_\beta^*)}{2(\Sigma w_j^2)^{\frac{1}{2}}} \Sigma w_j^2(Y_j^2 - 1).$$

We shall not investigate this further.

(iv) Some comments on two-sided tests

The discussion of locally most powerful tests has so far been for the one-sided case, using one tail of the distribution of the efficient score $U_.$. The derivation of this statistic in (37) shows, in effect, that the test is such that for the size α power, $\mathrm{pow}(\theta\,;\alpha)$, subject of course to $\mathrm{pow}(\theta_0\,;\alpha) = \alpha$, the derivative $\partial\,\mathrm{pow}(\theta_0\,;\alpha)/\partial\theta_0 = \mathrm{pow}'(\theta_0\,;\alpha)$ is maximized.

We can extend the argument to obtain the locally most powerful unbiased test. Here unbiasedness means that $\mathrm{pow}'(\theta_0\,;\alpha) = 0$ and maximum local power means that the second derivative $\mathrm{pow}''(\theta_0\,;\alpha)$ is maximized. We extend the argument leading to (37) by including terms of order δ^2. The solution is easily seen to be a critical region of the form

$$\left\{ y\,;\ \frac{\partial^2 f_Y(y\,;\theta_0)}{\partial\theta_0^2} \geqslant c_\alpha\,\frac{\partial f_Y(y\,;\theta_0)}{\partial\theta_0} + d_\alpha f_Y(y\,;\theta_0) \right\}$$

or equivalently

$$\left\{ y\,;\ \frac{\partial u_.(\theta_0)}{\partial\theta_0} + u_.^2(\theta_0) \geqslant c_\alpha u_.(\theta_0) + d_\alpha \right\}. \tag{74}$$

For details of these calculations, see Exercise 4.2.

Now it was argued in Section 4.7 that normally the best way to approach two-sided tests is in terms of two one-sided tests. We can do that here by using both tails of the distribution of $U_.$ and summarizing the conclusion by (34). One important difference between this and (74) is that in the former the nesting condition (8) is automatically satisfied, whereas this is not the case with (74); that is, with (74) we cannot rule out the possibility of data y being in, say, the 0.01 critical region, but not being in the 0.05 critical region.

Failure of the nesting condition means that significance probabilities cannot be deduced in a consistent way. To show that this situation can actually occur we outline an example of Chernoff (1951).

Example 4.17. Related normal mean and variance. Let Y_1 have the distribution $N\{\theta, 1/(1 + \theta)\}$, where $\theta > -1$. For a two-sided test of $H_0: \theta = 0$ versus $H_A: \theta \neq 0$, no uniformly most powerful unbiased test exists and we may use (74) to find a locally most powerful unbiased test. This is found to have critical regions of the form

$$\{y_1 ; z_1^4 + c z_1^2 + z_1 + d \geqslant 0 ; z_1 = \tfrac{1}{2}(y_1 - 1)\}. \tag{75}$$

Some calculation from (75) shows that, for suitably small α, the critical region is the union of three intervals $(-\infty, a_1)$, (a_2, a_3) and (a_4, ∞), where $a_1 < 1 < a_2 < a_3 < a_4$. As α decreases both a_2 and a_3 tend to one, thereby violating the nesting condition (8). Chernoff (1951) gives the details of these calculations.

The one-sided locally most powerful critical regions in this case are quite straightforward; for example, for alternatives $\theta > 0$, the region is

$$(y_1 - 1)^2 \leqslant k'_\alpha.$$

It can, however, be shown that this one-sided test has the unsavoury property that for quite moderate values of θ, e.g. $\theta > 3$ when $\alpha = 0.10$, the power is less than α. Of course this is an extreme example in that only one observation is involved. Difficulties are, however, to be expected whenever, as here, the cumulative distribution function of the test statistic is not monotonic in θ.

Asymptotic theory based on large values of n will be considered in detail in Chapter 9. Nevertheless it is convenient to make now some comments on the form of the present discussion for large n. For then U will be very nearly symmetrically distributed. Thus the combination of two one-sided tests is virtually equivalent to the measurement of significance by

$$\text{pr}\{U^2(\theta_0) \geqslant u^2_{\text{obs}}\}$$

or by critical regions

$$\{y ; u^2(\theta_0) \geqslant l_\alpha\}.$$

It is easily shown that this is approximately the same as (74), as is clear on general grounds, because the test based on the two tails of a symmetrically distributed random variable is bound to be unbiased.

Thus the inconsistency illustrated in Example 4.17 will not arise for large values of n.

4.9   Multidimensional alternatives

In the previous two sections it has been assumed not only that the null hypothesis is simple, but also that the alternatives are essentially one-dimensional, being characterized by a single parameter. Qualitatively this means that the alternatives are of one particular kind. Now in applications it will often happen that several quite different types of departure from H_0 are of particular interest. If the alternatives are specified parametrically then two or more independent parameters are required; we shall, for simplicity, deal here with the case of two parameters $\theta = (\phi, \chi)$. That is, under $H_0, \phi = \phi_0$ and $\chi = \chi_0$. To begin with, suppose that the alternatives are separately one-sided, i.e. that the full set of models contemplated has $\phi > \phi_0$ and $\chi > \chi_0$.

A simple example is that, working with the normal distribution, $N(\mu, \sigma^2)$, the null hypothesis might be $\mu = 0, \sigma = 1$, with alternatives $\mu > 0$, or $\sigma > 1$, or both. This problem arises, in particular, when the observations are test statistics obtained from independent sets of data in a preliminary stage of analysis.

It is clear that, except in degenerate cases, no uniformly most powerful test exists. Even if distributions within the exponential family hold, the density

$$\exp\{a_1(\phi, \chi)b_1(y) + a_2(\phi, \chi)b_2(y) + c(\phi, \chi) + d(y)\}$$

leads to a uniformly most powerful test if and only if the first two terms coalesce, which happens only when there is really just a single parameter.

There are several approaches to the problem. One is to take a particular direction in the parameter space in which high power is required. That is, we set $\chi = m(\phi)$, say, and then look either for a uniformly most powerful test or, failing that, for the locally most powerful test. For the latter the appropriate statistic is

$$\frac{\partial \log f_Y(Y; \phi_0, \chi_0)}{\partial \phi_0} + \frac{\partial \log f_Y(Y; \phi_0, \chi_0)}{\partial \chi_0} m'(\phi_0), \qquad (76)$$

a combination of the two components of the efficient score vector.

The drawback to this, and thus to any test based on a linear combination of the components of the efficient score vector, is that it will have negligible power in directions far from that chosen for maximum power. In an essentially one-sided version of the problem in which good power is required only in one quadrant, this is less serious than in the more common situation where it is required to detect any appreciable departure from H_0.

On the whole, a better approach is to use as test statistic a positive definite quadratic form in the components of the efficient score. Now under H_0, with obvious modification of the notation of (51),

$$U_{.} = \begin{bmatrix} U_{.\phi} \\ U_{.x} \end{bmatrix} = \begin{bmatrix} u_{.\phi}(Y; \phi_0, \chi_0) \\ u_{.x}(Y; \phi_0, \chi_0) \end{bmatrix} \tag{77}$$

has zero mean and covariance matrix $i_{.}(\phi_0, \chi_0)$.

In many ways, the most natural combined test statistic is

$$U_{.}^{T} i_{.}^{-1} U_{.}, \tag{78}$$

which has mean two under H_0 and a distribution which, for large n, can be treated as chi-squared with two degrees of freedom. One justification of (78) is as the largest of the standardized linear combinations

$$a^{T} U_{.}/\{\text{var}(a^{T} U_{.})\}^{\frac{1}{2}},$$

taken over all constant vectors a.

Composite statistics such as (78) have, however, the disadvantage that if clear evidence of discrepancy with H_0 is obtained, the test statistic by itself gives no indication of the nature of the departure. Further inspection of the data will always be necessary to interpret, even qualitatively, what has been found. A more fruitful approach is therefore often the following. We first parameterize so that ϕ and χ measure separate kinds of departure from H_0 which it is possible to interpret in distinct ways. Then we apply two separate tests one for $\phi = \phi_0$, the other for $\chi = \chi_0$. That is, if we work with locally most powerful tests, we calculate, in an obvious notation,

$$T_{\phi_0} = U_{.\phi}/\{i_{.\phi\phi}(\phi_0, \chi_0)\}^{\frac{1}{2}}, \quad T_{\chi_0} = U_{.x}/\{i_{.xx}(\phi_0, \chi_0)\}^{\frac{1}{2}}. \tag{79}$$

Under H_0 both have zero mean, unit variance and approximately normal distributions; the correlation coefficient under H_0 is $i_{.\phi x}/(i_{.\phi\phi} i_{.xx})^{\frac{1}{2}}$.

If we look at the more significant of T_{ϕ_0} and T_{χ_0}, an allowance for selection is necessary; see Section 3.4(iii). If we make the approximation that the two statistics have a bivariate normal distribution under H_0, then we need for given $t_{obs} = \max(|t_{\phi_0}|, |t_{\chi_0}|)$ to calculate

$$p_{obs} = \text{pr}\{\max(|T_{\phi_0}|, |T_{\chi_0}|) \geqslant t_{obs}; H_0\}$$

$$= 1 - \text{pr}(-t_{obs} \leqslant T_{\phi_0} \leqslant t_{obs}, -t_{obs} \leqslant T_{\chi_0} \leqslant t_{obs}; H_0) \quad (80)$$

and this is easily found from tables of the bivariate normal distribution.

As in all cases in which several test statistics are involved simultaneously, special care in interpretation is necessary. The level (80) is appropriate only when we look at both T_{ϕ_0} and T_{χ_0} and then pick out one just because it is the larger. Quite often it would be more sensible to look for a test of $\phi = \phi_0$ valid regardless of the value of χ, and for a test of $\chi = \chi_0$ valid regardless of the value of ϕ. The more significant of these would then be taken. This is equivalent to (80) only if the components of U are uncorrelated.

Example 4.18. Test of mean and variance of normal distribution.
Suppose that $Y_1, \ldots, Y_n$ are i.i.d. in $N(\mu, \sigma^2)$ and that the null hypothesis is $H_0: \mu = 0, \sigma = 1$. We test this against a family of alternatives in which either or both of μ and σ depart from their null values.

One practical interpretation of this is that the Y_j may be test statistics calculated from independent sets of data. Suppose, then, that the object is to detect departures from the distribution applying if a null hypothesis holds for all the sets of data. In practice, with an appreciable number of sets of data, we would plot the ordered Y_j against the expected normal order statistics, and regress the Y_j on explanatory variables describing the sets. The following tests may, however, be useful either with fairly small values of n or in borderline cases.

Now a shift in $E(Y_j)$ would be interpreted as a systematic effect present in all sets of data. On the other hand, if $E(Y_j) = 0$ for all j but $\text{var}(Y_j) > 1$, then the interpretation is that some kind of random effect is present, not systematically in one direction rather than another.

If we write $\sigma^2 = 1 + \chi$, we have, in an obvious notation, that

$$u._{\cdot\mu}(Y;0,0) = \left[\frac{\partial}{\partial\mu}\Sigma\log f_{Y_j}(Y_j;\mu,0)\right]_{\mu=0}$$

$$= \Sigma Y_j, \tag{81}$$

$$u._{\cdot\chi}(Y;0,0) = \left[\frac{\partial}{\partial\chi}\Sigma\log f_{Y_j}(Y_j;0,\chi)\right]_{\chi=0}$$

$$= \tfrac{1}{2}\Sigma(Y_j^2-1). \tag{82}$$

The information matrix is

$$i.(0,0) = \begin{bmatrix} i._{\cdot\mu\mu}(0,0) & i._{\cdot\mu\chi}(0,0) \\ i._{\cdot\chi\mu}(0,0) & i._{\cdot\chi\chi}(0,0) \end{bmatrix}$$

$$= \operatorname{diag}(n,\tfrac{1}{2}n)$$

and this is the covariance matrix of (81) and (82), as is easily verified directly.

Thus the single quadratic test statistic (78) is

$$\frac{1}{n}(\Sigma Y_j)^2 + \frac{1}{2n}\Sigma(Y_j^2-1)^2, \tag{83}$$

which has expectation two and approximately a chi-squared distribution with two degrees of freedom under H_0. This is a sensible test statistic, especially when we do not want to separate departures in mean from those in variance.

The separate test statistics for μ and for χ are uncorrelated and the test using the more significant of these is based on

$$\max\left\{\frac{|\Sigma Y_j|}{\sqrt{n}}, \frac{|\Sigma(Y_j^2-1)|}{\sqrt{(2n)}}\right\}, \tag{84}$$

which can be treated as having under H_0 approximately the distribution of the larger of two independent semi-normal random variables. That is, if p_{obs} is the significance level of the more significant statistic the level adjusted for selection is $1-(1-p_{\text{obs}})^2$, approximately.

In the light of the comment below (80), it will, except for rather small n, usually be more sensible to make the test for the mean not depend on the assumption about the variance, and *vice versa*; see Chapter 5. In this case that amounts to using the more significant of the Student t statistic and the chi-squared test for variance based on the corrected sum of squares, $\Sigma(Y_j - \bar{Y}_.)^2$.

Example 4.19. Agreement with an exponential distribution of known mean. A very similar discussion applies to testing agreement with a particular gamma distribution or, in a special case, with a particular exponential distribution, when one or both parameters of the gamma distribution may depart from the null values. Let $Y_1, \ldots, Y_n$ be i.i.d. with density

$$\frac{1}{\Gamma(\beta)} \left(\frac{\beta}{\mu}\right) \left(\frac{\beta y}{\mu}\right)^{\beta-1} e^{-\beta y/\mu}, \tag{85}$$

the null hypothesis being H_0: $\mu = 1$ and $\beta = 1$.

A possible interpretation and justification for this parameterization is as follows. We may have a series of independent estimates of variance, for simplicity each based on two degrees of freedom, and a value of the population variance derived from theory or from previous experience. We can thereby obtain a series of values which, under the null hypothesis, are exponentially distributed with unit mean; recall that a chi-squared random variable with two degrees of freedom is exponentially distributed with mean two. Now if the data $Y_1, \ldots, Y_n$, i.e. the estimates of variance, have the correct mean but the wrong shape of distribution, one interpretation is that the postulated variance is on the average correct, but that the individual data from which the Y_j are derived are non-normally distributed. If, however, the distributional form is correct but the mean is wrong (i.e. $\mu \neq 1$ and $\beta = 1$), the interpretation is that the theoretical variance is wrong.

It is routine to work through the details analogous to those of the previous example.

Bibliographic notes

Pioneer joint work of Neyman and Pearson is collected in a book (Neyman and Pearson, 1967); see also Pearson (1966) and Neyman (1967). Lehmann (1959) gives a systematic account of the theory. The approach taken in all these emphasizes an "accept-reject" decision-making role for tests, rather than the use of tail areas as measures of evidence; the introductory book of Ferguson (1967) emphasizes the decision-making aspect especially strongly.

The efficient score statistic (37) and the associated information function (41) occur in several contexts. They were introduced by Fisher (1925). Equation (45) is one justification for regarding (41) as a sensible measure of the amount of information about a parameter contributed by observing a random variable. A rather different notion of information has been extensively studied in connexion with communication theory; this is of the information contributed by observing a "message" when there is a known distribution over possible "messages". For a brief historical account, see Kendall (1973) and for a longer account and bibliography Papaioannou and Kempthorne (1971). Kullback (1968) gives an account of statistical theory in which likelihood based calculations are given an information-theoretical nomenclature and interpretation.

The discussion of the location family problem of Example 4.15 follows Fisher (1934); for another discussion, see Fraser (1968).

Further results and exercises

1. Formulate the choice of the best critical region of size α for testing $H_0: \theta = \theta_0$ versus $H_A: \theta = \theta_A$ as a problem in the calculus of variations, subject to a constraint. Apply the Euler equation to obtain the Neyman-Pearson lemma.

[Section 4.3]

2. The general non-randomized form of the Neyman-Pearson lemma is as follows. If $k_1(y), \ldots, k_{m+1}(y)$ are integrable functions over the sample space, not necessarily probability densities, then the region w which maximizes

$$\int_w k_{m+1}(y)dy,$$

subject to the constraints

$$\int_w k_j(y)dy = a_j \quad (j = 1, \ldots, m),$$

is defined by $\{y; k_{m+1}(y) \geqslant c_1 k_1(y) + \ldots + c_m k_m(y)\}$, provided that such constants $c_1, \ldots, c_m$ exist. Prove this result using the argument of Section 4.3 and use the result to verify the definitions of the critical regions of the locally most powerful one- and two-sided tests for a one-dimensional parameter.

[Sections 4.3, 4.7; Neyman and Pearson, 1933a, 1936, 1967]

3. To compare two simple hypotheses H_0 and H_1, observations are taken one at a time. After each observation the likelihood ratio is computed from all observations currently available; after, say, m observations denote this by $lr_{10}(y_1, \ldots, y_m) = lr_{10}(y^{(m)})$. Positive constants a and b are chosen, $a < 1 < b$ and

(i) if $a < lr_{10}(y^{(m)}) < b$, a further observation is taken;

(ii) if $a \geqslant lr_{10}(y^{(m)})$, no further observations are taken and H_0 is preferred to H_1;

(iii) if $b \leqslant lr_{10}(y^{(m)})$, no further observations are taken and H_1 is preferred to H_0.

Denote by d_i the conclusion that H_i is preferred ($i = 0, 1$).

By writing $lr_{10}(y^{(m)}) = f_1(y^{(m)})/f_0(y^{(m)})$ and summing over all terminal points in the infinite-dimensional sample space, show that

$$a \operatorname{pr}(d_0 ; H_0) \geqslant \operatorname{pr}(d_0 ; H_1), \quad b \operatorname{pr}(d_1 ; H_0) \leqslant \operatorname{pr}(d_1 ; H_1).$$

If, further, it can be assumed that the procedure terminates with probability one, show that $a \geqslant \beta/(1 - \alpha)$ and $b \leqslant (1 - \beta)/\alpha$, where α and β are the "probabilities of error".

Under what circumstances will we have $a \simeq \beta/(1 - \alpha)$, $b \simeq (1 - \beta)/\alpha$? Prove that if the observations are i.i.d., then the procedure terminates with probability one. Examine the form that the procedure takes for one exponential family problem, and show the relation with random walks.

Discuss critically the limitations on the practical usefulness of this formulation.

[Section 4.3; Wald, 1947]

4. In the comparison of two simple hypotheses H_0 and H_1 the observed likelihood ratio is lr_{obs}. The sampling scheme is unknown; for example, it might be that of Exercise 3 or it might use a fixed number of observations. Adapt the argument of Exercise 3 to show that the significance level for testing H_0, i.e. $\operatorname{pr}\{lr_{10}(Y) \geqslant lr_{obs} ; H_0\}$, is less than or equal to $1/lr_{obs}$.

[Section 4.3; Barnard, 1947]

5. Investigate in detail the test of Example 4.8 for a normal distribution with known coefficient of variation. Obtain the moment generating function under H_0 of the left-hand side of (22) and hence, or otherwise, find approximately the value of d_α. Discuss how to compare this test with those using only (a) $\Sigma(Y_j - \mu_0)$ and (b) $\Sigma(Y_j - \mu_0)^2$.

[Section 4.4]

6. Let Y_1 be a binary random variable, with $E(Y_1) = \text{pr}(Y_1 = 1) = \theta$. Consider $H_0: \theta = \epsilon$, $H_A: \theta = 1 - \epsilon$, with ϵ small. The "obvious" critical region consists of the single point $y_1 = 1$. What are the size and power?

Now let Y_1 and Y_2 be i.i.d. with the above distribution. A procedure for "accepting" and "rejecting" H_0 is defined by: accept H_0 for $y_1 = y_2 = 0$, accept H_A for $y_1 = y_2 = 1$, and otherwise randomize with equal probability between acceptance and rejection. Show that the size and power are identical to those with just one observation.

It is argued that this shows that, for comparing H_0 and H_A, "one binary observation is as good at two". Refute this by examining the levels of significance achievable in testing H_0.

[Section 4.5; Cohen, 1958]

7. The random variables $Y_1, \ldots, Y_n$ are i.i.d. in the exponential distribution with density $\rho e^{-\rho y}$ ($y \geq 0$). Obtain the uniformly most powerful test of $\rho = \rho_0$ against alternatives $\rho < \rho_0$, and derive the power function of the test.

To reduce dependence on possible outliers, it is proposed to reject the largest observation. Show that the representation (A2.8) of Appendix 2 can be used to obtain a simple uniformly most powerful test from the remaining observations and that the loss of power corresponds exactly to the replacement of n by $n - 1$.

Extend the result when the k largest values are omitted.

[Section 4.6]

8. Let $Y_1, \ldots, Y_n$ be i.i.d. with density proportional to $\exp(-\frac{1}{2}y^2 - \theta y^4)$. Obtain the uniformly most powerful test of $H_0: \theta = 0$ versus alternatives $\theta > 0$. Examine the distribution of the test statistic under H_0. For what practical situation might the test be useful?

[Section 4.6]

9. Let $Y_1, \ldots, Y_n$ be i.i.d. in $N(0, \sigma^2)$ and consider $H_0: \sigma^2 = 1$ versus $H_A: \sigma^2 \neq 1$. Obtain the most powerful unbiased test and compare its form numerically with the equitailed test (34) for the special case $\alpha = 0.1, n = 5$.

[Section 4.7]

10. Apply the procedure of (66) to find an improvement on the locally most powerful test for testing $H_0: \theta = 0$ versus $H_A: \theta > 0$ when $Y_1, \ldots, Y_n$ are i.i.d. in $N(\theta, 1 + a\theta^2)$, where a is a known positive constant.

[Section 4.8; Efron, 1974]

11. A multinomial distribution has four cells with probabilities respectively $\{\frac{1}{6}(1 - \theta), \frac{1}{6}(1 + \theta), \frac{1}{6}(2 - \theta), \frac{1}{6}(2 + \theta)\}$, as in Example 2.28. Show that for n independent trials the information is

$$i_.(\theta) = \frac{(2 - \theta^2)n}{(1 - \theta^2)(4 - \theta^2)}.$$

Show that there are two possible ancillary statistics and use the variance of the conditional information·to choose between them.

[Sections 4.8, 2.2(viii); Basu, 1964; Cox, 1971]

12. A finite Markov chain has transition probabilities $((p_{jk}(\theta)))$ depending on a scalar parameter θ. The chain is ergodic and at time zero the state has the stationary distribution of the chain. By examining the log likelihood, find the total information function $i_.(\theta)$ for observation at times $0, 1, \ldots, n$.

[Section 4.8]

13. Let $Y_1, \ldots, Y_n$ have independent Poisson distributions of means $\mu, \mu\rho, \ldots, \mu\rho^{n-1}$. Show how to obtain locally optimum tests (a) for $\mu = \mu_0$ when ρ is known and (b) for $\rho = \rho_0$ when μ is known.

[Section 4.8]

14. The independent random variables $Y_1, \ldots, Y_n$ are independently normally distributed with constant variance σ^2 and with $E(Y_i) = e^{\beta x_j}$, where $x_1, \ldots, x_n$ are known constants. Obtain the information matrix for $\theta = (\beta, \sigma^2)$. Find the locally most powerful test for $\beta = \beta_0$ versus $\beta \neq \beta_0$ with σ^2 known to be equal to σ_0^2.

The constants $x_1, \ldots, x_n$ can be chosen in $[0, 1]$, and it is proposed in fact to select them to maximize $i_.(\beta_0)$. Justify this and, assuming that all x_j's are taken equal, find the common value.

[Section 4.8]

15. Generalize the result of (45) to vector parameters by showing that $E\{U_.(\theta_0); \theta_0 + \delta\} = i_.(\theta)\delta + o(\delta)$, where δ is a column vector. Check this for the normal-theory linear model with known variance. A vector parameter θ is transformed into a new parameter ϕ by a one-one differentiable transformation. Show that the new information matrix is

$$\left(\frac{\partial \phi}{\partial \theta}\right)^{\mathrm{T}} i_.(\theta) \left(\frac{\partial \phi}{\partial \theta}\right) .$$

[Section 4.8(ii)]

16. Show that the integral in (23), namely

$$\int \log \left\{\frac{f_1(y)}{f_0(y)}\right\} f_0(y)dy,$$

is negative, unless the distributions concerned are identical, when it is zero. Hence show that a symmetrical positive distance between the two distributions is

$$\int \log \left\{\frac{f_1(y)}{f_0(y)}\right\} \{f_1(y) - f_0(y)\}dy.$$

Prove that if the distributions belong to the same regular parametric family, i.e. $f_0(y) = f(y; \theta)$ and $f_1(y) = f(y; \theta + \delta)$, then, as $\delta \to 0$, the distance measure is $i(\theta)\delta^2 + o(\delta^2)$.

[Section 4.8(ii)]

5 SIGNIFICANCE TESTS: COMPOSITE NULL HYPOTHESES

5.1 General

In Chapter 4 we dealt with a simple null hypothesis H_0, i.e. with one specifying completely the distribution of Y. Thus any suggested test statistic has a known distribution under H_0. In practice, however, in all but the simplest problems, null hypotheses are composite, i.e. do not specify the distribution of Y completely, and in the present Chapter we discuss the derivation of tests for such situations.

We again suppose that, in addition to the null hypothesis being tested, there is formulated one or more alternative hypotheses against which high power is required. The remarks of the previous Chapter about the unsymmetrical roles of null and alternative hypotheses continue to apply.

A fairly general formulation is to consider a null hypothesis giving Y the density $f_0(y; \xi)$ for $\xi \in \Omega_\xi$ and an alternative family of densities $f_A(y; \eta)$ for $\eta \in \Omega_\eta$. This covers a number of quite distinct situations which we now describe.

The first case is where we have a single parametric family of densities $f(y; \theta)$, where the parameter θ is partitioned into components $\theta = (\psi, \lambda)$ and where the parameter space is a Cartesian product $\Omega_\theta = \Omega_\psi \times \Omega_\lambda$. That is, any value of ψ can occur in combination with any value of λ. Suppose that the null hypothesis specifies $\psi = \psi_0$, whereas under the alternative $\psi \in \Omega_\psi - \{\psi_0\}$. Under both null and alternative hypotheses λ takes unspecified values in Ω_λ.

This is the most common situation and examples are widespread in applied statistics. For instance, in the normal-theory linear model our interest may be concentrated on one of the regression coefficients; then ψ is the component of interest, whereas λ is formed from the remaining regression coefficients and the unknown error variance. As a very special case, we have inference about the mean of a normal

131

distribution, the variance being unknown.

The second type of composite null hypothesis can again be represented in terms of a single family of distributions $f(y; \theta)$, the null hypothesis being $\theta \in \Omega_0 \subset \Omega_\theta$ and the alternative being $\theta \in \Omega_\theta - \Omega_0$. Of course, if Ω_0 can be represented as $(\psi_0, \lambda \in \Omega_\lambda)$ this reduces to the first situation, but there are many other possibilities. For instance, the null hypothesis may concern every component of θ, but still not specify the distribution completely. A simple example is where there is a single unknown parameter, the mean of a normal distribution. Two possible null hypotheses are that (a) the mean is in some interval (a_1, a_2) and that (b) the mean is an integer. In both cases the full parameter space could be the whole real line. It is arguable that null hypotheses should usually be formulated in terms of intervals of values with a common practical interpretation, rather than in terms of point values of parameters. However, this kind of situation is nearly always adequately handled for significance testing by concentrating attention on the boundaries. The case where the null hypothesis forms a disconnected set is more complicated, however.

A third type of problem with a composite null hypothesis arises when the family of alternative distributions has a parametric form different from that of the null hypothesis. Examples concerning i.i.d. random variables are where the null hypothesis is that the distribution is exponential, the alternative being a log normal distribution and where the null hypothesis is that the distribution is a Poisson distribution, the alternative being a geometric distribution (Cox, 1961, 1962).

As a more complicated example, suppose that, according to H_0, $Y_1, \ldots, Y_n$ have normal-theory simple linear regression, i.e. are independently normally distributed with expectations $\gamma + \beta x_1, \ldots, \gamma + \beta x_n$ and variance σ^2, whereas under H_A it is $\log Y_1, \ldots, \log Y_n$ that follow such a model. Obviously, for some parameter values, choice between H_0 and H_A will be easy: under H_A negative values of y are impossible, whereas under H_0 there may be an appreciable probability of such values. However, in the cases where the problem is likely to be of practical interest, it will usually happen that only for enormous values of n will choice on the basis of signs be a useful solution. One approach to this and similar problems is to take a comprehensive parametric model including both H_0 and H_A as special cases. For instance, we may consider a model in which for some unknown ψ, $(Y_j^\psi - 1)/\psi$ $(j = 1, \ldots, n)$ have normal-theory simple linear regression on x_j. This includes the log version as the limiting case $\psi \to 0$. Then the hypotheses H_0 and H_A correspond respectively to $\psi = 1$ and $\psi = 0$. This device reduces the third type of problem to the first, namely that of testing hypotheses about component parameters in the presence of nuisance parameters, the type of problem on which we shall now concentrate.

In such a problem, with ψ the parameter of interest and λ the nuisance parameter, a very naive approach would be to fix λ at an arbitrary value, say $\lambda^\dagger$. We can then test the null hypothesis $\psi = \psi_0$ by the methods of Chapter 4, thus obtaining a value of p_{obs} that will be a function of $\lambda^\dagger$. If for $\lambda^\dagger \in \Omega_\lambda$, this value of p_{obs} does not vary greatly, a conclusion can be drawn about the null hypothesis $\psi = \psi_0$. Unfortunately this is not often likely to lead to useful results, as the following example shows.

Example 5.1. Normal mean with unknown variance. Suppose that $Y_1, \ldots, Y_n$ are i.i.d. in $N(\mu, \sigma^2)$ and that we wish to test $H_0: \mu = 0$ versus $H_A: \mu > 0$, σ^2 under both hypotheses taking any positive value. In the general notation, μ becomes ψ and σ^2 becomes λ. Now if $\sigma = \sigma^\dagger$ is regarded as fixed, the best test is based on large values of $\bar{y}_. = \Sigma y_j/n$ and the one-sided significance level is

$$\mathrm{pr}(\bar{Y}_. \geqslant \bar{y}_.; H_0, \sigma^\dagger) = 1 - \Phi(\sqrt{n}\bar{y}_./\sigma^\dagger). \tag{1}$$

As $\sigma^\dagger$ varies over positive values, (1) varies between 1 and $\frac{1}{2}$ if $\bar{y}_. < 0$ and between 0 and $\frac{1}{2}$ if $\bar{y}_. > 0$; if $\bar{y}_. = 0$, (1) is $\frac{1}{2}$. Thus, we have the obvious conclusion that if $\bar{y}_. \leqslant 0$, there is no evidence against H_0; if $\bar{y}_. > 0$, we can draw no effective conclusion by this method.

The reason that this approach is of little use is that the nuisance parameter λ is allowed to take any value in its parameter space, whereas in fact the data will give information as to which are the relevant values of λ. We have to find some way of incorporating this information.

One possible modification of the above procedure is to form a "good" estimate $\tilde{\lambda}$ of λ and then to test the null hypothesis as if $\lambda = \tilde{\lambda}$, i.e. ignoring errors of estimation of λ. Unfortunately, while this gives a reasonably sensible answer in the special case (1), in general it gives quite misleading results. In particular, when H_0 is true, the chance of obtaining data apparently significant at level α may greatly exceed α. A second possibility, sometimes useful in

examining the sensitivity of the conclusions to simplifying assump-
tions, is to find a region of values $\lambda^\dagger$ within which the true λ is almost
certain to lie, and then to examine the dependence of significance
level on $\lambda^\dagger$ within that region.

Clearly, however, this is at best a rough and ready solution. In the
rest of this Chapter, we examine in detail two more satisfactory
general approaches to significance tests in the presence of nuisance
parameters. In both, we aim at procedures that are required to have
desirable properties whatever the value of the nuisance parameter λ.
In the first analysis, namely that of Section 5.2, we require the use
of so-called similar regions; this in effect means that whenever H_0 is
true the probability of falling in a size α critical region shall be
exactly α, whatever the value of λ. We shall achieve this property by
arguing conditionally on appropriate sufficient statistics; see Section
3.3. The second approach is by an appeal to the invariance principle
outlined in Section 2.3(vi). When this is applicable, it allows an
appreciable simplification of the problem, often to the point where
nuisance parameters have been effectively eliminated.

When neither of these techniques is applicable, we fall back on
approximate arguments based on limiting behaviour for a very large
number of observations. Discussion of this is postponed to Chapter 9.

Tests not explicitly based on a fully parametric formulation are
discussed separately in Chapter 6.

5.2   Similar regions

(i) Definitions

We begin by setting out more formally the ideas used in Section 3.3
in connexion with a pure test of significance of a composite null
hypothesis. Put in terms of critical regions of a given size α, the
central requirement is that the probability of falling in the critical
region shall equal the required value α for all values of the nuisance
parameter, when the null hypothesis is true. That is, we require that
for all $\lambda \in \Omega_\lambda$

$$\text{pr}(Y \in w_\alpha; \psi_0, \lambda) = \alpha. \tag{2}$$

A region satisfying (2) is called a *similar region of size* α and a nested
sequence of such regions for various values of α, defining as in Chapter 4 a test of significance, is called a *similar test*.

The term similar is short for similar to the sample space, the whole
sample space satisfying (2) for $\alpha = 1$.

We have already illustrated in Section 3.3 the procedure for finding similar regions. In fact, suppose that, given $\psi = \psi_0$, S_λ is sufficient for the nuisance parameter λ. Then the conditional distribution of Y given $S_\lambda = s$ does not depend on λ when H_0 is true, and therefore any region which has the required size in this conditional distribution for all s has the required property (2). Further, if S_λ is boundedly complete, then any similar region of size α must be of size α conditionally on $S_\lambda = s$, for almost all s; see Section 3.3. We call a critical region w_α with this property, namely

$$\mathrm{pr}(Y \in w_\alpha \,|\, S_\lambda = s\,;\, \psi_0) = \alpha \tag{3}$$

for all s, a region of *Neyman structure*. Clearly a region of Neyman structure satisfies (2), as is proved by averaging (3) with respect to the distribution of S_λ.

Assuming that S_λ is boundedly complete, we can use the results of Chapter 4 to find the similar test with maximum power for a particular alternative hypothesis $\psi = \psi_1, \lambda = \lambda_1$. For this we apply the Neyman-Pearson lemma to the conditional densities under H_0 and the alternative, thereby obtaining the best critical region of a particular size α as

$$\left\{ y \,;\, \frac{f_{Y|S_\lambda}(y\,|s\,;\,\psi_1, \lambda_1)}{f_{Y|S_\lambda}(y\,|s\,;\,\psi_0)} \geqslant c_\alpha \right\}. \tag{4}$$

If this same region applies to all alternatives being contemplated, i.e. to all ψ_1 and λ_1, then we call the region *uniformly most powerful similar*. Note that the conditional density $f_{Y|S_\lambda}(y\,|s\,;\,\psi, \lambda)$ will not in general be independent of λ when $\psi \neq \psi_0$.

In all the applications we shall consider, uniformly most powerful similar regions exist either for all s or for none, although there is clearly the theoretical possibility that such regions might exist only for certain special values of s. When such regions exist for all s and for all α, we call the corresponding test *uniformly most powerful similar*. It is then clear that the optimum power property holds unconditionally on s; indeed for any particular alternative this is true in that maximum unconditional power requires maximum conditional power for almost all s.

If uniformly most powerful similar regions do not exist, we use one or other of the approaches discussed in Section 4.6, applying them now to the conditional density $f_{Y|S_\lambda}(y\,|s\,;\,\psi, \lambda)$. For instance, one possibility is to consider locally most powerful similar tests, although some care is needed in applying the method of Section 4.8, as we shall see in part (ii) of this section (p. 146).

When the alternatives are two-sided, we follow the arguments of Section 4.7, merging two one-sided tests. The condition of unbiasedness, formulated and criticized in Section 4.7, could be adapted to the present problem by requiring that for all ψ and λ

$$\text{pr}(Y \in w_\alpha \mid S_\lambda = s ; \psi, \lambda) \geqslant \alpha. \tag{5}$$

Problems with multi-dimensional alternatives, arising if $\dim(\psi) \geqslant 2$ are essentially more difficult than those with two-sided alternatives. The multi-dimensional case will mainly be deferred to Sections 5.3 and 5.4.

A number of examples of the application of these definitions will now be given, a critical discussion of the definitions being deferred to (iii).

(ii) Some examples

We now give a number of relatively simple examples of the use of the similar region argument outlined in (i). Further examples are contained in Chapters 6 and 7. Several of the present examples concern the comparison of parameters in exponential family distributions. We begin with examples of the first type of problem mentioned in (i).

Example 5.2. Comparison of Poisson means. Suppose that Y_1 and Y_2 are independent Poisson random variables with means μ_1 and μ_2 and that H_0 is the hypothesis $\mu_1 = \psi_0 \mu_2$, where ψ_0 is a given constant. Here we may reparametrize so that $\psi = \mu_1/\mu_2$ and $\lambda = \mu_2$. One interpretation of this problem is that we observe independent sections of two Poisson processes, one for a time period $\psi_0 t$ and the other for a time period t, the random variables Y_1 and Y_2 representing the numbers of events in the two time periods. The null hypothesis H_0 then asserts that the rates of the two Poisson processes are identical.

The joint distribution of Y_1 and Y_2 is

$$f_{Y_1, Y_2}(y_1, y_2 ; \psi, \lambda) = \frac{e^{-\lambda(1+\psi)} \lambda^{y_1 + y_2} \psi^{y_1}}{y_1! \, y_2!}. \tag{6}$$

Under H_0: $\psi = \psi_0$, we have that $S_\lambda = Y_1 + Y_2$ is a complete sufficient statistic for λ; note that we have the special circumstance that the same S_λ holds for all ψ_0.

The conditional distribution of (Y_1, Y_2) given $S_\lambda = s$ is

$$f_{Y_1, Y_2|S_\lambda}(y_1, y_2 | s; \psi, \lambda) = \binom{s}{y_1}(1 + \psi)^{-s}\psi^{y_1}, \qquad (7)$$

because the distribution of S_λ is a Poisson distribution of mean $\mu_1 + \mu_2 = \lambda(\psi + 1)$. Alternatively, (7) can be specified by the binomial distribution of Y_1 given $S_\lambda = s$, the index s and the parameter $\psi/(1 + \psi)$. If the alternative to H_0 is one-sided, for example $\psi > \psi_0$, the likelihood ratio test rejects H_0 for large y_1. The significance level is thus the sum of (7) over values greater than or equal to the observed y_1, evaluated for $\psi = \psi_0$. The test is uniformly most powerful similar.

When s is reasonably large, a normal approximation with continuity correction can be used, the significance level being approximately

$$1 - \Phi\left[\frac{y_1 - s\psi_0/(1 + \psi_0) - \frac{1}{2}}{\{s\psi_0/(1 + \psi_0)^2\}^{\frac{1}{2}}}\right]. \qquad (8)$$

Example 5.3. Binary regression. Suppose that $Y_1, \ldots, Y_n$ are independent binary random variables, i.e. the possible values of each Y_j are zero and one. Correspondingly, there are values of an explanatory variable $x_1, \ldots, x_n$. To represent possible dependence of $E(Y_j) = \mathrm{pr}(Y_j = 1)$ on x_j, we consider the *logistic model*

$$\log\left\{\frac{\mathrm{pr}(Y_j = 1)}{\mathrm{pr}(Y_j = 0)}\right\} = \gamma + \beta x_j, \qquad (9)$$

$$\mathrm{pr}(Y_j = 1) = \frac{e^{\gamma + \beta x_j}}{1 + e^{\gamma + \beta x_j}}. \qquad (10)$$

The likelihood is

$$\frac{\exp(\gamma\Sigma y_j + \beta\Sigma x_j y_j)}{\Pi(1 + e^{\gamma + \beta x_j})}, \qquad (11)$$

so that the minimal sufficient statistic for (γ, β) is $(R = \Sigma Y_j, T = \Sigma x_j Y_j)$. Now consider the null hypothesis $\beta = \beta_0$, where often $\beta_0 = 0$ representing no dependence on the explanatory variable. For a given value β_0, the sufficient statistic for the nuisance parameter is $R = \Sigma Y_j$, the total number of ones. Hence similar regions are to be constructed by examining the conditional distribution of $T = \Sigma x_j Y_j$ given $\Sigma Y_j = r$;

note that the reduction to the consideration of T is possible because of sufficiency under the alternative hypotheses $\beta \neq \beta_0$.

It follows on summing (11) that

$$\text{pr}(R = r, T = t; \gamma, \beta) = \frac{c(r, t)e^{\gamma r + \beta t}}{\Pi(1 + e^{\gamma + \beta x_j})} \tag{12}$$

$$\text{pr}(R = r) = \frac{\Sigma_u c(r, u)e^{\gamma r + \beta u}}{\Pi(1 + e^{\gamma + \beta x_j})}, \tag{13}$$

where $c(r, t)$ is the number of distinct subsets of size r which can be formed from $\{x_1, \ldots, x_n\}$ and which sum to t. Formally $c(r, t)$ is the coefficient of $\zeta_1^r \zeta_2^t$ in the generating function

$$\prod_{j=1}^{n} (1 + \zeta_1 \zeta_2^{x_j}). \tag{14}$$

Thus, from (12) and (13), the required conditional distribution for arbitrary β is

$$\text{pr}(T = t \mid R = r; \beta) = c(r, t)e^{\beta t}/\Sigma_u c(r, u)e^{\beta u}. \tag{15}$$

It is now clear that the likelihood ratio for an alternative $\beta = \beta_A > \beta_0$ versus $\beta = \beta_0$ is an increasing function of t and that therefore the one-sided significance level for testing $\beta = \beta_0$ against $\beta > \beta_0$ is the upper tail probability

$$\sum_{u=t_{\text{obs}}}^{t_{\text{max}}} c(r, u)e^{\beta_0 u} / \sum_{u=0}^{t_{\text{max}}} c(r, u)e^{\beta_0 u}, \tag{16}$$

where, of course, $t_{\text{max}} \leqslant \Sigma x_j$.

When $\beta_0 = 0$, (15) simplifies and the statistic T is in effect the total of a random sample size r drawn without replacement from the finite population $\{x_1, \ldots, x_n\}$; see also Example 6.1. In particular, it follows that the conditional mean and variance of T under the null hypothesis $\beta = 0$ are

$$\frac{r\Sigma x_j}{n} \quad \text{and} \quad \frac{r(n-r)\,\Sigma(x_j - \bar{x}.)^2}{n(n-1)}. \tag{17}$$

A special case is the two-sample problem in which the first n_1 observations have, say, $\text{pr}(Y_j = 1) = \theta_1$, whereas the second group of $n_2 = n - n_1$ observations have a corresponding probability θ_2. This is covered, for example, by taking $x_1 = \ldots = x_{n_1} = 1$ and

$x_{n_1+1} = \ldots = x_n = 0$. Then

$$\beta = \log \frac{\theta_1}{1 - \theta_1} - \log \frac{\theta_2}{1 - \theta_2}, \tag{18}$$

and this is in some contexts a sensible scale on which to compare probabilities. In this case, $c(r, t)$ is the number of distinct ways of forming a set of size r containing t ones and $r - t$ zeroes from the finite population of x's and hence is equal to

$$\binom{n_1}{t} \binom{n_2}{r - t},$$

the sum over t being $\binom{n}{r}$.

Thus for the general null hypothesis the one-sided significance level (16) becomes

$$\sum_{u = t_{obs}}^{\min(r, n_1)} \binom{n_1}{u} \binom{n_2}{r - u} \beta_0^u \Big/ \sum_{u = \max(r - n_2, 0)}^{\min(r, n_1)} \binom{n_1}{u} \binom{n_2}{r - u} \beta_0^u$$

and for the special case $\beta = 0$ of no difference between groups this is

$$\sum_{u = t_{obs}}^{\min(r, n_1)} \binom{n_1}{u} \binom{n_2}{r - u} \Big/ \binom{n}{r}, \tag{20}$$

the tail area of the hypergeometric distribution. This test is often called Fisher's exact test for the 2×2 contingency table. It can, of course, be derived directly by extending the arguments of Example 5.2, the comparison of two Poisson means, to the comparison of two binomial parameters.

It is worth stating explicitly the optimal property of (20). For a given r', let $\alpha(r')$ be an achievable significance level, i.e. a possible value of (20) with $r = r'$. Then the only *exactly* similar regions of size $\alpha(r')$, i.e. regions w such that

$$\mathrm{pr}(Y \in w; \theta_1 = \theta_2 = \theta) = \alpha(r'), \quad 0 < \theta < 1,$$

are obtained as follows. Choose, for each r, a region having probability exactly $\alpha(r')$ under the conditional hypergeometric distribution for that r, the choice in general involving randomization.

Among all such regions the one with maximum power for all alternatives $\beta > 0$ for all $\theta_1 > \theta_2$ is based on the tail area (20). That is, for the particular r', the test gives exactly the unrandomized tail sum (20), whereas for other values of r the size $\alpha(r')$ critical region is formed by randomization, as in Example 4.9. We would, of course, use only those significance levels achievable for the particular observed r.

Example 5.4. Comparison of variances in bivariate normal distribution. Suppose that $(X_1, Y_1), \ldots, (X_n, Y_n)$ are i.i.d. pairs having a bivariate normal distribution of zero means and with $\text{var}(X_j) = \sigma_1^2$, $\text{var}(Y_j) = \sigma_2^2$ and $\text{cov}(X_j, Y_j) = \rho\sigma_1\sigma_2$. The restriction to zero means is inessential and is made to simplify some details. Suppose that it is required to test the null hypothesis $\sigma_2^2 = k_0\sigma_1^2$, with ρ and σ_1 as nuisance parameters and k_0 a given constant. We take for simplicity the one-sided alternatives $\sigma_2^2 > k_0\sigma_1^2$. It is known that if $\rho = 0$ the standard F test based on $(\Sigma Y_j^2)/(k_0\Sigma X_j^2)$ is applicable and can be shown to be optimal.

For general values of the parameters, the minimal sufficient statistic is $(\Sigma X_j^2, \Sigma X_j Y_j, \Sigma Y_j^2)$. Under H_0, however, this reduces to $\{\Sigma(k_0 X_j^2 + Y_j^2), \Sigma X_j Y_j\}$, which is complete. Thus optimal similar regions are based on the distribution of the full three-dimensional sufficient statistic conditionally on the reduced two-dimensional form. This is equivalent to the distribution of ΣY_j^2 given $\Sigma(k_0 X_j^2 + Y_j^2)$ and $\Sigma X_j Y_j$.

It is convenient to introduce a transformation to new variables that are independent under H_0 by writing

$$V_j = \sqrt{k_0}X_j + Y_j, \quad W_j = \sqrt{k_0}X_j - Y_j.$$

The conditional distribution required is that given $\Sigma(V_j^2 + W_j^2)$ and $\Sigma(V_j^2 - W_j^2)$, that is, given ΣV_j^2 and ΣW_j^2. Now, given these conditioning statistics, ΣY_j^2 is a monotone function of $\Sigma V_j W_j$, so that the test is equivalently based on the distribution of $\Sigma V_j W_j$ given ΣV_j^2 and ΣW_j^2, the random variables V_j and W_j being independently normally distributed. Under H_0, $\text{cov}(V_j, W_j) = 0$, whereas under alternatives $\sigma_2^2 > k_0\sigma_1^2$ the covariance is negative. The test is standardized by defining the sample correlation

$$R = \Sigma V_j W_j/(\Sigma V_j^2 \Sigma W_j^2)^{\frac{1}{2}}, \tag{21}$$

for which we have the result that

$$R\sqrt{(n-1)}/\sqrt{(1-R^2)} \tag{22}$$

has, under H_0, the Student t distribution with $n-1$ degrees of freedom independently of ΣV_j^2 and ΣW_j^2. For all alternatives $\sigma_2^2 > k_0\sigma_1^2$, R is stochastically smaller than under H_0, conditionally on ΣV_j^2 and ΣW_j^2, so that the lower tail of the conditional Student t distribution of (22) gives the significance level. The test is uniformly most powerful similar.

In the more general version of the problem in which $E(X_j)$ and $E(Y_j)$ are unknown, (21) and (22) are replaced by

$$R' = \Sigma(V_j - \bar{V}.)(W_j - \bar{W}.)/\{\Sigma(V_j - \bar{V}.)^2 \Sigma(W_j - \bar{W}.)^2\}^{\frac{1}{2}}$$

with

$$R'\sqrt{(n-2)}/\sqrt{(1-R'^2)}$$

having a Student t distribution with $n-2$ degrees of freedom. This test is due to Pitman (1939).

In the two-sample comparisons of Poisson and binomial parameters, Examples 5.2 and 5.3, and the two-parameter comparison of Example 5.4, quite simple optimal similar tests are available for null hypotheses specifying arbitrary values for

$$\frac{\mu_1}{\mu_2}, \quad \log\frac{\theta_1}{1-\theta_1} - \log\frac{\theta_2}{1-\theta_2} \quad \text{and} \quad \frac{\sigma_1^2}{\sigma_2^2},$$

respectively. When the null hypothesis specifies parameter equality, the particular parameterization is unimportant; for example, the null hypothesis $\mu_1/\mu_2 = 1$ and the family of alternatives $\mu_1/\mu_2 > 1$ are equivalent to the null hypothesis $h(\mu_1) - h(\mu_2) = 0$ and the family of alternatives $h(\mu_1) - h(\mu_2) > 0$, for any strictly increasing function $h(.)$. But for the general null hypothesis $\mu_1/\mu_2 = k_0$, with $k_0 \neq 1$, the particular parameterization is crucial; for example, no regions of Neyman structure are available for testing $\mu_1 - \mu_2 = l_0$, unless $l_0 = 0$. To explain these results in general terms, we relate them to the exponential family, taking for simplicity a comparison of two independent observations, each with one unknown parameter; an extended treatment is given in Section 7.3.

Suppose then that Y_1 and Y_2 are independent with Y_j having p.d.f.

$$\exp\{-\phi_j t_j + c^\dagger(\phi_j) + d^\dagger(y_j)\} \quad (j = 1, 2),$$

where ϕ is the natural parameter for the exponential family in question; see Section 2.2(vi). The likelihood is then

$$\exp\{-\phi_1 t_1 - \phi_2 t_2 + c^\dagger(\phi_1) + c^\dagger(\phi_2) + d^\dagger(y_1) + d^\dagger(y_2)\}.$$

Now for a null hypothesis $\phi_1 - \phi_2 = k_0$, $T_1 + T_2$ is sufficient for the nuisance parameter and is complete. For a null hypothesis $\phi_1 = l_0\phi_2$, $l_0 T_1 + T_2$ is sufficient for the nuisance parameter and is complete. However, for a general null hypothesis $\phi_1 = h(\phi_2)$, the likelihood under that null hypothesis is

$$\exp[-h(\phi_2)t_1 - \phi_2 t_2 + c^\dagger\{h(\phi_2)\} + c^\dagger(\phi_2) + d^\dagger(y_1) + d^\dagger(y_2)].$$

Therefore, unless $h(.)$ is linear, there is no reduction under H_0 of the dimensionality of the minimal sufficient statistic (T_1, T_2); hence no regions of Neyman structure are available.

Examples 5.2–5.4 concern hypotheses about differences of natural parameters. The corresponding tests for natural parameter ratios are often of less interest, partly for a reason to be discussed in Example 5.12

Example 5.5. Normal mean with unknown variance (ctd). Let $Y_1, \ldots Y_n$ be i.i.d. in $N(\mu, \sigma^2)$, both parameters being unknown. Consider the null hypothesis $H_0: \mu = \mu_0$, $0 < \sigma^2 < \infty$ with alternative $H_A: \mu > \mu_0$, $0 < \sigma^2 < \infty$. Under H_0, $V(\mu_0) = \Sigma(Y_j - \mu_0)^2$ is a complete sufficient statistic for σ^2. Further, also under H_0, the conditional distribution of the vector Y is uniform over the sphere with centre $(\mu_0, \ldots, \mu_0)$ and radius $\{v(\mu_0)\}^{\frac{1}{2}}$, given $V(\mu_0) = v(\mu_0)$. For a particular alternative $\mu_A > \mu_0$, the joint distribution of Y is constant on any sphere with centre at $(\mu_A, \ldots, \mu_A)$, the density conditional on $V(\mu_0) = v(\mu_0)$ thus increasing as $\Sigma(y_j - \mu_0)$ increases. Comparing this with the uniform conditional distribution under H_0, we find immediately that the likelihood ratio critical region for all alternatives $\mu_A > \mu_0$ takes the form

$$\{y \,; \Sigma(y_j - \mu_0)^2 \geqslant c_\alpha^{(n)} \{v(\mu_0)\}^{\frac{1}{2}}\}, \qquad (23)$$

a cap of the sphere $\Sigma(y_j - \mu_0)^2 = v(\mu_0)$. Now the ordinary Student t statistic can be written as

$$T_n = \frac{(\bar{Y}. - \mu_0)\{n(n-1)\}^{\frac{1}{2}}}{\{\Sigma(Y_j - \bar{Y}.)^2\}^{\frac{1}{2}}},$$

and it follows that (23) can alternatively be written as

$$\left\{y \,; \frac{t_n}{\sqrt{(t_n^2 + n - 1)}} \geqslant d_\alpha^{(n)}\right\},$$

i.e. as the region formed from all sufficiently large values of t_n. This is the ordinary one-sided Student t test, which is thus uniformly most powerful similar. If the alternatives are $\mu \neq \mu_0$, then we are lead similarly to the two-sided Student t test.

An appreciably longer argument along these lines shows that in the normal-theory linear model the standard Student t test is uniformly most powerful similar for testing a null hypothesis about one of the unknown regression parameters, when the other regression parameters and the variance are nuisance parameters. One proof proceeds by first reducing the problem by orthogonal transformation to the so-called canonical form (Scheffé, 1959, p. 21). Corresponding tests for two or more linear model parameters simultaneously are discussed in Sections 5.3 and 5.4.

Example 5.6. Two-sample problem with unequal variances. Suppose that we have two independent sets of i.i.d. normal random variables, the first set of n_1 variables having sample mean $\bar{Y}_{1.}$ and estimate of variance MS_1, with n_2, $\bar{Y}_{2.}$ and MS_2 being the corresponding values for the second set. Denote the two normal distributions by $N(\mu_1, \sigma_1^2)$ and $N(\mu_2, \sigma_2^2)$ and suppose that we wish to test the null hypothesis $H_0: \mu_1 - \mu_2 = \delta_0$, with σ_1^2 and σ_2^2 being independent nuisance parameters. This situation is not covered by the normal-theory linear model because of the presence of the two unrelated unknown variances σ_1^2 and σ_2^2.

Under H_0 the minimal sufficient statistic for the nuisance parameters is $(\bar{Y}_{1.}, \bar{Y}_{2.}, MS_1, MS_2)$. It is clear on dimensional grounds that this is not complete, and this is confirmed explicitly by noting that

$$(\bar{Y}_{1.} - \bar{Y}_{2.} - \delta_0)^2 - \frac{MS_1}{n_1} - \frac{MS_2}{n_2}$$

has zero expectation for all parameters values. Now completeness is connected solely with the uniqueness of the similar regions found by conditioning, not with their existence. On the other hand, in the present case, the sufficient statistic under H_0 is the same as that for arbitrary parameter values. Hence the distribution of Y conditionally on the sufficient statistic is always independent of all parameters, i.e. the conditional method totally fails. We return to this problem in Section 5.2(iv), and from a Bayesian viewpoint, in Example 10.5.

Example 5.7. Autoregressive time series. Unfortunately most problems of statistical inference connected with time series are too complicated to be handled exactly by the methods of the present Chapter. There are, however, some problems involving autoregressive processes which by some special approximations and modifications can be treated.

An mth order Gaussian autoregressive process of zero mean is defined by

$$Y_r = \beta_1 Y_{r-1} + \ldots + \beta_m Y_{r-m} + \epsilon_r,$$

where the ϵ_r's are i.i.d. in $N(0, \sigma^2)$. To begin with, suppose that initial conditions are specified by giving $Y_1, \ldots, Y_m$ either known values or values with a probability density $g(y_1, \ldots, y_m; \beta)$. Then, as in Exercise 2.2, we have the joint density of $Y_1, \ldots, Y_n$ as

$$g(y_1, \ldots, y_m; \beta) \frac{1}{(2\pi)^{\frac{1}{2}(n-m)} \sigma^{n-m}}$$

$$\exp\left\{ -\frac{\sum\limits_{r=m+1}^{n} (y_r - \beta_1 y_{r-1} - \ldots - \beta_m y_{r-m})^2}{2\sigma^2} \right\}.$$

The three main possibilities for the choice of the initial density $g(.)$ are

(a) to regard the initial values as given constants, in which case we simply omit the term $g(.)$;

(b) to give $g(.)$ that form which will make the series $Y_1, \ldots, Y_n$ stationary, and, in particular, will ensure that all Y_j have the same marginal density. This is the natural assumption if the system has been running some time before observation is started at an arbitrarily chosen time point;

(c) to make the process stationary around a circle, i.e. in effect to define $Y_{n+1} = Y_1$, etc. This is normally artificial, but has the advantage of simplifying some formulae.

Whichever form is taken, the exponent in the density is, except for end effects,

$$-\frac{(1 + \beta_1^2 + \ldots + \beta_m^2)}{2\sigma^2} \Sigma y_j^2 - \frac{(\beta_1 + \beta_1\beta_2 + \ldots + \beta_{m+1}\beta_m)}{\sigma^2} \Sigma y_j y_{j-1} - \ldots$$

$$-\frac{\beta_m}{\sigma^2} \Sigma y_j y_{j-m}$$

$$= -\phi_0 \Sigma y_j^2 - \phi_1 \Sigma y_j y_{j-1} - \ldots - \phi_m \Sigma y_j y_{j-m},$$

say. This suggests that simple formal results, at least indirectly relevant to the autoregressive process, are obtained by assigning $Y_1, \ldots, Y_n$ a joint density proportional to

$$\exp\left(-\phi_0 \sum_{j=1}^{n} y_j^2 - \phi_1 \sum_{j=2}^{n} y_j y_{j-1} - \ldots - \phi_m \sum_{j=m+1}^{n} y_j y_{j-m}\right),$$

or, more generally, the joint density proportional to

$$\exp\{-\phi_0 q_0(y) - \ldots - \phi_{m'} q_{m'}(y)\}, \tag{24}$$

where $q_0(y), \ldots, q_{m'}(y)$ are quadratic forms in the y's and $m' \geqslant m$.

Clearly uniformly most powerful similar tests are available about the natural parameters in (24), in particular of the null hypothesis, $\phi_{m'} = 0$, that the last term is null. A formidable distributional problem remains, namely that of finding the null hypothesis conditional distribution of the test statistic given the values of the conditioning statistics. Notice that if in (24) we have $m' > m$ the optimal similar test of $\phi_{m'} = 0$, with $\phi_0, \ldots, \phi_{m'-1}$ as nuisance parameters, may have no relevance to testing $\beta_m = 0$, because β_m may be specified by the nuisance parameters.

The simplest special case arises when y_1 is fixed and $m = 1$. It is then easy to show that the density is exactly of the form (24) with $m' = 2$ and with

$$\phi_0 = \frac{1}{2\sigma^2}, \quad \phi_1 = \frac{\beta^2}{2\sigma^2}, \quad \phi_2 = -\frac{\beta}{\sigma^2},$$

$$q_0(y) = y_2^2 + \ldots + y_n^2, \quad q_1(y) = y_1^2 + \ldots + y_{n-1}^2,$$

$$q_2(y) = y_1 y_2 + \ldots + y_{n-1} y_n.$$

An exact similar test is available if $y_1 = y_n$, when the problem involves reduced parameters $\phi_0' = \phi_0 + \phi_1$ and $\phi_1' = \phi_2$. Under any other assumption about y_1, no uniformly most powerful similar test is available concerning β_1.

Suppose that we wish to test the null hypothesis $\beta_1 = 0$ for arbitrary y_1. Under this hypothesis $Y_2^2 + \ldots + Y_n^2$ is sufficient for the nuisance parameter σ^2. As we have seen, no uniformly most powerful test exists, but if we examine local power, the terms in β_1^2 in the joint density can be ignored. Therefore, the locally most powerful similar test is given by examining the statistic $y_1 Y_2 + \ldots + Y_{n-1} Y_n$

conditionally on $Y_2^2 + \ldots + Y_n^2$; the distributional problem is simplified if $y_1 = 0$.

For a detailed discussion of tests of hypotheses connected with time series, see Anderson (1971, Chapter 5).

In the last example, we described without details how a locally most powerful similar test can be developed when a uniformly most powerful similar test does not exist. In principle, the necessary extension of the discussion in Section 4.8 is quite straightforward, but there is a point of possible ambiguity to be removed. Suppose, then, that (4) does not yield the same result for all ψ_1 and λ_1 in the alternative hypothesis, for fixed α. On considering the local alternative $\psi = \psi_0 + \delta$ for small δ and taking a specific $\lambda_1 \in \Omega_\lambda$, the likelihood ratio in (4) becomes

$$\frac{f_{Y|S_\lambda(\psi_0)}(y \,|\, s(\psi_0); \psi_0 + \delta, \lambda_1)}{f_{Y|S_\lambda(\psi_0)}(y \,|\, s(\psi_0); \psi_0)} =$$

$$1 + \delta \left[\frac{\partial \log f_{Y|S_\lambda(\psi_0)}(y|s(\psi_0); \psi_1, \lambda_1)}{\partial \psi_1} \right]_{\psi_1 = \psi_0} + o(\delta), \qquad (25)$$

where we have stressed the possible dependence of S_λ on ψ_0. The derivative is the locally most powerful test statistic for given λ_1. The explicit, if apparently clumsy, notation used in (25) is important when S_λ depends on ψ_0. For the above derivative is certainly not the same as

$$\frac{\partial \log f_{Y|S_\lambda(\psi_0)}(y \,|\, s(\psi_0); \psi_0)}{\partial \psi_0},$$

which would arise from the ratio

$$\frac{f_{Y|S_\lambda(\psi_0 + \delta)}(y \,|\, s(\psi_0 + \delta); \psi_0 + \delta)}{f_{Y|S_\lambda(\psi_0)}(y \,|\, s(\psi_0); \psi_0)}.$$

In the latter case, the likelihood ratio involves distributions conditioned to lie in different parts of the sample space. This likelihood ratio cannot be so powerful for the particular λ_1 as the properly conditioned ratio.

The crucial contrast between (25) and (26) is that the latter does not depend on λ_1, whereas the locally most powerful test derived from (25) may do so. Thus the locally most powerful similar test of

H_0 when $\lambda = \lambda_1$ may not be the same for all λ_1. No general discussion seems to be available; it is likely that (25), while good for the particular λ_1 used, can be poor for other values of λ, whereas (26), while inferior to (25) for the correct λ, has reasonably good properties for all λ. Therefore (26) is in general to be preferred to (25) when the latter depends on λ_1.

Example 5.8. Weighted estimates of a normal mean. Suppose that independent random variables in sets of sizes $n_1, \ldots, n_m$ are taken from the $N(\mu, \sigma_1^2), \ldots, N(\mu, \sigma_m^2)$ distribution, respectively. The sufficient statistic in each sample is $(\bar{Y}_{j.}, SS_j)$, where

$$SS_j = \Sigma(Y_{jk} - \bar{Y}_{j.})^2 \quad (j = 1, \ldots, m).$$

Consider testing the null hypothesis $H_0: \mu = \mu_0$, with one-sided alternative $H_A: \mu > \mu_0$.

Under H_0, the sufficient statistic for the variance in the jth sample is $S_j = SS_j + n_j(\bar{Y}_{j.} - \mu_0)^2$. For convenience, we introduce the notation $v_j = \bar{y}_{j.} - \mu_0$ and $\delta = \mu - \mu_0$. A simple calculation shows that the conditional p.d.f. of V_j given $S_j = s_j$ is

$$f_{V_j|S_j}(v_j|s_j; \delta, \sigma_j) \propto \frac{(s_j - n_j v_j^2)^{\frac{1}{2}(n_j - 2)} \exp\left\{-\left(\dfrac{s_j - 2n v_j \delta}{2\sigma_j^2}\right)\right\}}{\displaystyle\sum_{r=0}^{\infty} \frac{1}{r!}\left(\frac{n_j \delta^2}{2\sigma_j^2}\right)^r g_{n_j + 2r}\left(\frac{s_j}{2\sigma_j^2}\right)},$$

where $g_m(.)$ is the p.d.f. of the chi-squared distribution with m degrees of freedom. Evaluating the local likelihood ratio conditionally on $(S_1, \ldots, S_m)$, as in (25), we get the test statistic

$$\sum_{j=1}^{m} \frac{n_j(\bar{Y}_{j.} - \mu_0)}{\sigma_j^2}, \tag{27}$$

which, for given values of $\sigma_1^2, \ldots, \sigma_m^2$, is the best linear combination of the single-sample test statistics $\bar{Y}_{1.}, \ldots, \bar{Y}_{m.}$.

To test H_0 via (27), we have first to choose the combination of $(\sigma_1^2, \ldots, \sigma_m^2)$ for which maximum power is required; then the null distribution of (27) for any set of variances is derived from the conditional distribution of V_j for $\delta = 0$. It is clear on general grounds that if the ratios among the true σ_j^2 are very different from those used in defining (27), then the test will be poor.

In this case S_j does depend on μ_0 for each j and since

$$f_{V_j|S_j}(v_j\,|s_j\,;0,\sigma_j) \propto \left\{1 + \frac{n_j(\bar{y}_{j.} - \mu_0)^2}{SS_j}\right\}^{1-\frac{1}{2}n_j},$$

the statistic (26) is a multiple of

$$\sum_{j=1}^{m} \frac{(n_j - 2)n_j(\bar{Y}_{j.} - \mu_0)}{SS_j + n_j(\bar{Y}_{j.} - \mu_0)^2}, \tag{28}$$

which gives similar critical regions. The test statistic (28) was proposed by Bartlett (1936a) on the basis of (26). Notice that the statistic derived from (27) by substituting the sample variances $SS_j/(n_j - 1)$ bears a strong resemblance to (28), although the latter also uses information about σ_j^2 in $n_j(\bar{Y}_{j.} - \mu_0)^2$. At least for sufficiently large n, (28) is therefore very similar to (27) with the *correct* σ_j^2.

One peculiar feature of (28) is that samples of size two do not enter the test at all. This makes (28) non-existent if $n_1 = \ldots = n_m = 2$, although obviously there would be information available for testing H_0. How much power is lost by using (28) is not clear. For discussion of (28) from the point of view of asymptotic estimation theory, see Neyman and Scott (1948) and Exercise 9.3.

So far we have discussed examples where the parameter of interest, ψ, is a component of the factorized parameter $\theta = (\psi, \lambda)$ when observations have density $f(y; \theta)$, with $\Omega_\theta = \Omega_\psi \times \Omega_\lambda$. In principle we can apply the discussion to tests of separate families, the third type of problem described in Section 5.1. If S_λ is a complete sufficient statistic for λ under H_0, the conditional likelihood ratio test of H_A versus H_0 has critical regions of the form

$$\left\{y\,; \frac{f_{Y|S_\lambda}(y\,|s\,;H_A, \nu_1)}{f_{Y|S_\lambda}(y\,|s\,;H_0)} \geqslant c(\nu_1)\right\}$$

for any specific value ν_1 of the parameter ν determining the distribution under H_A; this is the direct generalization of (4). A uniformly most powerful test may exist, but if not, then ambiguity may arise as to how to proceed. Certainly we cannot use the approach of maximizing local power, because the term "local" has no meaning here when H_0 and H_A are separate. We shall discuss this problem in detail in Section 9.3(iv). Note that if S_λ is also the minimal sufficient

statistic for ν in H_A, then ν does not appear in the conditional likelihood ratio test, which is therefore uniformly most powerful.

Example 5.9. Geometric distribution versus Poisson distribution.
Suppose that under the null hypothesis H_0, $Y_1, \ldots, Y_n$ are i.i.d. in a
Poisson distribution with mean λ, the alternative H_A being that the
distribution is geometric with parameter ν. Then ΣY_j is sufficient for
λ in H_0 and also for ν in H_A. Simple calculation gives for $t = \Sigma y_j$ that

$$f_{Y|T}(y \,|\, t; H_0) \;=\; \frac{t!}{y_1! \ldots y_n!} \, n^{-t},$$

$$f_{Y|T}(y \,|\, t; H_A) \;=\; \left\{ \binom{n+t-1}{t} \right\}^{-1}$$

The likelihood ratio critical region therefore has the form

$$\{y; \Sigma \log y_j! \geqslant c_\alpha\}.$$

Direct enumeration is the only way to derive the exact distribution
of the test statistic under H_0, the upper point of the distribution
arising from configurations $(t, 0, \ldots 0), \ldots, (0, \ldots 0, t)$ for $(y_1, \ldots, y_n)$.
In this instance the moments required for a normal approximation
are difficult to obtain, and the discussion of large sample approximations will be postponed to Section 9.3(iv).

The final examples of the similar region approach are concerned with
the second type of problem defined in Section 5.1. The distinguish-
ing character of these problems is that under the null hypothesis the
joint distribution of the observations is indexed by the parameter θ
restricted to $\Omega_0 \subset \Omega_\theta$, where Ω_0 cannot be expressed in the form
$\{(\theta_0, \lambda), \lambda \in \Omega_\lambda\}$. Often the dimension of Ω_0 will be the same as
that of the full parameter space Ω_θ. In principle, the discussion in
(i) about similar tests does apply here. If a complete minimal sufficient
statistic exists under H_0, then all similar region tests of H_0 will have
Neyman structure, as defined by (3). But if S is also sufficient under
H_A when $\theta \in \Omega_\theta - \Omega_0$, then for $\theta \in \Omega_\theta$

$$\mathrm{pr}(Y \in w \,|\, S = s; \theta) \;=\; \text{const.} \tag{29}$$

That is, similar region tests are powerless and hence useless. This will
always be the case if $\dim(\Omega_0) = \dim(\Omega_\theta)$ and some other approach is

required to obtain a meaningful test. The following is in the spirit of our present discussion.

Suppose that Ω_0 and $\Omega_\theta - \Omega_0$ have a boundary Ω_0^*, and let H_0^* be the hypothesis $\theta \in \Omega_0^*$. Then often H_0^* will be either a simple hypothesis, or a composite hypothesis that reduces the dimension of θ, i.e. $\dim(\Omega_0^*) < \dim(\Omega_\theta)$. A suitable compromise to testing the original null hypothesis is to have a similar test of H_0^* with critical region w_α satisfying

$$\mathrm{pr}(Y \in w_\alpha ; \theta) \leqslant \alpha \quad (\theta \in \Omega_0 - \Omega_0^*). \tag{30}$$

Note that a similar region likelihood ratio test of H_0^* with particular alternative $\theta_1 \in \Omega_\theta - \Omega_0$ will satisfy (30) if $f(y ; \theta)$ has likelihood ratio monotonic in θ, because any interior point of Ω_0 is further away from θ_1 than some point in Ω_0^*. This approach will not work for a non-monotone likelihood ratio when the discussion of Section 9.3(iv) will be appropriate. When Ω_0^* is a single point, such difficulties, of course, do not arise.

Example 5.10. Normal mean with unknown variance (ctd). Let $Y_1, \ldots, Y_n$ be i.i.d. in $N(\mu, \sigma^2)$ and let the null hypothesis be H_0: $\mu \leqslant 0, 0 < \sigma^2 < \infty$, with alternative $H_A : \mu > 0, 0 < \sigma^2 < \infty$. Under H_0, $(\Sigma Y_j, \Sigma Y_j^2)$ is minimal sufficient and (29) is satisfied. Considering the approach outlined above, we see that Ω_0^* is the single point $\mu = 0$ and the uniformly most powerful similar test of this hypothesis against H_A is the one-tailed Student t test of Example 5.5, which satisfies (30).

A corresponding approach to testing the hypothesis that $a \leqslant \mu \leqslant b$ leads to a two-tailed Student t test.

Example 5.11. Circular bivariate normal distribution. Let (X_1, Y_1), $\ldots, (X_n, Y_n)$ be i.i.d. in the bivariate normal distribution with mean vector (θ_1, θ_2) and covariance matrix $\mathbf{I}$. Suppose that the null hypothesis asserts that $\theta_1^2 + \theta_2^2 \leqslant 1$, the alternative being $\theta_1^2 + \theta_2^2 > 1$. In this case the boundary between the hypotheses is $\Omega_0^* = \{(\theta_1, \theta_2); \theta_1^2 + \theta_2^2 = 1\}$, which is a composite hypothesis of the first type with nuisance parameter θ_1, say. Unfortunately the minimal sufficient statistic is $(\bar{X}, \bar{Y})$, which is not complete under the null hypothesis. Critical regions of Neyman structure will satisfy (30) and (29), but the lack of completeness indicates the existence of other similar regions. One often appropriate region is found by noticing that

$\bar{X}^2 + \bar{Y}^2$ has a distribution depending only on $\theta_1^2 + \theta_2^2$, so that a similar critical region is

$$\{(x, y); \; n(\bar{x}_.^2 + \bar{y}_.^2) \geqslant c_{2,\alpha}^*(n)\},$$

where $c_{2,\alpha}^*(n)$ is the upper α point of the chi-squared distribution with two degrees of freedom and non-centrality parameter n. This test satisfies (30) because the chi-squared distribution has monotone likelihood ratio. We shall see in Section 5.3 that the test is uniformly most powerful among a certain class of invariant tests.

All the examples in this section have dealt with one- and two-sample problems for which the alternative hypothesis is unambiguous, in the sense that departure from the null hypothesis being tested is in a specific direction in the parameter space. For example, when i.i.d. random variables are taken from $N(\mu_1, \sigma_1^2)$ and $N(\mu_2, \sigma_2^2)$ distributions and H_0 is that $\mu_1 - \mu_2 = k_0$, alternatives are described by values of $\mu_1 - \mu_2$ not equal to k_0. For m-sample problems with $m \geqslant 3$, different types of departure from the null hypothesis will typically exist, each with its entirely different optimal test. Examples of such problems are treated in Sections 5.3 and 5.4.

(iii) Discussion

The general idea of a similar test, the construction *via* distributions conditional on a sufficient statistic for the nuisance parameter, and the use of (4) to obtain most powerful similar tests, are of great theoretical importance. The argument provides a theoretical justification for many widely-used tests, and is constructively useful in handling new situations. Further, as we shall see in Chapter 7, the ideas are central to the obtaining of interval estimates for one parameter in the presence of nuisance parameters.

Unfortunately, despite the importance of the idea, there are serious difficulties about it. These are best illustrated by a simple, if extreme, example.

Example 5.12. A restricted conditional distribution. Suppose that Y_1 and Y_2 are independent Poisson random variables with unknown means μ_1 and μ_2. Write $\mu_1 = \mu_2^\psi$ and suppose that we require to test the null hypothesis $\psi = \psi_0$, μ_2 being a nuisance parameter. This hypothesis, while not normally physically sensible, could conceivably arise in practice. Also it is clear on general grounds that the hypothesis can be tested, even if only qualitatively. The hypothesis concerns the ratio of natural parameters, because $\psi = (\log \mu_1)/(\log \mu_2)$.

Now under H_0 the likelihood is

$$\exp(-\mu_2^{\psi_0} - \mu_2)\mu_2^{\psi_0 y_1 + y_2}/(y_1! y_2!), \tag{31}$$

in which the sufficient statistic for the nuisance parameter μ_2 is

$$S_{\mu_2} = \psi_0 Y_1 + Y_2. \tag{32}$$

That is, the only exactly similar regions are found by examining the conditional distribution over the non-negative integer pairs (y_1', y_2') such that

$$\psi_0 y_1' + y_2' = \psi_0 y_1 + y_2, \tag{33}$$

where (y_1, y_2) are the observed values. Thus

(a) if ψ_0 is irrational, (y_1, y_2) is the only point satisfying (33), so that no useful similar test exists;

(b) if ψ_0 is an integer or rational number, $\psi_0 = \gamma_0'/\gamma_0''$, where γ_0'' is not very large, there may for quite large (y_1, y_2) be an appreciable number of solutions to (33) and thus a reasonably rich set of achievable significance levels may exist;

(c) if ψ_0 is a rational number ξ_0'/ξ_0'', with ξ_0' or ξ_0'' large compared with y_1 and y_2, there will be very few solutions of (33) other than (y_1, y_2) itself, and very few achievable significance levels will exist.

Thus, as ψ_0 varies continuously over real values, the probability distribution to be used for constructing similar regions, and therefore also the significance levels corresponding to a given (y_1, y_2), vary violently; see Cox (1967) for further discussion.

The conclusion to be drawn from this example is that the requirement of *exact* similarity is in some instances too strong, placing an unnecessarily severe restriction on the tests available for consideration. From a practical point of view, approximate similarity, holding $\mathrm{pr}(Y \in w_\alpha ; H_0)$ reasonably close to α, would be entirely adequate. While a theory requiring only approximate similarity could certainly be formulated, it would lack the simple elegance of the theory based on exact similarity and would therefore be correspondingly less useful.

The most serious aspect is not so much the existence of cases where the attempt to find regions of Neyman structure totally fails, for then it is clear that some alternative, possibly approximate, *ad hoc* analysis has to be made; see Example 5.6 and the discussion in

(iv) below. Rather, the suspicion is raised that even when an apparently satisfactory exactly similar test has been discovered, there may be an appreciably more powerful test that is very nearly similar, but which has been excluded from consideration by the restriction to exact similarity. For this reason, it is natural to look for some alternative justification, preferably not applying to cases such as Example 5.12! One may try to do this by a direct appeal to a form of the conditionality principle of Section 2.3, such a principle being considered more basic than the idea of similarity. This approach is implied in Fisher's discussion of the "exact" test for the comparison of two binomial probabilities (Fisher, 1935), but attempts to formalize what is required (Cox, 1958; Kalbfleisch and Sprott, 1970; Barndorff-Nielsen, 1973b) seem to have covered satisfactorily only rather restricted cases. We shall not pursue the matter here, although the point is of considerable theoretical importance.

(iv) Two-sample problem with unequal variances

The discussion of Example 5.6 shows that conditioning on the minimal sufficient statistic under H_0 fails to produce a test for the difference of the means of two normal distributions, both variances being unknown. As an example of a quite different special argument, we now outline a formal solution of this problem due to Welch (1947a). We use the notation of Example 5.6.

The construction of a similar region based on the sufficient statistics is achieved if we can find a function $h_\alpha(\mathrm{MS}_1, \mathrm{MS}_2)$ such that for all σ_1^2 and σ_2^2

$$\mathrm{pr}\{Z - \delta < h_\alpha(\mathrm{MS}_1, \mathrm{MS}_2)\} = 1 - \alpha, \qquad (34)$$

where $Z = \bar{Y}_{1.} - \bar{Y}_{2.}$ and $\delta = \mu_1 - \mu_2$. Now the conditional probability of the event in (34), given MS_1 and MS_2, is

$$\Phi\left\{\frac{h_\alpha(\mathrm{MS}_1, \mathrm{MS}_2)}{(\sigma_1^2/n_1 + \sigma_2^2/n_2)^{\frac{1}{2}}}\right\} = \mathscr{P}_\alpha(\mathrm{MS}_1, \mathrm{MS}_2),$$

say, and therefore (34) can be written

$$E\{\mathscr{P}_\alpha(\mathrm{MS}_1, \mathrm{MS}_2)\} = 1 - \alpha,$$

where the expectation is with respect to the distributions of MS_1 and MS_2, which are independent of one another and of Z.

A Taylor expansion of $\mathscr{P}_\alpha(\mathrm{MS}_1, \mathrm{MS}_2)$ is now made about (σ_1^2, σ_2^2) by writing formally

$$\mathscr{P}_\alpha(\text{MS}_1, \text{MS}_2) = \exp\{(\text{MS}_1 - \sigma_1^2)\partial_1 + (\text{MS}_2 - \sigma_2^2)\partial_2\}\mathscr{P}_\alpha(x_1, x_2). \quad (35)$$

Here the exponential is to be expanded in a power series with, for example,

$$\partial_j^r \mathscr{P}_\alpha(x_1, x_2) = \left[\frac{\partial^r \mathscr{P}_\alpha(x_1, x_2)}{\partial x_j^r}\right]_{x_1 = \sigma_1^2, \, x_2 = \sigma_2^2}$$

Thus

$$\mathcal{W}\{\mathscr{P}_\alpha(x_1, x_2)\} = 1 - \alpha,$$

where the operator $\mathcal{W}$ is defined by

$$\mathcal{W} = E[\exp\{(\text{MS}_1 - \sigma_1^2)\partial_1 + (\text{MS}_2 - \sigma_2^2)\partial_2\}]. \quad (36)$$

Now, denoting the degrees of freedom of MS_1 and MS_2 by d_1 and d_2, we have that the expectation (36) is related to the formal moment generating function of the estimates of variance and therefore

$$\mathcal{W} = (1 - 2\sigma_1^2 \partial_1/d_1)^{-\frac{1}{2}d_1}(1 - 2\sigma_2^2 \partial_2/d_2)^{-\frac{1}{2}d_2}\exp(-\sigma_1^2\partial_1 - \sigma_2^2\partial_2)$$

$$= 1 + \left(\frac{\sigma_1^4 \partial_1^2}{d_1} + \frac{\sigma_2^4 \partial_2^2}{d_2}\right) + \ldots, \quad (37)$$

on expanding in powers of $1/d_1$ and $1/d_2$.

Thus a formal solution of the problem is provided by the equation

$$\mathcal{W} \; \Phi\left\{\frac{h_\alpha(x_1, x_2)}{(\sigma_1^2/n_1 + \sigma_2^2/n_2)^{\frac{1}{2}}}\right\} = 1 - \alpha. \quad (38)$$

That is, we carry out the partial differentiation with respect to x_1 and x_2, and then put $x_1 = \sigma_1^2, x_2 = \sigma_2^2$; this in principle determines $h_\alpha(\sigma_1^2, \sigma_2^2)$ and therefore also $h_\alpha(\text{MS}_1, \text{MS}_2)$, and hence the required similar region.

We solve by expansion in powers of the reciprocals of the degrees of freedom. The zeroth order solution is

$$h_{0\alpha}(\sigma_1^2, \sigma_2^2) = k_\alpha^*(\sigma_1^2/n_1 + \sigma_2^2/n_2)^{\frac{1}{2}},$$

where $\Phi(k_\alpha^*) = 1 - \alpha$.

If we introduce a differentiation operator Δ by

$$\Delta^r \Phi(v) = \left[\frac{d^r \Phi(v)}{dv^r}\right]_{v = k_\alpha^*},$$

we can write (38) in the form

$$\exp\left[\left\{\frac{h_\alpha(x_1, x_2)}{(\sigma_1^2/n_1 + \sigma_2^2/n_2)^{\frac{1}{2}}} - k_\alpha^*\right\}\Delta\right]\Phi(v) = \Phi(k_\alpha^*). \tag{39}$$

We now write $h_\alpha(.) = h_{0\alpha}(.) + h_{1\alpha}(.) + h_{2\alpha}(.) + \ldots$, where suffix indicates the order of magnitude in terms of powers of $1/d_1$ and $1/d_2$. On substitution and comparison of the terms of particular orders the functions $h_{i\alpha}(.)$ are determined and in particular

$$h_{1\alpha}(\sigma_1^2, \sigma_2^2) = \frac{k_\alpha^*(1 + k_\alpha^{*2})}{4} \frac{\left(\dfrac{\sigma_1^4}{n_1^2 d_1} + \dfrac{\sigma_2^4}{n_2^2 d_2}\right)}{\left(\dfrac{\sigma_1^2}{n_1} + \dfrac{\sigma_2^2}{n_2}\right)^{\frac{3}{2}}}. \tag{40}$$

Numerical tables based on this expansion are given, for example, in the *Biometrika* tables (Pearson and Hartley, 1970, Table 11). Determination of the higher order terms is extremely tedious.

This discussion raises a number of important points, bearing in part on the issues discussed in (iii). First, the above argument assumes that similar regions exist, i.e. that the satisfaction of (34) for all σ_1^2 and σ_2^2 is possible for some function $h(.)$. Also it is assumed that the particular method of solution is valid, for example that it converges. In fact, however, Linnik (1968) has shown that non-randomized similar regions using the sufficient statistics in a reasonably smooth way do not exist. That is, considered as an exact equation, (34) has no smooth solution. Next, in the light of the discussion of the previous subsection, we ask whether the regions actually tabulated from a finite form of the series expansion give regions that are nearly similar. Here the position seems entirely satisfactory; numerical studies have shown that the regions based on three or four terms of the expansion are very nearly of the required size. For instance, with $d_1 = d_2 = 7, \alpha = 0.05$, Welch (1956) has shown that the probability of falling in the tabulated critical region deviates from 0.05 by *at most* 0.0005

Finally, there is the question of whether the solution is satisfactory on general grounds, or whether the imposition of even near-similarity is so strong a requirement that it can only be met at the cost of introducing some undesirable feature. Now the similar region based on $h_{0\alpha}(.)$ is the "obvious" approximate solution based on neglecting the errors in estimating variances. This is likely to be satisfactory for large degrees of freedom and to the extent that the regions obtained here are close to those based on $h_{0\alpha}(.)$ it is unlikely that they will be dramatically unsatisfactory.

However, Fisher (1956) pointed out one disturbing feature of the
tabulated solution. He showed by numerical integration in the special
case $d_1 = d_2 = 6$, $\alpha = 0.1$ that conditionally on $MS_1/MS_2 = 1$ the probability of falling in the critical region uniformly exceeds 0.1 for all
parameter values. In fact, for Welch's tabulated values

$$\text{pr}\{Z - \delta \geqslant h_{0.1}(MS_1, MS_2)|MS_1 = MS_2; \sigma_1^2, \sigma_2^2\} \geqslant 0.108. \quad (41)$$

Note first that (41) in no way conflicts with the requirement of similarity; that involves probabilities taken over the distribution of MS_1
and MS_2 and therefore over the distribution of MS_1/MS_2. If one takes
similarity as the overwhelming requirement, then (41) is irrelevant;
from this point of view, it would be of concern if there were critical
regions of size at most the nominal level and with better power properties than given by Welch's method, but there is no evidence that
such regions exist.

Nevertheless, in the light of discussion of conditionality in Section
2.3(iii), (41) is relevant. If, in fact, we observe $MS_1 = MS_2$, we know
that we have observed a subfamily of cases in which the hypothetical
error rate is worse than the nominal one. Then the long-run of repetitions contemplated in the requirement of similarity is no longer
entirely relevant to the interpretation of the data under analysis.

Note, however, that although values of the ratio MS_1/MS_2 serve to
define sets within some of which the size is bounded away from the
nominal value, the statistic MS_1/MS_2 is not ancillary, at least in either
of the relatively simple senses discussed in Section 2.2(viii). Also,
conditioning on the observed value of MS_1/MS_2 does not lead directly
to similar regions; see Example 5.6.

At a practical level it is, of course, quite reasonable to argue that
the right-hand side of (41) is so close to 0.1 that the anomaly can be
disregarded. Buehler (1959) has shown that effects similar to (41)
can be obtained in many problems involving nuisance parameters,
including, for instance, the much simpler problem of testing a null
hypothesis about a single normal mean, discussed in Example 5.3.
Pierce (1973) has shown that conditioning to produce a probability
of inclusion bounded away from the nominal value is possible in
interval estimation problems whenever the procedure cannot be interpreted as corresponding to a Bayesian posterior distribution with respect to a proper prior distribution; see Chapter 10. While these are
conceptually disturbing results, it seems that numerically the effects
are very slight and that therefore at an immediate practical level they
can be disregarded.

5.3   Invariant tests

(i) General theory

A second approach to the construction of suitable tests for composite hypotheses is based on the invariance principle, which was outlined in Section 2.3(vi). This often yields the same tests as the requirement of similarity, but in some complicated situations the difficulties associated with similarity are bypassed. As noted in Section 2.3(vi), invariance arguments have greater appeal when the transformations considered have direct physical significance, for example connected with changes of units, than when the transformations arise solely from the formal mathematical structure of the problem. Simple standard normal-theory problems can all be handled by invariance considerations.

Suppose that Y has a probability density $f(y; \theta)$ with parameter space Ω, and that H_0 is the composite hypothesis that $\theta \in \Omega_0$, with alternative H_A that $\theta \in \Omega_A$. Here Ω_0 and Ω_A are disjoint subsets of Ω, and usually Ω_A is the complement of Ω_0, i.e. $\Omega_A = \Omega - \Omega_0$. Then the hypothesis testing problem is *invariant* under a group $\mathcal{G}$ of transformations acting on the sample space if for any transformation $g \in \mathcal{G}$ the probability distribution of gY belongs to the same subset of distributions as that of Y, so that both satisfy H_0 or both satisfy H_A. We put this more formally by saying that the probability distribution of gY is obtained from the distribution of Y by replacing θ by $g^*\theta$, such that the collection $\mathcal{G}^*$ of all such induced parameter transformations g^* is a group on the parameter space preserving both Ω_0 and Ω_A. Thus, for any given $g \in \mathcal{G}$ and all sets A in the sample space,

$$\text{pr}(gY \in A; \theta) = \text{pr}(Y \in A; g^*\theta) \tag{42}$$

for some $g^* \in \mathcal{G}^*$ satisfying

(i) $g^*\Omega_0 = \Omega_0$,
(ii) $g^*\Omega_A = \Omega_A$,
(iii) $g^*\Omega = \Omega$.

This clearly asserts that H_0 remains equally correct or incorrect if we observe gY rather than Y, for any $g \in \mathcal{G}$. One implication is that no subsets of Ω_0 and Ω_A which are not preserved by g^* are to be given special emphasis in a hypothesis test.

A test with critical region w_α is an *invariant test* if

$$Y \in w_\alpha \text{ implies } gY \in w_\alpha \text{ for all } g \in \mathcal{G}. \tag{43}$$

The appeal of an invariant test is that it is in accord with, and follows quite naturally from, the basic symmetry of the problem, expressed above in the notion of invariance. Thus the same inference should be made from gy as from y, because of (42) and (i) and (ii) above.

The appeal to invariance implies, among other things, that the coordinate system used for recording y has irrelevant features, such as the order of recording i.i.d. observations. The important implication in the parameter space, correspondingly, is that we can work with reduced forms of Ω_0 and Ω_A and so, hopefully, find quite simple tests. For all transformations $g^*\theta$ of a parameter value θ are equivalent under the invariance criterion, just as all transformations gy of an observation y are equivalent in the sense of (43).

The following example illustrates these ideas.

Example 5.13. Mean of a multivariate normal distribution. Let $Y_1, \ldots, Y_n$ be i.i.d. vectors having the $MN_p(\mu, \Sigma)$ distribution, i.e. the multivariate normal distribution in p dimensions with mean vector μ and covariance matrix Σ. Suppose that the null hypothesis to be tested is that $\mu = 0$, the alternatives being that $\mu \neq 0$, with Σ unspecified under each hypothesis. Now let $\mathcal{G}$ be the group of all non-singular $p \times p$ matrices $\mathbf{b}$, so that $gY_j = \mathbf{b}Y_j$ $(j = 1, \ldots, n)$. It is a simple matter to verify that the induced transformation on the parameter space is defined by

$$g^*(\mu, \Sigma) = (\mathbf{b}\mu, \mathbf{b}\Sigma\mathbf{b}^T),$$

because $\mathbf{b}Y_j$ has the $MN_p(\mathbf{b}\mu, \mathbf{b}\Sigma\mathbf{b}^T)$ distribution. Both Ω_0 and Ω_A are preserved by g^*, whatever the matrix $\mathbf{b}$. Therefore an invariant test will use a statistic $t(Y_1, \ldots, Y_n)$ which is invariant under all such non-singular matrix transformations of $Y_1, \ldots, Y_n$. Invariance here means that no direction away from the origin of μ should receive special emphasis.

The corresponding test for this problem is developed in Example 5.18.

Now suppose, instead, that the null hypothesis is the composite linear hypothesis H_0: $\mu_1 = \ldots = \mu_{p^{(1)}} = 0, -\infty < \mu_{p^{(1)}+1}, \ldots, \mu_p < \infty$, with the alternative that $(\mu_1, \ldots, \mu_{p^{(1)}}) \neq 0, -\infty < \mu_{p^{(1)}+1}, \ldots,$ $\mu_p < \infty$, again leaving Σ unspecified in each case. For simplicity, write $\psi = (\mu_1, \ldots, \mu_{p^{(1)}})$ and $\chi = (\mu_{p^{(1)}+1}, \ldots, \mu_p)$ with the corresponding partition $Y = (V, W)$; we follow our normal convention of treating ψ, χ, V, W and Y as column vectors. For any transforming matrix $\mathbf{b}$, we partition according to the partition of Y, so that

$$\mathbf{b} = \begin{bmatrix} \mathbf{b}_{vv} & \mathbf{b}_{vw} \\ \mathbf{b}_{wv} & \mathbf{b}_{ww} \end{bmatrix}.$$

Then

$$\mathbf{b}\mu = \begin{bmatrix} \mathbf{b}_{vv}\,\psi + \mathbf{b}_{vw}\,\chi \\ \mathbf{b}_{wv}\psi + \mathbf{b}_{ww}\chi \end{bmatrix},$$

so that if $gY = bY$, then the condition $g^*\Omega_0 = \Omega_0$ implies $\mathbf{b}_{vw} = 0$.

But we can also shift W by $a1$, an arbitrary multiple of the unit vector, because χ is unspecified, so that now the testing problem is invariant under non-singular transformations of the form

$$\begin{bmatrix} V \\ W \end{bmatrix} \rightarrow \begin{bmatrix} \mathbf{b}_{vv} & 0 \\ \mathbf{b}_{wv} & \mathbf{b}_{ww} \end{bmatrix} \begin{bmatrix} V \\ W \end{bmatrix} + \begin{bmatrix} 0 \\ a1 \end{bmatrix}.$$

Of course, given the group $\mathcal{G}$ under which the test statistic is to be invariant, there will typically be many such tests. Consider two possible observations y and y' from $f(y\,;\theta)$ with associated values $t(y)$ and $t(y')$ of some invariant statistic T. Then if $t(y) = t(y')$, we would make precisely the same inference regarding H_0 from both y and y'. The invariance of $t(.)$ also implies that $t(y) = t(gy)$ for any transformation $g \in \mathcal{G}$. Thus if y' is not equal to gy for any $g \in \mathcal{G}$, $t(.)$ must be invariant under some larger group of transformations than $\mathcal{G}$. Provided, then, that we take $\mathcal{G}$ to be the largest group under which the testing problem is invariant, we conclude, that an invariant test statistic is, in general undesirable unless

$$t(y) = t(y') \text{ implies that } y' = gy \text{ for some } g \in \mathcal{G}. \quad (44)$$

If (44) holds, $t(.)$ is called a *maximal invariant* with respect to $\mathcal{G}$.

Any test based on a maximal invariant satisfies the definition (43) of an invariant test. For if $t(y) = t(y')$, then $y' = gy$, so that if y is in a critical region w_α, so too is y'.

As the various transformations in $\mathcal{G}$ are applied to a given sample point y, an *orbit* of points is traced out in the sample space. On each orbit a maximal invariant statistic $t(.)$ is constant, the value necessarily being unique to that orbit because of (44). In other words, $t(.)$ effects a partition of the sample space in much the same way as does a sufficient statistic. The natural question to ask is: how are maximal invariants and sufficient statistics related? The answer, roughly speaking, is that the two methods of reduction complement each other. If we start with y and the group $\mathcal{G}$, there may be several maximal invariants. For any particular maximal invariant $t'(.)$, there will exist a sufficient reduction $s'\{t'(.)\}$, which by definition is also a maximal invariant. Now suppose that $s(.)$ is a sufficient reduction of y in the whole parameter space, and that for the group $\mathcal{G}$ we find a maximal invariant $t\{s(.)\}$ of $s(.)$. Then in most, but not all, cases the two reductions $s'\{t'(.)\}$ and $t\{s(.)\}$ lead to the same statistic. Thus it is normally best to begin with the sufficient reduction $s(y)$ and then to apply the appropriate invariance argument, because this is computationally more convenient.

Although we have defined a maximal invariant statistic, it may not be immediately clear how we should proceed to find a maximal invariant. However, the above notion of orbits of sample values traced out by $\mathcal{G}$ indicates that the maximal invariant $t(.)$ uniquely characterises each orbit. Thus for any given sample value y we determine the orbit obtained by applying all the transformations in $\mathcal{G}$ and then characterize that orbit uniquely by a function of y. Note that to verify that a statistic $t(y)$ is a maximal invariant, we must check that (44) is satisfied. Thus, given any two samples y and y' which have the same value of t, we must show that there is a transformation $g \in \mathcal{G}$ such that $y' = gy$.

We illustrate these ideas with two simple examples.

Example 5.14. Standard deviation of a normal distribution. Suppose that $Y_1, \ldots, Y_n$ are i.i.d. in $N(\mu, \sigma^2)$ and let the null hypothesis be that $\sigma = \sigma_0$, with alternatives $\sigma > \sigma_0$, the mean μ being unknown in each case, $-\infty < \mu < \infty$. Then the appropriate transformations g are location shifts defined by

$$gy_j = y_j + a \quad (j = 1, \ldots, n),$$

where a is an arbitrary fixed constant. One maximal invariant is $t_1(y) = (y_2 - y_1, \ldots, y_n - y_1)$, corresponding to the location parameter ancillary statistic for arbitrary location distributions; it reduces the dimension of the sample by one. To verify that $t_1(y)$ is a maximal invariant, suppose that y' is another possible sample vector with $t_1(y') = t_1(y)$. Then we can write $y' = y + a1$ where $a = y_1' - y_1$, so that (44) is satisfied.

A direct calculation of the distribution of $T_1 = t_1(Y)$ shows that the sufficient reduction gives

$$s(T_1) = \Sigma(Y_j - \bar{Y})^2;$$

this is obtained also from such other maximal invariants as $t_2(y) = (y_1 - y_3, \ldots, y_n - y_3)$.

On working in reverse order, the sufficient statistic for (μ, σ^2) is $S' = \{\Sigma Y_j, \Sigma(Y_j - \bar{Y})^2\}$. As successive transformations in $\mathcal{G}$ are applied, the orbits in two dimensions are lines parallel to the Σy_j axis with constant values of $\Sigma(y_j - \bar{y})^2$. This latter statistic, therefore, indexes the orbits and is a maximal invariant.

Notice that S' could be obtained as a maximal invariant directly from $(Y_1, \ldots, Y_n)$ by considering the group of translations and orthogonal rotations, which effectively anticipates the normal form of the distribution involved.

Example 5.15. Index of gamma distribution. Let $Y_1, \ldots, Y_n$ be i.i.d. in the gamma distribution with index ψ and scale parameter λ, where the parameter space is $\Omega_\psi \times \Omega_\lambda = [1, \infty) \times [0, \infty)$. Suppose that the null hypothesis is $\psi = \psi_0$, with alternative $\psi > \psi_0$, λ being arbitrary. This problem is invariant under scale transformations $gy = by$, for positive constants b.

Taking the maximal invariant reduction first, we find, on applying successive scale transformations, that the orbits are lines through the origin $y = 0$. The set of angular tangents characterizes these lines, so that a maximal invariant is

$$T = (Y_2/Y_1, \ldots, Y_n/Y_1).$$

Now a straightforward calculation shows that the distribution of $T = (T_2, \ldots, T_n)$ has density

$$\frac{\Gamma(n\psi)}{\{\Gamma(\psi)\}^n} \frac{\prod\limits_{j=2}^{n} t_j^{\psi-1}}{(1 + \sum\limits_{j=2}^{n} t_j)^{n\psi}}.$$

The minimal sufficient reduction of T is therefore

$$S = s(T) = \sum_{j=2}^{n} \log T_j - n \log(1 + \sum_{j=2}^{n} T_j)$$

$$= \sum_{j=1}^{n} \log Y_j - n \log(\sum_{j=1}^{n} Y_j),$$

or equivalently the ratio of geometric and arithmetic means.

The same statistic is derived more easily by applying the sufficiency reduction first to give $(\Sigma \log Y_j, \Sigma Y_j)$ and then applying the invariance argument in two dimensions. Thus, taking the more convenient form $(n^{-1} \Sigma \log Y_j, \log \Sigma Y_j)$ for the sufficient statistic, the linear orbit traced out by the transformed points $(n^{-1} \Sigma \log Y_j + \log b, \log \Sigma Y_j + \log b)$ is characterized by the intercept, which is $n^{-1} \Sigma \log Y_j - \log \Sigma Y_j$, as before.

In certain very rare cases, the "sufficiency plus invariance" and "invariance plus sufficiency" reductions do not give a unique statistic; conditions for uniqueness are discussed by Hall, Wijsman and Ghosh (1965). For the examples considered here unique reductions are always obtained. In any case, it seems natural to us to apply the sufficiency reduction first, this being the main unifying principle of statistical inference. In the rest of this Section we shall assume that an initial reduction is made to minimal sufficient statistics.

The invariance principle yields not only maximal invariants of the data y, but also of the parameter θ, because θ and $g^*\theta$ are equivalent for all g^* in the group $\mathcal{G}^*$ induced by $\mathcal{G}$. Denoting the maximal invariant of θ by $\phi = \phi(\theta)$, we see that the distribution of the maximal invariant $t(Y)$ depends only on ϕ. It is the reduction to ϕ that reduces or removes the nuisance parameter in the testing problem, for the null hypothesis that $\theta \in \Omega_0$ is reduced to a hypothesis about ϕ, often to a simple hypothesis.

As mentioned in Section 5.2, H_0 may sometimes implicitly specify a boundary in the parameter space Ω_θ; for example, $H_0 : \theta \leqslant \theta_0$ has as boundary with $H_A : \theta > \theta_0$ the single point θ_0. In terms of the reduced parameter ϕ, H_0 will specify a new boundary and it will often be appropriate to test that the reduced parameter lies on this boundary. The ideal situation is where the new boundary is a single point, as in Example 5.17.

Now suppose that H_0 and H_A are rephrased in terms of the reduced parameter ϕ. Suppose also that, possibly in its boundary form, H_0 is a simple hypothesis $\phi = \phi_0$. Then, for a particular alternative ϕ_1, the best invariant test will be determined by the likelihood ratio, the critical region at level α based on the maximal invariant sufficient statistic T being given by

$$w_\alpha = \left\{ t : \frac{f_T(t; \phi_1)}{f_T(t; \phi_0)} \geqslant c_\alpha \right\}. \tag{45}$$

Should it turn out that this critical region does not depend on ϕ_1, the test is *uniformly most powerful invariant*. This will happen if the likelihood ratio is monotonic and the alternatives one-dimensional and one-sided in the reduced parameter space Ω_ϕ. If w_α does depend on ϕ_1, some other basis for the selection of a test is required such as maximization of local power. The discussion of Section 4.8 is directly relevant here, now being applied to $f_T(t; \phi)$, rather than to $f_Y(y; \theta)$.

If H_0 contains more than one point, even in its boundary form, some modification to the above arguments is necessary to ensure desirable test properties, for instance unbiasedness, but this is usually unnecessary.

One particular feature of invariant tests is that they frequently remove the difficulties associated with the existence of several different types of alternative, insofar as strong requirements of symmetry are imposed on the parameter space and the structure of the parameter space thereby is simplified. Therefore even when there is no uniformly most powerful similar test a uniformly most powerful invariant test may exist. This is illustrated by Examples 5.17 and 5.18 below, where directional coordinates in the original parameter space are removed by the appeal to invariance. We emphasize this property because in particular applications it may be undesirable; that is, the relevance of invariance needs critical examination, especially when it arises from the mathematical structure of the problem.

(ii) Some examples
We now examine some applications of the basic invariance ideas discussed in (i).

Example 5.16. Normal mean with unknown variance (ctd). Let $Y_1, \ldots, Y_n$ be i.i.d. in $N(\mu, \sigma^2)$ with both μ and σ^2 unknown, and consider testing the null hypothesis H_0: $\mu = 0, 0 < \sigma^2 < \infty$, with two-sided alternative H_A: $\mu \neq 0, 0 < \sigma^2 < \infty$. The problem is scale invariant, i.e. invariant under transformations $y_j \to by_j$ for arbitrary constants b, and $g^*(\mu, \sigma) = (b\mu, b\sigma)$.

The sufficient statistic here is $S = (\Sigma Y_j, \Sigma Y_j^2)$, and, as successive scale transformations are applied to any sample point s, the orbit traced out in the plane is a parabola through the origin indexed by the curvature, so that a maximal invariant is $t(Y) = (\Sigma Y_j)^2/\Sigma Y_j^2$. The corresponding maximal invariant of $\theta = (\mu, \sigma^2)$ is $\phi = \mu^2/\sigma^2$. The null hypothesis is therefore reduced to the single point $\phi = 0$. Note that $t(Y)$ is a monotone function of the square of the Student t statistic of Example 5.5, so that invariance and similarity lead to the same test statistic in this case. Critical regions are formed by large values of the square of the Student t statistic. The test is uniformly most powerful invariant.

If H_0 were that $\mu \leqslant 0$, with alternative that $\mu > 0$, the above calculations would no longer apply. For this problem the constant b must be positive, so that the maximal invariant is $T = \Sigma Y_j / (\Sigma Y_j^2)^{\frac{1}{2}}$ and $\phi = \mu/\sigma$. This leads to the one-sided Student t test of Example 5.5, if we reduce the null hypothesis to its boundary $\phi = 0$.

Example 5.17. Circular bivariate normal mean (ctd). Let $(X_1, Y_1), \ldots, (X_n, Y_n)$ be i.i.d. bivariate normal with mean (μ_1, μ_2) and covariance matrix $\sigma^2 I$ and suppose that H_0 is that $\mu_1^2 + \mu_2^2 \leqslant 1$ and $\sigma^2 = 1$, with alternative H_A that $\mu_1^2 + \mu_2^2 > 1$ and $\sigma^2 = 1$. We immediately take H_0^*, that $\mu_1^2 + \mu_2^2 = 1$ and $\sigma^2 = 1$, as the working null hypothesis. In Example 5.11 we found that the similar region approach did not work for this problem because the minimal sufficient statistic is not reduced under H_0. Now the problem is invariant under rigid rotation, with

$$g\begin{bmatrix} x \\ y \end{bmatrix} = \begin{bmatrix} \sin a & \cos a \\ \cos a & -\sin a \end{bmatrix} \begin{bmatrix} x \\ y \end{bmatrix} \quad (0 < a \leqslant 2\pi).$$

The sufficient statistic is $S = (\bar{X}_., \bar{Y}_.)$, and the orbit traced out by applying successive rotations to s is a circle, centred at the origin. The maximal invariant statistic is therefore $T = \bar{X}_.^2 + \bar{Y}_.^2$, the squared radius of the circular orbit, with corresponding maximal invariant parameter $\phi = \mu_1^2 + \mu_2^2$.

Because $\bar{X}_.\sqrt{n}$ and $\bar{Y}_.\sqrt{n}$ are independently $N(\mu_1\sqrt{n}, 1)$ and $N(\mu_2\sqrt{n}, 1)$, $U = nT$ has a non-central chi-squared distribution with two degrees of freedom and p.d.f.

$$f_U(u;\phi) = \tfrac{1}{2}\exp(-\tfrac{1}{2}u^2 - \tfrac{1}{2}n\phi^2)I_0\{u\sqrt{(n\phi)}\}.$$

This has monotone likelihood ratio. We therefore "reject" H_0 for large values of u and the significance probability is

$$\text{pr}(U \geqslant u; H_0) = \tfrac{1}{2}e^{-\tfrac{1}{2}n} \int_u^\infty e^{-\tfrac{1}{2}v^2} I_0(v\sqrt{n})dv.$$

If the variance σ^2 is unspecified by H_0 and by H_A, then the maximal invariant becomes

$$T' = \frac{T}{\Sigma(X_j - \bar{X}_.)^2 + \Sigma(Y_j - \bar{Y}_.)^2},$$

which is the maximal invariant reduction of $\{T, \Sigma(X_j - \bar{X}.)^2,$
$\Sigma(Y_j - \bar{Y}.)^2\}$ under scale transformations. The test statistic $U' = n(2n - 2)T'$ has a non-central F distribution with 2 and $2n - 2$
degrees of freedom and non-centrality parameter $n\phi/\sigma^2$. This particular test is well-known in time series analysis as the Schuster test
for the amplitude of a given periodogram component (Anderson,
1971, p. 118).

Note that the reduction by invariance would be less compelling if
there were reason to think that the x and y axes had been specially
chosen, for example so that alternatives close to the axes were particularly likely. Note also that the invariance is destroyed if the
bivariate distribution is not circular.

Example 5.18. Hotelling's test. Let $Y_1, \ldots, Y_n$ be independent
p-dimensional random variables with the multivariate normal distribution $MN_p(\mu, \Sigma)$. Suppose that Σ is unspecified and that $H_0: \mu = 0$,
with alternative $H_A: \mu \neq 0$. In Example 5.13 we saw that this problem is invariant under all non-singular transformations. The sufficient
statistic is $S = (\bar{Y}., \text{SS})$, where, for $u, v = 1, \ldots, p$,

$$\text{SS} = ((\text{SS}_{uv})) = \left(\left(\sum_{j=1}^{n} (Y_{uj} - \bar{Y}_{u.})(Y_{vj} - \bar{Y}_{v.}) \right) \right)$$

is the sample cross-product matrix. The maximal invariant reduction
of S is

$$T = \bar{Y}.^{\mathrm{T}} \text{SS}^{-1} \bar{Y}., \tag{46}$$

with corresponding reduction of the parameter to

$$\phi = \mu^{\mathrm{T}} \Sigma^{-1} \mu.$$

To find the maximal invariant statistic, we note that the same
statistic must be obtained after the particular transformations
$Y \to \pm \text{SS}^{-\frac{1}{2}} Y$, which reduce SS to the identity matrix. Conversely,
(44) can be verified by noting that T is the only non-zero eigenvalue of
$\bar{Y}.\bar{Y}.^{\mathrm{T}} \text{SS}^{-1}$ and that equality of eigenvalues for two matrices B_1 and
B_2 occurs only if $B_2 = AB_1A^{\mathrm{T}}$, for some nonsingular matrix A.

The statistic T is the multivariate analogue of the Student t statistic, and is usually referred to as Hotelling's statistic. It is not difficult to show that $\{n(n - p)T\}/\{p(n - 1)\}$ has the F distribution with
$(p, n - p)$ degrees of freedom and non-centrality parameter $n\phi$
(Anderson, 1958, Chapter 5). Therefore large values of the statistic

are significant, and the significance probability of an observed value t_{obs} is

$$\text{pr}(T \geq t_{obs}; H_0) = \text{pr}\left\{F_{p,n-p} \geq \frac{n(n-p)}{p(n-1)} t_{obs}\right\}.$$

An important feature of the test is that it reduces (μ, Σ) to a single number ϕ and a uniformly most powerful invariant test results. Clearly the test is in a sense equally powerful in all directions of the μ space, which is a strong condition. A uniformly most powerful similar test will not exist, because any specific alternative μ_1 indicates a preferred direction in which the Student t test based on $\mu_1^T \bar{Y}$ is uniformly most powerful. In fact, the invariant Hotelling statistic is the largest of all such directional Student t statistics. Therefore if a particular type of alternative to H_0 is relevant, the invariant test is not particularly appropriate, because it looks for significant departures from H_0 in irrelevant as well as in relevant directions.

Example 5.19. Hotelling's test on a subspace. To continue with the setting of the previous example, suppose that the null hypothesis specifies only a subset of the vector μ. For simplicity, suppose that μ is partitioned into $(\mu^{(1)}, \mu^{(2)})$ and that the null hypothesis is H_0: $\mu^{(1)} = 0$ with alternatives H_A: $\mu^{(1)} \neq 0$, both $\mu^{(2)}$ and Σ being unspecified throughout. The induced partitions of $\bar{Y}$, SS and Σ will be denoted by

$$\begin{bmatrix} \bar{Y}^{(1)} \\ \bar{Y}^{(2)} \end{bmatrix}, \begin{bmatrix} SS_{11} & SS_{12} \\ SS_{12}^T & SS_{22} \end{bmatrix}, \begin{bmatrix} \Sigma_{11} & \Sigma_{12} \\ \Sigma_{12}^T & \Sigma_{22} \end{bmatrix},$$

respectively. Then a generalization of the argument in Example 5.13 shows the testing problem to be invariant under non-singular transformations

$$gy = \begin{bmatrix} b_{11} & 0 \\ b_{21} & b_{22} \end{bmatrix} \begin{bmatrix} y^{(1)} \\ y^{(2)} \end{bmatrix} + \begin{bmatrix} 0 \\ a^{(2)} \end{bmatrix}. \tag{47}$$

The maximal invariant reduction of $(\bar{Y}, SS)$ is found by applying the transformation defined in (47) to the partitioned forms of $\bar{Y}$ and SS, and then imposing the invariance requirement on H_0 and H_A. This gives the Hotelling statistic

$$T^{(1)} = (\bar{Y}^{(1)})^T SS_{11}^{-1} \bar{Y}^{(1)}, \tag{48}$$

corresponding to (46). Note that no information is contributed from $\bar{Y}^{(2)}$, because all its information goes into inference about the nuisance parameter $\mu^{(2)}$.

Suppose now, however, that H_0 and H_A both specify the value of $\mu^{(2)}$, so that information in $\bar{Y}^{(2)}$ is made available. Without loss of generality take $\mu^{(2)} = 0$. Now the problem is invariant under the arbitrary transformation (47) only if $a^{(2)} = 0$, and one can verify by applying the transformation that a maximal invariant is

$$T = (\bar{Y}_.^{T} \mathbf{ss}^{-1} \bar{Y}_., \bar{Y}_.^{(2)T} \mathbf{ss}_{22}^{-1} \bar{Y}_.^{(2)}), \tag{49}$$

instead of (48). The question now arises of how to construct the test of H_0 from T. Notice that the second component $T^{(2)}$ of T is an ancillary statistic; it has a central F distribution when $\mu^{(2)} = 0$, as we mentioned in Example 5.18. Therefore, because of the conditionality principle, we test H_0 using the distribution of T conditionally on $T^{(2)} = t^{(2)} = \bar{y}_.^{(2)T} \mathbf{ss}_{22}^{-1} \bar{y}_.^{(2)}$. Some simplification is achieved by replacing $\bar{y}_.^{(1)}$ with the residual from its regression on $\bar{y}_.^{(2)}$, namely

$$Z = \bar{Y}_.^{(1)} - \mathbf{ss}_{12} \mathbf{ss}_{22}^{-1} \bar{Y}_.^{(2)},$$

which is distributed independently of $\bar{Y}_.^{(2)}$. By decomposing the quadratic form $\bar{Y}_.^{T} \mathbf{ss}^{-1} \bar{Y}_.$, we find that the maximal invariant (49) is equal to

$$\{Z^{T}(\mathbf{ss}_{11} - \mathbf{ss}_{12}\mathbf{ss}_{22}^{-1}\mathbf{s}_{12}^{T})Z + T^{(2)}, T^{(2)}\}.$$

The likelihood ratio test, conditionally on $t^{(2)}$, therefore uses the conditional distribution of the quadratic form in z. An extension of the multivariate distribution theory used for the Hotelling test statistic in Example 5.18 shows that the standardized statistic

$$T^{(1.2)} = \frac{Z^{T}(\mathbf{ss}_{11} - \mathbf{ss}_{12}\mathbf{ss}_{22}^{-1}\mathbf{ss}_{12}^{T})Z}{1 + nt^{(2)}}$$

is distributed independently of $T^{(2)}$ under H_0, and is proportional to an F statistic with $p^{(1)}$ and $n - p$ degrees of freedom and non-centrality parameter

$$n\phi = n\mu^{(1)T}(\mathbf{\Sigma}_{11} - \mathbf{\Sigma}_{12}\mathbf{\Sigma}_{22}^{-1}\mathbf{\Sigma}_{12}^{T})^{-1}\mu^{(1)}.$$

The essential character of $T^{(1.2)}$ is most easily seen by looking at the case $p = 2$, when $\mathbf{ss}_{12}\mathbf{ss}_{22}^{-1}$ is the ordinary sample regression coefficient, b say. Then $T^{(1.2)}$ becomes

$$\frac{(\bar{Y}_{.}^{(1)} - b\bar{Y}_{.}^{(2)})^2 \{\Sigma(Y_j^{(1)} - \bar{Y}_{.}^{(1)})^2 - b^2 \Sigma(Y_j^{(2)} - \bar{Y}_{.}^{(2)})^2\}^{-1}}{1 + n(\bar{Y}_{.}^{(2)})^2 \{\Sigma(Y_j^{(2)} - \bar{Y}_{.}^{(2)})^2\}^{-1}}.$$

This can be derived directly as the usual F statistic for analysis of covariance. Thus we adjust $\bar{Y}_{.}^{(1)}$ for regression on $\bar{Y}_{.}^{(2)}$ to give $\bar{Y}_{.}^{(1)} - b\bar{Y}_{.}^{(2)}$ whose variance conditionally on $\bar{Y}_{.}^{(2)} = \bar{y}_{.}^{(2)}$ is

$$\text{var}(Y^{(1)}|Y^{(2)} = y^{(2)}) \left[\frac{1}{n} + \frac{(\bar{y}_{.}^{(2)})^2}{\Sigma(y_j^{(2)} - \bar{y}_{.}^{(2)})^2}\right].$$

The variance of $Y^{(1)}$ given $Y^{(2)}$ is estimated by the usual residual sum of squares about the regression line.

Note that we have argued conditionally on the ancillary statistic $T^{(2)}$ and, as in Example 4.6, this results in our test not being uniformly most powerful invariant unconditionally.

As we have seen, one advantage of appeal to invariance is that multidimensional alternatives may be reduced to one-dimensional alternatives. As a final example of this reduction we give a simple application in analysis of variance.

Example 5.20. One-way analysis of variance. Consider the balanced one-way linear model

$$Y_{jk} = \mu_j + \epsilon_{jk} = \mu + \alpha_j + \epsilon_{jk} \quad (j = 1, \ldots, m; k = 1, \ldots, r)$$

where the ϵ_{jk} are i.i.d. in $N(0, \sigma^2)$ and the parameters α_j represent contrasts satisfying $\Sigma\alpha_j = 0$. Let the null hypothesis be $H_0: \alpha_1 = \ldots = \alpha_m = 0$, μ and σ being unspecified, with alternative hypothesis that the α_j are not all zero. The sufficient statistic is $(\bar{Y}_{1.}, \ldots, \bar{Y}_{m.}, \text{SS}_w)$, where $\text{SS}_w = \Sigma\Sigma(Y_{jk} - \bar{Y}_{j.})^2$. Under H_0 there is the further reduction to the overall mean $\bar{Y}_{..}$ and

$$\text{SS}_0 = \Sigma\Sigma(Y_{jk} - \bar{Y}_{..})^2.$$

Conditionally on $\bar{Y}_{..} = \bar{y}_{..}$ and on SS_0, the Y_{jk} lie on the intersection of the hyperplane $\Sigma\Sigma Y_{jk} = mr\bar{y}_{..}$ and the hypersphere with centre $\bar{y}_{..}$ and radius $\sqrt{\text{SS}_0}$. By generalizing the result of Example 5.5, it may be shown that the likelihood ratio tests for the specific alternatives $H_{A,1}: \mu_1 \neq \mu_2 = \ldots = \mu_m$ and $H_{A,m}: \mu_1 = \ldots = \mu_{m-1} \neq \mu_m$ have critical regions defined respectively by large absolute values of $\bar{y}_{1.} - \bar{y}_{..}$ and by $\bar{y}_{m.} - \bar{y}_{..}$. Thus no uniformly most powerful similar test exists for $m > 2$. Applying the invariance argument, we see that the problem is invariant under suitable sets of orthogonal transformations.

The maximal invariant is, by generalization of the result of Example 5.16, $\Sigma(\bar{Y}_{j.} - \bar{Y}_{..})^2/\text{SS}_w$, which is equivalent to the usual F ratio.

Similar results apply to more complicated balanced analyses of variance, but they do depend entirely on the balance. Of course, in practice, the primary purpose of analysis of variance is rarely the testing of null hypotheses such as the one considered here.

So far we have not examined how invariance arguments might apply to tests of separate families of hypotheses, as defined in Section 5.1. The following example indicates what can be done when the nuisance parameters have the same form under each hypothesis.

Example 5.21. Two separate location and scale families. Let $Y_1, \ldots,$ Y_n be i.i.d. each with p.d.f. $f_Y(y)$, the null hypothesis being that $f_Y(y) = \tau^{-1}g\{(y - \mu)/\tau\}$ and the alternative hypothesis that $f_Y(y) = \tau^{-1}h\{(y - \mu)/\tau\}$; both μ and τ are unrestricted in each hypothesis. The problem is invariant under transformations $gy = ay + b$, for arbitrary a and b, and so a maximal invariant is the ancillary statistic with respect to μ and τ, namely

$$T = \left(\frac{Y_3 - Y_2}{Y_2 - Y_1}, \ldots, \frac{Y_n - Y_2}{Y_n - Y_1} \right) = (T_3, \ldots, T_n).$$

Now write $y_2 = u$ and $y_2 - y_1 = v$, so that the p.d.f. of T under H_0 is

$$f_T(t; H_0) = \iint_{-\infty}^{\infty} v^{n-2} g(v) g(u - v) \prod_{j=3}^{n} g(vt_j + u) \, du \, dv; \qquad (50)$$

under H_A, we simply replace $g(.)$ by $h(.)$ in (50). Notice that the alternative hypothesis is simple after the invariance reduction, so that the ratio of (50) to its counterpart under H_A is the most powerful invariant test statistic. An obvious difficulty with this test is the calculation of the distribution of the likelihood ratio. Frequently some approximation will be necessary, possibly a numerical approximation based on Monte Carlo simulation of the likelihood ratio. More often a large-sample approximation will be convenient, in which case the discussion of Section 9.3(iv) will apply.

(iii) Discussion

In the examples used in (ii) to illustrate the application of the invariance principle, we have obtained tests that are reasonable on general grounds. The principle is, apparently, particularly useful in multi-dimensional problems, because it takes direct advantage of any lack of preferred direction; this is well illustrated by Example 5.20. However, the mathematical symmetry of a problem may not correspond to the symmetry of the hypothesis being tested. Thus in Example 5.17, invariance would be of no help if the null hypothesis were that $\mu_1^2 + b\mu_2^2 \leqslant 1$, with $b \neq 1$, because this hypothesis is not rotation invariant.

Of course, we must expect that for some problems the invariance principle will not be applicable, and that for some other problems it will not give much reduction of the original sample. In the latter case, initial reduction by sufficiency followed by application of the invariance principle will often lead to greater reduction than an initial application of the invariance principle, as in the case of Example 5.5, because the choice of maximal invariant is not unique. This is an argument of convenience and it is not clear whether there are other arguments for preferring to make the reduction in this order.

Finally, note that even when the invariance principle is applicable, it can lead to useless tests. An often-quoted illustration of this is the following example due to C. Stein.

Example 5.22. A useless invariant test. Let $X = (X_1, X_2)$ and $Y = (Y_1, Y_2)$ have independent bivariate normal distributions with means zero and covariance matrices Σ and $\psi\Sigma$, where both ψ and Σ are unknown. If the null hypothesis is $H_0: \psi = 1$, with alternative $\psi > 1$, then the problem is invariant under common non-singular transformations of X and Y.

Now the matrix

$$\mathbf{Z} = \begin{bmatrix} X_1 & Y_1 \\ X_2 & Y_2 \end{bmatrix}$$

is itself non-singular with probability one. For any two values z_1 and z_2, there is a non-singular matrix $\mathbf{a}$ such that $z_1 = \mathbf{a}z_2$, namely $\mathbf{a} = z_1 z_2^{-1}$. Thus all possible sample points are on the same orbit. Therefore no effective maximal invariant statistic exists under non-singular transformations. The only invariant level α test of H_0 is to "reject" H_0 randomly with probability α, independently of the data.

However, among tests using X_1 and Y_1 alone, there is a uniformly most powerful test, namely that with critical regions of the form

$$\{x, y ; y_1^2/x_1^2 \geqslant c_\alpha\}.$$

The distribution of $\psi X_1^2/Y_1^2$ is F with $(1, 1)$ degrees of freedom, so that the power always exceeds the test level for $\psi > 1$. Similarly a test can be constructed based on X_2^2/Y_2^2, but there is no invariant combination of them.

5.4   Some more difficult problems

The similar region and invariant test procedures are central to the solution of composite hypothesis testing problems, but not all problems can be solved completely without further reduction. One reason may be that sufficiency and invariance do not provide a reduction of the problem anyway; another may be that we are left with several statistics and no unique way to combine them in a single test. This often happens in multi-sample or multivariate problems. We now give some examples to illustrate this situation, and outline some methods for forming suitable test statistics.

Example 5.23. Comparison of two covariance matrices. Suppose that $Y_1, \ldots, Y_{n_1}$ are i.i.d. vectors from the $MN_p(0, \Sigma_1)$ distribution, and that $Z_1, \ldots, Z_{n_2}$ are i.i.d. vectors from the $MN_p(0, \Sigma_2)$ distribution, the two sets being independent. Then $SS_1 = Y_1 Y_1^T + \ldots + Y_{n_1} Y_{n_1}^T$ and $SS_2 = Z_1 Z_1^T + \ldots + Z_{n_2} Z_{n_2}^T$ are sufficient statistics, each having a Wishart distribution. One hypothesis of interest in this situation is $H_0: \Sigma_1 = \Sigma_2$, with the global alternative $H_A: \Sigma_1 \neq \Sigma_2$.

This problem is invariant under any non-singular transformation applied to both samples, and the maximal invariant is the set of eigenvalues of $SS_1 SS_2^{-1}$, i.e. the solutions $l_1, \ldots, l_p$ of

$$|SS_1 SS_2^{-1} - l\mathbf{I}| = 0.$$

This is derived by noting that there exists a non-singular matrix $\mathbf{a}$ such that $\mathbf{a} SS_1 \mathbf{a}^T = \text{diag}(l_1, \ldots, l_p)$ and $\mathbf{a} SS_2 \mathbf{a}^T = \mathbf{I}$. The maximal invariant is unchanged by this transformation and so is necessarily a function of these eigenvalues. It is easy to verify that the eigenvalues themselves are maximal invariant.

Just one function of the maximal invariant is the product of the eigenvalues, $|SS_1|/|SS_2|$, the ratio of generalized variances, but there are many other possible functions. In general, a strong reason for not selecting the ratio of generalized variances is that it is invariant under orthogonal transformations of the sets of random variables, and hence is useless for testing the angles between principal axes of the two concentration ellipsoids.

Under the null hypothesis, $SS_1 + SS_2$ is sufficient for the common covariance matrix, so that similar region tests can be constructed from the distribution of SS_1 given $SS_1 + SS_2$. This conditional distribution over the $\frac{1}{2}p(p + 1)$ variables in SS_1 is complicated, and no uniformly most powerful similar test exists because of the many possible directions of departure from H_0; see Example 5.20.

Where the sort of ambiguity illustrated in the last example exists, the derivation of a single test statistic is bound to be rather arbitrary, unless some extra considerations are introduced. One approach that achieves a formal unification is the maximum likelihood ratio method, which extends the criterion of the Neyman-Pearson lemma as follows. When H_0 specifies more than one point in the parameter space, and no points are preferred to any others, take the maximum possible value of the likelihood and compare it with the maximum possible value under the alternative. Specifically, if $H_0: \theta \in \Omega_0$ and the alternative is $H_A: \theta \in \Omega_A$, use the criterion

$$e^{\frac{1}{2}w'} = \frac{\sup_{\Omega_A} \mathrm{lik}(\theta; y)}{\sup_{\Omega_0} \mathrm{lik}(\theta; y)} \tag{51}$$

where $\mathrm{lik}(\theta; y) = f_Y(y; \theta)$, as defined in Section 2.1. If $\Omega_A = \Omega - \Omega_0$, as is usually the case, it is more convenient to use the statistic

$$e^{\frac{1}{2}w} = \frac{\sup_{\Omega} \mathrm{lik}(\theta; y)}{\sup_{\Omega_0} \mathrm{lik}(\theta; y)}, \tag{52}$$

which is related to w' by $w = \max(w', 0)$. Large values of the maximum likelihood ratio test statistic are significant. The definitions (51) and (52) are in exponential form for later convenience.

Unfortunately, the strong optimum properties associated with the likelihood ratio method for simple hypotheses are not carried over to composite hypothesis problems in general. We shall see in Chapter 9 that in large samples a certain optimality is achieved, but this is weak compared to that obtained in earlier sections. Further, from a practical point of view, the likelihood ratio often has the disadvantage that significance probabilities are difficult to calculate without large-sample approximations.

Notice from (51) that the maximum likelihood ratio test depends only on the sufficient statistic S for $\theta \in \Omega$. Moreover, if the testing problem is invariant under the groups $\mathcal{G}$ and $\mathcal{G}^*$ operating on the sample space and on Ω respectively, then w' is invariant, because $\mathrm{lik}(g^*\theta \, ; y) = \mathrm{lik}(\theta \, ; gy)$, and the suprema involved in (51) are preserved by the transformations $g^* \in \mathcal{G}^*$. Therefore, if a uniformly most powerful invariant test exists, then the maximum likelihood ratio test can be no better; it may be worse.

The advantage of the maximum likelihood ratio test, by its construction, lies in reducing otherwise irreducible problems to a single test statistic. Thus, in Example 5.23 it is easily shown that $e^{\frac{1}{2}W}$ is proportional to $|SS_1|^{n_1} |SS_2|^{n_2}/|SS_1 + SS_2|^{n_1 + n_2}$, a single function of the invariant eigenvalues. Here W simultaneously examines differences in concentration, shape and rotation of the two multivariate populations. For a discussion of W and other invariant test statistics in this particular problem, see Pillai and Jayachandran (1968).

A method of reduction widely used in multivariate analysis is Roy's union-intersection method (Roy, 1953; Roy, Gnanadesikan and Srivastava, 1971), which in essence is designed to generalize optimal univariate tests while retaining invariance. The idea is most conveniently described for the one-way analysis of variance for multivariate data.

Example 5.24. Multivariate one-way analysis of variance. Suppose that $Y_{j1}, \ldots, Y_{jr}$ are i.i.d. in $\mathrm{MN}_p(\mu_j, \Sigma)$ independently for each $j = 1, \ldots, m$. Let the null hypothesis be H_0: $\mu_1 = \ldots = \mu_m$, with alternative that the means are not all equal, the covariance matrix Σ being unrestricted throughout. This is the multivariate extension of Example 5.20, for which the F test was uniformly most powerful invariant. Thus for any linear combination $l^T Y$, where l is a $p \times 1$ vector of constants, there exists a uniformly most powerful invariant test of $H_0^{(l)}$: $l^T\mu_1 = \ldots = l^T\mu_m = 0$, which is the projected version of H_0. The sufficient statistic for $(\mu_1, \ldots, \mu_m, \Sigma)$ is $(\bar{Y}_{1.}, \ldots, \bar{Y}_{m.}, SS_w)$, where SS_w is the pooled within-samples cross-product matrix $\Sigma\Sigma(Y_{jk} - \bar{Y}_{j.})(Y_{jk} - \bar{Y}_{j.})^T$.

The univariate invariant test of H_0 using the $l^T Y_{jk}$ is based on the statistic

$$F^{(l)} = \frac{nl^{\mathrm{T}}\{\Sigma(Y_{j.} - Y_{..})(Y_{j.} - Y_{..})^{\mathrm{T}}\}l}{l^{\mathrm{T}}\mathrm{SS}_w l} = \frac{l^{\mathrm{T}}\mathrm{SS}_b l}{l^{\mathrm{T}}\mathrm{SS}_w l},$$

say. The idea of Roy's method, then, is that a multivariate critical region for testing H_0 should be the union over all l of critical regions determined by $F^{(l)}$. Thus in effect we use as test statistic the maximum possible $F^{(l)}$. By imposing $l^{\mathrm{T}}\mathrm{SS}_w l = 1$ as a normalizing constraint on the length of l, it is easy to show that the maximum of $F^{(l)}$ over all l is the largest eigenvalue of $\mathrm{SS}_b \mathrm{SS}_w^{-1}$.

Since the testing problem is invariant under transformations

$$gy = \mathbf{d}y + k$$

for non-singular matrices $\mathbf{d}$ and arbitrary vectors k, it follows that the eigenvalues of $\mathrm{SS}_b \mathrm{SS}_w^{-1}$ are maximal invariant. Thus Roy's method gives one of several invariant tests. Another invariant test is the maximum likelihood ratio with statistic

$$e^{\frac{1}{2}W} = |\mathbf{I} + \mathrm{SS}_b \mathrm{SS}_w^{-1}|,$$

which involves all the eigenvalues of $\mathrm{SS}_b \mathrm{SS}_w^{-1}$. A third statistic is the unweighted sum of eigenvalues, i.e. the trace of $\mathrm{SS}_b \mathrm{SS}_w^{-1}$, which puts equal emphasis on several directional statistics $F^{(l)}$. For a discussion of these statistics, see Roy, Gnanadesikan and Srivastava (1971).

One practical advantage of Roy's method is that the vector l which maximizes $F^{(l)}$ is the eigenvector corresponding to the largest eigenvalue, and is therefore a descriptively valuable by-product of the calculation of the test statistic. Further eigenvectors and eigenvalues are also useful descriptively, and would not be available if W alone were calculated. Note that when $m = 2$ there is only one non-zero eigenvalue of $\mathrm{SS}_b \mathrm{SS}_w^{-1}$ and this is proportional to Hotelling's statistic, the uniformly most powerful invariant and maximum likelihood ratio test statistic.

Bibliographic notes

For general references, see the notes to Chapters 2 and 4.

The theory of similar tests is quite closely tied to that of the exponential family; see Lehmann (1959) and, for more mathematical accounts, Barndorff-Nielsen (1973a) and Linnik (1968).

For further development of the binary regression model of

Example 5.3, see Cox (1970). Rasch (1960) has done extensive work on models of this general type with reference, in particular, to item analysis. The two-sample problem of Section 5.2(iv) has an extensive literature. The treatment given here follows Welch (1947a); see also Welch (1947b, 1951, 1956), James (1956, 1959) and Wijsman (1958). An alternative analysis *via* the idea of fiducial probability leads to the Behrens-Fisher test which is, however, not similar; see also Chapter 7 and Example 10.5. For a numerical comparison of different procedures, see Mehta and Srinivasan (1970), and for a more theoretical discussion Pfanzagl (1974).

Introductory accounts of invariance theory are given by Lehmann (1959) and Zacks (1971).

Further results and exercises

1. The random variables $Y_1, \ldots, Y_{n_1}$ are i.i.d. in a geometric distribution of parameter θ_1 and independently $Y_{n_1+1}, \ldots, Y_{n_1+n_2}$ are i.i.d. in a geometric distribution of parameter θ_2. Obtain a uniformly most powerful similar test of $\theta_1 = \theta_2$ versus alternative $\theta_1 > \theta_2$. What more general null hypothesis can be tested similarly? Suggest a test of the adequacy of the model.

[Section 5.2]

2. Events occur in a time-dependent Poisson process of rate $\rho e^{\beta t}$, where ρ and β are unknown. The instants of occurrence of events are observed for the interval $(0, t_0)$. Obtain a uniformly most powerful similar test of $\beta = \beta_0$ versus alternatives $\beta > \beta_0$. Examine in more detail the form of the test in the special case $\beta_0 = 0$.

[Section 5.2; Cox and Lewis, 1966, Chapter 3]

3. There are available m independent normal-theory estimates of variance, each with d degrees of freedom. The corresponding parameters are $\sigma_1^2, \ldots, \sigma_m^2$. If $\sigma_j^2 = 1/(\lambda + \psi j)$ $(j = 1, \ldots, m)$, obtain a uniformly most powerful similar test of $\psi = 0$ versus alternatives $\psi > 0$. Suggest a simpler, although less efficient, procedure based on the log transformation of the variances.

[Section 5.2]

4. Observations are obtained on a two-state Markov chain for times $0, \ldots, n$; the notation of Example 2.5 is used, and the state i_0 at time

zero is regarded as fixed. Prove that if $\log(\theta_{01}/\theta_{00}) = \gamma$ and $\log(\theta_{11}/\theta_{10}) = \gamma + \delta$ then the likelihood is

$$\frac{\exp(\gamma m_{.1} + \delta m_{11})}{(1 + e^{\gamma})^{m_{0.}} \cdot (1 + e^{\gamma + \delta})^{m_{1.}}},$$

not quite of the exponential family form leading to a uniformly most powerful similar test of the null hypothesis $\delta = \delta_0$. Show further, however, that the conditional distribution of M_{11} given the number r of occupancies of state one, and both the initial and final states, i_0 and i_n, is

$$\frac{c_{i_0 i_n}(r, m_{11}) e^{\delta m_{11}}}{\Sigma c_{i_0 i_n}(r, t) e^{\delta t}},$$

where the combinatorial coefficient in the numerator is the number of distinct binary sequences with the given initial and final states, with r ones and $n + 1 - r$ zeroes and with the required m_{11}.

[Section 5.2; Cox, 1970, Section 5.7]

5. Let $Y_1, \ldots, Y_n$ be i.i.d. in an exponential distribution with lower terminal θ_1 and mean $\theta_1 + \theta_2$. Construct appropriate similar or invariant tests of the hypotheses (a) $H_0: \theta_1 = 0$ against $H_A: \theta_1 > 0$, (b) $H_0: \theta_2 = 1$ against $H_A: \theta_2 \neq 1$.

Examine the corresponding two-sample homogeneity tests.

[Section 5.2, 5.3]

6. Let $Y_1, \ldots, Y_n$ be independent random variables such that some unknown permutation of them have gamma distributions whose densities

$$\rho_j(\rho_j y)^{\beta - 1} e^{-\rho_j y} / \Gamma(\beta)$$

depend on two unknown parameters ψ and λ through the regression relationship $\rho_j = \exp(\lambda + \psi z_j)$, where the z_j's are known. Verify that the problem of testing $H_0: \psi = 0$ against the alternative $\psi \neq 0$ is invariant under permutation and scale change of the observations, and that a maximal invariant is $(Y_{(1)}/Y_{(n)}, \ldots, Y_{(n-1)}/Y_{(n)})$. Show that the locally most powerful invariant test has rejection regions determined by large values of $(\Sigma Y_j^2)/(\Sigma Y_j)^2$.

[Section 5.3; Ferguson, 1961]

7. Let Y_{jk} $(j = 1, \ldots, m; k = 1, \ldots, r)$ follow a normal-theory components of variance model, i.e. $Y_{jk} = \mu + \eta_j + \epsilon_{jk}$, where the η_j's and ϵ_{jk}'s are independently normally distributed with zero mean and with variances respectively σ_b^2 and σ_w^2. Show that the minimal sufficient statistic is $(\bar{Y}_{..}, SS_b, SS_w)$ where $SS_b = r\Sigma(\bar{Y}_{j.} - \bar{Y}_{..})^2$ and $SS_w = \Sigma(Y_{jk} - \bar{Y}_{j.})^2$. Two independent sets of data of this structure are available with the same values of m and r. It is required to test the null hypothesis that the ratio σ_b^2/σ_w^2 is the same for both sets, all other parameters being arbitrary. Formulate a relevant group of transformations for the application of invariance theory and by a reduction to the minimal sufficient statistic, followed by the calculation of the maximal invariant, show that the appropriate test statistic is the ratio of the two values of SS_b/SS_w. Suggest a simple approximation to the null distribution of the test statistic, based on a log transformation. For what kind of practical situation may this problem be relevant?

[Section 5.3; Dar, 1962]

8. Suppose that $Y_1, \ldots, Y_n$ are i.i.d. random variables either in $N(\mu, \sigma^2)$ or in the uniform distribution on $(\mu - \frac{1}{2}\sigma, \mu + \frac{1}{2}\sigma)$, the parameters being unknown. Show that the most powerful test between these two hypothetical distributions which is invariant under location and scale changes has test statistic $(Y_{(n)} - Y_{(1)})/\{\Sigma(Y_j - \bar{Y}_.)^2\}^{\frac{1}{2}}$.

[Section 5.3; Uthoff, 1970]

9. Observations are taken independently from a multivariate normal distribution with unknown mean and covariance matrix. The null hypothesis H_0 asserts that the first q coordinates are independent of the remaining $p - q$ coordinates, the alternative hypothesis being completely general. Find a suitably wide group of transformations such that the maximal invariant statistic is the set of canonical correlations. That is, if the sample cross-product matrix SS is partitioned into $SS_{11}, SS_{12}, SS_{21}$ and SS_{22} corresponding to the first q and last $p - q$ coordinates, then the maximal invariant is the set of solutions of $|SS_{11} - l\,SS_{12}\,SS_{22}^{-1}\,SS_{21}| = 0$. What function of the maximal invariant is used in the maximum likelihood ratio test?

[Sections 5.3, 5.4; Anderson, 1958, Chapter 12]

10. The sample space of the bivariate random variable Y consists of two concentric circles and their centre point. Under the null hypothesis, H_0, the centre point has known probability p_0 and the inner and outer circles have total probabilities respectively $1 - 2p_0$ and p_0, distributions being uniform over each circle. Under the composite alternative hypothesis, the centre point has known probability $p_A > p_0$, the inner circle has zero probability and there is an arbitrary bounded distribution over the outer circle. The problem is thus invariant under rotation of the plane about the centre point. Find the maximal invariant statistic and hence describe the uniformly most powerful invariant test of H_0. Compare this test with that based on the maximum likelihood ratio.

[Sections 5.3, 5.4; Lehmann, 1959, Section 6.12]

11. Let $Y_1, \ldots, Y_{n_1}$ be i.i.d. in $MN_p(\mu, \Sigma)$ and $Y_1', \ldots, Y_{n_2}'$ be i.i.d. in $MN_p(\mu', \Sigma')$. Derive an invariant test of the hypothesis $H_0 : \Sigma = \Sigma'$ against the general alternative by finding the most significant of the test statistics $F(a)$ used in testing the corresponding hypotheses about variances of the scalar combinations $a^T Y$ and $a^T Y'$.

[Section 5.4]

6 DISTRIBUTION-FREE AND RANDOMIZATION TESTS

6.1 General

In previous Chapters we have considered tests in which the distribution under the null hypothesis is either completely specified or is given except for a finite number of unknown parameters. In this Chapter we deal primarily with situations in which the null hypothesis involves, explicitly or implicitly, arbitrary and usually unknown densities. The words distribution-free and nonparametric are used broadly for the resulting techniques; in fact, however, there are a number of distinct ideas involved. Clarification is hindered by the fact that many procedures can be regarded from several points of view.

Suppose first that the null hypothesis is that the random variables $Y_1, \ldots, Y_n$ are i.i.d. in some unknown density. The type of departure to be tested may, for example, be (a) trend with serial order, (b) dependence on some explanatory variable, (c) stochastic difference between observations $1, \ldots, n_1$ and observations $n_1 + 1, \ldots, n$.

In this situation we call a test *distribution-free* if the distribution of the test statistic under the null hypothesis is the same for some family of densities more general than a finite parameter family, e.g. if it is the same for all continuous densities. By the arguments of Sections 3.3 and 5.2, this can be achieved only by conditioning on the complete minimal sufficient statistic, that is by regarding the order statistics as fixed and using the consequence that under the null hypothesis all permutations of the ordered values are equally likely. For this reason such tests are often called *permutation tests*.

Next, we contrast tests that use the values of the observations as recorded on some scale, with tests that use only ranks, i.e. the serial numbers of the observations when the observations are arranged in increasing order. These latter are called *rank tests*. In some applications only the ranks may be available; in other cases it may be reasonable to look for a test invariant under arbitrary monotonic changes of the scale of measurement. Such a very strong invariance requirement dictates the use of rank tests.

In obtaining test statistics, whether for rank tests or not, we may follow the approach of Chapter 3, leading to a pure significance test. That is, we simply define a test statistic that is a reasonable indicator of the type of departure we want to detect. Alternatively, we may formulate parametric alternatives against which we want high power and then find that distribution-free test with maximum power against the particular alternatives. The purpose of the parametric formulation is here purely to indicate a suitable test statistic; it plays no part in determining the distribution under the null hypothesis.

Next, in Section 6.4, we consider tests whose justification lies in the randomization used in allocating treatments to experimental units in an experiment. We call such tests *randomization tests*. Formally such tests are the same as permutation tests, but the conceptual basis is quite different.

Finally, it is convenient to consider in Section 6.5 tests derived by a quite different kind of argument. Here, in the simplest case, the null hypothesis specifies the distribution of a set of i.i.d. random variables. To test that the distribution has the specified form, we define a distance measure between two distribution functions and then proceed to estimate the distance between the distribution function of the sample and the postulated form. Such procedures are called *distance tests*. In some cases the null hypothesis distribution of the distance test statistic may be independent of the particular null hypothesis under test; that is why some of the previous distribution-free tests can be regarded also as distance tests.

The arguments in favour of distribution-free tests are clear; it is always appealing to relax assumptions and because, for example, we can test for trend without postulating the functional form of the density under the null hypothesis it might seem that we should as a rule do this, avoiding an apparently unnecessary assumption. While distribution-free procedures certainly have a role in situations where careful testing of significance is of primary importance and no parametric assumption is strongly indicated, there are a number of considerations that limit the practical importance of such techniques. These are as follows:

(a) If in fact a formulation in terms of a parametric model, i.e. one involving distributions known except for a finite number of parameters, is appropriate, there will be some loss of efficiency or power in using a distribution-free test instead of the most efficient parametric test. On the whole, however, at least in simple common problems, this loss of efficiency is quite small, provided that the best distribution-free test is used, so that this is not so strong an argument against distribution-free tests as might be thought.

(b) While quite simple distribution-free tests are available in fairly unstructured problems, such questions as testing the parallelism of regression lines, the absence of interactions and the adequacy of non-linear regression models, are not so readily tackled.

(c) On the whole, provided that data are screened for outliers, the results of distribution-free tests are not often very different from those of analogous parametric tests. Therefore the extra work that is often involved in setting up a distribution-free procedure is frequently not really justified. This applies particularly to the calculation of distribution-free interval estimates, which we shall see in Chapter 7 follow in principle fairly directly from the tests to be discussed in the present Chapter.

(d) The main emphasis in distribution-free tests is on avoiding assumptions about distributional form. In many applications, however, the most critical assumptions are those of independence.

(e) Most importantly, the main objective in analysing data, especially fairly complicated data, is to achieve a description and understanding of the system under investigation in concise and simple terms. This nearly always involves a model having as small a number of unknown parameters as possible, each parameter describing some important aspect of the system. That is, at least in complex problems, the demands of economy in final description point towards parametric specification. Note, however, that even when a full formulation in terms of a limited number of unknown parameters is used, consideration of the possible effect of outliers will always be required. Either the data must be screened for possible aberrant values, or so-called robust techniques used. For the latter, see Sections 8.6 and 9.4; in fact, some useful robust techniques are based on tests of the type to be considered in the present Chapter.

When, however, great importance attaches to the cautious testing of significance, for example of the difference between two groups, use of distribution-free methods deserves careful consideration.

6.2   Permutation tests

Suppose that $y_1, \ldots, y_n$ are scalar observations. Under the null hypothesis, H_0, the corresponding random variables $Y_1, \ldots, Y_n$ are i.i.d. with an unknown p.d.f. $f(.)$. Some type of departure from the null hypothesis is defined, such as that $Y_1, \ldots, Y_{n_1}$ are i.i.d. with a different density from the i.i.d. set $Y_{n_1+1}, \ldots, Y_n$. Under H_0, the order statistics $Y_{(1)}, \ldots, Y_{(n)}$ are sufficient for $f(.)$. Therefore similar region tests of H_0 can be obtained by conditioning on the observed order statistics, i.e. on the event $Y_{(.)} = y_{(.)}$. Under H_0, all $n!$ permutations of y are equally likely; see Example 2.11.

That is, if we have chosen a test statistic $T = t(Y) = t(Y_1, \ldots, Y_n)$, then for any permutation $(i_1, \ldots, i_n)$ of $(1, \ldots, n)$, we have that

$$\mathrm{pr}\{T = t(y_{i_1}, \ldots, y_{i_n}) | Y_{(.)} = y_{(.)}\} = \frac{1}{n!}, \qquad (1)$$

provided that the $n!$ values of T are all distinct, there being an obvious modification if the values of the T are not all different. This is the basis of all permutation tests, there being an extension to the multivariate case.

The simplest way to choose a test statistic is often to take the statistic used in some analogous finite parameter problem. This is illustrated by the two-sample location problem.

Example 6.1. Permutation t test. Suppose that the alternative to homogeneity is that $Y_1, \ldots, Y_{n_1}$ have p.d.f. $f(y - \theta)$, whereas $Y_{n_1+1}, \ldots, Y_n$ have p.d.f. $f(y)$, the null hypothesis therefore being as before, i.e. that $\theta = 0$. The function $f(.)$ is unknown; we write $n_2 = n - n_1$. We take for simplicity the one-sided case, where we wish to test that the first set of observations tends to be larger than the second, i.e. that $\theta > 0$.

The optimal normal-theory test statistic for the problem is the Student t statistic

$$T = (\bar{Y}_1 - \bar{Y}_2) \Big/ \left[\frac{n}{n_1 n_2 (n-2)} \{(n_1 - 1)\mathrm{MS}_1 + (n_2 - 1)\mathrm{MS}_2\} \right]^{\frac{1}{2}}, \quad (2)$$

where $(\bar{Y}_1, \bar{Y}_2)$ are the means and $(\mathrm{MS}_1, \mathrm{MS}_2)$ are the usual estimates of variance. Large values of t are evidence against H_0. We therefore calculate the permutation significance level corresponding to t_{obs} by

$$p_{\text{obs}} = \text{pr}\{t(Y) \geq t_{\text{obs}} \,|\, Y_{(.)} = y_{(.)}; H_0\} = \frac{k(y)}{n!}, \qquad (3)$$

where $k(y)$ is the number of permutations of $\{1, \ldots, n\}$ giving values of the test statistic greater than or equal to the observed value.

Two important simplifications are possible. First, only distinct pairs of samples need be considered, so that there are only $n!/(n_1!n_2!)$ distinct possibilities to enumerate rather than $n!$, and a corresponding simplification of $k(y)$ is possible. Secondly, the test statistic itself can be drastically simplified. Given $Y_{(.)} = y_{(.)}$ we have that

$$n_1 \bar{Y}_1 + n_2 \bar{Y}_2 = y_{(1)} + \ldots + y_{(n)} = \text{const},$$

$$(n_1 - 1)\text{MS}_1 + (n_2 - 1)\text{MS}_2 + n_1 \bar{Y}_1^2 + n_2 \bar{Y}_2^2 = y_{(1)}^2 + \ldots + y_{(n)}^2 = \text{const},$$

so that T is a monotonic function of $\bar{Y}_1$, or equivalently of $Y_1 + \ldots + Y_{n_1}$. Thus to evaluate (3) we have only to find the proportion of all permutations, or equivalently of all samples, giving $\bar{Y}_1$ greater than or equal to its observed value. This feature, that because of the conditioning familiar test statistics take on a much simpler form than usual, arises quite generally.

Even with the above simplifications, enumeration of all distinct samples to determine (3) is tedious or impracticable when n_1 and n_2 are appreciable. Then we look for approximate versions of the permutation distribution, usually found by calculating the moments of the test statistic in the permutation distribution and then approximating by some convenient distribution with the same moments. An alternative is computer sampling of the permutation distribution until the significance level is estimated with adequate precision.

Calculation of moments is particularly easy in the present case, because under the null hypothesis the random variable $\bar{Y}_1$ is the mean of a random sample of size n_1 drawn randomly without replacement from the finite population $\{y_1, \ldots, y_n\}$. The moments have been evaluated, in fact up to a high order (Kendall and Stuart, 1967-69, Vol. 1) and the most relevant results are that

$$E(Y_1) = K_1, \quad \text{var}(Y_1) = a_1 K_2, \quad E(Y_1 - K_1)^3 = (a_2 - 3a_1/n)K_3, \quad (4)$$

where, in the usual notation of finite sampling theory,

$$a_r = (1/n_1^r - 1/n^r), \quad K_1 = \Sigma y_j/n, \quad K_2 = \Sigma(y_j - K_1)^2/(n-1),$$

$$K_3 = n\Sigma(y_j - K_1)^3/\{(n-1)(n-2)\}.$$

Very often a normal approximation will be adequate; where judged necessary, this can be refined by corrections based on the third and fourth moments, as indicated in Appendix 1.

We shall not discuss in detail the connexion between the permutation significance limits for T and those based on, for example, normal theory. Note, however, that the unconditional distribution for T when Y has any distribution is obtained by averaging the conditional distributions. Thus the permutation significance points are distributed around the corresponding values for the unconditional distribution of T. In large samples this unconditional distribution is approximately standard normal for all $f(.)$ with finite variance. This can be shown to imply that, at least in large samples, there is unlikely to be much difference between the permutation test and a "parametric" procedure.

Example 6.2. Test of linear dependence. Suppose that $(y_1, z_1), \dots, (y_n, z_n)$ are n independent pairs of values. The following analysis applies whether

(a) the data correspond to i.i.d. bivariate random variables $(Y_1, Z_1), \dots, (Y_n, Z_n)$, or

(b) the $z_1, \dots, z_n$ are fixed constants and $Y_1, \dots, Y_n$ are random variables.

The null hypothesis is that in (a) Y_j and Z_j are independent with arbitrary densities and that in (b) the Y_j are i.i.d. and therefore have a distribution not depending on z_j.

If we are interested in testing linear dependence, it is natural to take as the observed test statistic $t = \Sigma y_j z_j$. In case (a) the null hypothesis distribution is formed by conditioning on the order statistics of the y's and on the order statistics of the z's; conditionally on these, all $(n!)^2$ permutations are equally likely. It is easily seen that this is equivalent to fixing, say, the z's and taking the $n!$ permutations of the y's. This is exactly the permutation distribution appropriate for case (b) in which the z's are given constants.

Thus the null hypothesis distribution of the test statistic is that of a random variable

$$T = \Sigma y_{I_j} z_j,$$

where $(I_1, \dots, I_n)$ is a random permutation of $(1, \dots, n)$. In particular, we have that for $j, k = 1, \dots, n$ and for $j_1 \neq j_2, k_1 \neq k_2 = 1, \dots, n$

$$\text{pr}(I_j = k) = \frac{1}{n}, \quad \text{pr}(I_{j_1} = k_1, I_{j_2} = k_2) = \frac{1}{n(n-1)}, \quad (5)$$

etc. The permutation moments of T can now be calculated fairly easily and are

$$E(T) = \frac{1}{n}\Sigma y_j \Sigma z_j, \quad \text{var}(T) = \frac{1}{n-1}\Sigma(y_j - \bar{y}.)^2 \Sigma(z_j - \bar{z}.)^2,$$
$$(6)$$
$$E\{T - E(T)\}^3 = \frac{n^2}{(n-1)(n^2 - 2n + 2)}\Sigma(y_j - \bar{y}.)^3 \Sigma(z_j - \bar{z}.)^3.$$

There are several ways of calculating these moments. The most direct is to use the expression for T in terms of the permutation random variables $I_1, \ldots, I_n$ together with the properties (5) and their higher order analogues. An approach involving less detailed calculation hinges on the symmetry and algebraic structure of the permutation expectations. Thus $E(T)$ is linear in the y's and in the z's and is invariant under an arbitrary permutation and is symmetric; it therefore must be of the form $a_n \Sigma y_j \Sigma z_j$, where a_n is a constant. Consideration of the special case when all y's and z's are equal to one shows that $a_n = n^{-1}$. A similar argument establishes that

$$\text{var}(T) = b_n \Sigma(y_j - \bar{y}.)^2 \Sigma(z_j - \bar{z}.)^2, \quad (7)$$

where b_n is a constant. Any special case can be used to determine b_n. The simplest approach is to take $y_1 = z_1 = 1$, $y_j = z_j = 0(j = 2, \ldots, n)$. Then $\text{pr}(T = 1) = 1/n$, $\text{pr}(T = 0) = 1 - 1/n$, from which $\text{var}(T)$, $\Sigma(y_n - \bar{y}.)^2$, $\Sigma(z_n - \bar{z}.)^2$ are easily determined and substituted in (7) to prove that $b_n = (n-1)^{-1}$.

It can be shown, under weak assumptions about the y's and the z's, that the permutation distribution of T is approximately normal for large values of n.

Note that in the conditional distribution considered here the test statistic is equivalent to the sample correlation coefficient or to either sample regression coefficient; we have chosen the simplest form. For comparison with the normal-theory test of linear dependence, which can of course be expressed equivalently in terms of any of the above three statistics, we choose the correlation coefficient

$$r = \frac{\Sigma y_j z_j - n\bar{y}.\bar{z}.}{\{\Sigma(y_j - \bar{y}.)^2 \Sigma(z_j - \bar{z}.)^2\}^{\frac{1}{2}}}.$$

Thus the permutation moments of R are, from (6),

$$0, \frac{1}{n-1} \text{ and } \frac{n^2 \Sigma(y_j - \bar{y}.)^3 \Sigma(z_j - \bar{z}.)^3}{(n-1)(n^2 - 2n + 2)\{\Sigma(y_j - \bar{y}.)^2 \Sigma(z_j - \bar{z}.)^2\}^{\frac{3}{2}}} \sim \frac{g_{1y} g_{1z}}{n^2},$$

$$(8)$$

where g_{1y} and g_{1z} are standardized third moments formed from the y's and the z's. For comparison, the normal-theory moments are

$$0, (n-1)^{-1} \text{ and } 0. \tag{9}$$

Note that the standardized third moment ratio for the permutation distribution is, for large n, $g_{1y} g_{1z}/\sqrt{n}$.

Comparison between (8) and (9) illustrates the connexion between permutation distributions and those arising directly from the original specification in terms of i.i.d. random variables, i.e. the unconditional or "infinite model" distributions. If in fact one or both variables y and z have symmetrical sample arrangements, the agreement is exact for the first three moments; of course, a more refined discussion would bring in the fourth moments.

If both Y and Z have normal distributions, g_{1y} and g_{1z} are of order $n^{-\frac{1}{2}}$ in probability and approximately symmetrically distributed. This not only confirms that the permutation distribution averages to the "infinite model" distribution, but shows that, because the permutation standardized third moment is of order $n^{-\frac{3}{2}}$, the permutation distribution is likely to be very nearly symmetrical. If, however, the distributions of both Y and Z are skew, both g_{1y} and g_{1z} will be of order one and the standardized third moment of the permutation distribution will typically be of order $n^{-\frac{1}{2}}$. That is, when we attempt more than the crudest normal approximation to the distribution, the permutation distribution will give different results from normal theory if both Y and Z have unsymmetrical distributions.

So far we have chosen test statistics either by the "pure significance test" approach of Chapter 3 or by analogy with parametric theory, in the latter case aiming at good power properties if in fact the parametric assumptions hold. Another way of looking at test statistics is that they are based on empirical measures of departure from H_0 towards some alternative. This idea may be used more explicitly, as in Section 6.5 and Examples 3.2 and 3.3.

Note that while permutation tests are constructed to be unaffected by changes in the functional form of distributions when the null hypothesis is true, we have left entirely open the question of behaviour under the alternative hypothesis. In the next section we discuss this in connexion with rank tests.

6.3   Rank Tests

(i) General remarks

An important particular class of permutation test are those defined in terms of the sample ranks $(r_1, \ldots, r_n)$, where $y_j = y_{(r_j)}$; thus $r_j = 1$ if y_j is the smallest observation, etc. An important property of the vector r is that it is the maximal invariant (Section 5.3(i)) under monotonic transformations of the original y. Statistics that involve the observations only through r are therefore particularly appropriate if the scale of measurement is so arbitrarily defined that it does little more than define the ordering of the different values.

If $Y_1, \ldots, Y_n$ are i.i.d. with an arbitrary continuous density and if the possibility of ties, i.e. exact equality of observations, can be disregarded, then the rank vector R and the order statistics $Y_{(.)}$ are independent with

$$f_R(r) = \frac{1}{n!}, \tag{10}$$

where r is any permutation of $(1, \ldots, n)$. Thus, if in (1) $t(Y) = t(R)$, then the statement of the condition $Y_{(.)} = y_{(.)}$ is redundant. The problem of ties will be ignored in the rest of the discussion, except for some brief comments in Section 6.3(v); the complications they introduce are minor.

Rank tests are concerned with essentially comparative problems. In the simplest case the null hypothesis H_0 is that $Y_1, \ldots, Y_n$ are i.i.d. with some arbitrary and unknown density. The possible departures from H_0 include multi-sample differences, y being the overall sample, and also regression-like dependencies on an explanatory variable $z_1, \ldots, z_n$. We exclude such problems as single-sample tests of location for which information additional to r is essential; a single-sample test is described in Exercise 6.1.

We start with the comparison of two samples, a problem which best illustrates the essential features of rank tests.

(ii) Two-sample location tests

Suppose that we have two sets of data $y_1, \ldots, y_{n_1}$ and $y_{n_1+1}, \ldots, y_n$, which we suspect may be from distributions differing in location. The null hypothesis is again that $Y_1, \ldots, Y_n$ are i.i.d. with an unknown p.d.f. For simplicity we consider the one-sided case in which the first set of observations is suspected of being the larger. We therefore expect under the alternative hypothesis that in the ranking of the whole data the ranks $r_1, \ldots, r_{n_1}$ of the observations in the first set will tend to be large. One possibility is to take $R_1 + \ldots + R_{n_1}$ as test statistic, large values to be significant. Because the components of the vector R are fixed, this is equivalent to the two-sample Student t statistic calculated from the ranks; see Example 6.1.

Obviously many other choices of test statistic are possible, and to get a more formal method of obtaining a test statistic, we aim for good power properties when the densities for the two sets of data are $g(y - \theta)$ and $g(y)$, where now $g(.)$ will affect the power properties but not the size of the critical region under the null hypothesis that the full set of random variables are i.i.d. We continue to look at the one-sided case, $\theta > 0$.

Now for general values of θ the distribution of the ranks is

$$f_R(r; \theta) = \int_{\{y_j = y_{(r_j)}\}} \prod_{k=1}^{n_1} g(y_k - \theta) \prod_{l=n_1+1}^{n} g(y_l) dy. \qquad (11)$$

Write $x_{r_j} = y_{(r_j)}$ $(j = 1, \ldots, n)$; then (11) becomes

$$f_R(r; \theta) = \int_{x_1 < \ldots < x_n} \left\{ \prod_{j=1}^{n} g(x_{r_j}) \right\} \left\{ \prod_{k=1}^{n_1} \frac{g(x_{r_j} - \theta)}{g(x_{r_j})} \right\} dx.$$

But because $n! \prod_{j=1}^{n} g(x_{r_j})$ is the p.d.f. of $X_{(1)}, \ldots, X_{(n)}$ when $X_1, \ldots, X_n$ are i.i.d. with density $g(x)$, we have that

$$f_R(r; \theta) = \frac{1}{n!} E \left\{ \prod_{j=1}^{n_1} \frac{g(Y_{(r_j)} - \theta)}{g(Y_{(r_j)})} ; 0 \right\}. \qquad (12)$$

From (12) we can obtain the optimal rank test for any specified alternative θ by calculating the likelihood ratio. In general, there is no uniformly most powerful test and one possibility is therefore to use the locally most powerful procedure; see Section 4.8. Thus the critical regions are formed from large values of

$$\left[\frac{\partial f_R(r; \theta)}{\partial \theta} \right]_{\theta = 0}.$$

This leads to the statistic

$$t(R) = -\sum_{j=1}^{n_1} E\left\{\frac{g'(X_{(R_j)})}{g(X_{(R_j)})}; 0\right\} \tag{13}$$

$$= \sum_{k=1}^{n_1} w_{gn}(R_k),$$

where

$$w_{gn}(j) = -E\left\{\frac{g'(Y_{(j)})}{g(Y_{(j)})}; 0\right\} \quad (j = 1, \ldots, n). \tag{14}$$

A convenient approximation to (14) is obtained from the fact that $E\{G(Y_{(j)}); 0\} = j/(n + 1)$ and that for large n, $\mathrm{var}\{G(Y_{(j)}); 0\} = O(n^{-2})$; see Appendix 2. This leads to

$$w_{gn}(j) \simeq w_{gn}^*(j) = -\frac{g'\left\{G^{-1}\left(\dfrac{j}{n+1}\right)\right\}}{g\left\{G^{-1}\left(\dfrac{j}{n+1}\right)\right\}}, \tag{15}$$

which is often more easily calculated than (14).

The null distribution does not depend on the particular density of the observations; the purpose of $g(.)$ is to determine a suitable test statistic. It is therefore of interest to examine how a particular $t(R)$ behaves for different alternatives, such as the normal, Cauchy, logistic and double exponential. The theory of a method for doing this will be given in (iii).

One important special case of (14) is when $g(.)$ is the standard normal density. Then

$$w_{gn}(j) = E(Y_{(j)}),$$

the expected value of the jth largest observation out of n from a standard normal distribution; see Appendix 2.

Example 6.3. Wilcoxon test. To derive the weights for a logistic distribution we take

$$g(y) = \frac{e^y}{(1 + e^y)^2},$$

when the expression (14) gives $w_{gn}(j) = -1 + 2j/(n + 1)$. Linear transformations of the test statistic being irrelevant, we can take the statistic to be $R_1 + \ldots + R_{n_1}$, which was earlier suggested as, in effect, the Student t statistic calculated directly from the ranks. The resulting test is called the two-sample Wilcoxon or Mann-Whitney statistic.

In principle, the exact distribution of the general test statistic (13) under the null hypothesis is found by enumerating all permutations. The moments of the test statistic can be found from the theory of sampling a finite population without replacement and, in particular,

$$E(T; H_0) = \frac{n_1}{n} \sum_{j=1}^{n} w_{gn}(j) = \bar{w},$$

(16)

$$\text{var}(T; H_0) = \frac{n_1 n_2}{n(n-1)} \Sigma \{w_{gn}(j) - \bar{w}\}^2.$$

This suggests the use of a normal approximation based on

$$\frac{T - E(T; H_0)}{\sqrt{\text{var}(T; H_0)}}$$

(17)

with a continuity correction when, as for the Wilcoxon statistic, the set of achievable values has a known spacing. A formal proof of limiting normality is sketched in (iv).

Problems in which it is required to obtain maximum power for scale changes of distribution of known location can be tackled by (14), after logarithmic transformation.

Example 6.4. Comparison of exponential means. Suppose that maximum power is required in a two-sample rank test when the two samples correspond to exponential distributions with different means. Now if Y has an exponential distribution of mean μ, $U = \log Y$ has the extreme value density

$$\exp(u - \gamma - e^{u - \gamma}),$$

(18)

where $\gamma = \log \mu$. The rank test based on the observed U's is identical with that based on the observed Y's, so that we can apply (14) with $g(.)$ given by (18). We have that $g'(u)/g(u) = -e^u = -y$, so that we base the test on the scores determined from the expected order statistics in sampling the exponential distribution of unit mean. In fact if $Y_{(1)}, \ldots, Y_{(n)}$ are the order statistics corresponding to random variables i.i.d. with density e^{-y}, it is shown in (A2.9) that

$$E(Y_{(j)}) = \frac{1}{n} + \ldots + \frac{1}{n - j + 1}.$$

(19)

Thus the test consists in finding the ranks of the observations in the first set, assigning each its score from (19) and using the total score as test statistic, large values being evidence that the first sample has the larger mean.

The general formulae (16) and (17) can be used to approximate to the null hypothesis distribution; a better approximation is obtained by noting that the efficient parametric test is based on the ratio of the sample means, which has an F distribution under the null hypothesis. The rank test is equivalent to one using the ratio of the sample mean scores and an F distribution with adjusted degrees of freedom can be used to approximate to its distribution.

(iii) More complicated problems

The discussion of two-sample location rank tests can, in principle, be extended to more complicated problems such as multi-sample comparisons, regression relationships, and the detection of serial correlation. The tests take the form that there are a set of constant "scores" $w(1), \ldots, w(n)$ and the observations $y_1, \ldots, y_n$ are replaced first by ranks $r_1, \ldots, r_n$ and then by the scores $w(r_1), \ldots, w(r_n)$. We then use these scores to obtain a distribution-free rank test of the null hypothesis having good power for the alternatives in question. Often we arrive at the more general form

$$t(R, c) = \sum_{j=1}^{n} c_j w(R_j), \qquad (20)$$

with $c_1, \ldots, c_n$ depending on the particular nature of the alternative.

Corresponding to the normal approximation (17) in the two-sample problem we can, under mild conditions on the c's, treat linear combinations such as (20) as approximately normal for large n. This induces normal and chi-squared approximations for the distributions of most rank test statistics arising in applications.

Example 6.5. Rank analysis of variance. Suppose that we have m sets of data $(y_1, \ldots, y_{n_1}), (y_{n_1}, \ldots, y_{n_1 + n_2}), \ldots$ with sample sizes $n_1, \ldots, n_m$, the total number of observations being $n = \Sigma n_l$. A one-way analysis of variance can be defined based on the rank scores by setting

$$T_l = t_l(R) = \frac{1}{n_l} \sum_{j = n_1 + \ldots + n_{l-1} + 1}^{n_1 + \ldots + n_l} w(R_j) \quad (l = 1, \ldots, m),$$

$$\bar{T}. = \frac{1}{n} \Sigma n_l t_l(R) = \frac{1}{n} \sum_{j=1}^{n} w(j) = \bar{w}.,$$

$$b(R) = \Sigma n_l \{t_l(R) - \overline{T}_.\}^2.$$

Here we are using m linear rank statistics, subject to one linear constraint. The large-sample normal approximation to the null hypothesis distribution of $t_l(R)$, valid for large n_l's, involves the moments

$$E\{t_l(R); H_0\} = \overline{w}_., \operatorname{var}\{t_l(R); H_0\} = \frac{(n - n_l)}{nn_l} \Sigma\{w(j) - \overline{w}_.\}^2/(n-1);$$

these results are easily derived from the formulae for random sampling from a finite population. It follows from standard results for quadratic forms that

$$(n-1)b(R)/\Sigma\{w(j) - \overline{w}_.\}^2$$

has a large-sample chi-squared distribution with $m-1$ degrees of freedom.

(iv) Large-sample properties of linear rank tests

We now discuss briefly the distribution of linear rank statistics, both under the null hypothesis and under alternatives. In particular, we derive large-sample approximations. Large-sample theory in general is treated in Chapter 9, but it is convenient to treat linear rank statistics here, albeit rather sketchily.

The arguments depend on a number of approximations and a careful treatment would involve showing that for sufficiently large values of n the neglected terms are indeed negligible. We do not attempt such a treatment here.

To deal with the statistic (20), we suppose that the weights $w(j)$ are defined by

$$w(j) = E[a\{G(X_{(j)})\}] = E\{a(U_{(j)})\}$$

for some function $a(.)$, supposing $G(.)$ to be the cumulative distribution function of $X_1, \ldots, X_n$, so that $U_1, \ldots, U_n$ are i.i.d. with uniform density over $(0, 1)$. This corresponds exactly to the $w_{gn}(j)$ of (14), and approximately to the $w_{gn}(j)$ of (15) for large n. Now, because

$$E(U_{(j)}) = \frac{j}{n+1}, \quad \operatorname{var}(U_{(j)}) = O\left(\frac{1}{n^2}\right),$$

under the null hypothesis we may use the first-order approximations

$$w(R_j) \simeq a(U_j), \quad w(1 + [un]) \simeq a(u).$$

Therefore the linear rank statistic (20) is given to first order by

$$t(R, c) = \Sigma c_j w(R_j) \simeq \Sigma(c_j - \bar{c}_.)a(U_j) + n\bar{c}_.\bar{w}_., \quad (21)$$

Notice that this holds independently of the distribution for the i.i.d. random variables from which R is derived. We then apply the central limit theorem to (21), under suitable conditions on c and $a(.)$. Under H_0, when (10) holds, the exact permutation moments of $t(R, c)$ are

$$E(T; H_0) = n\bar{c}_.\bar{w}_., \quad \mathrm{var}(T; H_0) = \Sigma(c_j - \bar{c}_.)^2 \Sigma\{w(j) - \bar{w}_.\}^2/(n - 1),$$

but it is sometimes convenient to use for the variance the approximation

$$\Sigma(c_j - \bar{c}_.)^2 \int_0^1 \{a(u) - \bar{a}_.\}^2 du. \quad (22)$$

We now examine the behaviour under the alternative hypothesis, that is the power, or distribution of the observed significance level, when the null hypothesis is not true. The object is to have some fairly simple procedure for comparing the relative merits of different test statistics. The direct and exact approach is to use (12), but this rarely leads to simple answers. We now derive a large-sample normal approximation in which we consider alternatives close to the null hypothesis; in so doing we are effectively finding the derivative of the power function at the null hypothesis.

Consider location-type alternatives in which $Y_1, \ldots, Y_n$ are independent with Y_j having the density $f_0(y - \theta_j)$; here $f_0(.)$ is not necessarily the same density as $g(.)$, for which optimal power properties are desired. The null hypothesis of homogeneity is that $\theta_j = \bar{\theta}_.$; in considering alternatives we suppose that $\theta_j - \bar{\theta}_. = O(n^{-\frac{1}{2}})$.

Define

$$L = l(Y) = \log\left|\frac{\Pi f_0(Y_j - \theta_j)}{\Pi f_0(Y_j - \bar{\theta}_.)}\right|,$$

the log likelihood ratio for testing H_0 as a simple hypothesis versus the particular alternative, using the observations *directly* rather than through the ranks.

The essential step in the calculation of the distribution of $t(R, c)$ under an alternative hypothesis is the result that

$$\mathrm{pr}(T \leqslant t_0; H_A) = \int_{\{y; T \leqslant t_0\}} \exp\{l(Y)\} \Pi f_0(y_j - \theta) dy.$$

$$= \int_{-\infty}^{t_0} \int_{-\infty}^{\infty} e^l f_{T, L}(t, l; H_0) dt\, dl, \quad (23)$$

where $f_{T,L}(t, l; H_0)$ is the joint p.d.f. of $t(R, c)$ and $l(Y)$ under H_0. We already have a useful approximation to T from (21); corresponding to this we get the Taylor expansion

$$l(Y) \simeq -\sum_{j=1}^{n}(\theta_j - \bar{\theta}_.)\frac{f_0'(Y_j - \bar{\theta}_.)}{f_0(Y_j - \bar{\theta}_.)} - \frac{1}{2}i_{f_0}\Sigma(\theta_j - \bar{\theta}_.)^2,$$

where i_{f_0} is the information or intrinsic accuracy for the density $f_0(.)$. For the joint density we use a bivariate normal approximation, the relevant permutation means and covariance matrix being easily calculated. The resulting approximation to (23) is that

$$\text{pr}\left\{\frac{T - E(T; H_0)}{\sqrt{\text{var}(T; H_0)}} \leqslant z; \theta\right\} \simeq \Phi(z - \lambda_{TL}),\tag{24}$$

where

$$\lambda_{TL} = \frac{\text{cov}(T, L; H_0)}{\sqrt{\text{var}(T; H_0)}} \simeq \frac{\Sigma(c_j - \bar{c}_.)(\theta_j - \bar{\theta}_.)\int_0^1 a(u)b(u)du}{[\Sigma(c_j - \bar{c}_.)^2 \int_0^1 \{a(u) - \bar{a}_.\}^2 du]^{\frac{1}{2}}},\tag{25}$$

and $b(u) = f_0'\{F_0^{-1}(u)\}/f_0\{F_0^{-1}(u)\}$.

The ratio λ_{TL} is a covenient tool for comparison of alternative forms of linear rank statistics for different parent distributions, although it is, of course, restricted to large-sample comparisons near the null hypothesis.

Example 6.6. Efficiency of two-sample tests. Consider again the two-sample problem, where, without loss of generality, we can now write $\theta_1 = \ldots = \theta_{n_1} = \delta$ and $\theta_{n_1+1} = \ldots = \theta_n = 0$. It is easily shown from (25) that the previous choice $c_j \propto \theta_j$ is optimal. The non-centrality parameter then simplifies to

$$\lambda_{TL} \simeq \delta\left\{\frac{n_1 n_2}{n}\right\}^{\frac{1}{2}}\frac{\int_0^1 a(u)b(u)du}{\left[\int_0^1 \{a(u) - \bar{a}_.\}^2 du\right]^{\frac{1}{2}}}.\tag{26}$$

Two alternative one-sided tests based on statistics T_1 and T_2 can be compared by the ratio of sample sizes necessary to get the same power. To within the approximations made here, this implies that the efficiency of T_1 relative to T_2 is

$$\text{eff}(T_1, T_2) = \left(\frac{\lambda_{T_1 L}}{\lambda_{T_2 L}}\right)^2.\tag{27}$$

For example, we may take f_0 to be the normal density, T_1 to be the Wilcoxon statistic with $a(u) = u$, and T_2 to be the best linear rank test when $b(u) = \Phi^{-1}(u)$. Then simple calculation shows that $\mathrm{eff}(T_1, T_2) = 0.955$, i.e. that the Wilcoxon test has only slightly poorer power properties than the best rank test for the problem, in large samples.

A second important deduction from (26) is that for any given $f_0(.)$ the maximum value of λ_{TL} is $\delta(n_1 n_2 i_{f_0}/n)^{\frac{1}{2}}$. This, taken in conjunction with (24) and (4.45), shows that asymptotically the best rank test has the same power as the locally most powerful test, i.e. the best non-rank test. Of course, this leaves open the question of how large n has to be before these results give a good approximation to the actual power.

We have outlined here the main results about linear rank statistics. The full mathematical treatment is much more extensive (Hájek and Sidák, 1967). One strengthening of the results is to include nonlocal power comparisons (Chernoff and Savage, 1958).

(v) Occurrence of ties
When the underlying random variables are continuous, the probability that any two observations are equal is zero and the possibility of tied ranks can be disregarded. On the other hand, in practice ties arise quite often. The effect of ties on the choice of optimal statistic is hard to assess, but is likely to be small unless the proportion of tied observations is high. So far as the null distribution of test statistics is concerned, ties cause no difficulty in principle. We argue condition-ally on the order statistics, which take account of the multiplicities of tied observations. Thus, in the Wilcoxon two-sample test, we have just to decide how tied observations are to be scored; an obvious possibility is to assign all the observations in a tied group the average rank. Then the full set of ranks for the n individuals is fixed. The test statistic is again the total score in, say, the first sample and under the null hypothesis this is again a random sample of size n_1 drawn ran-domly without replacement from a known finite population. Detailed formulae for moments are altered because the finite population is no longer $\{1, ..., n\}$, but the principle of obtaining the null hypothesis distribution is in no way affected. For some rank statistics, formulae for moments in the presence of ties are given by Hájek (1969).

6.4 Randomization tests

We now turn to a quite different kind of argument which, in particular, leads again to some of the tests considered in the previous sections. To take as simple a version of the argument as possible, consider an experiment to compare two alternative treatments A and B. Let $2m$ experimental units be available and suppose that a physical randomizing device is used to choose m of them to receive A, the remaining m units receiving treatment B. That is, a physical randomizing device is used to choose one of the $(2m)!/(m!)^2$ designs in which each treatment occurs the same number of times. The chosen design is applied to the experimental units and we suppose that the resulting observations are denoted by $y'_1, \ldots, y'_m$ for treatment A, and $y''_1, \ldots, y''_m$ for treatment B.

Now consider the null hypothesis that the response observed on any experimental unit is the same whichever treatment is applied to that unit and is unaffected also by the allocation of treatments to other units. This is a deterministic hypothesis; note that it involves no reference to experimental units other than those used in the experiment.

It follows that for any of the designs that might have been used we can, under the null hypothesis, find exactly the observations that would have been obtained; this is done simply by permuting the data. That is, for any test statistic we can find the exact null hypothesis distribution. With the particular design used here we take all permutations of the $2m$ observations and treat them as equally likely.

Thus any of the two-sample tests of Sections 6.2 and 6.3 can be applied, in particular a permutation Student t test or some form of two-sample rank test.

In the previous discussion these tests were justified in that the null hypothesis specified the physical system under investigation as generating i.i.d. values. Now, however, the justification is that the randomization used in design makes all permutations of the data equally likely under the very strong null hypothesis.

The argument can be applied very generally whenever a design, i.e. an allocation of treatments to experimental units, is selected at random from some set of possible designs, by a physical randomizing device. Under the deterministic null hypothesis that the response on any experimental unit is the same whatever treatment is applied to that and any other experimental unit, the null hypothesis distribution of any test statistic can be generated. In principle, we write down the set of data we would have obtained for each possible design, and then assign each such data set its appropriate probability. Nearly always designs are drawn from some set of desirable designs, giving each design equal chance of selection.

An important general point is that we are here interpreting data from a given design by hypothetical repetitions over a set of possible designs. In accordance with general ideas on conditionality, this interpretation is sensible provided that the design actually used does not belong to a subset of designs with recognizably different properties from the whole set. For example, if the $2m$ experimental units in the completely randomized design just discussed were arranged in sequence in time, then the possible design in which the first m units all receive A is an extreme instance of a design with different properties from most of the reference set of designs. If the relevance of the serial order is realized at the design stage, the completely randomized design becomes inappropriate; if it is not realized until after the data have been obtained one can only abandon the approach *via* randomization theory and attempt an *ad hoc* model to represent treatment effect and time trend.

The most important application of randomization theory is not so much in connexion with tests of significance as in the development of an appropriate linear model for the analysis of relatively complicated experimental designs, such as Latin squares, split plot experiments, incomplete block designs, etc. On the face of it, the conventional discussion of such designs by the least squares analysis of a suitable linear model requires the special choice of an appropriate model for each design. In fact, however, when the analysis is based on the randomization, a second-order analysis can be made from the same assumptions throughout. This is a specialized topic, however, rather outside the main scope of this book, and will not be developed further.

We conclude this section with another simple application of the randomization argument for significance testing.

Example 6.7. Matched pair experiment. We again consider the comparison of two treatments, this time not in a completely randomized design but in a matched pair experiment. That is, the experimental units are grouped into m pairs. One unit in each pair receives treatment A while the other receives treatment B, the choice being made independently at random for each pair, the two possible assignments having equal probability. Denote the differences between the response to the unit receiving A and that receiving B in the same pair by $x_1, \ldots, x_m$.

Now, under the null hypothesis considered previously that response is unaffected by the treatment allocation, it follows that if on, say, the jth pair the opposite allocation had been adopted, then the resulting difference would have been $-x_j$. That is, under the null hypothesis the sets of differences that would have been obtained with alternative designs have the form

$$\pm x_1, \ldots, \pm x_m$$

and, moreover, under the particular design used, the 2^m possibilities have equal probability. Thus the null hypothesis distribution of any test statistic is determined.

One obvious test statistic is the Student t statistic based on the differences and this is equivalent over the distribution used here to the sum of the differences. Denote the random variables representing the differences by $X_1, \ldots, X_m$; under the null hypothesis they are independent and each has a two-point distribution. We can thus easily obtain the null hypothesis moments of

$$U = \Sigma X_j,$$

and in fact all the odd order moments vanish, whereas

$$E(U^2; H_0) = \Sigma x_j^2,$$

$$E(U^4; H_0) = \Sigma x_j^4 + 6 \underset{j > k}{\Sigma} x_j^2 x_k^2$$

$$= 3(\Sigma x_j^2)^2 - 2\Sigma x_j^4.$$

Usually a normal approximation to the distribution of U will be adequate.

6.5   Distance tests

All the tests described so far in this Chapter have the property that the null hypothesis distribution is known exactly for some very general null hypothesis; this is achieved in Sections 6.2 and 6.3 by arguing conditionally on the order statistics. The tests are essentially ones involving comparison between sets to detect inhomogeneity. In discussing criteria of optimality, we returned to the approach of earlier chapters requiring, subject to the distribution-free character of the test, maximum power for some alternative expressed in terms of a few unknown parameters. The general idea behind this is clear; we shall often have some notion of the type of parametric assumption that is reasonable even though we want the size of the test not to depend on this assumption.

A different approach, also not calling for a specification with a finite number of parameters, is to define some quite general measure of distance between two cumulative distribution functions and to estimate this distance from data; the object may be to test agreement with some given distribution or family of distributions, or to test the difference between two or more samples.

Tests of this type have already been mentioned in Chapter 3. The simplest one uses the Kolmogorov statistic based on the measure of distance

$$\sup|F_1(y) - F_2(y)|,$$

$F_1(.)$ and $F_2(.)$ being any two cumulative distribution functions. To test the consistency of data with a given distribution function $F_0(.)$, we take $\sup|\tilde{F}_n(y) - F_0(y)|$, where $\tilde{F}_n(.)$ is the sample cumulative distribution function.

Quite often we can argue as follows. Let $Y_1, \ldots, Y_n$ be i.i.d. with some unknown cumulative distribution function $F(.)$ and suppose that the null hypothesis H_0 is that $F(.) = F_0(.)$, this being possibly composite. Now suppose that for some $m \leqslant n$, we can define a function $g_0(y_1, \ldots, y_m)$, such that

$$\Delta_0(F) = \int .. \int g_0(y_1, \ldots, y_m) dF(y_1) \ldots dF(y_m) \qquad (28)$$

is a suitable measure of the distance of $F(.)$ from $F_0(.)$ for the alternatives to H_0 that are of interest; we might define $\Delta_0(F)$ so that $\Delta_0(F_0) = 0$, but this is not essential. The functional $\Delta_0(F)$ is linear, so that the Kolmogorov measure $\sup|F(y) - F_0(y)|$ is excluded.

An obvious statistic for testing the null hypothesis is $\Delta_0(\tilde{F}_n)$, i.e. the estimate of $\Delta_0(F)$ based on the sample distribution function. But often we can find a simple analogue of this statistic as follows. Suppose that for some $r < n$, we can find a symmetric function $k(Y_1, \ldots, Y_r)$, not depending on $F(.)$ or $F_0(.)$, such that

$$E\{k(Y_1, \ldots, Y_r)\} = \Delta_0(F). \qquad (29)$$

Now, effectively, we have $n!/\{(n-r)!r!\}$ replicates of the unbiased estimate $k(Y_1, \ldots, Y_r)$, corresponding to the number of subsets of size r in $(Y_1, \ldots, Y_n)$. We therefore take as a test statistic the average

$$U = \frac{1}{\binom{n}{r}} \Sigma k(Y_{j_1}, \ldots, Y_{j_r}), \qquad (30)$$

summation being over all selections of r distinct integers from $\{1, \ldots, n\}$. The formal motivation for U is that it is the average of $k(Y_1, \ldots, Y_r)$ conditionally on $Y_{(.)}$, the sufficient statistic, and, as we shall show in Section 8.4(ii), this implies that U has the smallest variance among all unbiased estimates of $\Delta_0(F)$, whatever the true distribution $F(.)$.

The U statistic is very similar to the estimate $\Delta_0(\tilde{F}_n)$, because (29) is nothing more than a re-expression of (28) in terms of a symmetric kernel function $k(.)$ not depending on $F(.)$ or $F_0(.)$. The estimate $\Delta_0(\tilde{F}_n)$, then, can be written as

$$\Delta_0(\tilde{F}_n) = \int k(y_1, \ldots, y_r) d\tilde{F}_n(y_1) \ldots d\tilde{F}_n(y_r)$$

$$= \frac{1}{n^r} \sum_{j_1=1}^{n} \ldots \sum_{j_r=1}^{n} k(Y_1, \ldots, Y_r).$$

A simple illustration is the variance of Y, for which we might take as the initial representation

$$\Delta_0(F) = \int \{y_1 - \int z dF(z)\}^2 dF(y_1),$$

corresponding to (28), which can be expressed as $E\{\frac{1}{2}(Y_1 - Y_2)^2\}$, corresponding to (29).

It can be shown quite generally that U and $\Delta_0(\tilde{F}_n)$ are identical to first order and have the same limiting distribution for large n, the essential difference being that $\Delta_0(\tilde{F}_n)$ is biased if $r > 1$. Hoeffding (1948) obtained many properties of the statistic (30) and its generalizations, in particular their asymptotic normality. In fact to first order the variance of (30) is

$$\text{var}_{Y_1} [E\{k(Y_1, \ldots, Y_r)| Y_1\}]/n. \qquad (31)$$

Because U has the smallest variance among unbiased estimates of $\Delta_0(F)$, it is clearly the best large-sample test statistic based on the given distance measure. However, for particular alternative distributions, the test based on U may not be very efficient. The advantages of the U statistic are that it provides an estimate of the discrepancy with $F_0(.)$, and generally has simpler distributional properties than $\Delta_0(\tilde{F}_n)$.

We give two examples to illustrate the use of the U statistic.

Example 6.8. Test for constant hazard. When, as in reliability theory, we have a non-negative random variable representing failure-time, an important way of specifying the distribution is by the hazard function or age-specific failure rate. This is defined by

$$\rho_Y(y) = \lim_{\Delta y \to 0+} \frac{\text{pr}(y < Y \leqslant y + \Delta y \,|\, y < Y)}{\Delta y} = \frac{f_Y(y)}{1 - F_Y(y)}.$$

For the exponential distribution, $\rho_Y(y)$ is constant, an important alternative to this being that the individuals "age" in the sense that $\rho_Y(y)$ is monotone increasing. In terms of the cumulative distribution function this implies that for any y and z

$$1 - F_Y(y + z) > \{1 - F_Y(y)\}\{1 - F_Y(z)\},$$

so that a suitable non-parametric measure of distance from the null hypothesis of an exponential distribution is

$$\Delta_0(F) = \iint [1 - F_Y(y + z) - \{1 - F_Y(y)\}\{1 - F_Y(z)\}] dF_Y(y) dF_Y(z).$$

This can be expressed in the form (29) by noting that, apart from a constant, it is equal to $\text{pr}(Y_1 > Y_2 + Y_3)$ where Y_1, Y_2 and Y_3 are i.i.d. with distribution function $F_Y(.)$. Therefore, with $r = 3$, a suitable symmetric function $k(.)$ is

$$k(y_1, y_2, y_3) = \text{hv}(y_1 - y_2 - y_3) + \text{hv}(y_2 - y_1 - y_3) +$$
$$\text{hv}(y_3 - y_1 - y_2) - \tfrac{1}{4},$$

where $\text{hv}(x) = 1 \ (x \geqslant 0)$, $\text{hv}(x) = 0 \ (x < 0)$, the unit Heaviside function.

Under the null hypothesis of an exponential distribution, the U statistic has zero mean and the large-sample variance (31) is, without loss of generality taking Y_1 to have unit mean,

$$\text{var}(Y_1 e^{-Y_1})/(9n) = \frac{5}{3888n}.$$

The standardized statistic $U\sqrt{(3888n/5)}$ is approximately $N(0, 1)$

under the null hypothesis, and large values are evidence of positive
ageing.

A discussion of the U statistic and its competitors for this particular problem is given by Proschan and Hollander (1972).

The next example illustrates the possibility of extending the definition of the U statistic to cover situations more complicated than
those involving single samples of i.i.d. variables.

Example 6.9. Wilcoxon test (ctd). The two-sample Wilcoxon test of
Example 6.3 can be regarded as based on a simple distance measure.
Let Y' and Y'' denote generic random variables from the two groups.
Consider the null hypothesis H_0 to be that $\mathrm{pr}(Y' > Y'') = \frac{1}{2}$, and take
for general distributions the measure of departure from H_0

$$\mathrm{pr}(Y' > Y'') - \tfrac{1}{2} = \int F_{Y'}(y)dF_{Y''}(y) - \tfrac{1}{2}.$$

An unbiased estimate of this based on two sets of i.i.d. random variables $Y_1', \ldots, Y_{n_1}'$ and $Y_1'', \ldots, Y_{n_2}''$ is

$$\sum_{i=1}^{n_1} \sum_{j=1}^{n_2} D_{ij}/(n_1 n_2) - \tfrac{1}{2}, \qquad (32)$$

where $D_{ij} = 1 \ (Y_i' > Y_j'')$, $D_{ij} = 0 \ (Y_i' \leqslant Y_j'')$, and this is equivalent
to the Wilcoxon statistic. The statistic (32) corresponds to the U
statistic (30) but with the two-sample kernel function $k(Y_1', \ldots, Y_{r_1}';$
$Y_1'', \ldots, Y_{r_2}'')$, where in this particular case $r_1 = r_2 = 1$.

Note that this approach is conceptually quite different from the
derivation as an optimal rank test for location. The present approach
makes no appeal to finite parameter formulations and has the advantage of giving an interpretation of the test statistic when the null
hypothesis is false.

There is no reason in general why a distance test should be distribution-free; indeed, to achieve that, the null hypothesis distribution
will have, explicitly or implicitly, to be taken as the appropriate conditional one.

Bibliographic notes

Some of the ideas connected with distribution-free tests can be
traced back to the latter half of the 19th century in the work of
Sir Francis Galton and to the beginning of the 20th century in the

work of psychologists such as C. Spearman. Results on optimality are, however, much more recent.

Fisher (1966, 1st ed. 1935) established both the possibility of an "exact" test of significance based on randomization used in allocating treatments, and the implications for the correct choice of error term in the analysis of variance. For a review of the former, see Kempthorne and Doerfler (1969) and, for the second, see Anscombe (1948), Kempthorne (1952), Nelder (1965a, b), and Scheffé (1959, Chapter 9). For the group-theoretic implications, see Grundy and Healy (1950).

Pitman (1937, a, b, c) examined permutation versions of the common normal-theory tests of significance with emphasis on the magnitude of the discrepancies between normal-theory and permutation significance limits; see also Box and Andersen (1955).

Early work on rank tests concentrated on measures of rank correlation and their use for testing null hypotheses of independence (Kendall, 1962). The use of expected normal order-statistics as a basis for rank tests is due to Fisher and Yates (1963, 1st ed. 1938), although similar ideas occur in the work of K. Pearson. The formal discussion of optimality was first given in the 1950's; see the textbook of Lehmann (1959). A comprehensive modern treatment of rank tests is given by Hájek and Šidák (1967).

Multivariate distribution-free tests are discussed systematically by Puri and Sen (1971). Puri (1970) has edited a collection of papers of theoretical work in distribution-free methods. The asymptotic theory is based largely on the papers by Hoeffding (1948) and von Mises (1947).

Distance methods based on the Kolmogorov statistic and its generalizations have a long history. Recent work has concentrated largely on a rigorous approach to asymptotic theory using the idea of weak convergence; see Billingsley (1968) and Durbin (1973).

Further results and exercises

1. Suppose that $x_1, \ldots, x_n$ are i.i.d. with a density $f(x - \theta)$, where $f(.)$ is an unknown continuous density symmetric about zero, and consider testing the null hypothesis $H_0: \theta = 0$ against the alternative that $\theta > 0$. Let the positive x values be denoted by $y_1, \ldots, y_{n_1}$ and the negative values by $-y_{n_1+1}, \ldots, -y_n$. Show that H_0 is equivalent to a hypothesis comparing the samples $y_1, \ldots, y_{n_1}$ and $y_{n_1+1}, \ldots, y_n$,

and hence construct a rank-type test of H_0.

[Section 6.2]

2. Use the arguments outlined in Example 6.2 to obtain the first four moments of the sample mean in random samples of size n_1 drawn without replacement from the finite population $\{y_1, \ldots, y_n\}$. By examining the third and fourth cumulant ratios, show that the rate of approach to normality as $n \to \infty$, $n_1/n \to a$, $0 < a < 1$, is faster than that for sampling with replacement. Give some explicit results for the finite population $\{1, \ldots, n\}$ and show the application to the two-sample Wilcoxon test.

[Section 6.2]

3. In the matched pair experiment of Example 6.7, suppose that the observed differences between treatment A and treatment B are $8, 5, 4, -2, -2, 0$ and 1. Show, by enumeration, that the exact two-sided significance level for testing the null hypothesis of treatment equivalence is $1/4$. Compare this with the result of (a) applying the "ordinary" one-sample Student t test; (b) a normal approximation to the permutation distribution, with a continuity correction; (c) an approximation to the permutation distribution using the first four moments.

[Section 6.2]

4. Let $(Y_1, Z_1), \ldots, (Y_n, Z_n)$ be n i.i.d. bivariate random variables with an arbitrary continuous distribution. Show that the rank statistic

$$T = \frac{1}{n(n-1)} \sum_{j > k} \text{sgn}\{(Y_j - Y_k)(Z_j - Z_k)\} + \tfrac{1}{2},$$

equivalent to Kendall's τ, has expectation equal to the probability that $Y - Y'$ and $Z - Z'$ have the same sign, where (Y, Z) and (Y', Z') are independent pairs with the distribution in question. Show that under the null hypothesis that Y and Z are independent, the permutation distribution of T has mean $\tfrac{1}{2}$ and variance $(2n + 5)/\{18n(n - 1)\}$.

Prove that if the bivariate pairs have a bivariate normal distribution of correlation coefficient ρ, then the expectation of T is $\tfrac{1}{2} + \pi^{-1}\sin^{-1}\rho$.

[Section 6.3; Kendall, 1962]

5. Let $Y_1, \ldots, Y_{n_1}$ be i.i.d. with density $h(y)$, and $Y_{n_1+1}, \ldots, Y_n$ be independent i.i.d. with density $\tau^{-1}h\{(y - \mu)/\tau\}$. Develop the theory corresponding to that of Section 6.3(ii) for constructing rank tests of the null hypothesis $H_0: \tau = 1, \mu = 0$ against the alternatives (a) $H_A: \tau > 1, \mu > 0$ and (b) $H_A': \tau > 1, \mu < 0$.

[Section 6.3]

6. Examine in more detail the null hypothesis distribution of the optimal two-sample rank test for exponential distributions in Example 6.4.

Suppose that under the null hypothesis $Y_1, \ldots, Y_n$ are i.i.d. and exponentially distributed, whereas under the alternative hypothesis Y_j has mean $1/(\gamma + \beta z_j)$, where $z_1, \ldots, z_n$ are known. Obtain optimum parametric and rank tests, and suggest an approximation to the null hypothesis distribution of the latter.

[Sections 6.3, 5.2]

7. A randomized block design is to be used to compare m treatments. The experimental units are set out as mr units in each of b blocks, each treatment occurring exactly $r > 1$ times in each block, and subject to this the treatment allocations are made at random. Let x_{ijk} be the response that would be observed on the jth unit of the ith block if treatment k were applied. Population linear contrasts are defined as $\mu = \bar{x}_{...}, \beta_i = \bar{x}_{i..} - \bar{x}_{...}, \tau_k = \bar{x}_{..k} - \bar{x}_{...}, \gamma_{ik} = \bar{x}_{i.k} - \bar{x}_{i..} - \bar{x}_{..k} + \bar{x}_{...}$, and error terms are $e_{ij} = \bar{x}_{ij.} - \bar{x}_{i..}$. In addition, assume that $x_{ijk} - \bar{x}_{ij.} \equiv \bar{x}_{i.k} - \bar{x}_{i..}$. Now suppose that y_{ikl} is the response on the lth replicate in block i with treatment k. The sample $\{y_{ikl}\}$ is related to the population solely in terms of an indicator random variable saying which treatment is applied to each unit. Derive the linear model for $\bar{y}_{ik.}$, and discuss the randomization analogue of the normal-theory test of $H_0: \tau_1 = \ldots = \tau_m = 0$, when the interactions γ_{ik} are assumed to be zero.

[Section 6.4]

8. Prove that the null hypothesis distribution of the Kolmogorov distance statistic is unaffected by an arbitrary monotonic transformation and hence that, without loss of generality, the null hypothesis to be tested can be taken as the uniform distribution over $(0, 1)$. In this case, let $\tilde{F}_n(y)$ be the proportion of the first n observations less than or equal to y. Prove that the first and second moment properties

of $\tilde{F}_n(y) - y$ are, as $n \to \infty$, the same as those of a tied Brownian motion (sometimes called Brownian bridge), i.e. Brownian motion starting at $(0, 0)$ and conditioned to go through $(1, 0)$. Hence interpret the limiting null hypothesis distribution in terms of a tied Brownian motion with absorbing barriers and show that, as $n \to \infty$,

$$\lim (D_n^+ \sqrt{n} \leqslant x) = 1 - e^{-2x^2},$$

and

$$\lim (D_n \sqrt{n} \leqslant x) = 1 + 2 \sum_{k=1}^{\infty} (-1)^k e^{-2k^2 x^2}.$$

[Sections 6.5 and 3.2; Bartlett, 1966]

9. To test whether random variables Y and Z are independent is to test whether their joint cumulative distribution function is of the form $F_{Y,Z}(y, z) = G_Y(y)K_Z(z)$, where $G_Y(.)$ and $K_Z(.)$ are the marginal cumulative distribution functions. Construct a distribution-free permutation test of monotonic dependence based on the measure of dependence

$$\Delta = \iint \{F_{Y,Z}(y, z) - G_Y(y)K_Z(z)\} dG_Y(y)dK_Z(z),$$

giving the sampling distribution of the test under the null hypothesis of independence.

[Section 6.5]

7 INTERVAL ESTIMATION

7.1 Introduction

In Chapters 4 and 5 we were concerned with a fully specified parametric family of models in which, however, the relation between the different possible parameter values is not symmetrical. The null hypothesis is clearly specified from scientific or technological considerations, whereas the alternative hypotheses serve only to indicate the types of departure from the null hypothesis thought to be of particular importance. As has been emphasised repeatedly in previous chapters, null and alternative hypotheses are on a different logical footing.

We now turn to situations in which all members of the family of distributions are to be treated as of equal importance. Provisionally we assume that one of the distributions in the family is the true distribution, an assumption which is, of course, always tentative. What can we say about the true parameter value? This can reasonably be regarded as the central problem of statistical inference, and the results of Chapters 4–6, while important in their own right, are in a sense a preliminary to the present Chapter.

The approach to be taken may be viewed as summarizing inferences about which distributions in the given family are consistent with the data, as judged by significance tests. That is to say, we do not in this Chapter deal with post-data probability statements about parameter values. This can only be done with the extra structure of a prior probability distribution of some type over the family of distributions, and this situation is dealt with separately in Chapters 10 and 11. In view of the earlier statement about the equal importance attached to distributions in the family, in certain situations the statements derived from the theory of this Chapter may be virtually indistinguishable from the probability statements obtained from uniform prior distributions. Nevertheless, it will become quite clear that, in general, interval estimates cannot be taken as probability statements about parameters, and foremost is the interpretation "such and such parameter values are consistent with the data."

We shall deal in turn with situations in which interest is concentrated on

  (a) a scalar parameter;
  (b) one parameter in the presence of nuisance parameters;
  (c) a vector parameter;
  (d) a further observation taken from the same random system as the data.

Even for problems with a single unknown parameter, there are a number of aspects that need careful discussion. Therefore, even though most practical problems involve nuisance parameters, a good part of the Chapter is taken up with the discussion of the single parameter situation.

7.2   Scalar parameter

(i) Confidence limits

Suppose that the observed vector random variable Y has p.d.f. $f_Y(y; \theta)$ depending on the scalar parameter θ. In a non-statistical problem involving an unknown constant which cannot be determined exactly, it will often be sensible to try to find upper and lower bounds for the unknown. To generalize this idea to deal with a statistical problem, we introduce the idea of upper and lower confidence limits.

For $\alpha > 0$, let $T^\alpha = t^\alpha(Y)$ be a statistic, i.e. a function of Y, such

$$\text{pr}(T^\alpha \geqslant \theta; \theta) = 1 - \alpha \quad (\theta \in \Omega); \tag{1}$$

if $\alpha_1 > \alpha_2$ and T^{α_1} and T^{α_2} are both defined, we require that

$$T^{\alpha_1} \leqslant T^{\alpha_2}. \tag{2}$$

We then call T^α a $1 - \alpha$ *upper confidence limit* for θ. Since in principle we rarely want to consider just one value of α, the requirement (2) is needed. We are essentially concerned with the system of limits T^α and cannot tolerate a situation in which a nominally less stringent limit, say with $\alpha = 0.1$, is larger than a nominally more stringent limit, say with $\alpha = 0.01$. In practice, however, we rarely work explicitly with more than a few values of α.

Note that if $g(.)$ is strictly increasing, $g(T^\alpha)$ is an upper $1 - \alpha$ confidence limit for $g(\theta)$.

The key requirement (1) gives a physical interpretation to the confidence limits. For fixed α, if we were to assert for $Y = y$, $T^\alpha = t^\alpha$, that

$$\theta \leqslant t^\alpha, \tag{3}$$

then we would be following a procedure that would be wrong only in a proportion α of cases, in hypothetical repeated applications, whatever may be the true value θ. Note that this is a hypothetical statement that gives an empirical meaning, which in principle can be checked by experiment, rather than a prescription for using confidence limits. In particular, we do not recommend or intend that a fixed value α_0 should be chosen in advance and the information in the data summarized in the single assertion $\theta \leqslant t^{\alpha_0}$.

Further, in order that the long-run property should be relevant to the interpretation of a particular set of data under analysis, the probability property (1) should hold conditionally on the values of any ancillary statistics for the problem.

In many ways a system of confidence limits calculated from the data corresponds to treating the parameter as if, in the light of the data, it is a random variable Θ with cumulative distribution function implicitly defined by

$$\mathrm{pr}(\Theta \leqslant t^\alpha) = 1 - \alpha. \tag{4}$$

The distinction, emphasized in the standard treatments of confidence intervals, is that in the hypothetical repetitions that give physical meaning to $\mathrm{pr}(X \leqslant x)$, where X is a random variable, we regard x as a fixed arbitrary constant; more generally we consider the probability that X falls in some *fixed* set. On the other hand, in (3) it is t^α that varies in hypothetical repetitions, not θ: hence it is not legitimate to treat θ as a random variable. Nevertheless, for many purposes the distinction appears to be a rather legalistic one; we return to this very important point in Section 7.2 (vii) and (viii).

Equations (1) and (2) define upper confidence limits. The definition of $1 - \alpha$ lower confidence limits T_α is entirely analogous, the condition corresponding to (1) being

$$\mathrm{pr}(T_\alpha \leqslant \theta; \theta) = 1 - \alpha \quad (\theta \in \Omega). \tag{5}$$

Except for the possibility that $T_\alpha = \theta$, a lower $1 - \alpha$ limit is an upper α limit; nearly always, however, we work with values of α appreciably less than $\frac{1}{2}$.

If, instead of (1) and (5), we have

$$\text{pr}(T^\alpha \geqslant \theta; \theta) \geqslant 1 - \alpha, \quad \text{pr}(T_\alpha \leqslant \theta; \theta) \geqslant 1 - \alpha, \tag{6}$$

the confidence limits are called *conservative*.

Example 7.1. Normal mean with known variance (ctd). If $Y_1, \ldots, Y_n$ are i.i.d. in $N(\mu, \sigma_0^2)$, where the variance σ_0^2 is known, it is easily shown that

$$\bar{Y}_. - k_{1-\alpha}^* \frac{\sigma_0}{\sqrt{n}} = \bar{Y}_. + k_\alpha^* \frac{\sigma_0}{\sqrt{n}} \tag{7}$$

is a $1 - \alpha$ upper confidence limit for μ. Note that we have expressed the upper limit in its general form, followed by the special form resulting from the symmetry of the normal distribution; if $\alpha < \frac{1}{2}$, $k_\alpha^* > 0$.

If $\tilde{Y}$ is the sample median and we make the approximation, which is very slight for all except extremely small values of n, that $\tilde{Y}$ is normally distributed with mean and variance $(1.25)^2 \sigma_0^2/n$, as shown in Section A2.3, we have similarly that

$$\tilde{Y} - 1.25k_{1-\alpha}^* \frac{\sigma_0}{\sqrt{n}} = \tilde{Y} + 1.25k_\alpha^* \frac{\sigma_0}{\sqrt{n}} \tag{8}$$

is a $1 - \alpha$ upper confidence limit. The approximation can be avoided by replacing $k_{1-\alpha}^*$ by a suitable constant derived from the exact distribution of the median in samples from the standard normal distribution.

Example 7.2. Cauchy distribution (ctd). If $Y_1, \ldots, Y_n$ are i.i.d. with density

$$\frac{1}{\pi\{1 + (y - \theta)^2\}},$$

we may want confidence limits for θ. Let T be any statistic with density of the form $h(t - \theta)$; examples are the median and, more generally, linear functions of the order statistics with sum of weights one, and, more generally still, any location statistic. Let h_α^* be the upper α point of the distribution with density $h(x)$. Then

$$\text{pr}(T - \theta \geqslant h_{1-\alpha}^*; \theta) = 1 - \alpha,$$

so that if

$$T^\alpha = T - h^*_{1-\alpha}, \tag{9}$$

the defining properties (1) and (2) for upper confidence limits are satisfied. For a statistic with a symmetric distribution, $-h^*_{1-\alpha} = h^*_\alpha$.

A similar argument gives us a large number of ways of obtaining confidence limits for general location parameter problems in which the p.d.f. of the Y_j is of the form $h(y - \theta)$. Some examples which are particularly relevant when $h(.)$ is an unknown symmetry density are given in Section 9.4.

We have, however, in this analysis ignored the requirement that the probability statements are to be conditional on the ancillary statistics for the problem. We shall see in Example 7.7 that when we take account of this, an essentially unique answer is obtained.

Example 7.3. Pivotal quantities. It is sometimes convenient to derive confidence limits from the distribution of a pivot, this being defined as a function of the data and the parameter having a fixed distribution the same for all parameter values. That is, if $P(Y, \theta)$ has a continuous distribution with p.d.f. $f_P(.)$ not involving θ, we can find a constant $p^*_{1-\alpha}$ such that

$$\mathrm{pr}\{P(Y, \theta) \geqslant p^*_{1-\alpha}; \theta\} = 1 - \alpha. \tag{10}$$

If now for each Y, $P(Y, \theta)$ is a monotonic decreasing function of θ, the inequality $P(Y, \theta) \geqslant p^*_{1-\alpha}$ is equivalent to $\theta \leqslant T^\alpha$, where T^α is a function of $p^*_{1-\alpha}$ and Y. If the pivot is increasing in θ rather than decreasing, lower confidence limits result; to obtain upper confidence limits it is necessary only to change the sign of the pivot.

Both Examples 7.1 and 7.2 can be expressed in this form; for instance in Example 7.1 both $\bar{Y}_\cdot - \mu$ and $\tilde{Y} - \mu$ are pivots, and $p^*_{1-\alpha}$ is proportional to $k^*_{1-\alpha}$.

We assume that the reader is familiar with other simple examples of confidence limits.

(ii) Optimality requirements
Quite often there may be several ways of obtaining upper confidence limits and it is required to have some way of comparing them and, possibly, picking out one method as optimum. The rough idea is that given some small α, the smaller the value of the upper limit the better; ideally T^α would be only very slightly greater than the parameter θ

for which it is intended as an upper bound. There are several ways in which this idea might be expressed formally. For example, the statement that T_1 is preferable to T_2 for $\alpha < \frac{1}{2}$ if

$$E(T_1^\alpha ; \theta) \leqslant E(T_2^\alpha ; \theta), \tag{11}$$

for all $\theta \in \Omega$, with inequality for some θ, would express in a reasonable way the preferability of the limit (7) using the normal mean to the limit (8) using the median. A disadvantage of this approach, even apart from its mathematical clumsiness in many applications, is that it is not invariant under transformations of the parameter θ to a monotone function of θ, even though, as noted below (2), the notion of confidence limits does have such invariance.

An alternative and preferable approach was suggested in 1937 by J. Neyman. Subject to the primary requirement (1) we want $T^\alpha - \theta$ to be stochastically small. Thus if θ' is an arbitrary value larger than θ, $\theta' \in \Omega$, we want the probability of including θ' below the upper limit on θ, i.e.

$$\text{pr}(T^\alpha \geqslant \theta' ; \theta), \tag{12}$$

to be small, for all pairs θ and θ' such that $\theta' > \theta$. Here α is taken to be small. If we are concerned with values of α near one, it will be more sensible to think in terms of lower confidence limits. Minimizing (12) makes the upper bound sharp; ideally we would like (12) to be zero.

If we can minimize (12) simultaneously for all pairs $\theta, \theta' \in \Omega$, $\theta' > \theta$, a strong optimum property has been achieved. Otherwise some weaker requirement, such as the use of a local property when θ' is near θ, may be necessary. A mathematical advantage of this approach is that it establishes a strong link with the theory of tests.

(iii) General procedure

It would be possible to develop a theory of the construction of confidence limits from first principles but in fact, as mentioned above, it is convenient to use the mathematical formulation of the theory of tests. The extent to which confidence limits are regarded as conceptually dependent on the idea of tests of significance is partly a matter of taste.

The following procedure yields "good" confidence limits for an important class of situations. Let θ_0 be an arbitrary value of $\theta \in \Omega$. Consider a size α critical region for testing $\theta = \theta_0$ having "good" properties for alternatives $\theta = \theta_{A0} < \theta_0$. Now suppose such a critical region $w_\alpha(\theta_0)$ set up for each possible $\theta_0 \in \Omega$. For any given $Y = y$ we take the set of all θ values not "rejected" at level α, i.e.

$$\{\theta; y \notin w_\alpha(\theta)\}. \qquad (13)$$

Under some natural monotonicity conditions, and provided that tests for the different null hypotheses are of like form, the set (13) may turn out to be equivalent to

$$\{\theta; \theta \leqslant t^\alpha(y)\}. \qquad (14)$$

An upper limit, rather than a lower limit, is obtained because of the direction of the alternatives to θ_0 used in the tests. Thus for sufficiently large values of θ_0 the data will indicate that θ is in fact smaller than θ_0. We shall discuss in Section 7.2(v) the situation in which (13) is not equivalent to (14). If the equivalence does hold the defining property of an upper confidence limit is satisfied. For, by the property defining the size α critical regions,

$$\mathrm{pr}\{Y \in w_\alpha(\theta); \theta\} = \alpha,$$

i.e. $$\mathrm{pr}\{Y \notin w_\alpha(\theta); \theta\} = 1 - \alpha, \qquad (15)$$

so that the probability that a random Y satisfies (13), and hence (14), for the true θ is $1 - \alpha$.

Further, if the critical regions are most powerful for all null and alternative values, the required optimum property of the confidence limits holds. To see this, let $\theta^\dagger < \theta^{\dagger\dagger}$ be two arbitrary values. Then the property of maximum power means that

$$\mathrm{pr}\{Y \in w_\alpha(\theta^{\dagger\dagger}); \theta^\dagger\} \qquad \text{is a maximum,}$$

i.e. $$\mathrm{pr}\{Y \notin w_\alpha(\theta^{\dagger\dagger}); \theta^\dagger\} \qquad \text{is a minimum,}$$

i.e. $$\mathrm{pr}\{\theta^{\dagger\dagger} \leqslant T^\alpha; \theta^\dagger\} \qquad \text{is a minimum.} \qquad (16)$$

Now take $\theta^\dagger = \theta$ to be an arbitrary possible true value of θ and $\theta^{\dagger\dagger} = \theta'$ to be any other value such that $\theta' > \theta$. Then (16) is exactly the required optimum property (12), asserting that a false value that is too large is included within the upper confidence limit with minimum probability. If the critical regions have some weaker optimality property, then the confidence limits will have some corresponding property.

Finally, the nesting property of critical regions will imply a corresponding property for the confidence limits corresponding to different values of α, and this is (2).

Confidence limits for a normal mean based on the sample mean, as given in Example 7.1, illustrate these ideas. The uniformly most powerful size α critical region for testing $\mu = \mu_0$ against alternatives $\mu < \mu_0$ is

$$\{y; \bar{y}_. < \mu_0 - k_\alpha^* \sigma_0 / \sqrt{n}\}$$

and the set of parameter values not satisfying this is

$$\{\mu; \mu \leqslant \bar{y}_. + k_\alpha^* \sigma_0 / \sqrt{n}\}.$$

Further, any value above the true mean is below the upper limit with minimum probability.

This discussion can be summed up in the simple rule:

to obtain "good" $1 - \alpha$ upper confidence limits, take all those parameter values not "rejected" at level α in a "good" significance test against lower alternatives.

There is an obvious modification for obtaining "good" lower confidence limits. Note that in principle different types of test statistic could be used for each null hypothesis, but in general this would lead to violation of (2). It would, in any case, seem reasonable to use the same type of statistic throughout and usually (16) will determine which one.

This argument is important in at least three ways. First, it gives a valuable constructive procedure for finding confidence limits in non-standard cases. Secondly, it provides a criterion of optimality and a basis for comparing alternative procedures. Thirdly, it suggests that we can regard confidence limits as specifying those parameter values consistent with the data at some level, as judged by a one-sided significance test.

One fairly general situation for which optimum confidence limits are available by this method is when we observe i.i.d. continuous random variables having the exponential family density

$$\exp\{a(\theta)b(y) + c(\theta) + d(y)\}, \tag{17}$$

where $a(.)$ is monotonic. In this situation uniformly most powerful tests are given by the likelihood ratio, as we pointed out in Section 4.6.

(iv) Discrete distributions

Just as with testing problems, there are some difficulties in the formal theory when we work with discrete distributions, arising because it is not in general possible to find a critical region with exactly the required size. Except in extreme cases, this is, however, not a serious practical difficulty. Rather than discuss the problem in generality, we again take a special case.

Example 7.4. Poisson distribution (ctd). Let Y have a Poisson distribution of unknown mean μ, and, to be specific, suppose that we require a 0.95 upper confidence limit for μ. For a given μ_0, the best critical region against smaller values of μ is formed from small values of y. Because of the discreteness of the distribution it will be possible to find critical regions of exact size 0.05 only for isolated values of μ_0. We discard the possibility of randomized regions. One approach is to use conservative regions, i.e. for a given μ_0 we look for the largest $y_{0.95}^*(\mu_0)$ such that

$$\mathrm{pr}\{Y < y_{0.95}^*(\mu_0); \mu_0\} \leqslant 0.05. \tag{18}$$

There results the set of μ values not in the relevant critical region, namely

$$\{\mu; y \geqslant y_{0.95}^*(\mu)\}$$

and, for each possible observed y, this is equivalent to

$$\mu \leqslant t^{0.05}(y). \tag{19}$$

Pearson and Hartley (1970, Table 40) give numerical values. The general form of the solution can be best seen by examining small values of μ_0. For very small μ_0, the critical region (18) is null. However, when $\mu = \mu_0^{(1)}$, where

$$e^{-\mu_0^{(1)}} = 0.05, \quad \mu_0^{(1)} \simeq 3.00,$$

it becomes possible to include the value zero in the critical region, i.e. $y_{0.95}^*(\mu_0^{(1)}) = 1$. Similarly, as μ_0 increases further, at the value $\mu_0^{(2)}$ where

$$e^{-\mu_0^{(2)}} + \mu_0^{(2)} e^{-\mu_0^{(2)}} = 0.05, \quad \mu_0^{(2)} \simeq 4.74,$$

the critical region can be expanded to include the point 1, i.e. $y_{0.95}^*(\mu_0^{(2)}) = 2$, etc.

Thus, if we observe $y = 0$, this is in the critical region if $\mu > 3.00$, so that 3.00 is the 0.95 upper confidence limit, etc.

One way of assessing the conservative nature of the confidence limits is to plot

$$\text{pr}(\mu \leqslant T^{0.05} ; \mu) \tag{20}$$

as a function of μ. This is done in Fig. 7.1. In the continuous case this would be constant and equal to 0.95. Note, for example, that (20) is one for $\mu < 3.00$, because whatever the value of y the confidence limit is at least 3.00.

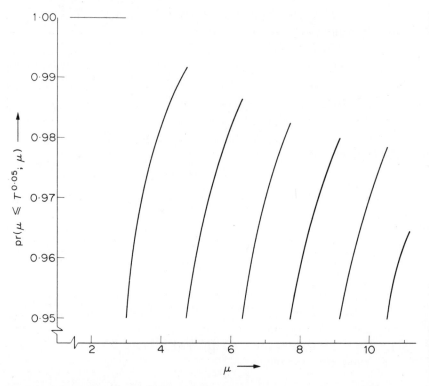

Fig. 7.1. Probability that conservative upper 0.95 confidence limit, $T^{0.05}$, is greater than true value μ, as a function of μ.

From one point of view the conservative nature of the limits is not very serious. We use the confidence limits to give a general idea of the uncertainty in μ. If, however, we are interested in the relation with some particular μ_0, we can always look at the actual significance level achieved in testing conformity with that value and, as has been argued previously, the use of some preassigned level is unnecessary. On the other hand, if we were using the confidence limits as a general guide,

perhaps to the graphical analysis of large sets of data presumed to
come from Poisson distributions, then it would be more sensible to
arrange that the function (20) plotted in Fig. 7.1 averages in some
sense around 0.95 rather than having 0.95 as its lower bound.

We shall not discuss in detail how this might be achieved. When we
approximate to the Poisson distribution by a normal distribution, the
use of a continuity correction leads to an approximation to the probability (18), and hence eventually to confidence limits which are in a
reasonable sense approximately conservative. On the other hand, it
can be shown that, if we omit the continuity correction, then limits
for which (20) has approximately the correct average value are obtained. Thus in significance testing the continuity correction should
be included when a normal approximation is used, although for the
rarer type of application mentioned at the end of the last paragraph
the continuity correction should be omitted.

(v) Confidence limits, intervals and regions

So far, we have tacitly assumed that the specification of the information about the unknown parameter is best done in terms of upper
and lower limits at various levels. By taking a lower limit and an
upper limit in combination we form an interval, called a *confidence
interval*, and in some treatments the idea of a confidence interval is
taken as the primary concept; the interval $[T_., T^{.}]$ is called a $1 - \alpha$
confidence interval if, for all $\theta \in \Omega$,

$$\text{pr}(T_. \leqslant \theta \leqslant T^{.}; \theta) = 1 - \alpha. \tag{21}$$

At least in the continuous case, a combination of upper and lower
limits at levels α_1 and α_2, with $\alpha_1 + \alpha_2 = \alpha$, will satisfy (21), so that
there are many ways of obtaining a $1 - \alpha$ interval, even when we use
only one particular type of confidence limit. It is possible to set up a
criterion of unbiasedness to choose between these. The criterion corresponding to (4.32) is that the probability of covering an arbitrary
parameter value θ' should be a maximum when θ' is equal to the true
parameter value θ. For essentially the reasons put forward in Section
4.7, this is quite unnecessary in the majority of cases. In principle we
are concerned with all levels α_1 and α_2 simultaneously, and from this
point of view we can best regard the upper and lower limits as the
primary quantities. If we do want to give intervals, one rather simple
convention is to use the limits $T_{\frac{1}{2}\alpha}$ and $T^{\frac{1}{2}\alpha}$, but in unsymmetric situations and cases with multimodal likelihoods this has the disadvantage

of including in the interval parameter values with lower likelihood than some values outside. The problem is best viewed as that of summarizing characteristics of distributions and the solution will depend on the quantitative nature of the distribution concerned. A more generally satisfactory convention is given after (22).

The optimum property (16) has a strong physical interpretation in certain location parameter problems. To take a simple illustration, suppose that T_1 and T_2 are respectively $N(\theta, \sigma_1^2)$ and $N(\theta, \sigma_2^2)$, where $\sigma_1^2 < \sigma_2^2$. Then T_1 is more powerful than T_2 for testing hypotheses about θ. It is clear from Example 7.1 that the confidence intervals for θ based on T_1 and T_2 have widths proportional to σ_1 and σ_2, so that the more powerful statistic gives a physically more precise, i.e. shorter, interval. This argument applies approximately to the mean and median in Example 7.1, and, as follows from Section 9.3(viii), there is a quite general equivalence between the optimality criterion (16) and lengths of confidence intervals in large samples.

In the very occasional application where the statement of a single interval is all that is required, special consideration would be necessary of the basis for the choice of the interval; probably the decision-theoretic formulation of Section 11.7 would be called for.

There is one situation in which even a specification by upper and lower limits can be rather misleading. This is when the data are consistent with parameter values in two or more disjoint intervals, but highly inconsistent with parameter values in between. While a full set of confidence limits would reveal the situation, the limits become a clumsy way of specifying the information.

These remarks lead us to generalize (21) to the idea of a confidence region. We call the nested set of regions $\mathcal{R}_\alpha(Y)$, depending on the random variable Y, a $1 - \alpha$ *confidence region* if for all $\theta \in \Omega$

$$\text{pr}\{\theta \in \mathcal{R}_\alpha(Y); \theta\} = 1 - \alpha,$$

$$\mathcal{R}_{\alpha_1}(Y) \subset \mathcal{R}_{\alpha_2}(Y) \quad (\alpha_1 > \alpha_2). \tag{22}$$

A rather arbitrary convention may be necessary in forming such regions, and will need to be made clear in each application; a commonly used criterion is to arrange that all parameter values in $\mathcal{R}_\alpha$ have higher likelihood than those outside. We then call $\mathcal{R}_\alpha$ a *likelihood-based confidence region*.

In the majority of applications the dependence on the parameter is sufficiently monotone for confidence limits to be entirely adequate.

We now, however, give three examples where this is not so.

Example 7.5. Sinusoidal regression. Let $Y_1, \ldots, Y_n$ be independently normally distributed with known variance σ_0^2 and with $E(Y_j) = \cos(\omega j)$, where ω is unknown but can be taken without loss of generality to satisfy $0 \leqslant \omega < 2\pi$. The log likelihood is, except for a constant,

$$-\frac{1}{2\sigma_0^2} \Sigma\{y_j - \cos(\omega j)\}^2. \tag{23}$$

There is no sufficient statistic of low dimension for this problem, and there is no strongly optimal set of confidence regions. One approach is to note that under the null hypothesis $\omega = \omega_0$ the distribution of

$$\frac{\Sigma\{Y_j - \cos(\omega_0 j)\}^2}{\sigma_0^2} \tag{24}$$

is the chi-squared distribution with n degrees of freedom, and that large values of (24) indicate lack of fit. Hence

$$\{\omega; \Sigma\{y_j - \cos(\omega j)\}^2 \leqslant c_{n,\alpha}^* \sigma_0^2\} \tag{25}$$

forms a $1 - \alpha$ confidence region for ω. The region may consist of an interval, the union of several intervals or it may be null. We return to the interpretation of the last case in Section 7.2(vii).

It is clear on general grounds that the test statistic (24) is not sensitive for changes in ω away from ω_0, except possibly for very small values of n. The test statistic for local departures from ω_0 is, by the theory of Section 4.8,

$$\Sigma j \, Y_j \sin(\omega_0 j). \tag{26}$$

When $\omega = \omega_0$ the distribution of (26) is normal with mean and variance, respectively,

$$\Sigma j \cos(\omega_0 j) \sin(\omega_0 j) \quad \text{and} \quad \sigma_0^2 \Sigma j^2 \sin^2(\omega_0 j).$$

This suggests the confidence region

$$\{\omega; |\Sigma j \sin(\omega j)\{y_j - \cos(\omega j)\}| \leqslant k_{\frac{1}{2}\alpha}^* \sigma_0 \{\Sigma j^2 \sin^2(\omega j)\}^{\frac{1}{2}}\}. \tag{27}$$

In general, neither (25) nor (27) yields intervals of ω values.

Example 7.6. Uniform distribution of known range (ctd). Let $Y_1, \ldots, Y_n$ be independently distributed in the uniform distribution of range

$(\theta - \frac{1}{2}, \theta + \frac{1}{2})$. The likelihood is, as in Example 4.7,

$$
\begin{cases}
1 & (\theta - \frac{1}{2} < \min(y_j) \leqslant \max(y_j) < \theta + \frac{1}{2}), \\
0 & \text{otherwise}
\end{cases}
\tag{28}
$$

and the minimal sufficient statistic is $\{\min(Y_j), \max(Y_j)\}$. Now it is clear from invariance considerations that the range

$$
W = \max(Y_j) - \min(Y_j)
$$

has a distribution not depending on θ and hence is ancillary. Thus critical regions, and therefore confidence regions, to be optimal and relevant, must be based on the conditional distribution of

$$
M = \frac{1}{2}\{\max(Y_j) + \min(Y_j)\}
$$

given $W = w$.

It is easily shown that conditionally on $W = w$, M is uniformly distributed over the range

$$
(\theta - \frac{1}{2} + \frac{1}{2}w, \theta + \frac{1}{2} - \frac{1}{2}w).
\tag{29}
$$

If now we follow through the general procedure, we need to test the null hypothesis $\theta = \theta_0$ and for this we examine the likelihood ratio for θ_0 versus an alternative θ_{A0}. This likelihood ratio can be computed from the conditional distribution of M given $W = w$ and it follows immediately from (29) that the ratio is infinity, one or zero depending on whether the values θ_0 and θ_{A0} are inside or outside the interval

$$
(m - \frac{1}{2} + \frac{1}{2}w, m + \frac{1}{2} - \frac{1}{2}w).
\tag{30}
$$

The significance tests therefore take the degenerate form that values of θ are either consistent or inconsistent with the null hypothesis and thus the only achievable significance levels in a likelihood ratio test are one and zero. Therefore, if we keep to confidence regions based on likelihood ratio tests, the only statement possible is that all values in the interval (30) are entirely consistent with the data.

This example illustrates two important points. First, it is necessary to condition on the ancillary statistic if silly answers are to be avoided. If we used the marginal distribution of M to obtain inferences, we would be overlooking the information that, for example, a value of w near one tells us that θ can be determined almost exactly. The second point is that it would be possible to obtain upper and lower confidence limits from the distribution of M given $W = w$. This would be done by taking critical regions in the tail of the uniform distribution. However, if we did this, we would be excluding values of θ which as judged by a likelihood ratio test are as consistent with the data as values included.

Example 7.7. Location parameter (ctd). Suppose that $Y_1, \ldots, Y_n$ are continuous i.i.d. with p.d.f. $h(y - \theta)$, where $h(.)$ is a known density function and θ is an unknown parameter. In Example 7.1 we illustrated the calculation of confidence limits for θ when $h(.)$ is a normal density and in Example 7.2 we gave some rather *ad hoc* methods for the Cauchy density, which are, in fact, quite generally applicable. They did not, however, take account of the ancillary statistics for the problem and we now show that when this is done, an effectively unique solution is reached. The discussion has much in common with that in Example 4.15.

In general the minimal sufficient statistic is the set of order statistics $Y_{(1)} \leqslant \ldots \leqslant Y_{(n)}$ whose joint density, where it is nonzero, is

$$n! \, h(y_{(1)} - \theta) \ldots h(y_{(n)} - \theta). \tag{31}$$

Now the set of random variables

$$C_2 = Y_{(2)} - Y_{(1)}, \ldots, C_n = Y_{(n)} - Y_{(1)} \tag{32}$$

has a joint distribution not depending on θ and is therefore ancillary. If we write $T = Y_{(1)}$, then the inference about θ is to be based on the conditional distribution of T given $C = c$. Incidentally, note that, given $C = c$, any other location statistic, such as the sample mean, differs from T by a mere constant.

The Jacobian of the transformation from $Y_{(1)}, \ldots, Y_{(n)}$ to T, $C_2, \ldots, C_n$ is one and therefore the joint density of the new variables is

$$n! \, h(t - \theta) h(t + c_2 - \theta) \ldots h(t + c_n - \theta). \tag{33}$$

The conditional density of T given $C = c$ is therefore

$$f_{T|C}(t|c; \theta) = \frac{n! \, h(t - \theta) h(t + c_2 - \theta) \ldots h(t + c_n - \theta)}{\int n! \, h(t' - \theta) h(t' + c_2 - \theta) \ldots h(t' + c_n - \theta) dt'}. \tag{34}$$

This conditional distribution still has the location form, so that we have now reduced the problem to one in which a single observation is taken from a density of the form $h^+(t - \theta)$ and in which θ is the unknown parameter of interest. The form in which the conclusions are best expressed depends on the form of $h^{\dagger}(.)$, but essentially θ is formally distributed in the density $h^{\dagger}(t - \theta)$. Thus, if the distribution $h^{\dagger}(.)$ is unimodal and confidence limits are required, then these can be obtained as in (9). If the distribution is multimodal and likelihood-based confidence regions are required, we can take a cut-off level at a constant value $h^{\dagger\alpha}$ chosen to include a proportion $1 - \alpha$ of the distribution and then take

$$\{\theta; h^{\dagger}(t - \theta) \geqslant h^{\dagger\alpha}\} \tag{35}$$

as the $1 - \alpha$ confidence region.

These results, due to Fisher (1934), can be restated in the following equivalent form. Find the likelihood function

$$h(y_1 - \theta) \dots h(y_n - \theta). \tag{36}$$

Take this as a function of θ and normalize so that its total area is one; note that this is not typically meaningful with likelihood functions. This gives

$$\frac{h(y_1 - \theta) \dots h(y_n - \theta)}{\int h(y_1 - \theta') \dots h(y_n - \theta')d\theta'}, \tag{37}$$

which is clearly formally identical with the p.d.f. (34). This is, therefore, a very special case in which the normalized likelihood function can be interpreted as a sampling distribution. Confidence limits, or likelihood-based confidence regions, can be read off directly from the normalized likelihood function.

For the normal distribution this leads, by a rather roundabout route, to the familiar limits based on the sample mean. For other distributions it is not easy to get simple analytical results, although the method lends itself to numerical work on the computer.

(vi) Local properties
In Section 4.8 we saw that in regular problems for which no uniformly most powerful tests are available it is possible to get simple and general results by maximizing local power, leading to the use of the efficient score statistic

$$U = U_.(\theta_0) = \partial \log f_Y(Y; \theta_0)/\partial\theta_0$$

$$= \Sigma \, \partial \log f_{Y_j}(Y_j; \theta_0)/\partial\theta_0 = \Sigma \, U_j(\theta_0), \tag{38}$$

the form (38) applying when Y has independent components. While conceptually the emphasis on local properties is questionable, the method is a valuable one, especially with moderate or large amounts of data; in such cases the good power properties are likely to hold over a useful range and also central limit approximations to the distribution of (38) will be good.

The same idea can be used in connexion with confidence intervals and indeed has already been illustrated in Example 7.5. The method is just to form the critical region needed for testing the arbitrary null hypothesis $\theta = \theta_0$ from the upper and lower tails of the distribution of the efficient score (38). Where confidence limits are appropriate, one-sided tests based on the efficient scores can be used.

In most cases a normal approximation to the distribution of the efficient score will be in order, but there is no need in principle for this to be used. If available, the exact distribution can be used, or in other cases more refined approximations involving higher moments can be employed. A brief discussion of these approximations is given in Section 9.3(vi). Here we give just one example of the approximate procedure based on (38).

Example 7.8. Truncated exponential distribution. Suppose that $Y_1, \ldots, Y_n$ are independently distributed with the truncated exponential density

$$
\begin{cases}
\theta e^{-\theta y}/(1 - e^{-\theta y_0}) & (0 < y < y_0), \\
0 & \text{otherwise.}
\end{cases}
\tag{39}
$$

A direct calculation gives, in the notation of Section 4.8, that

$$
U_j(\theta_0) = \frac{1}{\theta_0} - Y_j - \frac{Y_j e^{-\theta_0 y_0}}{1 - e^{-\theta_0 y_0}},
\tag{40}
$$

$$
U(\theta_0) = \Sigma U_j(\theta_0),
$$

$$
i_j(\theta_0) = E\left\{-\frac{\partial^2 \log f_{Y_j}(Y_j; \theta_0)}{\partial \theta_0^2}\right\}
$$

$$
= \frac{1}{\theta_0^2} - \frac{y_0^2 e^{-\theta_0 y_0}}{(1 - e^{-\theta_0 y_0})^2}.
\tag{41}
$$

Thus, when $\theta = \theta_0$, the statistic $U = \Sigma U_j$ has zero mean and variance $i(\theta_0) = ni_j(\theta_0)$. If we want an upper confidence limit for θ we test against lower alternatives and for this small values of U are significant.

The set of θ not "rejected" in such a test is, with a normal approximation,

$$\{\theta \; ; u_.(\theta) > -k_\alpha^* \sqrt{i_.(\theta)}\} \tag{42}$$

and this will in most cases give a region of the form $\theta \leqslant t^\alpha$. In principle, the exact distribution of $U_.$ can be found, for example by inverting the moment generating function of ΣY_j, which can easily be calculated. The third and fourth cumulants of $U_.$ can be found either directly, or by using the general formulae of Bartlett (1953a), and then higher terms of an Edgeworth expansion used to refine the approximation (42). Some details are given in Section 9.3(vi). Note that ΣY_j is sufficient, so that more than local optimum properties hold. Calculation *via* (40)–(42) is, all the same, convenient.

(vii) Restricted parameter space

Sometimes there may be a restriction of the allowable parameter values to a subset of the mathematically natural ones. For example, it may be known that the mean of a normal distribution is
   (a) non-negative, or
   (b) an integer.

When we apply the previous arguments some complications arise, although these seem essentially minor. First, it may happen that at any particular level all parameter values are inconsistent with the data; this possibility arose also in the problem of sinusoidal regression, Example 7.5. Secondly, procedures may be conservative at particular parameter values.

We can illustrate these problems by considering Example 7.1, concerning the mean of a normal distribution, with the restricted parameter spaces:

$$\Omega_1 : \mu \text{ an integer;}$$

$$\Omega_2 : \mu \geqslant 0;$$

$$\Omega_3 : \mu > 0.$$

In some ways it would be physically natural to suppose that parameter spaces are closed at boundary points and hence to exclude Ω_3.

In the first case there are two extreme situations depending on whether $\sigma_0/\sqrt{n} \ll 1$ or $\sigma_0/\sqrt{n} \gg 1$. In general, except in the latter case, there are difficulties if confidence limits are calculated and we therefore concentrate on likelihood-based confidence regions. For each integer μ_0, we can form the symmetrical size α critical region and the $1 - \alpha$ confidence region for μ is the set of integers not "rejected", and is therefore

$$\{\mu; \mu \in \Omega_1, \; \bar{y}_. - k^*_{\frac{1}{2}\alpha}\sigma_0/\sqrt{n} < \mu < \bar{y}_. + k^*_{\frac{1}{2}\alpha}\sigma_0/\sqrt{n}\}. \qquad (43)$$

This set may be null, i.e. there may be no integer consistent with the data at the level in question. Of course, for some sufficiently small values of α the integers near to $\bar{y}_.$ will be included. Note that the true value of μ is included within the $1 - \alpha$ region in a proportion $1 - \alpha$ of trials in the long run.

One's attitude to this and similar phenomena depends on the attitude to formalized inferences of the kind we are discussing. At a practical level the answer that none of the contemplated parameter values is consistent with the data at a reasonable level is entirely acceptable and in extreme cases will be the "common-sense" conclusion. We would, in such cases, summarize this aspect of the analysis by giving the level of significance attaching to the integer nearest to $\bar{y}_.$; this is the level at which the confidence regions become non-null. This does bring out quite forcibly the fact that confidence region statements cannot in general be regarded as probability statements about parameters. In this particular case such a probability statement could be made if μ were restricted *a priori* to Ω_1 with a discrete probability distribution, but (43) has no such prior restriction. The confidence region (43) implicitly incorporates a significance test of the hypothesis that μ is in Ω_1, whereas in a Bayesian solution of the problem this test would be handled separately. The Bayesian approach is covered in Chapter 10. In the present situation, if we wish to end with a conclusion within the parameter space Ω_1, we should include the nearest integer to $\bar{y}_.$ within the confidence region, thereby making the region conservative.

A second criticism sometimes made of procedures that can lead to null confidence regions rests on the idea that a $1 - \alpha$ confidence region procedure is a rule for making assertions of the form

$$\mu \in R_\alpha(y),$$

a proportion $1 - \alpha$ of such assertions to be true in the long run. From this point of view the assertion that μ lies in a null region is certainly false, under the assumption of Ω_1. What is the point of making assertions known to be false? Further, if the confidence region is not null, the probability of its including μ must exceed

$1 - \alpha$, although typically not uniformly so for all values of μ in Ω_1. This line of argument seems to us to confuse the hypothetical statements that give confidence intervals their operational meaning with the kind of inference actually drawn from data. It is not a false or useless statement to say that the data deviate at such and such a significance level from all specified parameter values. Again, in Example 7.13 in the next section we shall have a situation where sometimes the confidence region is the whole parameter space. Viewed as a single statement this is trivially true, but, on the other hand, viewed as a statement that all possible parameter values are consistent with the data at a particular level it is a strong statement about the limitations of the data.

It must, then, be borne in mind that (43) in particular, and (22) in general, are entirely satisfactory statements about consistency of parameter values with observed data but not necessarily about the credibility of parameter values in the probabilistic sense as derived from use of Bayes's theorem.

When we have the parameter space $\Omega_2 : \mu \geqslant 0$, upper confidence limits are obtained by testing $\mu = \mu_0$ versus lower alternatives. Thus, if $\mu_0 \neq 0$, then the usual region is obtained, whereas if $\mu_0 = 0$, then no critical region can be formed against smaller alternatives, at least so long as we take alternatives within Ω_2. Hence the $1 - \alpha$ upper confidence limit is $[\bar{y}. + k_\alpha^* \sigma_0/\sqrt{n}]^+$, where $x^+ = \max(0, x)$. The lower confidence limit is $[\bar{y}. - k_\alpha^* \sigma_0/\sqrt{n}]^+$ and is taken to be zero when the usual value is negative, because only values of μ in Ω_2 need to be considered. It is easily verified that if $\mu \neq 0$ the confidence limits have the defining property (1); if, however, $\mu = 0$, the limits are certain to be correct, i.e. the limits are conservative at this single parameter point.

When the parameter space is $\Omega_3 : \mu > 0$, the argument is slightly different. In calculating upper limits there is no value that cannot be "rejected" in favour of a lower value and the set of values in Ω_3 not "rejected" at level α is

$$\begin{cases} 0 < \mu \leqslant \bar{y}. + k_\alpha^* \sigma_0/\sqrt{n} & (\bar{y}. + k_\alpha^* \sigma_0/\sqrt{n} > 0), \\ \text{null} & \text{otherwise.} \end{cases} \tag{44}$$

The lower limit is the same as for Ω_2. In this case the defining property (1) holds for all $\mu \in \Omega_3$.

From a practical point of view the distinction between the parameter spaces closed and open at zero seems very minor and the distinction between the answers is immaterial.

Similar arguments apply much more generally both to non-normal problems and to situations in which the parameter space is more complicated, as for example when the parameter may take values in any of several disjoint intervals.

(viii) Further comments on interpretation

We now return to the interpretation of confidence limits and intervals, a matter discussed briefly in subsections (ii) and (vii) of the present Section. For simplicity, consider the case where upper and lower confidence limits are a convenient way of specifying the conclusions. From the mathematical point of view, the interpretation of the confidence limit statement (3), i.e. $\theta \leqslant t^{\alpha}$, is entirely unambiguous. It is calculated by a rule which, assuming the correctness of the model, will be wrong only in a proportion α of a long run of hypothetical repetitions. Our attitude toward the statement $\theta \leqslant t^{\alpha}$ might then be taken to be the same as that to other uncertain statements for which the probability is $1 - \alpha$ of their being true, and hence we might say that the statement $\theta \leqslant t^{\alpha}$ is virtually indistinguishable from a probability statement about θ. However, this attitude is in general incorrect, in our view, because the confidence statement can be known to be true or untrue in certain unexceptional cases, such as those described in (vi) and Example 7.13. The system of confidence limits simply summarizes what the data tell us about θ, given the model. If external qualitative information about θ is available, this can be combined qualitatively with the information in the confidence limits. In effect, this is how we arrive at confidence statements such as (43) and (44) in the previous subsection. The very special case where external information can be formulated probabilistically is handled by the Bayesian methods to be discussed in Chapter 10.

To take the distinction between confidence limit statements and direct probability statements about θ a stage further, there are two kinds of manipulation that are in general entirely invalid:

(a) it is wrong to combine confidence limit statements about different parameters as though the parameters were random variables,

(b) it is wrong to examine probabilistically the connexion of the parameter with particular values determined from external sources.

An exception to (a) is, of course, the situation corresponding to simple multiplication of confidence levels for the intersection of two independent confidence statements.

As an example of (b), suppose that we are concerned with the mean of a normal distribution and are particularly interested in the value $\mu = 0$. Even though the confidence interval statements about μ have the formal structure that μ is apparently normally distributed with mean and variance $\bar{y}.$ and σ_0^2/n, we cannot therefrom conclude that the probability that $\mu = 0$ is zero, or that the probability that μ is negative is

$$\Phi(-\bar{y}.\sqrt{n}/\sigma_0). \tag{45}$$

When we examine the set of hypothetical repetitions that give the confidence interval its frequency interpretation, the value 0 is not simply related to them and a physical interpretation to (45) is not directly available as a probability. There is no clear sense within our framework in which a long-run proportion (45) of values of μ is less than zero; see, however, Exercise 7.6. ·

To summarize, when considered as the end-point of an analysis, confidence limits can be thought of as virtually equivalent to probability statements about what can be learned from the data, but statements about different parameters must not be combined by the laws of probability, nor must statements about the relation with externally specified values of the parameter be made in probability terms.

7.3   Scalar parameter with nuisance parameters

(i) Similar and invariant procedures

We now consider what in practice is the most important situation, namely where there is a single parameter ψ of interest and a nuisance parameter λ, in general a vector.

The arguments are a combination of those of the previous sections with those of Chapter 5, which dealt with significance tests of composite hypotheses. The two main ideas of that Chapter for handling tests with nuisance parameters were those of

(a) similar regions, in which probability properties are required to hold for all values of the nuisance parameter;

(b) invariant procedures, in which attention is restricted to tests having an appropriate kind of invariance.

In view of the central principle of the previous section that confidence limits, intervals or regions are formed from those values not "rejected" in good tests, it is clear that both (a) and (b) are immediately applicable to the present problem. We shall concentrate largely on (a), the use of similar regions.

We call T^{α} a $1 - \alpha$ upper confidence limit for ψ in the presence of the nuisance parameter λ if for all ψ and λ

$$\text{pr}(\psi \leqslant T^{\alpha}; \psi, \lambda) = 1 - \alpha. \tag{46}$$

To construct such limits, and more generally to obtain intervals or regions, we test the null hypothesis that $\psi = \psi_0$ using a size α similar critical region which has good power for alternatives $\psi < \psi_0$. The main technique for finding similar critical regions is that given in Section 5.2, namely the use of distributions conditional on a complete sufficient statistic for λ when $\psi = \psi_0$.

The optimum properties of Section 7.2(ii) generalize to this situation. Rather than give a series of formal definitions which largely duplicate what has been said already, we concentrate on examples which illustrate how the method is used and also its limitations.

(ii) Some examples
The first examples to be given concern relatively simple problems involving distributions within the exponential family.

Example 7.9. Normal mean with unknown variance (ctd). Let $Y_1, \ldots,$ Y_n be i.i.d. in $N(\mu, \sigma^2)$, both parameters being unknown. Suppose that confidence limits are required for μ with σ^2 as a nuisance parameter; that is, in the general notation $\mu = \psi$, $\sigma^2 = \lambda$.

It was shown in Example 5.5 that the uniformly most powerful similar test of $\mu = \mu_0$ versus alternatives $\mu < \mu_0$ is formed from small values of the Student t statistic, i.e. has the form, in the usual notation,

$$\left\{ y; \frac{\bar{y}. - \mu_0}{\sqrt{(\text{MS}/n)}} < -t^*_{n-1, \alpha} \right\}. \tag{47}$$

Upper confidence limits are formed by taking the parameter values for which the sample is not included in the critical region, that is

$$\{\mu; \mu \leqslant \bar{y}. + t^*_{n-1, \alpha} \sqrt{(\text{MS}/n)}\}. \tag{48}$$

This procedure is invariant under the group of scale changes. It can also be derived by the procedure of Example 7.3, noting that the quantity

$$\frac{\bar{Y}_. - \mu}{\sqrt{(\text{MS}/n)}} \tag{49}$$

is a pivot, i.e. has a distribution which is the same for all parameter values. In general, confidence procedures with the similarity property (46) can be obtained whenever a pivot can be discovered involving only the parameter of interest, although further arguments are necessary to establish optimality.

A similar argument from the most powerful similar regions leads to confidence limits for σ^2 when μ is a nuisance parameter, based on the chi-squared distribution of the pivotal quantity

$$(n-1)\text{MS}/\sigma^2. \tag{50}$$

Example 7.10. Normal-theory linear model (ctd). Although the details are more complicated, essentially the same arguments as used in Example 7.9 deal with the calculation of confidence limits for a component regression parameter in the general normal-theory linear model of Example 2.4. The most powerful similar critical regions are based on the Student t distribution of the pivotal quantity

$$\frac{\text{least sq. est} - \text{parameter value}}{\text{est. st. dev. of least sq. est.}}. \tag{51}$$

Example 7.11. Exponential family problems. The general form of a number of important problems can best be seen by considering a multi-parameter exponential family situation in which the dimensionality of the minimal sufficient statistic equals that of the parameter. The likelihood can be written, as in Section 2.2(vi),

$$\exp\left\{-\sum_{r=1}^{q} \phi_r s_r + c^\dagger(\phi) + d^\dagger(y)\right\}, \tag{52}$$

where the ϕ_r's are the natural parameters and the S_r's the minimal sufficient statistic.

Suppose first that the parameter of interest is one of the ϕ_r, without loss of generality ϕ_1. That is, in the general notation $\psi = \phi_1$, $\lambda = (\phi_2, \ldots, \phi_q)$. Then the most powerful similar regions are obtained by conditioning on $(S_2, \ldots, S_q)$, i.e. what in effect we have to do is to compute the conditional distribution of S_1 given $S_2 = s_2, \ldots, S_q = s_q$, and this is, at least in principle, simple, the answer depending only on ϕ_1. Note that if the parameter of interest is a linear combination of the ϕ_r's, a simple reparameterization will make ϕ_1 the parameter of interest.

A second possibility is that, perhaps after an initial linear transformation of the ϕ_r's, the parameter of interest is a ratio, say $\psi = \phi_1/\phi_2$. When we test the null hypothesis $\psi = \psi_0$, the sufficient statistic for the nuisance parameter is obtained by writing $\phi_1 = \psi_0\phi_2$ in (52) and is

$$(\psi_0 S_1 + S_2, S_3, ..., S_q) \tag{53}$$

and depends on ψ_0. We need the distribution of S_1 given (53). This situation is conceptually and mathematically less clear than the first; from some points of view the fact that we are getting confidence regions by, in effect, conditioning on many different sufficient statistics makes the solution logically more complex.

Example 7.9, concerned with the normal distribution with both parameters unknown, illustrates both the above possibilities. The likelihood is, when written in the form (52),

$$\exp\left\{-\frac{\Sigma y_j^2}{2\sigma^2} + \frac{n\bar{y}\,\mu}{\sigma^2} - \frac{n\mu^2}{2\sigma^2} - n\log\sigma - \tfrac{1}{2}n\log(2\pi)\right\}, \tag{54}$$

so that the natural parameters are $1/\sigma^2$ and μ/σ^2. Thus inference about $1/\sigma^2$, i.e. in effect about σ^2, is mathematically the more straightforward to examine, requiring the conditional distribution of ΣY_j^2 given $\bar{Y} = \bar{y}.$; because under this condition

$$\Sigma Y_j^2 = \Sigma(Y_j - \bar{Y})^2 + \text{const} \tag{55}$$

and $\Sigma(Y_j - \bar{Y})^2$ is distributed independently of $\bar{Y}.$, the usual results follow immediately. On the other hand, the mean μ is the ratio of two natural parameters and for $\mu = \mu_0$, the conditioning statistic is $\Sigma Y_j^2 - 2n\bar{Y}.\mu_0$ or equivalently $\Sigma(Y_j - \mu_0)^2$. Because of the special features of this problem the conditional distribution has a simple form and leads, as we have seen in Example 7.9, to the limits based on the Student t distribution.

Finally, returning to the general model (52), suppose that the parameter of interest is neither a natural parameter nor a ratio of such. If $\psi = \psi(\phi_1, ..., \phi_q)$ we can, when $\psi = \psi_0$, solve for say ϕ_1 in the form $\phi_1 = \phi_1^0(\phi_2, ..., \phi_q)$. Thus the likelihood under $\psi = \psi_0$ has the form

$$\exp\{-s_1\phi_1^0(\phi_2, ..., \phi_q) - \sum_{r=2}^{q} s_r\phi_r + c^\dagger(\phi) + d^\dagger(y)\}, \tag{56}$$

and the minimal sufficient statistic is in general of dimension q. Then the reduction of dimensionality that enables similar regions to be constructed in the space of the full set of sufficient statistics does not hold. Thus either no similar regions exist, and approximate arguments have to be used, or similar regions have to be found by some other approach. These results extend the discussion of Section 5.2(ii).

Thus in the case of the normal distribution the present methods apply when the parameter of interest is σ^2 or μ/σ^2 or μ. In fact, however, similar regions based on the non-central Student t distribution may be found when the parameter of interest is $(\mu - a)/\sigma$, where a is an arbitrary constant, as is clear by applying invariance arguments.

Example 7.12. Comparison of Poisson means (ctd). Let Y_1 and Y_2 have independent Poisson distributions with means μ_1 and μ_2. Tests of composite null hypotheses in this situation have been examined in Examples 5.2 and 5.12 and the results are immediately applicable to interval estimation. The natural parameter for the Poisson distribution is the log mean, and therefore simple confidence intervals are available for $\log \mu_1 - \log \mu_2$, or equivalently for $\psi = \mu_1/\mu_2$. Example 5.2 shows how to test the null hypothesis, $\psi = \psi_0$, *via* the binomial distribution of Y_1 given $Y_1 + Y_2 = s$, which has parameter $\psi_0/(1 + \psi_0)$. Hence the calculation of confidence limits for ψ is reduced to the consideration of limits for a binomial parameter from a single observation.

We are led to this solution once we make the strong requirement that the defining probability property of the confidence limit must be valid whatever the value of the nuisance parameter. The use of the conditional distributions given $Y_1 + Y_2$ is, in any case, intuitively reasonable because it has some of the features of an ancillary statistic. It gives by itself no information about ψ, but it does enable the precision of the inference to be specified; see Section 2.2(viii).

The results of Example 5.12 show that while confidence limits for $\log \mu_1/\log \mu_2$ can formally be obtained by a similar argument, the answer is unacceptable.

Example 7.13. Ratio of two normal means. Suppose that we have two independent sets of data of sizes n_1 and n_2 drawn from normal distributions of unknown means μ_1 and μ_2 and known variances σ_1^2 and σ_2^2. By sufficiency, inference about any aspect of μ_1 and μ_2 will involve the data only through the means $\bar{Y}_{1.}$ and $\bar{Y}_{2.}$. Suppose that the parameter of interest is the ratio $\rho = \mu_1/\mu_2$. Because this is the ratio of two natural parameters, we know from the previous discussion in Example 7.11 that similar regions can, in principle, be found.

Given the null hypothesis $\rho = \rho_0$, it is easily shown that the sufficient statistic for the nuisance parameter is

$$\rho_0 \frac{\bar{Y}_{1.}}{\sigma_1^2/n_1} + \frac{\bar{Y}_{2.}}{\sigma_2^2/n_2} . \tag{57}$$

We have therefore to examine the distribution of $(\bar{Y}_{1.}, \bar{Y}_{2.})$ given (57) and one way to achieve this is to note that (57) and the statistic

$$\bar{Y}_{1.} - \rho_0 \bar{Y}_{2.} \tag{58}$$

are independently distributed. We can therefore base the size α critical region on extreme values of the unconditional distribution of the statistic (58) and take it to be of the form

$$\left\{ y ; |\bar{y}_{1.} - \rho_0 \bar{y}_{2.}| > k_{\frac{1}{2}\alpha}^* \left(\frac{\sigma_1^2}{n_1} + \rho_0^2 \frac{\sigma_2^2}{n_2} \right)^{\frac{1}{2}} \right\} . \tag{59}$$

Thus the corresponding confidence region is

$$\{ \rho : |\bar{y}_{1.} - \rho \bar{y}_{2.}| \leqslant k_{\frac{1}{2}\alpha}^{*2} (\sigma_1^2/n_1 + \rho^2 \sigma_2^2/n_2)^{\frac{1}{2}} \} . \tag{60}$$

To examine the form of the confidence region, note that its end points are the solutions of the quadratic equation in ρ'

$$\left(\bar{y}_{1.}^2 - \frac{\sigma_1^2}{n_1} k_{\frac{1}{2}\alpha}^{*2} \right) - 2\bar{y}_{1.}.\bar{y}_{2.}.\rho' + \left(\bar{y}_{2.}^2 - \frac{\sigma_2^2}{n_2} k_{\frac{1}{2}\alpha}^{*2} \right) \rho'^2 = 0. \tag{61}$$

The roots of this equation are either
  (a) both real, in which case they lie on opposite sides of the point $\bar{y}_{1.}/\bar{y}_{2.}$; or
  (b) both complex, in which case for all real ρ' strict inequality in (60) is satisfied, since it is satisfied for $\rho' = \bar{y}_{1.}/\bar{y}_{2.}$.

In case (a), the region (60) is the interval between the two roots if $\bar{y}_{2.}^2 - k_{\frac{1}{2}\alpha}^{*2} \sigma_1^2/n_1 \geqslant 0$, and the complement of this interval if $\bar{y}_{2.}^2 - k_{\frac{1}{2}\alpha}^{*2} \sigma_1^2/n_1 < 0$. In case (b), the confidence region is the whole real line, which is to say that, at the level in question, all values of ρ are consistent with the data. This situation is complementary to that discussed in Section 7.2(vi) where the confidence region was null, and the general comments on interpretation there apply here also. The occurrence of the whole real line as confidence region does not mean that a vacuous statement is being made; it will arise when μ_1 and μ_2 are probably both very close to zero. Indeed, for some applications it might be argued that if both means could be zero at the level in question, we should summarize the situation in terms of a joint confidence statement about (μ_1, μ_2) rather than about ρ. This is discussed for the more general problem of ratios of linear model parameters by Scheffé (1970). Notice that, conditionally on the confidence region not being the entire parameter space, the level of the confidence region is less than the unconditional level $1 - \alpha$, but not uniformly so. To our way of thinking, the occurrence of the whole real line as confidence region seems, on common-sense grounds, entirely proper and indeed inevitable.

It is, of course, important here as elsewhere to remember that the consistency between data and parameter values is measured by a particular significance test, in this case the optimal similar test. By using a less efficient test, we might for a particular set of data avoid phenomena of the type discussed above, but we would be ignoring some information about the parameter in question and violating (16). Difficulty only arises if we incorrectly attempt to deduce a probability statement about the parameter from the confidence region statement.

If the variances are equal and unknown and an estimate MS is available having the usual normal-theory distribution with d degrees of freedom, independently of $\bar{Y}_{1.}$ and $\bar{Y}_{2.}$, then it can be shown that essentially the above argument leads to the confidence region

$$\{\rho : |\bar{y}_{1.} - \rho \bar{y}_{2.}| \leqslant t^*_{d, \frac{1}{2}\alpha}\{\mathrm{MS}(1/n_1 + \rho^2/n_2)\}^{\frac{1}{2}}\}, \tag{62}$$

with properties similar to those of (60). The argument can be generalized to deal with the ratio of any two parameters in the normal-theory linear model, such as a simple regression intercept.

Most of the previous discussion applies to optimal procedures based on sufficient statistics, which give confidence region statements that are as precise as possible in the sense of minimizing the probability of including either all or some incorrect parameter values. However, the method of constructing confidence regions from significance tests applies generally to include the use of tests obtained by *ad hoc* arguments, and also the randomization and permutation tests of the type discussed in Chapter 6. If a non-optimal test is used, the interpretation of the resulting confidence statements is somewhat less clear. In a sense, the result is a conservative statement about consistency of parameter values with the data. This is quite natural in the situations where randomization and permutation tests are used, because the statements are valid in a much more general setting than a single parametric family. We discuss just one example.

Example 7.14. Matched pair experiment (ctd). In Example 6.7 we considered a significance test for the difference between two treatments based on a paired comparison experiment in which

(a) a very strong null hypothesis is considered in which the response observed on a particular individual is in no way affected by the treatment applied to that individual;

(b) the treatment allocation is randomized independently for each pair of individuals.

The same approach can be used to produce confidence limits for a treatment effect. We replace the null hypothesis by the more general assumption that there is an unknown parameter δ, measuring the difference between the treatments, such that the response observed on any particular individual if A is applied is exactly δ more than it would have been had B been applied to that individual. Treatments are allocated at random in a matched pair design as before.

Confidence limits and intervals are in principle obtained as follows. To test an arbitrary null hypothesis, $\delta = \delta_0$, subtract δ_0 from all observations on individuals receiving treatment A, i.e. from the differences analysed in Example 6.7. The new "observations" are, if $\delta = \delta_0$, derived from a situation to which the original null hypothesis of complete equivalence applies. Hence the distribution under the null hypothesis of a suitable test statistic $T(\delta_0)$, say, can be found in virtue of the randomization, and either an exact level of significance computed or some convenient approximation used.

This can be done in principle for all δ_0 and hence confidence regions, typically intervals, found. For confidence limits, one-sided tests are used. In practice, the level of significance is calculated for a few suitably chosen values of δ_0 and the approximate limiting values for a given α found by interpolation. In the one-sided case, with $p_{obs}(\delta_0)$ denoting the level of significance achieved in testing $\delta = \delta_0$, linear interpolation of $\Phi^{-1}\{p_{obs}(\delta_0)\}$ against δ_0 will often give good results. There is a wide choice of test statistics available.

Note that while the hypothetical frequency properties of confidence regions would be retained if a different form of test statistic were used at different values of δ_0, the nesting requirement (2) of the limits would in general be violated. Hence we do not regard the use of a different form of test statistic at different values of δ_0 as permissible.

The same numerical procedure can be viewed as giving a distribution-free set of confidence limits based on the quite different assumption of permutation theory, namely that for the jth pair the responses to treatments A and B are independent random variables with distribution functions $F_j(x - \delta)$ and $F_j(x)$, where $F_j(.)$ is an arbitrary and unknown distribution function.

An analogous procedure can be used for some other problems. For example, with a regression problem with dependent variable Y and explanatory variable x, it might be assumed that the effect of assigning a particular individual to a value x' is to increase the associated value of Y by exactly $\beta x'$, as compared with what it would have been with a zero value of the explanatory variable. For any particular value β_0, we subtract $\beta_0 x$ from each Y and then apply a randomization test for the null hypothesis of zero regression.

7.4   Vector parameter

(i) Confidence regions for a vector parameter

We now consider the situation where the parameter of interest is a vector; there may or may not be nuisance parameters. One new point in such situations is that no uniformly most powerful critical regions will exist and therefore we have to use some compromise procedure with reasonable sensitivity against alternatives concerning all parameters. Often it will be reasonable to look for likelihood-based regions, i.e. ones such that all parameter values within a particular confidence region have greater likelihood than those outside and, therefore for which the boundary is a surface of constant likelihood.

Two further practical aspects are, first, that confidence regions in more than two dimensions are difficult to describe in easily grasped form. One approach is to show two-dimensional sections and this will, of course, be completely appropriate when the region is ellipsoidal. A second, and more fundamental, point is that on general grounds it is, wherever feasible, sound to separate distinct aspects of the problem into separate parameters, i.e. to give each component parameter a distinct physical interpretation. Then confidence regions for vector parameters will not be required. This does, however, assume that a rather clear formulation of the problem has been achieved and it is therefore, on the whole, in the more exploratory stages of an investigation that confidence regions for vector parameters are most likely to be useful.

The most important example of a confidence region for a vector parameter is that involving the normal-theory linear model.

Example 7.15. Partitioned parameter in normal-theory linear model. Suppose that the $n \times 1$ random vector Y has components that are independently normally distributed with variance σ^2 and with expectation

$$E(Y) = \mathbf{x}\beta = (\mathbf{x}_1 : \mathbf{x}_2) \begin{pmatrix} \beta_1 \\ \vdots \\ \beta_2 \end{pmatrix}, \qquad (63)$$

where the $q_x \times 1$ parameter β has been decomposed into the two components β_1 and β_2 of sizes $q_{x_1} \times 1$ and $q_{x_2} \times 1$, and all matrices are of full rank. If interest centres on β_1, we use the invariant test of the null hypothesis $\beta_1 = \beta_1^0$ with the F statistic

$$\frac{(\text{SS for fitting } \beta - \text{SS for fitting } \beta_2 \text{ with } \beta_1 = \beta_1^0)/q_{x_1}}{(\text{Residual SS for fitting } \beta)/(n - q_x)}$$

and hence the confidence region has the form

$$\left\{ \beta_1 : \frac{(\hat{\beta}_1 - \beta_1)^{\text{T}} \{\mathbf{x}_1^{\text{T}} \mathbf{x}_1 - \mathbf{x}_1^{\text{T}} \mathbf{x}_2 (\mathbf{x}_2^{\text{T}} \mathbf{x}_2)^{-1} \mathbf{x}_2^{\text{T}} \mathbf{x}_1\} (\hat{\beta}_1 - \beta_1)}{(q_{x_1} \, \text{SS}_{\text{res}})/(n - q_x)} \leqslant F^*_{q_{x_1}, \, n - q_x; \alpha} \right\},$$
$$(64)$$

an ellipsoid centred on $\hat{\beta}_1$ whose bounding contours have constant marginal likelihood.

The next example is essentially the simplest special case of this, but is worth treating separately.

Example 7.16. Several normal means. Suppose that there are m independent groups each of r independent observations all normally distributed with unknown variance σ^2. Denote the observed means by $\bar{y}_1, \ldots, \bar{y}_m$. corresponding to expected values $\mu_1, \ldots, \mu_m$, with MS_w the independent estimate of σ^2 based on the within-group variation.

The rotation-invariant analysis of variance test procedure leads to the confidence region for the vector μ

$$\left\{ \mu; \frac{r\Sigma(\bar{y}_{j.} - \mu_j)^2}{m\,\mathrm{MS}_w} \leqslant F^*_{m, mr-m;\alpha} \right\},\tag{65}$$

a hypersphere centred on the sample means, where the constant is the appropriate upper limit of the F distribution.

The confidence region (65) is likelihood-based in that the bounding contours have constant likelihood for any fixed σ^2.

Of course, the confidence region (65) summarizes what the data tell us about the vector $\mu = (\mu_1, \ldots, \mu_m)$ and we cannot infer directly exactly what the data tell about an individual component such as μ_1. However, conservative statements can be made, as we shall see in Example 7.17.

We can obtain formal confidence regions with the right probability of covering μ and with shape entirely different from (65). In particular, we can produce a rectangular confidence region from individual confidence intervals for each μ_j. Thus in the case where σ^2 is known, a formal $1 - \alpha$ confidence region is given by

$$\left\{ \mu; |\bar{y}_{j.} - \mu_j| \leqslant b_j^* \frac{\sigma}{\sqrt{r}} \quad (j = 1, \ldots, m) \right\},\tag{66}$$

where

$$\prod_{j=1}^{m} \{2\Phi(b_j^*) - 1\} = 1 - \alpha.\tag{67}$$

Yet it is clear that, especially in rather extreme cases, this is quite inadequate as a summary of the information provided by the data. To see this we have only to consider the limiting case

$$b_1^*, \ldots, b_{m-1}^* \to \infty, \quad b_m^* \to k^*_{\frac{1}{2}\alpha}.\tag{68}$$

It is not too difficult to see that for any fixed α, the area of the rectangular region (66) is greater than the area of (65), even in the symmetric case $b_1^* = \ldots = b_m^*$. Of course, (66) is likelihood-based separately on each coordinate axis.

While the use of likelihood-based regions is a little arbitrary, and not always possible, any regions that depart strongly from the likelihood base will contain points with much lower likelihood than some excluded points and this cannot be acceptable. This comment applies to (66), where it would be possible to exclude, say, $\mu = 0$ when in fact such a value is quite consistent with the data. We shall see in the next subsection a situation where rectangular regions arise in a slightly different setting.

(ii) Multiple confidence regions for a set of parameter functions

We now turn to a type of problem which mathematically is closely connected with the calculation of confidence regions for vector parameters, but which conceptually is different. We discuss only a simple special case to illustrate the ideas. For a more extensive treatment, see Miller (1966).

The problem is concerned with a number of scalar combinations of the parameters which are of physical interest, it being required to calculate intervals for all of them such that the probability that they are simultaneously correct is some pre-assigned value $1 - \alpha$. Such problems are often referred to as *multiple comparison problems*.

Exercise 7.17. Simultaneous limits for linear combinations of means. Suppose that we have the same data and model as in Example 7.16. For an individual unrestricted linear combination $\Sigma a_j \, \mu_j$, it follows, either from first principles or by results on the normal-theory linear model, that equi-tailed confidence limits are

$$\Sigma a_j \, \bar{y}_{j.} \pm t^*_{mr-m, \, \frac{1}{2}\alpha}(\mathrm{MS}_w \, \Sigma a_j^2 / r)^{\frac{1}{2}}, \tag{69}$$

where MS_w is the usual residual mean square.

Suppose now that we are interested in all such linear combinations. If we restrict ourselves to symmetric intervals that have lengths a fixed multiple of the standard error, it follows that we must try to find a constant b^*_α such that for all $\mu_1, \ldots, \mu_m$

$$\mathrm{pr} \left\{ \Sigma a_j \, \bar{Y}_{j.} - b^*_\alpha \left(\frac{\mathrm{MS}_w \, \Sigma a_j^2}{r} \right)^{\frac{1}{2}} \leqslant \Sigma a_j \mu_j \leqslant \Sigma a_j \bar{Y}_{j.} + b^*_\alpha \left(\frac{\mathrm{MS}_w \, \Sigma a_j^2}{r} \right)^{\frac{1}{2}} ; \right.$$

$$\left. \text{all } a_1, \ldots, a_m \right\} = 1 - \alpha. \tag{70}$$

There are various ways of relating this to the confidence region (65) for μ. Notice that there is an arbitrary scale factor in (69) and (70); that is, multiplication of all $a_1, \ldots, a_m$ by a fixed constant leaves the statements unaltered. Thus, without loss of generality, set $\Sigma a_j^2 = 1$. Then a quite simple way of relating (70) to (65) is to notice that the limits in (70) are pairs of planes in the m-dimensional space of $(\mu_1, \ldots, \mu_m)$ vectors at a constant distance $b^*_\alpha(\mathrm{MS}_w / r)^{\frac{1}{2}}$ from, and parallel to, the planes through the point $(\bar{y}_1, \ldots, \bar{y}_m)$. That is, the limits form a pair of planes tangential to a sphere of radius $b_\alpha^*(MS_w/r)^{\frac{1}{2}}$ centred at $(\bar{y}_1, \ldots, \bar{y}_m)$. As $a_1, \ldots, a_m$ take on all possible values, these tangent planes describe precisely the boundary of a sphere. Therefore the inequalities in (70) hold simultaneously only for points $(\mu_1, \ldots, \mu_m)$ inside the sphere, so that (70) is exactly equivalent to the confidence region (65) for μ and necessarily

$$b_\alpha^* = (mF_{m, mr-m;\alpha}^*)^{\frac{1}{2}}. \tag{71}$$

Essentially the argument is that any set of statements implied by μ lying within the $1 - \alpha$ confidence region can be made at a level at least $1 - \alpha$ and a set of such statements sufficiently rich to be equivalent to the confidence region has level exactly $1 - \alpha$.

In problems of this type, we may be interested in a particular subset of linear combinations. If the subset spans a subspace of the full parameter space, there is a straightforward modification of (71). To illustrate this, suppose that we are interested in contrasts among the means, i.e. in linear combinations $\Sigma a_j \mu_j$ with $\Sigma a_j = 0$. We shall derive the simultaneous confidence statement corresponding to (70) by a slightly different argument.

The standardized deviation of a sample contrast from its expected value is

$$z_a = \frac{\Sigma a_j(\bar{y}_j. - \mu_j)}{(MS_w \Sigma a_j^2/r)^{\frac{1}{2}}}. \tag{72}$$

If now, for a given vector y, we take the supremum of z_a^2 over all contrasts we have, for example using the method of Lagrange multipliers, that $a_j \propto (\bar{y}_j. - \mu_j) - (\bar{y}.. - \bar{\mu}.)$ and

$$\sup z_a^2 = \frac{r\Sigma\{(\bar{y}_j. - \bar{y}..) - (\mu_j - \bar{\mu}.)\}^2}{MS_w} = (m-1)F(\mu), \tag{73}$$

where $F(\mu)$ is the ordinary F ratio for testing agreement with the basic set of contrasts $\mu_1 - \bar{\mu}., \ldots, \mu_m - \bar{\mu}.$, considered as a null hypothesis. Thus, since the maximum squared standardized deviation is less than $(m-1)F_{m-1, mr-m;\alpha}^*$ with probability $1 - \alpha$, it follows that for all contrasts simultaneously

$$\mathrm{pr}\left[\Sigma l_j \bar{Y}_{j.} - \left\{(m-1)F^*_{m-1,mr-m;\alpha}\frac{\Sigma l_j^2}{r}\mathrm{MS}_w\right\}^{\frac{1}{2}} \leqslant \Sigma l_j \mu_j\right.$$

$$\left. \leqslant \Sigma l_j \bar{Y}_{j.} + \left\{(m-1)F^*_{m-1,mr-m;\alpha}\frac{\Sigma l_j^2}{r}\mathrm{MS}_w\right\}^{\frac{1}{2}}\right] = 1 - \alpha. \quad (74)$$

Thus, because the dimension spanned by linear combinations is now $m-1$ rather than m, (70) holds with b_α^* given by $\{(m-1)F^*_{m-1,mr-m;\alpha}$ instead of (71). If we maximize z_a^2 over a space of lower dimension, we correspondingly obtain (70) with appropriately changed degrees of freedom.

This discussion can be extended directly to the general normal-theory linear model of Example 7.15. For a simple linear regression, the simultaneous confidence statement has the physical interpretation of the usual confidence band for the linear regression function. The discussion also extends to the multivariate linear model. One simple special case of this is that of simultaneous limits for linear combinations $a^T \mu$ of a multivariate normal mean; the simultaneous aspect of the corresponding significance test for this was mentioned in Example 5.18.

Mathematically, the idea of simultaneous or multiple confidence regions is quite natural, and is a reasonable way of interpreting the confidence region for the appropriate vector parameter, as in the last example. Physically, it is unlikely that one would want to characterize uncertainty by the probability that a very large number of statements are all simultaneously correct. In the example we have discussed here, one motivation might be that conservative statements can be made about linear combinations that are indicated to be of interest by looking at the data.

If in a large analysis one is going to consider many distinct statements about pre-selected combinations it is often more natural to consider the hypothetical probability of error associated with individual statements. If, however, all the distinct statements bear on the same issue, then it is better to think of the problem as one of the estimation of a single vector parameter.

(iii) Selected comparisons
We have now examined estimation of a vector parameter from two different points of view. In the first and most direct approach attention

is concentrated directly on a region for the vector parameter. In the second formulation, we consider a number of statements about scalar parameters, the probability of simultaneous correctness being required. In this final subsection we consider a third interpretation of what are still essentially the same mathematical results.

In the situation of Example 7.17, it may quite often be that the contrasts in terms of which the conclusions are most usefully expressed are determined only after inspection of the data. When the primary emphasis is on significance testing of null hypotheses andd the contrast selected is the one most significant among some set of possible contrasts, an adjustment to the significance level is necessary; see Section 3.4(iii). Similar remarks apply to confidence intervals.

No matter how we select the contrast or contrasts in terms of which the conclusions are finally presented, the confidence limits (74) can be used, because they apply to all possible contrasts. In practice, however, this solution tends to be very conservative, i.e. to produce unnecessarily wide intervals. There are two reasons for this:

(a) it will be unusual for selection to take place over all possible contrasts, so that the inflation of the multiples from $t^*_{mr-m, \alpha}$ to $\{(m-1)F^*_{m-1, mr-m; \alpha}\}^{\frac{1}{2}}$ is too great. If the method of selection used can be specified, then in principle a better solution can be found;

(b) often selection will be of contrasts which deviate most significantly from zero. The use of the F distribution is thus correct in a significance test, but is conservative for finding a confidence interval. This is qualitatively obvious from the consideration that if the selected contrast is very highly significant and all contrasts orthogonal to it are small then little selection is involved and the usual confidence limits based on Student's t will be very nearly appropriate at the usual levels. A full discussion of this problem seems hard.

The general point, then, is that if we have a simultaneous confidence region for a collection of parametric functions, and one or more particular functions are selected in the light of the data, conservative confidence regions for the selected parameters are implied by the simultaneous region.

7.5   Estimation of future observations

In the whole of the previous work, it has been supposed that the objective is an unknown parameter in the probability model assumed to have generated the data. Depending somewhat on the nature of the application, the unknown parameter represents in a concise form all future observations that might be obtained from the same random system, or some notional "true" value free of random errors of determination. Occasionally, however, it is required to predict explicitly values of one or more future observations taken from the same random system.

Suppose then that we observe a random vector Y with p.d.f. $f(y; \theta)$ and that interest is focussed on an as yet unobserved random vector $Y^\dagger$ independent of Y with p.d.f. $g(y; \theta)$.

Very often $Y^\dagger$ will be a scalar with distribution the same as that of one of the components of Y; if $Y^\dagger$ and Y are dependent their joint distribution has to be specified. We call a region $\mathcal{P}_\alpha(Y)$ in the sample space of $Y^\dagger$ a $1 - \alpha$ *prediction region* for $Y^\dagger$ if for all θ

$$\mathrm{pr}\{Y^\dagger \in \mathcal{P}_\alpha(Y); \theta\} = 1 - \alpha, \tag{75}$$

where the probability is calculated over the distribution of Y and $Y^\dagger$. When $Y^\dagger$ is a scalar, there is an obvious extension to upper and lower prediction limits and to prediction intervals. A nesting condition analogous to (2) is required.

An extended discussion will not be given. It is clearly essential to find a function of Y and $Y^\dagger$ having a distribution not involving θ. This is possible if, denoting for the moment the parameters governing the random variables Y and $Y^\dagger$ by θ and $\theta^\dagger$, a similar test can be found of the null hypothesis $\theta = \theta^\dagger$. Suppose that w_α is a size α similar critical region in the sample space of $(Y, Y^\dagger)$. If now for $Y = y$, we define a region in the sample space of $Y^\dagger$ by

$$\mathcal{P}_\alpha(y) = \{y^\dagger; (y, y^\dagger) \notin w_\alpha\}, \tag{76}$$

it is clear that

$$\mathrm{pr}\{Y^\dagger \in \mathcal{P}_\alpha(Y); \theta\} = 1 - \alpha. \tag{77}$$

The introduction of the notional parameter $\theta^\dagger$ is, in the first place, just a device for establishing the mathematical·connexion with the problem of constructing similar regions. However, the use of optimum critical regions leads to prediction regions having minimum probability under alternative distributions; in some cases, this is a reasonable way of requiring the prediction region to be as small as possible subject to the defining condition (75). If θ is a vector parameter, it may be appropriate to take $\theta^\dagger$ as possibly different from θ only in certain components.

By taking various kinds of critical region, we can, at least in simple cases, obtain upper limits, lower limits and intervals. An alternative approach is to look directly for a function of Y and $Y^\dagger$ having a distribution not involving θ. Subject to some conditions of monotonicity, a region for $Y^\dagger$ can then be obtained by inversion. The function used here is analogous to a pivotal quantity in the study of confidence regions and it is therefore reasonable to extend the term pivot to cover this case.

Example 7.18. Normal distribution. The simplest illustration is when Y consists of n i.i.d. random variables from $N(\mu, \sigma_0^2)$, where μ is unknown and σ_0^2 is known. Let $Y^\dagger$ be a further independent random variable with the same distribution. Critical regions for the notional null hypothesis that $\theta = \theta^\dagger$ are formed *via* the standardized normal deviate

$$\frac{Y^\dagger - \bar{Y}}{\sigma_0(1 + 1/n)^{\frac{1}{2}}}, \tag{78}$$

the pivot for the problem.

This leads to upper and lower limits and, in particular, to equitailed prediction limits

$$[\bar{Y} - k^*_{\frac{1}{2}\alpha}\sigma_0(1 + 1/n)^{\frac{1}{2}}, \ \bar{Y} + k^*_{\frac{1}{2}\alpha}\sigma_0(1 + 1/n)^{\frac{1}{2}}]. \tag{79}$$

The interpretation of these limits depends on the joint distribution of Y and $Y^\dagger$. In particular, the limits (79) do *not* have the interpretation that, for given $Y = y$, a proportion $1 - \alpha$ of repeat observations on $Y^\dagger$ lie within the limits.

If it is required to obtain limits containing with specified probability all or most of a set of future measurements, then the problem must be reformulated; for example, one might consider a set $Y_1^\dagger, \ldots,$ $Y_m^\dagger$ of future observations and require prediction limits containing them all with specified probability $1 - \alpha$. This is, in principle, easily accomplished by finding upper and lower $1 - \frac{1}{2}\alpha$ limits for, respectively, $\max(Y_j^\dagger)$ and $\min(Y_j^\dagger)$.

In the case of unknown variance, an interval estimate corresponding to (78) for a single random variable $Y^\dagger$ cannot be deduced from a test of the notional null hypothesis $\mu = \mu^\dagger$, $\sigma = \sigma^\dagger$ because $\mu^\dagger$ and $\sigma^\dagger$ are not identifiable from one observation. However, if we assume $\sigma = \sigma^\dagger$ and set up a test of $\mu = \mu^\dagger$, we are led to

$$\left[\bar{Y}. - t^*_{n-1, \frac{1}{2}\alpha} \sqrt{\left\{ MS \left(1 + \frac{1}{n}\right)\right\}} , \ \bar{Y}. + t^*_{n-1, \frac{1}{2}\alpha} \sqrt{\left\{ MS \left(1 + \frac{1}{n}\right)\right\}}\right]. \tag{80}$$

With more than one observation to be predicted, the possibilities are more diverse.

This discussion is easily extended to prediction limits for future observations on a normal-theory linear model. The results can, of course, be combined with those of Section 7.4 to form prediction bands for observations on regression models.

Example 7.19. Poisson distribution. Suppose that Y has a Poisson distribution of mean μ and $Y^\dagger$ a Poisson distribution of mean $a\mu$, where a is a known constant. Having observed Y, it is required to predict $Y^\dagger$. One interpretation of this is that a Poisson process has been observed for a time t_0, Y being the number of events. It is required to predict the number of events in the Poisson process in some future time interval of length at_0.

In accordance with the general discussion, we suppose that $Y^\dagger$ has mean $a\mu^\dagger$ and test the null hypothesis $\mu^\dagger = \mu$. The critical regions are formed from the conditional binomial distribution of, say, $Y^\dagger$ given that $Y + Y^\dagger = t$.

Thus for any $Y = y$ and for any $Y^\dagger = y^\dagger$, we can in principle determine whether $y^\dagger$ lies in the conservative critical region of size at most α, using one-sided or equi-tailed regions depending on the form of prediction required. If we use a normal approximation without a continuity correction we base the critical region on the fact that

$$\left(Y^\dagger - \frac{at}{1+t}\right) \bigg/ \left\{\frac{at}{(1+t)^2}\right\}^{\frac{1}{2}} \tag{81}$$

is approximately a standardized normal deviate. Thus, on writing $t = y + y^\dagger$, an approximate $1 - \alpha$ region for $y^\dagger$ is defined by

$$\left\{y^\dagger - \frac{a(y + y^\dagger)}{(1 + a)}\right\}^2 \bigg/ \left\{\frac{a(y + y^\dagger)}{(1 + a)^2}\right\} \leqslant k^{*2}_{\frac{1}{2}\alpha}. \tag{82}$$

On solving for $y^\dagger$, the limits are found as the roots of a quadratic equation.

Bibliographic notes

The formal theory of confidence intervals and regions stems from Neyman(1937). The mathematical techniques being those of testing hypotheses, the references of Chapters 4–6 are relevant. Confidence statements are not the same as probability statements and alternative approaches attempt to derive probability statements for unknown parameters. Of these, the Bayesian approach to be studied in Chapter 10 is straightforward conceptually once prior distributions are accepted and found numerically.

The method of fiducial probability aims to get probability statements without the use of Bayes's theorem; probabilities concerning pivotal quantities are inverted into formal statements about parameters. This method predates confidence intervals, being introduced by Fisher in the early 1930's. In the late 1930's strong disagreement between the fiducial and confidence interval approaches emerged centering partly on the physical interpretation of the answers and more specifically on the legitimacy of manipulating the answers as ordinary probabilities. A crucial instance is that of Example 5.6, the two-sample normal problem with unequal variances. The fiducial solution, the so-called Behrens-Fisher distribution, does not have the frequency property required for confidence intervals; see also Example 10.5.

For the difficulties of linking fiducial theory in general with Bayes's theorem, see Lindley (1958). Fiducial theory is a continuing field of study although, in our opinion, no account satisfactorily justifying the Behrens-Fisher distribution has yet emerged. For an ingenious frequency interpretation of fiducial distributions, see Fraser (1961); see also Exercise 7.6.

A different approach to these problems is *via* structural probability (Fraser, 1968, 1971). This is related mathematically to conditioning on ancillary statistics, but the terminology and emphasis are different, stress being laid on the nature of the model and the role of error.

O'Neill and Wetherill (1971) have given a review and bibliography on simultaneous confidence intervals; for an extensive account, see Miller (1966).

Guttman (1970) has reviewed estimation for as yet unobserved values; see also Aitchison and Sculthorpe (1965). Emphasis on future observations rather than parameters occurs in the philosophical work of de Finetti (1937) and in work by P. Martin-Löf, unpublished at the time of writing.

Further results and exercises

1. The random variable Y has a binomial distribution of parameter $\frac{1}{2}$ and with an unknown number θ of trials. Obtain confidence limits for θ.

[Section 7.2]

2. Let $Y_1, \ldots, Y_n$ be i.i.d. with p.d.f. $h(y - \theta)$, where $h(y)$ is an unknown density symmetrical about zero. Explain how the null hypothesis $\theta = 0$ can be tested by using the statistic ΣY_j. To obtain a confidence region for θ, this test is applied to $\Sigma(Y_j - \theta_0)$. Show that the inversion of this test at any fixed level of significance yields an interval of θ values.

[Section 7.2; Kempthorne and Doerfler, 1969]

3. Let $Y_1, \ldots, Y_n$ be i.i.d. with the unknown continuous cumulative distribution function $F_Y(y)$. For $0 < p < 1$, the p-quantile ξ_p of the distribution is defined by $p = F_Y(\xi_p)$; any reasonable convention can be used to make ξ_p unique should this equation not have a single solution. Show how to test the null hypothesis $\xi_p = \xi_{p0}$ against (a) one-sided and (b) two-sided alternatives by counting the number of observations exceeding ξ_{p0}. Show that for large n and for one-sided alternatives $\xi_p > \xi_{p0}$, the data fall in the size α critical region if and only if the rth order statistic exceeds ξ_{p0}, where

$$r = [np - k_\alpha^* \{np(1 - p)\}^{\frac{1}{2}} + \tfrac{1}{2}].$$

Hence obtain confidence limits for ξ_p.

[Section 7.2]

4. Let $Y_{(1)}, \ldots, Y_{(n)}$ be order statistics from the uniform distribution on $(\theta, \theta + 1)$. Using the appropriate uniformly most powerful one-sided test, show that the optimum lower confidence limit is $T_\alpha = \max(Y_{(1)} - c, Y_{(n)} - 1)$, where $(1 - c)^n = 1 - \alpha$. Comment on the fact that

$$\mathrm{pr}(T_\alpha \leqslant \theta \mid Y_{(n)} - Y_{(1)} \geqslant 1 - c) = 1$$

and suggest a more relevant lower confidence limit for θ.

[Section 7.2; Pierce, 1973]

5. Following the discussion of Example 7.7, involving a single unknown location parameter, suppose that Y_1 and Y_2 are i.i.d. in the Cauchy distribution. Show that the likelihood is bimodal if and only if $c_2 = y_{(2)} - y_{(1)} > 2$. Examine the forms of conditioned confidence regions for the location parameter when $c_2 \leqslant 2$ and when $c_2 > 2$, and comment on the value of the mean as a location statistic.

[Section 7.2(v); Barnard and Sprott, 1971]

6. Show that if the scalar random variable Y has the distribution $N(\mu, 1)$ and if we observe $Y = y$, then μ is distributed in $N(y, 1)$ in the following frequency sense. Imagine hypothetical repetitions $(y^{(1)}, \mu^{(1)}), (y^{(2)}, \mu^{(2)}), \ldots$, each $y^{(j)}$ being the observed value of a random variable $Y^{(j)}$ having the distribution $N(\mu^{(j)}, 1)$ and the $\mu^{(j)}$ being arbitrary. Suppose now that all pairs are translated so that the observed value is y. That is, the jth pair is translated to $(y, \mu^{(j)} + y - y^{(j)}) = (y, \mu^{(j)\prime})$, say. Then the $\mu^{(j)\prime}$ have the frequency distribution $N(y, 1)$; this is the hypothetical frequency distribution of true means in a set of repetitions all having the observed value y. Use a similar argument starting from $Y_1, \ldots, Y_n$ i.i.d. in $N(\mu, \sigma^2)$ to show that μ has a Student t distribution, when samples are rescaled to have the observed mean and mean square.

[Section 7.2(viii); Fraser, 1961]

7. The parameter μ is an unknown integer and is observed with an error equally likely to be $+1$ or -1. On the basis of an observation y it is claimed that

$$\text{pr}(\mu = y - 1) = \text{pr}(\mu = y + 1) = \tfrac{1}{2}.$$

Show that this "probability" statement does not have the full properties of an ordinary probability statement in that it can be defeated by the following betting strategy:

if $y \geqslant 1$, bet even money that $\mu = y - 1$;
if $y < 1$, refuse to bet.

Prove that if $\mu = 0$ this strategy "wins", whereas if $\mu \neq 0$ it breaks even. Note that if the "cut-off" is chosen to be $y = m$ with probability $\pi_m > 0$, the strategy wins for all m.

[Section 7.2(viii); Buehler, 1971]

8. Obtain confidence intervals corresponding to the testing problems of Exercises 5.1 and 5.7.

[Section 7.3]

9. Suppose that it is required to obtain a $1 - \alpha$ confidence interval of preassigned width $2l$ for the unknown mean μ of a normal distribution, the variance also being unknown. To achieve this, n_0 observations are taken; let MS_0 be the estimate of variance from these observations. Now take further observations so that the total number of observations is

$$\max(n_0, [4t^{*2}_{n_0-1, \frac{1}{2}\alpha} MS_0/l^2] + 1),$$

where $[x]$ is the integer part of x. Show, by first arguing conditionally on MS_0, that, if $\bar{Y}$ is the mean of the combined data, then $(\bar{Y} - l, \bar{Y} + l)$ defines a conservative $1 - \alpha$ confidence interval for μ. Outline how, by suitable randomization, the confidence interval can be made exact. Comment on more efficient uses of the data obtained by such a procedure.

[Section 7.3; Stein, 1945]

10. The random variables $Y_1, \ldots, Y_n$ are independently normally distributed with unknown variance σ^2 and with $E(Y_j) = \gamma + \beta e^{-\rho x_j}$, where $x_1, \ldots, x_n$ are known constants and γ, β and ρ are unknown parameters. Show that there is no minimal sufficient statistic of low dimension. Consider the null hypothesis $\rho = \rho_0$ and show that for local departures $\rho = \rho_0 + \delta$, the model becomes $E(Y_j) = \gamma + \beta e^{-\rho_0 x_j} - \beta \delta x_j e^{-\rho_0 x_j}$. Hence, show that an exact locally most powerful test of $\rho = \rho_0$ is obtained by considering the model $E(Y_j) = \gamma + \beta e^{-\rho_0 x_j} + \psi x_j e^{-\rho_0 x_j}$ and by applying the normal-theory linear model test of $\psi = 0$ using Student's t. Thence obtain a $1 - \alpha$ confidence region for ρ. Give a general formulation of the argument of which this exercise is a special case.

[Section 7.3; Williams, 1962]

11. Independent binary trials are made with constant but unknown probability of success. In n trials, r successes are observed. What can be said about the number of successes to be observed in a further m trials?

[Section 7.5]

8 POINT ESTIMATION

8.1 General

Suppose that we have a fully specified parametric family of models. Denote the parameter of interest by θ. In previous chapters we have considered significance tests of hypotheses about θ and confidence intervals or regions for θ, the latter specifying ranges of values of θ consistent with the data. Suppose now that we wish to calculate from the data a single value representing the "best estimate" that we can make of the unknown parameter. We call such a problem one of *point estimation*.

Superficially, point estimation may seem a simpler problem to discuss than that of interval estimation; in fact, however, any replacement of an uncertain quantity by a single value is bound to involve either some rather arbitrary choice or a precise specification of the purpose for which the single quantity is required. Note that in interval estimation we explicitly recognize that the conclusion is uncertain, whereas in point estimation, as it is conceived here, no explicit recognition of uncertainty is involved in the final answer.

There are at least four situations where point estimation may seem relevant, although the first of these discussed below is in fact not point estimation in the sense understood here. For simplicity, we discuss the estimation of a scalar parameter of interest.

First, it quite often happens that, at least approximately, the $1 - \alpha$ equitailed confidence interval for θ is of the form $(t - k^*_{\frac{1}{2}\alpha}\sigma_t,\ t + k^*_{\frac{1}{2}\alpha}\sigma_t)$, where t and σ_t are functions of the data. It is then very convenient in reporting on data to give not the confidence limits but the values t and σ_t. If the limits are only approximately of the above form it may be enough to give values t and σ_t that will reproduce approximately, say, the 95% limits. However, the choice of a t with these properties in mind is best not regarded as a genuine problem of point estimation but merely as an important convenient device for simplifying the presentation of confidence intervals. Note that if the intervals are very unsymmetric nonlinear transformation of the parameter may be helpful. Failing that, however, presentation via t and σ_t is likely to be misleading.

Secondly, it may happen that the uncertainty in the final conclusion arising from random error is very small. In that case the main objective in collecting and analysing data must be to ensure that the systematic error in the final estimate is also small. This case is particularly likely to arise in very precise work in the physical sciences. Even here, however, some indication of precision is likely to be of interest, i.e. the problem is really then one of the first type. If the estimate t is represented by a random variable T, at least approximately normally distributed, then the absence of appreciable systematic error might be formalized by the very weak requirement that

$$|E(T;\theta) - \theta| \not\gg \text{st.dev.} (T;\theta),\qquad(1)$$

i.e. that the systematic error must not greatly exceed the random error. On the other hand, the requirement that confidence intervals can be conveniently summarized is that

$$|E(T;\theta) - \theta| \ll \text{st.dev.} (T;\theta),\qquad(2)$$

where it is supposed that st.dev.$(T;\theta)$ is small.

The third possibility, and the first that is a "genuine" point estimation problem in the sense discussed here, is where a decision has to be made, for example to settle the size of an adjustment to a control mechanism, to settle the quantity of material for a store, etc. Whatever the uncertainty, a definite single number has to be specified. We shall discuss decision problems in detail in Chapter 11; no completely satisfactory solution can, however, be given until fairly explicit information is available both about the quantitative consequences of errors of estimation and also about relevant information other than the data under analysis. When such information is available in the form of a utility function and a prior probability distribution a definitive solution of the point estimation problem can be given and this in a real sense is the only point estimation problem for which an entirely satisfactory solution is possible.

Finally, point estimation problems arise in the analysis of complex data, whenever it is wise to proceed in stages. Suppose that there are a number of sets of data arranged in some kind of structure; a particular example is an experiment laid out, perhaps, in several Latin squares and with a group of observations in each cell of the design. The first stage of the analysis may then be the estimation of one or more parameters from each group, and the second stage will be the analysis of these estimates, in the example by the usual methods for a Latin square. Now, in principle, it might be thought best to set up a single model for the whole data and to develop a method of analysis for this from first principles. In practice, however, in sufficiently complicated cases, it is undoubtedly useful to proceed in the two-stage way sketched above. The requirements that it is sensible to impose on the point estimates obtained in the first stage of the analysis depend on what is to be done in the second stage. If, as will often be the case, the second stage methods are linear, the most crucial requirement on the point estimate t calculated from a set of data with parameter value θ is that

$$|E(T;\theta) - \theta| \tag{3}$$

is small. How small (3) needs to be depends on the number of data sets to be combined as well as on the dispersion of the individual T's. Subject to (3) being satisfactorily small, it is desirable that the random dispersion of T should be as small as possible.

8.2   General considerations on bias and variance

Suppose that we have a scalar parameter θ, and let T be a statistic and t its observed value. When estimating θ, T is sometimes called an *estimator*, t an *estimate*; we shall use the single term estimate. Write

$$E(T;\theta) - \theta = b(\theta), \tag{4}$$

where $b(\theta)$ is called the *bias*. The *mean squared error* of T is

$$\mathrm{mse}\,(T;\theta) = E\{(T-\theta)^2;\theta\}$$

$$= \mathrm{var}\,(T;\theta) + \{b(\theta)\}^2. \tag{5}$$

A simple and elegant theory is obtained by restricting attention to unbiased estimates, i.e. by requiring that $b(\theta) = 0$, and subject to this minimizing $\mathrm{var}(T;\theta)$. In the light of the discussion of Section 8.1, it is clear that this formulation is at best a rough representation of what is required. It most closely corresponds to the fourth type of problem listed in Section 8.1, although the restriction to exactly unbiased estimates is too severe and, as shown in Example 8.1 below, may have unexpectedly drastic consequences. An estimate with a small bias and a small variance will be preferable to one with a zero bias and large variance; the more estimates there are to be combined in the second stage of the analysis the more important will control of bias be.

For the other kinds of problem, the search for unbiased estimates of minimum variance seems, at best, indirectly relevant. For a decision problem in which the consequences of errors of estimation are measured approximately by the squared error $(t - \theta)^2$, the mean squared error (5) is important. Note, however, that no useful progress can be made by trying to minimize it without further constraint. For example, the degenerate estimate $t \equiv \theta_0$ minimizes mean squared error at $\theta = \theta_0$ and has very good properties if $\theta \simeq \theta_0$. Some condition is usually needed to ensure that the estimate has sensible properties for all possible parameter values and while unbiasedness is one expression of this idea it has, in a decision-making context, no very particular justification. The connexion between unbiasedness and simplified confidence interval estimation is about equally tenuous. Note also that the dispersions of two distributions of different shapes are not necessarily best compared in terms of variances, so that, even accepting the requirement of unbiasedness, the minimization of variance is not necessarily completely appropriate.

As a final warning against overemphasizing the importance of exactly unbiased estimates, we give an example where there is only one unbiased estimate, not a sensible one.

Example 8.1. Unique unbiased estimate. Let Y have the geometric distribution

$$f_Y(r;\theta) = (1 - \theta)^{r-1}\theta \quad (r = 1,2,...). \tag{6}$$

If $T = t(Y)$ is an unbiased estimator of θ, then

$$\sum_{r=1}^{\infty} t(r)\,\phi^{r-1}(1 - \phi) = 1 - \phi,$$

where $\phi = 1 - \theta$. Thus, equating coefficients of powers of ϕ, we see that $t(1) = 1$ and $t(r) = 0$ $(r = 2,3, \ldots)$. From most points of view this is a poor estimate.

8.3   Cramér-Rao inequality

Let the data be represented by a vector random variable Y, having density $f_Y(y;\theta)$ depending on the scalar parameter θ. We introduce as in Chapter 4 the efficient score

$$U_. = U_.(\theta) = u_.(Y;\theta) = \frac{\partial \log f_Y(Y;\theta)}{\partial \theta} \qquad (7)$$

with, for a regular problem, which we assume,

$$i_. = i_.(\theta) = E(U_.^2;\theta) = \text{var}(U_.;\theta). \qquad (8)$$

We often use the abbreviated notation $U_.$ and $i_.$ when no ambiguity can arise. If the component random variables are independent, both $U_.$ and $i_.$ are sums of contributions.

We now show that if $T = t(Y)$ is an estimate with bias $b(\theta)$, then

$$\text{var}(T;\theta) \geqslant \{1 + b'(\theta)\}^2/i_.(\theta). \qquad (9)$$

In particular, if T is unbiased,

$$\text{var}(T;\theta) \geqslant 1/i_.(\theta). \qquad (10)$$

Further, in (9) and (10), equality holds if and only if T is a linear function of $U_.(\theta)$.

The right-hand sides of (9) and (10) are both called the *Cramér-Rao lower bound*. The form (10) is, in particular, useful as a standard against which to judge the performance of unbiased estimates.

To prove these results, we start from

$$E(T;\theta) = \theta + b(\theta) = \int t(y)f_Y(y;\theta)dy.$$

Because the problem is assumed regular, differentiation with respect to θ is in order and gives

$$1 + b'(\theta) = \int t(y)\frac{1}{f_Y(y;\theta)}\frac{\partial f_Y(y;\theta)}{\partial \theta}f_Y(y;\theta)dy$$

$$= E\{TU_.(\theta);\theta\}$$

$$= \text{cov}\{T, U_.(\theta);\theta\},$$

because $E\{U_.(\theta);\theta\} = 0$.

But, by the Cauchy-Schwarz inequality (see Appendix 3),

$$\text{var}(T;\theta)\,\text{var}\{U_.(\theta);\theta\} \geqslant [\text{cov}\{T,U_.(\theta);\theta\}]^2, \qquad (11)$$

with equality if and only if T and $U.(\theta)$ are linear functions of one another. This proves (9) and (10) and the condition for equality.

Example 8.2. The exponential distribution. Suppose that $Y_1, \ldots, Y_n$ are independently distributed with density $\mu^{-1} e^{-y/\mu} (y \geq 0)$. Then, with U_j denoting the efficient score from the jth value, we have

$$U_j = \frac{\partial}{\partial \mu} \left(-\log \mu - \frac{Y_j}{\mu} \right) = -\frac{1}{\mu} + \frac{Y_j}{\mu^2}, U. = -\frac{n}{\mu} + \frac{\Sigma Y_j}{\mu^2}, i. = \frac{n}{\mu^2}. \quad (12)$$

Thus, by (10), any unbiased estimate of μ has variance at least μ^2/n. But $\text{var}(\bar{Y}) = \mu^2/n$, so that $\bar{Y}$ is the minimum variance unbiased estimator of μ; note that it is a linear function of $U.$.

Suppose now that the exponential density is written in the form $\rho e^{-\rho y}$ and that we consider estimation of ρ. If we temporarily denote quantities referring to the new parameter with a dash, we have that

$$U_j' = \frac{\partial}{\partial \rho} (\log \rho - \rho Y_j) = \frac{1}{\rho} - Y_j, U_.' = \frac{n}{\rho} - \Sigma Y_j, i_.' = \frac{n}{\rho^2}. \quad (13)$$

Thus any unbiased estimate of ρ has variance at least ρ^2/n. Further no observable linear function of $U_.'$, i.e. of $\bar{Y}.$, can have expectation $\rho = 1/\mu$, so that the lower bound cannot be achieved.

For the moment we continue the discussion of this example in a rather *ad hoc* way. Consider estimates of the form $c/\Sigma Y_j$, where c is to be chosen to make the estimate unbiased. Now ΣY_j has density

$$\rho(\rho x)^{n-1} e^{-\rho x}/(n-1)!,$$

from which it follows that $E(1/\Sigma Y_j) = \rho/(n-1)$. Hence we take $c = n - 1$ and a further simple calculation gives that

$$\text{var}\left(\frac{n-1}{\Sigma Y_j}\right) = \frac{\rho^2}{n-2}. \quad (14)$$

It will be shown in Section 8.4 that this is the smallest variance attainable by an unbiased estimate of ρ.

Similar results are obtained for the one-parameter exponential family with one-dimensional sufficient statistic.

Suppose now that θ is a vector parameter and that we are interested in a component, say θ_1. If T_1 is an unbiased estimate of θ_1, exactly the previous argument is applicable and we have that

$$\operatorname{var}(T_1;\theta) \geqslant 1/i_{.11}(\theta).$$

This bound can, however, be improved. For we can introduce the vector of efficient scores with components

$$U_{.s}(\theta) = \frac{\partial \log f_Y(Y;\theta)}{\partial \theta_s} \quad (s = 1, \ldots, q).$$

Then differentiation of the equation

$$\theta_1 = \int t_1(y) f_Y(y;\theta) dy$$

in turn with respect to $\theta_1, \ldots, \theta_q$ leads to the equations

$$\operatorname{cov}\{T_1, U_{.1}(\theta);\theta\} = 1, \operatorname{cov}\{T_1, U_{.s}(\theta);\theta\} = 0 \ (s = 2, \ldots, q).$$
$$(16)$$

A generalization of the Cauchy-Schwarz inequality, given in Appendix 3, now shows that

$$\operatorname{var}(T_1;\theta) \geqslant i^{11}(\theta), \tag{17}$$

with equality if and only if T_1 is a linear function of $U_{.1}(\theta), \ldots, U_{.q}(\theta)$. This is, in general, an improvement on (15).

Example 8.3. Normal distribution. Suppose that $Y_1, \ldots, Y_n$ are i.i.d. in $N(\mu, \tau)$. Then, on regarding μ and τ as the components of the vector θ, straightforward calculation gives that the two components of $U_.$ are

$$U_{.1} = \Sigma(Y_j - \mu)/\tau, \ U_{.2} = -\frac{n}{2\tau} + \Sigma(Y_j - \mu)^2/(2\tau^2),$$

and that

$$i_. = \begin{bmatrix} n/\tau & 0 \\ 0 & n/(2\tau^2) \end{bmatrix}, i_.^{-1} = \begin{bmatrix} \tau/n & 0 \\ 0 & 2\tau^2/n \end{bmatrix}.$$

The lower bound (17) for an unbiased estimate of μ is thus $\tau/n = \sigma^2/n$, whereas that for an unbiased estimate of $\tau = \sigma^2$ is $2\tau^2/n$, not quite attained by the standard unbiased estimate which has variance $2\tau^2/(n-1)$. Attainment and non-attainment of the lower bounds is explained by the structure of the efficient score vector. Note that the lower bounds (15) and (17) are equal because the score components are uncorrelated.

Example 8.4. Normal-theory linear model (ctd). We consider the standard normal-theory linear model with

$$E(Y_j) = \sum_{s=1}^{q_x} x_{js} \beta_s. \tag{18}$$

Then the log density of Y_j is

$$-\tfrac{1}{2} \log (2\pi\sigma) - (Y_j - \Sigma x_{js}\beta_s)^2/(2\sigma^2), \tag{19}$$

so that

$$U_{.s} = (\Sigma_j x_{js} Y_j - \sum_{t=1}^{q_x}\Sigma_j x_{js} x_{jt} \beta_t)/\sigma^2 \quad (s = 1, \ldots, q_x). \tag{20}$$

The least squares estimates of the β's are linear combinations of the efficient score vector $U_.$ and it follows without further calculation that, being unbiased, they are of minimum variance. Note that this result applies whether or not the variance is known.

Example 8.5. Linear model with non-normal error. Suppose that $Y_1, \ldots, Y_n$ are independently distributed with expectations as in Example 8.4. The errors are assumed to be independently and identically distributed with density $g(z ; \lambda)$ of mean zero, where the vector parameter λ is a specification of scale and shape of the distribution. In practice most linear models contain the general mean and then we can suppose that in (18)

$$x_{j1} = 1, \Sigma x_{js} = 0 \ (s = 2, \ldots, q_x). \tag{21}$$

The log likelihood is

$$\Sigma \log g (Y_j - \mu_j ; \lambda),$$

where μ_j is the expected value of Y_j as given by the model (18).

It is now a fairly routine calculation (Cox and Hinkley, 1968) to find the second derivatives with respect to the parameters and, by taking expectations, to evaluate the matrix $i_.$. With the additional assumption (21), it can be shown that any element involving one parameter from the set (λ, β_1) and one from the set $(\beta_2, \ldots, \beta_{q_x})$, is zero. Further, the part of the information matrix involving the second set of parameters has the form

$$((\Sigma x_{js} x_{jt})) \int g(z ; \lambda) \frac{\partial^2 \log g (z ; \lambda)}{\partial z^2} dz, \tag{22}$$

where the integral is, for fixed λ, the intrinsic accuracy of the density $g(.)$ as defined in Example 4.14, in connexion with a location parameter problem.

The Cramér-Rao lower bounds for the variances are the diagonal elements of the inverse of (22), whereas the variances of the least squares estimates are of similar form, with the variance of the density $g(.)$ replacing the reciprocal of the intrinsic accuracy.

The final conclusion is thus that the ratio of the variance of the least squares estimate to the Cramér-Rao lower bound does not depend on the design matrix x, subject to (21).

8.4   Achievement of minimum variance and removal of bias

(i) General
In the previous section we saw that in certain rather exceptional cases the Cramér-Rao lower bound can be attained and that therefore an unbiased estimate of minimum variance is available. We need, however, to complete the theoretical discussion, to deal with the case where the Cramér-Rao bound is not attainable; for example, is the final estimate of Example 8.2 of minimum variance?

It turns out that the discussion of this is closely related with the problem of reducing the bias of estimates. Therefore, in this section we first discuss the existence of minimum variance unbiased estimates and then go on to develop further methods for the approximate removal of bias.

(ii) Existence of unbiased estimates of minimum variance
Let V be any unbiased estimate for θ, or in the vector parameter case for the component of interest. Let S be sufficient for the whole parameter. Define $T = E(V|S)$, a function only of S. Note that the sufficiency of S is crucial in order that the conditional distribution, given S, shall not depend on θ.

Then, for any given θ,

$$E(T;\theta) = E_S\{E(V|S);\theta\} = E(V;\theta) = \theta,$$

and

$$\text{var}(V;\theta) = \text{var}_S\{E(V|S);\theta\} + E_S\{\text{var}(V|S);\theta\}.$$

Thus T is unbiased and

$$\text{var}(T;\theta) \leqslant \text{var}(V;\theta),$$

with equality if and only if var $(V|S) = 0$, i.e. if and only if V is a function of S with probability one.

Suppose now that we start with two different unbiased estimates V_1 and V_2 and apply the above construction leading, say, to T_1 and T_2. Then T_1 and T_2 are functions of S such that for all θ

$$E(T_1 - T_2; \theta) = 0.$$

If S is complete, it follows that $T_1 = T_2$.

An immediate deduction is that any function of a complete sufficient statistic is the unique minimum variance unbiased estimate of its expectation. In particular the estimate of ρ in Example 8.2 has the stated property. This argument is due to Rao (1945) and Blackwell (1947).

The above construction is mainly valuable as a theoretical argument, but it is occasionally useful also for constructing unbiased estimates. One example is the U statistic estimate of a distance measure in Section 6.5. We now give one other example.

Example 8.6. Probability of zero in a Poisson distribution. Let $Y_1, \ldots, Y_n$ be i.i.d. in a Poisson distribution of mean μ and suppose that $\theta = e^{-\mu}$ is the parameter of interest. Now $\overline{Y}$ is the minimum variance unbiased estimate of μ and our problem is to find a function of Y with expectation $e^{-\mu}$. The obvious estimate $e^{-\overline{Y}}$, while satisfactory for many purposes, is, of course, slightly biased.

To follow through the general construction, we start with any unbiased estimate, no matter how inefficient and improve it in one step. Let

$$V = \begin{cases} 1 & (Y_1 = 0), \\ 0 & (Y_1 \neq 0). \end{cases}$$

This is obviously an unbiased estimate of θ. We require $T = E(V|S)$; conditionally on S, Y_1 has a binomial distribution of parameter $1/n$ and index S and therefore

$$T = \left(1 - \frac{1}{n}\right)^S \tag{23}$$

Note that $S = n \overline{Y}$ and that if $\overline{Y}$ is of order one and if n is large, then we can write

$$T = e^{-\bar{Y}} \left\{ 1 - \frac{\bar{Y}}{2n} + o_p \left(\frac{1}{n} \right) \right\}, \tag{24}$$

where the term $o_p(1/n)$ is $o(1/n)$ with high probability and hence negligible.

It is, however, only in relatively simple problems that this device can be used for an explicit analytical answer and we now turn to methods which, while only approximate, are more widely useful.

(iii) Reduction of bias by series expansion

A careful treatment of the methods to be discussed in this and the following subsection really involves asymptotic considerations which we examine in more detail in Chapter 9. The basic ideas are, however, conveniently treated now.

Very often we have a natural estimate which is, however, slightly biased. For example, if T is a simple and good unbiased estimate of μ, and if it is required to estimate $\theta = g(\mu)$, then the natural estimate $g(T)$ will nearly always be slightly biased. This may be entirely unimportant, but for the fourth kind of problem mentioned in Section 8.1 bias removal may be required. Examples 8.2 and 8.6 illustrate the kind of bias likely to be involved; normally it will be quite small.

If the random variation of T around μ is relatively small, for example if var (T) is of order n^{-1}, we can use the first two terms of a Taylor series for $g(.)$, i.e. we can write

$$g(T) \simeq g(\mu) + (T-\mu)g'(\mu) + \tfrac{1}{2}(T-\mu)^2 g''(\mu)$$

and therefore, on taking expectations, we have the approximation

$$E\{g(T)\} \simeq g(\mu) + \tfrac{1}{2} \text{ var} (T) g'' (\mu). \tag{25}$$

The second term, which we suppose to exist, is small by virtue of the small dispersion assumed for T. If, therefore, we can estimate the second term in (25) we can produce an estimate more nearly unbiased. If var (T) is a known constant, we have only to replace $g''(\mu)$ by $g''(T)$; when var (T) is of order n^{-1}, the expression (25) is the same to that order, with $g''(T)$ in place of $g''(\mu)$. Often var (T) will also have to be estimated. Some other approximations of the type (25) will be described in Chapter 9; see also Exercise 8.12.

Example 8.6 can be used to illustrate the procedure. Here $g(\mu) = e^{-\mu}$, $T = \bar{Y}$ and var $(T) = \mu/n$. Thus the expansion (25) becomes

$$E(e^{-\bar{Y}_.}) \simeq e^{-\mu} \left(1 + \frac{\mu}{2n}\right).$$

The bias term is estimated by $\bar{Y}_. e^{-\bar{Y}_.}/(2n)$, so that

$$e^{-\bar{Y}_.} \left(1 - \frac{\bar{Y}_.}{2n}\right) \tag{26}$$

has expectation $e^{-\mu} + O(1/n^2)$. The exact solution is (23) and we have in (26) recovered the leading terms in its series expansion; see (24).

The argument could, in principle, be extended to give higher terms, although in practice this is rarely necessary. Transformation of variables is, for example, widely used in analysis of variance and regression and if it is required to express the conclusions back on the original scale of measurement, the above results may be useful.

(iv) Sample-splitting

The method of the previous subsection requires that the estimate is a function of a statistic whose expectation and variance can be calculated to a sufficient order of accuracy. We now discuss a method that can be used purely numerically, without detailed analytical calculations, and which is therefore particularly suitable for complex problems.

The method, which is called the *jackknife*, depends on the qualitative idea that some aspects of the stability of an estimate can be judged empirically by examining how much the estimate changes as observations are removed.

Let T_n be an estimate calculated from $Y_1, \ldots, Y_n$ and let $T_{n-1,j}$ be the estimate of the same form calculated from the set of $n - 1$ random variables obtained by omitting Y_j. Write $\bar{T}_{n-1,.}$ for the average of the $T_{n-1,j}$ over $j = 1, , \ldots, n$.

Now suppose that for all sufficiently large m, in particular for $m = n - 1$ and n,

$$E(T_m) = \theta + a_1(\theta)/m + a_2(\theta)/m^2 + O(1/m^2). \tag{27}$$

Then

$$E(\bar{T}_{n-1,.}) = \theta + a_1(\theta)/(n-1) + O(1/n^2),$$

$$E(T_n) = \theta + a_1(\theta)/n + O(1/n^2).$$

From the last two equations it is clear that we can find a linear combination of $\bar{T}_{n-1,.}$ and T_n that will have bias of order $1/n^2$. In fact

$$T_n^J = nT_n - (n-1)\bar{T}_{n-1,.} \tag{28}$$

$$= nT_n - \frac{n-1}{n}\Sigma T_{n-1,j}$$

has $E(T_n^J) = \theta + O(1/n^2)$. Often T_n^J is referred to as the jackknifed estimate corresponding to T_n.

Note that if T_n is an average,

$$T_n = \Sigma l(Y_j)/n, \tag{29}$$

say, then $T_n = \bar{T}_{n-1,.} = T_n^J$.

The argument requires only the knowledge that the leading term in the bias is of order $1/n$, although, as with other asymptotic arguments, the implications of this numerically for any particular n in principle need exploring. It is theoretically possible that $a_1 = 1$, $a_2 = 10^{10}$, in which case the above arguments are relevant only for enormous values of n.

If the leading term in the expansion of $E(T_n)$ is, say, of order $1/n^2$ a modification to (27) and (28) is needed. Of course, bias of order one cannot be removed. As explained above, the main use of the jackknife is in relatively complicated problems in which direct analytical work is not feasible. Nevertheless, it is instructive to look at the result of applying the formulae to relatively simple cases. In Example 8.6, an estimate

$$ne^{-\bar{Y}.} - \frac{n-1}{n}\Sigma e^{-\bar{Y}_j'}$$

is obtained, where $\bar{Y}_j'$ is the mean of the data omitting Y_j. This seems inferior to the estimate obtained in the two previous subsections; for example, it is not a function of the minimal sufficient statistic $\bar{Y}.$.

Another elementary example concerns the estimation of variance starting from

$$T_n = \frac{\Sigma (Y_j - \bar{Y}.)^2}{n} ;$$

some direct calculation, which we leave as an exercise, shows that

$$T_n^J = \frac{\Sigma (Y_j - \bar{Y}.)^2}{n-1}.$$

As noted above, the more useful applications are probably in more complex fields, such as survey analysis, time series analysis and multivariate analysis. As an example illustrating some extensions of the sample-splitting idea, we consider the estimation of error probabilities in normal-theory discriminant analysis.

Example 8.7. Error rate in discriminant analysis. We make the usual tentative assumptions of two-sample normal-theory discriminant analysis that the two sets $\{Y_j\}$ and $\{Z_j\}$ of i.i.d. vector variables have multivariate normal distributions with the same covariance matrix Σ. The sufficient statistics are the mean vectors $\bar{Y}$ and $\bar{Z}$ and the usual pooled sample covariance matrix MS, which are all unbiased estimates. The estimated optimum linear discriminant is found in the usual way and a score $(\bar{Y} - \bar{Z})^T \text{MS}^{-1} x$ assigned to an arbitrary vector measurement x. Suppose that the discriminant is used to classify future observations in a symmetrical way, so that an individual is classified as belonging to that population whose sample mean score is closer to its own score; thus the cut-off point is the mid-point of the scores corresponding to the sample means, namely $\frac{1}{2}(\bar{Y} - \bar{Z})\text{MS}^{-1}(\bar{Y} + \bar{Z})$. The unconditional probabilities of error, i.e. that a future individual from one population is classified as belonging to the other, are equal. Suppose that this common error probability is to be estimated.

One way is to use the linear discriminant to classify all individuals in the data and to find the proportion misclassified. It is fairly clear that this will tend to underestimate the probability of error and numerical work (Lachenbruch and Mickey, 1968) shows that the underestimation can be severe.

One adaptation of the sample-splitting idea to this problem is to take each individual in turn, and to classify it using the discriminant function based on the remaining data. When this is repeated for all individuals, the proportion so misclassified is an almost-unbiased estimate of the probability of misclassification.

An alternative procedure is to use the properties of the normal distribution to estimate the error probability. While this is more efficient if the normality assumptions hold, the previous estimates have, of course, the major advantage of robustness. When the discriminant function is scaled to have unit estimated variance the two sample means differ by $D = \{(\bar{Y} - \bar{Z})^T \text{MS}^{-1} (\bar{Y} - \bar{Z})\}^{\frac{1}{2}}$ on the discriminant scale, and the error probability may be estimated by $\Phi(-\frac{1}{2}D)$. This estimate also is liable to be too low. We modify the argument of (27) − (29) to deal with this two-sample situation.

Suppose that in a problem with two groups of data T_{mn} denotes an estimate using m observations from the first group and n from the second. We delete in turn one observation from the first group, the corresponding estimates being denoted $T_{m-1,n;j0}$ $(j = 1, ..., m)$. Similarly by deleting observations from the second group we obtain estimates $T_{m,n-1;0j}$. Under the assumption that

$$E(T_{mn}; \theta) = \theta + \frac{a_1(\theta)}{m} + \frac{a_2(\theta)}{n} + ... ,$$

it is easily shown that the estimate

$$(m + n - 1)T_{mn} - \frac{m-1}{m} \sum T_{m-1,n;j0} - \frac{n-1}{n} \sum T_{m,n-1;0j} \qquad (30)$$

has a lower order of bias and this is the natural generalization of the previous estimate.

There is a second aspect of sample-splitting techniques which, while rather outside the scope of this chapter, it is convenient to mention here. This is the use to assess precision by giving approximate confidence intervals. In its crudest form the data are split into independent sections which are analysed separately, the precision of the overall average being assessed from the dispersion between sections. The method based on (27) and (28) is a more elaborate version of this.

We argue as follows. If the estimate were an average of the form (29) we would have

$$\text{var}(T_n) = \text{var}\{l(Y)\}/n$$

and this can be calculated directly, or estimated by

$$\sum\{l(Y_j) - T_n^J\}^2/\{n(n-1)\}.$$

Further, in this special case,

$$l(Y_j) = nT_n - (n-1)T_{n-1,j}.$$

This suggests that if the estimate is only approximately of the form (29), then we write

$$T_j^P = nT_n - (n-1)T_{n-1,j}, \tag{31}$$

and estimate $\mathrm{var}(T_n^J)$ by

$$\Sigma(T_j^P - T_n^J)^2/\{n(n-1)\}; \tag{32}$$

note from (28) that $T_n^J = \Sigma T_j^P/n$. Of course (32) also estimates $\mathrm{var}(T_n)$. The quantities (31) are called *pseudo-values*. Subject to approximate normality, confidence intervals can be found in the usual way. If the estimate is radically non-linear, e.g. is a single order statistic, the method fails.

One distinct advantage that (32) has is that it is free of distributional assumptions, and so may be more reliable than a highly parametric estimate of the standard error of T_n^J.

Some further discussion of the jackknife is given in Section 9.4.

The devices of sample-splitting described above are all relatively sophisticated. The much simpler idea of making independent parallel analyses of two halves of the data can be very useful, especially with extensive data, and has of course much appeal in presenting conclusions nontechnically.

(v) Estimation within a restricted class

In the previous discussion we have not put any artificial restriction on the mathematical form of the estimate. Often, however, it is useful to look for estimates within some restricted family and, in particular, when unbiasedness is relevant, to look for unbiased estimates of minimum variance within a special class. There are two reasons for doing this. One is that there may be an appreciable gain in simplicity. The other is that it may, by such a restriction, be possible to relax some other assumptions in the model.

A familiar example of the second feature is that for the linear model, if only estimates linear in the observations are considered, the least squares estimates are minimum variance unbiased under much weaker assumptions about the random error; independence and normality can be replaced by zero correlation.

We shall not here discuss further particular examples in detail. The following are some important special cases:

(a) in estimating scale and location it may be useful to consider estimates linear in the order statistics;

(b) in estimating the variance in linear models we may consider the family of estimates quadratic in the observations.

The special case (a) arises in the theory of robust estimation, described briefly in Section 8.6 and more fully in Section 9.4.

8.5   Estimates of minimum mean squared error

In the discussion of Section 8.1 it was mentioned that in many contexts the criterion of unbiasedness is not particularly relevant; one reason is that an estimate of small bias and small variance will for most purposes be preferable to one with no bias and appreciable variance. This naturally raises the possibility of using mean squared error as the criterion to be minimized. A full discussion requires the decision theoretical formulation of Chapter 11 but sometimes useful estimates can be obtained by minimizing mean squared error within some natural family of estimates determined, for example, by invariance arguments. We illustrate this by some examples.

Example 8.8 Normal variance with unknown mean. Let $Y_1, \ldots, Y_n$ be i.i.d. in $N(\mu, \sigma^2)$. If we consider the estimation of σ^2 using the minimal sufficient statistic $(\Sigma Y_j, \Sigma Y_j^2)$ and insist on invariance under scale and location changes, then we are led to estimates of the form

$$a\Sigma(Y_j - \bar{Y})^2.$$

More generally, in a normal-theory linear model problem we can consider an estimate of σ^2 of the form $b\mathrm{MS}$, where MS is the residual mean square based on d degrees of freedom. In the special case, $d = n - 1$ and $b = a(n-1)$.

A direct calculation now gives that the mean squared error is

$$E\{(b\mathrm{MS} - \sigma^2)^2\} = \frac{2b^2\sigma^4}{d} + (b-1)^2\sigma^4,$$

and this is clearly minimized by $b = d/(d+2)$. Thus the estimate of σ^2 is the residual sum of squares divided by the degrees of freedom plus two.

Ought one then to use this estimate routinely in applications in place of the conventional unbiased estimate? The following points are relevant:

(a) When the estimate is required for use in assessing the precision of a mean or regression coefficient *via* the Student t distribution it is entirely a matter of convention what value of b is taken.

(b) When what are really required are confidence intervals for σ^2, it is again purely conventional what value of b is used, or indeed whether or not one considers a point estimate to be a useful intermediate step in the calculation.

(c) If a further analysis of a group of sample estimates of variance is proposed it is quite likely that a linear model in terms of log σ will be used and in that case an approximately unbiased estimate of that parameter will be useful; the minimization of mean squared deviation from σ^2 is not then likely to be relevant.

(d) In a decision problem, the above calculation is likely to be more appropriate, although the particular criterion is probably usually less appealing than mean squared error in terms of σ or of log σ.

To summarize, while the above estimate illustrates an interesting kind of argument, it is unlikely that the particular estimate obtained here will often be useful.

The comparison of alternative estimates *via* mean squared error needs critical care for at least four reasons:

(a) If the distributions of the estimates are of very different shapes, then the mean squared error may not be a fair basis of comparison, except for a very specific decision-making situation. In particular, if one or more of the estimates has infinite variance, then the comparison needs particular caution; an estimate with infinite variance may be entirely acceptable if the divergence is produced by an effect of very small probability.

(b) If several estimates of the same type are to be combined in some way, e.g. averaged, then the mean squared error of a single component is not relevant, bias becoming relatively more important.

(c) Particular care is needed when the mean squared error depends on the value of the unknown parameter, because it is always possible to produce a good mean squared error over a particular part of the range by "shrinking" the estimate towards that part of the range. This will be sensible if there is genuine prior information about the parameter value which it is required to include, but in a general comparison of estimates in the absence of specific prior knowledge the shrinkage can be misleading.

(d) The possible existence of ancillary statistics should be considered.

Example 8.9 Inverse estimation in regression. Some of the above points are well illustrated by an important problem connected with linear regression and arising particularly in connexion with calibration experiments. Suppose, first, that we have the usual situation of normal-theory linear regression with independent data represented by $(x_1, Y_1), \ldots, (x_n, Y_n)$, where the usual distributional assumptions are made.

Suppose, further, that a new individual independent of the previous ones has an unknown value $\chi^\dagger$ of the explanatory variable and that the corresponding dependent variable $Y^\dagger$ is observed. It is required to make inferences about $\chi^\dagger$.

The answer to this problem is appreciably affected by whether any supplementary assumptions are reasonable. For example, in some contexts it might be sensible to assume that the explanatory variable, X, also is random and that the pairs (X, Y) have a bivariate normal distribution. If, further, it can be assumed that the new individual is drawn from the same bivariate distribution as the initial data, the problem is a straightforward one. The regression equation of X on Y can be fitted and, in particular, an unbiased estimate of $\chi^\dagger$ of minimum variance is

$$\bar{X}_. + \frac{\Sigma(X_j - \bar{X}_.)(Y_j - \bar{Y}_.)}{\Sigma(Y_j - \bar{Y}_.)^2}(Y^\dagger - \bar{Y}_.) = \bar{X}_. + b_{XY}(Y^\dagger - \bar{Y}_.), \quad (33)$$

say.

In the majority of applications, however, this extra assumption is not reasonable; the x values in the original data would be systematically chosen to span the range of interest.

If no additional assumptions are legitimate we can argue as follows, working first in terms of interval estimation or significance testing. If we want confidence regions for $\chi^\dagger$ using the sufficient statistics for the problem, we consider a test of the null hypothesis $\chi^\dagger = \chi_0^\dagger$ similar with respect to the nuisance parameters and, as in Example 7.13, this is based on the quantity

$$Z(\chi_0^\dagger) = \frac{Y^\dagger - \left\{\bar{Y}. + \dfrac{\Sigma(x_j - \bar{x}.)(Y_j - \bar{Y}.)}{\Sigma(x_j - \bar{x}.)^2}(\chi_0^\dagger - \bar{x}.)\right\}}{MS^{\frac{1}{2}}\{1 + n^{-1} + (\chi_0^\dagger - \bar{x}.)^2/\Sigma(x_j - \bar{x}.)^2\}^{\frac{1}{2}}}$$

$$= \frac{Y^\dagger - \{\bar{Y}. + b_{Yx}(\chi_0^\dagger - \bar{x}.)\}}{\text{est.st.dev.}},$$

which has a Student t distribution with $n - 2$ degrees of freedom. Thus the $1 - \alpha$ confidence for $\chi^\dagger$ is

$$\{\chi^\dagger ; Z^2(\chi^\dagger) \leqslant t_{n-2, \frac{1}{2}\alpha}^{*2}\}. \tag{34}$$

The form is closely related to that of Example 7.13, as is to be expected since we are concerned implicitly with $(Y^\dagger - \alpha)/\beta$, a ratio of parameters. In practice, we shall nearly always be working with the case where n is fairly large, so that the regression coefficients are quite precisely estimated, and where clearcut regression is present. In that case the region (34) consists of an approximately symmetrical interval centred on the zero of the numerator, namely on

$$\bar{x}. + \frac{1}{b_{Yx}}(Y^\dagger - \bar{Y}.). \tag{35}$$

The geometrical interpretation of (35) is to be contrasted with that of (33). Both are the values of x corresponding to the point $Y = Y^\dagger$ on a fitted regression line, (33) being for the regression of X on Y, and (35) being for the regression of Y on x.

One interpretation of the difference between them is that (33) is shrunk towards $\bar{X}.$, corresponding to the information that the new individual is drawn from the same source as the original set of data. The estimate (35) is the sensible one as a centre of confidence intervals, in the absence of supplementary information. It is natural, therefore, to examine it from other points of view. Although in the circumstances under study, of strong regression and large n, the distribution of (35) is very nearly normal with mean $\chi^\dagger$, the mean squared error of the full sampling distribution is infinite, because of the behaviour of (35) near zero values of the sample regression coefficient. This is a striking example of the dangers of using mean squared error as a criterion for comparing distributions of very different shapes.

The singularity can be avoided, for example by conditioning on the event that the sample regression coefficient of Y on x is not too near zero. We shall not give details; the conclusions are what would be expected in view of the relation between the two estimates. If the unknown $\chi^\dagger$ is close to $\bar{x}$. the estimate (33) has smaller mean squared error, whereas if $\chi^\dagger$ is appreciably different from $\bar{x}$., and in particular if it lies outside the original data, (33) has the larger mean squared error.

When a regression model is used in this way it would be common for repeated applications to be made, with a series of individuals for calibration. Of course, the confidence interval statement refers to a single application and would need modification if, for example, it were required that the probability of error referred to the simultaneous correctness of a whole string of statements; see Section 7.4 (ii). More relevant to the present discussion is the fact that the estimate (33) would be bad if a whole series of individuals were involved with about the same value of $\chi^\dagger$ and if interest were focused in part on the average $\chi^\dagger$.

This example illustrates a great many of the points arising in the discussion of point estimates.

8.6   Robust estimation

The derivation of optimum statistical procedures usually requires a rather strong specification of the distribution of the random variables involved, for example that they are normal, exponential, etc. This is, in practice, at best a good approximation and the inevitable idealization involved raises problems. To some extent we can rely on studies of robustness, showing, as just one example, that some types of small departure from normality have relatively little effect on the efficiency or on the long-run frequency properties of the method of least squares. That is, as in any piece of applied mathematics, one pauses at the end of an application to consider which of the various assumptions made for developing the analysis are likely to have a critical effect on the answer.

One particular type of departure from the simplified assumptions that we make is the presence of aberrant observations, i.e. there may be present a fairly small proportion of observations subject to gross errors of recording or measurement, or in some other way untypical of the random system under investigation. With fairly small sets of data, it is feasible to inspect the data for such outliers, and to

investigate, correct and replace individual observations where
appropriate. With large quantities of data, however, this is not
practicable and one looks instead for procedures which are insensitive
to the presence of moderate degrees of contamination by outliers,
while, hopefully, doing reasonably well under the ideal conditions of
normality or whatever. Such procedures have been developed, on the
whole by rather *ad hoc* methods, largely for the point estimation of
the mean of a symmetrical distribution from independent and identically distributed random variables.

A different, although closely related, problem is to produce an
estimate of the mean of a symmetrical distribution of unknown form
which is almost as good as the estimate that would be used if the
functional form were known.

A reason for posing these problems in terms of symmetrical distributions is that it is not clear out of context what aspect of an
unsymmetrical distribution should be estimated. In principle these
considerations are not particular to point estimation rather than to
interval estimation or significance testing; at the time of writing,
however, most work concerns point estimation. A defence of this is
that the potentially most useful application is probably in the first
stage of the reduction of complex sets of data.

Suppose that $Y_1, \ldots, Y_n$ are independent and identically distributed in a continuous symmetrical density $g(y - \mu)$, with
$g(y) = g(-y)$. One important class of estimates is that of trimmed
means, defined as follows. Let $Y_{(1)}, \ldots, Y_{(n)}$ be the order statistics
and for $0 \leqslant p < \frac{1}{2}$ define

$$\bar{Y}_p = \frac{1}{n - 2[np]} \sum_{j=[np]+1}^{n-[np]} Y_{(j)}. \tag{36}$$

In the crudest version of the use of these, p is chosen in advance, in
the light of the amount of contamination expected. It can be shown
that if $g(.)$ is either normal or is a normal density symmetrically
contaminated, then p can be chosen so that reasonably high efficiency
is retained if in fact the distribution is normal, calling really for $p = 0$;
whereas, if contamination is present, then there is substantial gain
from the rejection of the more extreme values.

The estimate (36) is a linear combination of order statistics and
clearly there is much scope for making such combinations more
general than (36).

The analytical study of these and similar estimates requires the use of asymptotic arguments, i.e. the study of limiting behaviour for large n. A detailed account is therefore postponed to Section 9.4. We note here, however, that (36) can be extended by making p depend on the data. One way of choosing p would be rather subjectively, taking the smallest value that would exclude any suspiciously outlying values; of course, this would be no good for large-scale data handling. One precisely defined approach, which does, however, call for appreciable computing, is to choose p to minimize the estimated variance. For this we need a result which follows directly from (9.99), namely that the asymptotic variance of $\bar{Y}_p$ is

$$\frac{1}{n(1-2p)^2}\left[\int_{G^{-1}(p)}^{G^{-1}(1-p)} x^2 g(x)\,dx + p\{G^{-1}(p)\}^2 + p\{G^{-1}(1-p)\}^2\right],$$

$$(37)$$

which is estimated by substituting $\tilde{F}_n(y + \bar{Y}_p)$ for $G(y)$, i.e. by

$$\frac{1}{n(1-2p)^2}\frac{1}{n}\left[\sum_{j=[np]+1}^{n-[np]}(Y_{(j)}-\bar{Y}_p)^2 + p\{(Y_{([np]+1)}-\bar{Y}_p)^2 + (Y_{(n-[np])}-\bar{Y}_p)^2\}\right].$$

Thus (38) can be computed for selected values of p and finally the mean estimated by $\bar{Y}_{p'}$, where p' minimizes (38). A similar idea can, in principle, be used for more complicated linear combinations of order statistics. Notice that (38) could be replaced by the jackknife estimate (32), which, as we shall see in Section 9.4, is closely related to (38).

Bibliographic notes

Unbiased estimation has been widely considered both in the context of the linear model and more generally, although the limitations of exact unbiasedness as a criterion are widely recognized. The central inequality (10) is essentially due to Fisher (1925), although it was first given in its present form by Cramér (1946) and Rao (1945); see also Aitken and Silverstone (1942).

Further bounds involving higher derivatives are due to Bhattacharya (1950); generalizations to nonregular problems are given by Chapman and Robbins (1951) and Kiefer (1952) and to sequential estimation by Wolfowitz (1947).

The use of sample splitting to eliminate bias was suggested by Quenouille (1949, 1956) and in the form discussed here by Tukey

(1958), who suggested the name of jackknife. Miller (1974) has reviewed the subject; see also the book of Gray and Shucany (1972).

For the discussion of quadratic estimates of variances and variance components based on fourth order moment assumptions, see Rao (1973), Atiqullah (1962) and Rao and Mitra (1971).

A careful account of inverse regression is given by Finney (1964). For discussion and comparison of different approaches to this problem, see Williams (1969a, b), Tallis (1969), Krutchkoff (1967) and Halperin (1970).

For further references on, and more discussion of, robust estimation, see Section 9.4.

Further results and exercises

1. Suppose for simplicity that there is a scalar parameter. Let the likelihood attain its largest value at a point where the second derivative exists; call the value at which the maximum is attained the maximum likelihood estimate. For sufficiently large α, the $1 - \alpha$ likelihood-based confidence interval is approximately symmetrical about the maximum likelihood estimate and according to the view that point estimates are sometimes useful as indicating the centres of confidence intervals this provides a special status to the maximum likelihood estimate. What further property of the likelihood function would make this more compelling?

[Section 8.1]

2. For continuous distributions, an estimate T is called *median unbiased* for θ if, for all θ,

$$\mathrm{pr}(T < \theta\,;\theta) = \mathrm{pr}(T > \theta\,;\theta) = \tfrac{1}{2}.$$

Prove that if $g(.)$ is strictly monotonic, $g(T)$ is median unbiased for $g(\theta)$. Show that if SS_d is a normal-theory sum of squares based on d degrees of freedom, then the median unbiased estimate of variance is SS_d/c_d, where $c_d \doteq d - \tfrac{2}{3}$. Show also that the same value of c_d applies approximately if the estimate is that $\tilde{\theta}$ such that $\mathrm{pr}(T \leqslant t_{\mathrm{obs}}\,;\tilde{\theta}) = \tfrac{1}{2}$, leading to the "most likely value" estimate. Discuss critically the circumstances under which these requirements might be relevant.

[Section 8.2]

3. Show that if $Y_1, \ldots, Y_n$ are i.i.d. with density

$$\exp\{a(\theta)b(y) + c(\theta) + d(y)\},$$

then $\frac{1}{n}\Sigma b(Y_j)$ attains the Cramér-Rao lower bound for an unbiased estimate of its expectation.

[Section 8.3]

4. Explain qualitatively the presence of the term $\{1 + b'(\theta)\}^2$ in the general form of the Cramér-Rao inequality, as follows:
   (a) explain why $b'(.)$ rather than $b(.)$ occurs;
   (b) explain why the bound must vanish when $b'(\theta) = -1$;
   (c) explain why the term is squared rather than of degrees one.
Discuss why the vanishing of $b(.)$ at an isolated θ_0 is not enough to justify the simplified version of the bound at θ_0.

[Section 8.3]

5. Subject to some regularity conditions, the equation $k(Y, \tilde{\theta}) = 0$, considered as an equation for $\tilde{\theta}$, is called an *estimating equation* for θ if $E\{k(Y, \theta); \theta\} = 0$, for all θ. Such an equation is called optimum if it has the smallest possible value of the index

$$E\{k^2(Y, \theta); \theta\}/E[\{\partial k(Y, \theta)/\partial\theta\}^2; \theta].$$

Defend this definition. Show that the value of the index is at least $1/i(\theta)$, the lower bound being attained if and only if $k(.,.)$ is proportional to the efficient score. Generalize to the multiparameter case.

[Section 8.3; Godambe, 1960]

6. Examine the special case of Exercise 5 holding when the estimate is given explicitly by the equation $g_1(Y) - \tilde{\theta} g_2(Y) = 0$ with the requirement that, for all θ, $E\{g_1(Y); \theta\} = \theta E\{g_2(Y); \theta\}$. The equation is then called an *unbiased estimating equation*. Given several sets of data with a common value of θ how would a combined estimate be formed?

[Section 8.3; Durbin, 1960]

7. In the first order autoregressive process

$$Y_{j+1} - \theta Y_j = \epsilon_{j+1},$$

where $Y_0 = y_0$ is a constant and $\epsilon_1, \ldots, \epsilon_n$ are i.i.d. with normal distributions of zero mean. Show that

$$\Sigma Y_j \, Y_{j+1} - T \Sigma Y_j^2 \, = \, 0$$

is an unbiased estimating equation attaining the lower bound of Exercise 5.

[Section 8.3]

8. Prove that the relations between the efficient scores and information for the alternative parameterizations of the exponential distribution in Example 8.2 are a consequence of the general relations developed in Section 4.8.

[Section 8.3]

9. Show that in the multiparameter case an unbiased estimate of the linear combination $\Sigma l_r \, \theta_r = l^T \theta$ has variance at least

$$l^T \{i_.(\theta)\}^{-1} l.$$

[Section 8.3]

10. Let $Y_1, \ldots, Y_n$ be i.i.d. in the exponential density $\rho e^{-\rho y}$. Show that the minimum variance unbiased estimate of the cumulative distribution function is

$$1 - \{1 - yk(y \, ; 0, \, \Sigma Y_j) / \Sigma Y_j\} + k(y \, ; \Sigma Y_j, \infty),$$

where $k(y \, ; a, b)$ is 1 or 0 according as $y \in [a, b)$ or not.

[Section 8.4 (ii)]

11. In connexion with the estimation of the variance σ^2 of a normal distribution, obtain an unbiased estimator of $\log \sigma$ using (a) exact, and (b) approximate arguments. When would the results be relevant? Suggest also an application to the analysis of data from Poisson processes.

[Section 8.4 (iii)]

12. Apply the argument of (25) to show that if $E(T \, ; \theta) = \theta$, then

$$\text{var}\{g(T) \, ; \theta\} \simeq \{g'(\theta)\}^2 \, \text{var}(T \, ; \theta),$$

and hence suggest how independent estimates of several parameters $\theta_1, \ldots, \theta_m$ with variances $v(\theta_1), \ldots, v(\theta_m)$ can be transformed to have approximately constant variability.

[Section 8.4 (iii)]

13. The random variable Y is said to be lognormally distributed if $Z = \log Y$ is normal with, say, mean λ and variance τ^2. Prove that $E(Y) = \exp(\lambda + \frac{1}{2}\tau^2)$. For a set of i.i.d. variables $Y_1, \ldots, Y_n$ with the above distribution, the sufficient statistic is the estimates of mean and variance $\bar{Z}$. and MS_z from the logs. Thus an approximately unbiased estimate of $E(Y)$ is $\exp(\bar{Z}. + \frac{1}{2} MS_z)$. Examine bias-removing techniques as applied to this problem.

[Section 8.4 (iii)]

14. Examine the sample-splitting estimate of (28) as applied to the range of the uniform distribution $(0, \theta)$. Take T_n to be the maximum observation in the data. Compare the estimate with that based on the sufficient statistic. Show that the point estimate is very inadequate as a summary of the confidence regions in such a case.

[Section 8.4 (iv)]

15. Prove that in estimating a variance starting from the biased estimate $\Sigma(Y_j - Y)^2/n$, the sample-splitting estimate (28) exactly eliminates bias. Examine the usefulness of (32) as an estimate of precision.

[Section 8.4 (iv)]

16. Observations y_{jk} $(j = 1, \ldots, m; k = 1, \ldots, r)$ are represented by random variables Y_{jk} with unknown location parameters μ_j and unknown constant dispersion. The values of the μ_j's being possibly close together, the linear combination

$$\tilde{\mu}_j(a) = a\bar{y}_{j.} + (1-a)\bar{y}_{..}$$

is considered as a suitable estimate for μ_j. Show that the squared prediction error criterion $\Sigma\Sigma\{y_{jk} - \mu_j(a)\}^2$ is minimized when $a = 1$.

Denote the estimate $\tilde{\mu}_j(a)$ based on the data with y_{jk} omitted by $\tilde{\mu}_{j,k}(a)$. Then the prediction error for y_{jk} may be measured by the "pseudo-value" $y_{jk} - \tilde{\mu}_{j,k}(a)$, leading to the overall squared error criterion $\Sigma\Sigma\{y_{jk} - \tilde{\mu}_{j,k}(a)\}^2$. Show that the value of a which minimizes this criterion is a monotone increasing function of MS_b/MS_w. Comment on the explicit form of the resulting estimate for μ_j and the method of derivation.

[Section 8.4 (iv), 8.5; Stone, 1973]

17. Consider a stationary Poisson process with rate ρ, and suppose that on the basis of observation for a time t we wish to estimate $\theta = \exp(-\rho x)$. If the number of events in $(0, t)$ is $N(t)$, then a "natural" estimate of θ is

$$\hat{\theta} = \exp\{-N(t)x/t\}.$$

Now split the time interval up into r equal sub-intervals of length t/r and calculate the jackknifed estimate $\tilde{\theta}_r$ of θ obtained by successively deleting single sub-intervals. Show that as r increases, with t fixed,

$$\tilde{\theta}_r \to \hat{\theta}\{1 - N(t)(e^{x/t} - 1 - x/t)\};$$

that is, with very fine subdivision of $(0,t)$ the jackknifed estimate $\tilde{\theta}_r$ is equivalent to $\hat{\theta}$ with first-order bias correction.

[Section 8.4 (iv); Gaver and Hoel, 1970]

18. Under the conditions of Section 8.4 (iv), let T_n be an estimate from the full set of n observations and let $T'_{\frac{1}{2}n}$ and $T''_{\frac{1}{2}n}$ denote the corresponding estimates from the two halves when the data are split at random into two. Show that $2T_n - (T'_{\frac{1}{2}n} + T''_{\frac{1}{2}n})$ has reduced bias.

[Section 8.4 (iv); Quenouille, 1949, 1956]

19. The linear model $E(Y) = x\beta$ is called quadratically balanced if the diagonal elements of the matrix $x(x^T x)^{-1} x^T$ are all equal. Let the errors be independent with constant variance and with constant kurtosis γ_2. Prove that of all quadratic forms that are unbiased estimates of the error variance, the usual residual mean square has minimum variance, provided that either the model is quadratically balanced or that $\gamma_2 = 0$.

[Section 8.5; Atiqullah, 1962]

20. Prove Gauss's theorem that for a linear model with uncorrelated errors of constant variance the least squares estimators are of minimum variance among all linear unbiased estimates. Compare this with the property of Example 8.4. Show that a linear estimate that is not unbiased has unbounded mean squared error over the whole parameter space and that therefore the requirement of unbiasedness in Gauss's theorem could be replaced by that of bounded mean squared error.

[Section 8.5; Barnard, 1963]

21. Two random variables X and Y are such that the marginal distribution of Y and the conditional distribution of Y given X can be observed. Show that these two distributions determine the conditional distribution of X given Y if and only if a certain Fredholm integral equation has a unique solution. If the conditional distribution of Y given $X = x$ is normal with mean $\gamma + \beta x$ and constant variance, show that the conditional distribution of X given Y is always determined and examine in more detail the special case when the marginal distribution of Y too is normal. Discuss the possible relevance of these results to the inverse estimation problem of Example 8.9.

[Section 8.5; Tallis, 1969]

9 ASYMPTOTIC THEORY

9.1 Introduction

In Chapters 3–8 we have discussed significance tests, interval estimates and point estimates and have shown how to obtain procedures optimal according to various criteria. While that discussion enables us to deal with important practical problems, there remain many, including for example the majority of those connected with time series, for which the previous methods are ineffective. This is sometimes because the distributional calculations required, for example in connexion with significance tests, are too complex to be readily handled. A more serious difficulty is that the techniques of Chapters 5 and 7 for problems with nuisance parameters are of fairly restricted applicability.

It is, therefore, essential to have widely applicable procedures that in some sense provide good approximate solutions when "exact" solutions are not available. This we now discuss, the central idea being that when the number n of observations is large and errors of estimation correspondingly small, simplifications become available that are not available in general. The rigorous mathematical development involves limiting distributional results holding as $n \to \infty$ and is closely associated with the classical limit theorems of probability theory; full discussion involves careful attention to regularity conditions and this we do not attempt. Practical application, on the other hand, involves the use of these limiting results as an approximation for some finite n. The numerical adequacy of such an approximation always needs consideration; in some cases, we shall be able to extend the discussion to investigate higher-order approximations.

It helps in following the detailed discussion to have a few general points in mind.

In the previous discussion, the likelihood function has played a central role. Now, for regular problems, it can be shown that the log likelihood has its absolute maximum near to the true value; it is then easy to show that the second derivative at the maximum, being in the independent case the sum of n terms, is of order n in probability. It follows from this that the log likelihood is negligible compared with its maximum value as soon as we move a distance from the maximum large compared with $n^{-\frac{1}{2}}$. It then follows that over the effective range of parameter values the log likelihood is approximately described by the position of the maximum and the second derivative there. These, or quantities asymptotically equivalent to them, are therefore in a sense approximately sufficient statistics.

There is a close connexion between the position of the maximum, i.e. the so-called maximum likelihood estimate (m.l.e), and the efficient score

$$U_.(\theta) = u_.(Y;\theta) = \frac{\partial}{\partial \theta} \log f(Y;\theta).$$

For we have used repeatedly the fact that in regular problems

$$E\{U_.(\theta);\theta\} = 0.$$

In the regular case under discussion, the m.l.e. $\hat{\theta}$ satisfies the closely related equation

$$U_.(\hat{\theta}) = 0.$$

The close connexion between efficient scores, the second derivative of the log likelihood function at or near its maximum, and the behaviour of the likelihood function itself at or near its maximum is used repeatedly in the subsequent discussion. Efficent scores were introduced in Section 4.8 in connexion with locally most powerful tests and there is a strong link between such tests and asymptotically optimal tests.

A final general comment is that in asymptotic theory estimates and tests defined in apparently quite different ways may be equivalent in the limit. To give a unified treatment, we shall concentrate on procedures, such as maximum likelihood estimation, which have a strong direct link with the likelihood function; in particular problems, however, there may be alternative procedures that are computationally much simpler but which in the limit are equivalent to, say, maximum likelihood estimation. The wide availability of computers has made such alternatives of reduced practical importance and this is added justification for emphasizing maximum likelihood methods. In later sections we shall consider briefly second-order calculations which can in some cases give a basis for choosing between alternative asymptotically equivalent procedures.

The simple relation between maximum likelihood theory and efficient scores holds for regular problems. In fact a relatively straightforward treatment of asymptotic maximum likelihood theory is possible under the following regularity conditions on $f_Y(y; \theta) = \text{lik}(\theta; y)$:

(a) the parameter space Ω has finite dimension, is closed and compact, and the true parameter value is interior to Ω;

(b) the probability distributions defined by any two different values of θ are distinct;

(c) the first three derivatives of the log likelihood $l(\theta; Y)$ with respect to θ exist in the neighbourhood of the true parameter value almost surely. Further, in such a neighbourhood, n^{-1} times the absolute value of the third derivative is bounded above by a function of Y, whose expectation exists;

(d) the identity of (4.52), namely

$$E\{U(\theta)U^T(\theta); \theta\} = E\{-\frac{\partial}{\partial\theta} U^T(\theta); \theta\} = i(\theta),$$

holds for the total information $i(\theta)$, which is finite and positive definite in the neighbourhood of the true parameter value.

These conditions can be used to justify Taylor expansions and other similar techniques. We shall use such expansions in a fairly informal manner. The discussion will not be limited to these assumptions, but exceptions will be clearly stated. The reader interested in the mathematical aspects of the asymptotic theory should refer, for example, to LeCam (1953, 1956, 1970).

The distributional results of this Chapter concern limiting behaviour as the number n of observations tends to infinity. More specifically, the majority of the discussion will be concerned with showing that some standardized statistic T_n has a probability distribution that is, for very large n, indistinguishable from the probability distribution of a random variable T, i.e.

$$\text{pr}(T_n \leqslant t) \to \text{pr}(T \leqslant t)$$

as n tends to infinity for any finite value t. To do this, we show that

$$T_n = T + o_p(1),$$

the general notation $o_p(n^a)$ meaning a random variable Z_n such that for any $\epsilon > 0$

$$\lim_{n \to \infty} \text{pr}(n^{-a}|Z_n| > \epsilon) = 0.$$

The notation $O_p(n^a)$ similarly denotes a random variable Z_n which is $O(n^a)$ with high probability. The calculus of $o_p(.)$ and $O_p(.)$ is simply a stochastic version (Mann and Wald, 1943) of that of $o(.)$ and $O(.)$ used for sequences of fixed numbers.

Often we discuss moments, especially means and variances, of limiting distributions, and we shall then use the rather loose notation $E(T_n)$, etc. This strictly denotes either a convergent approximation to the mean of the limiting distribution, or the mean of a truncated version of T_n which has that same limiting distribution. The later discussion of higher-order properties, in effect, relates to construction of an asymptotic expansion of, say, $E(T_n)$, which is a series of the form $\mu_0 + \mu_1 n^{-1} + \ldots + \mu_r n^{-r}$, which for fixed r converges at the rate n^{-r+1} to the mean of the relevant limiting distribution. The series need not converge, of course, especially if $E(T_n)$ does not exist. The object of using higher terms in such series is primarily to obtain an improved approximation to $\text{pr}(T_n \leqslant t)$ in terms of the limiting distribution. Care is needed with such expansions to see that all terms involving the requisite power of n^{-1} have been retained. The Edgeworth expansion of Appendix 1 is often a valuable guide; for example, if an improved mean and variance are used in a normal approximation, adjustments for skewness and kurtosis will usually be required too.

The connexion between significance tests and interval estimates, described in detail in Chapter 7, has encouraged us to describe in Section 9.3 common material under the heading of significance tests. The reader can make his own interpretations of the results in terms of large-sample confidence statements, with reference to the early parts of Chapter 7 for any points of interpretation or methodology. This has been done for convenience of exposition and, of course, does not mean that interval estimation is to be considered as less important than significance testing.

The final section of this Chapter deals with the large-sample theory of robust estimation, particularly of the location parameter of a symmetric distribution. This complements the discussion in Section

8.6 on robustness, and forms also the basis of large-sample interval estimation in the absence of definite assumptions about the probability model.

9.2 Maximum likelihood estimates

(i) General remarks
A central role is played in large-sample parametric inference by the maximum likelihood estimate (m.l.e.) of the parameter of interest. In the terminology of Section 2.1, the likelihood function corresponding to an observed vector y from the density $f_Y(y; \theta)$ is written

$$\mathrm{lik}_Y(\theta'; y) = f_Y(y; \theta'),$$

whose logarithm is denoted by $l_Y(\theta'; y)$. If the parameter space is Ω, then the maximum likelihood estimate $\hat{\theta} = \hat{\theta}(y)$ is that value of θ' maximizing $\mathrm{lik}_Y(\theta'; y)$, or equivalently its logarithm $l_Y(\theta'; y)$, over Ω. That is,

$$l_Y(\hat{\theta}; y) \geqslant l_Y(\theta'; y) \quad (\theta' \in \Omega). \tag{1}$$

Where it is necessary to make the distinction, we use θ to denote the true parameter value and θ' as the general argument of the likelihood function. The distinction is mainly required for the general arguments of (ii) and (iii), and is normally unnecessary in particular examples.

In this section we derive the asymptotic properties of the m.l.e., aiming to outline general results that will cover most practical situations. For much of the discussion it is convenient to suppose that y corresponds to a vector Y of i.i.d. random variables, but in fact we show later that most of the results extend easily to non-homogeneous and non-independent random variables.

In many problems the most convenient way to find the m.l.e. is to examine the local maxima of the likelihood. In terms of log likelihood this means solving the *likelihood equation*

$$\frac{\partial l_Y(\theta'; Y)}{\partial \theta'_s} = 0 \quad (s = 1, \ldots, q). \tag{2}$$

The relevant solutions being local maxima, we can restrict attention to those θ' satisfying (2) that are in Ω and for which the matrix of second derivatives is negative definite. For notational simplicity, we shall denote the column vector of first derivatives by $U(\theta') = \partial l_Y(\theta'; Y)/\partial$

and the symmetric matrix of second derivatives by $\partial U^{T}(\theta')/\partial\theta' = \partial^{2}l_{Y}(\theta' ; Y)/\partial\theta'^{2}$. It is important in applications to check that the chosen solution of (2) corresponds to an overall maximum of the likelihood in Ω.

The following examples illustrate some of the points that can arise.

Example 9.1. Normal mean and variance. Let $Y_{1}, \ldots, Y_{n}$ be i.i.d. in $N(\mu, \sigma^{2})$. Then

$$l_{Y}(\mu,\sigma ; Y) = -n\log\sigma - \frac{1}{2\sigma^{2}}\{\Sigma(Y_{j}-\bar{Y})^{2}+n(\bar{Y}-\mu)^{2}\}. \qquad (3)$$

We exclude the very special situation when all observations are exactly equal, when the likelihood tends to infinity as $\sigma \to 0$ for the appropriate value of μ. Otherwise, for $n \geqslant 2$, the likelihood is bounded, tends to zero as $|\mu|$ or σ becomes infinite or as $\sigma \to 0$ and is differentiable everywhere.

The likelihood equation (2) has the single solution

$$\hat{\mu} = \bar{Y}, \hat{\sigma}^{2} = \frac{1}{n}\Sigma(Y_{j}-\bar{Y})^{2}, \qquad (4)$$

which is a local maximum. Because the likelihood has no other turning points and is continuously differentiable (4) defines the m.l.e. provided that the parameter space is $(-\infty, \infty) \times [0, \infty)$. If, however, μ is restricted, say to be non-negative, then (4) is appropriate only if $\bar{Y} \geqslant 0$. For $\bar{Y} < 0$, it is easy to see from (4) that $l_{Y}(0,\sigma ; Y) > l_{Y}(\mu,\sigma;Y)$ for all $\mu < 0$. Thus $\hat{\mu} = \max(0,\bar{Y})$ and $\hat{\sigma}^{2} = \Sigma(Y_{j} - \hat{\mu})^{2}/n$.

Example 9.2. Uniform distribution. If $Y_{1}, \ldots, Y_{n}$ are i.i.d. in the uniform density on $(0, \theta)$, the likelihood is

$$\theta^{-n} \, \text{hv}(\theta - Y_{(n)}),$$

where $Y_{(n)}$ is the largest value and $\text{hv}(.)$ is the unit Heaviside function. Therefore the likelihood function is a monotone decreasing function of θ beyond $\theta = Y_{(n)}$, at which point it is a maximum. The likelihood is discontinuous at this point; therefore the likelihood derivative does not exist there. In fact the likelihood derivative is only zero in the range $(0, Y_{(n)})$ where the likelihood is identically zero. Thus we have a situation in which the maximum likelihood estimate is not a solution of the likelihood equation (2). In fact, whenever the range of the random variable depends on the parameter it is important to see whether the maximum of the likelihood is achieved at an extreme point rather than at a solution of (2).

Example 9.3. Double exponential distribution. If $Y_1, \ldots, Y_n$ are i.i.d. with the double exponential density

$$\tfrac{1}{2} \exp \left(-|y - \theta| \right), \quad (-\infty < y < \infty ; -\infty < \theta < \infty),$$

then the log likelihood function is

$$l_Y(\theta ; Y) = -n \log 2 - \Sigma |Y_j - \theta|.$$

A special feature here is that the likelihood is differentiable everywhere except at the values $\theta = Y_{(1)}, \ldots, Y_{(n)}$, and is continuous everywhere. Location of $\hat{\theta}$ is simplified by writing

$$l_Y(\theta ; Y) = l_Y(Y_{(1)};Y) - 2 \sum_{j=1}^{m} Y_{(j)} - (n-2m)\theta \;\; (Y_{(m)} \leqslant \theta < Y_{(m+1)});$$

the likelihood is increasing for $\theta \leqslant Y_{(1)}$ and decreasing for $\theta \geqslant Y_{(n)}$. Examination of the linear sections of the log likelihood now shows that, for odd values of n, $\hat{\theta}$ is the median $Y_{(\frac{1}{2}n+\frac{1}{2})}$, whereas for even values of n the log likelihood is maximized by any value of θ in the interval $(Y_{(\frac{1}{2}n)}, Y_{(\frac{1}{2}n+1)})$, including the conventional median $\tfrac{1}{2} Y_{(\frac{1}{2}n)} + \tfrac{1}{2} Y_{(\frac{1}{2}n+1)}$. Thus the m.l.e. is given by the likelihood equation (2) only for even values of n, and then not uniquely.

Example 9.4. Special bivariate normal distribution. Let $Y_1, \ldots, Y_n$ be i.i.d. vectors with the bivariate normal distribution of zero mean, unit variances and unknown correlation coefficient θ. That is, the p.d.f. of a single vector is

$$\frac{1}{2\pi(1 - \theta^2)^{1/2}} \exp \left\{ -\frac{1}{2(1 - \theta^2)} y^{\mathrm{T}} \begin{bmatrix} 1 & -\theta \\ -\theta & 1 \end{bmatrix} y \right\}.$$

If we write

$$\Sigma Y_j Y_j^{\mathrm{T}} = \begin{bmatrix} \mathrm{SS}_{11} & \mathrm{SS}_{12} \\ \mathrm{SS}_{21} & \mathrm{SS}_{22} \end{bmatrix},$$

then the likelihood equation becomes

$$n\hat{\theta}^3 - \mathrm{SS}_{12}\,\hat{\theta}^2 + (\mathrm{SS}_{11} + \mathrm{SS}_{22} - n)\hat{\theta} - \mathrm{SS}_{12} = 0.$$

This equation has three solutions, two of which may be complex. Thus careful checking is necessary in applying the method to see that the right solution of (2) is taken.

It has been shown numerically (Barnett, 1966) that likelihood functions with several maxima arise with appreciable probability in estimating the location parameter of a Cauchy distribution from a small number of i.i.d. random variables. Examples of likelihood functions with saddle points are not common, but Solari (1969) has shown that this does occur in connexion with structural equations; see Exercise 9.1.

The phenomenon of Example 9.4 does not arise if the covariance matrix is completely unknown, when the problem is within the exponential family with three unknown parameters and a three-dimensional minimal sufficient statistic, there then being a unique solution to the likelihood equation. This is a particular case of the following general result for the exponential family, although here, for simplicity, we keep to the one-parameter case. In this, for i.i.d. random variables with the density

$$\exp\{a(\theta)b(y) + c(\theta) + d(y)\},$$

the likelihood equation (2) becomes

$$a'(\hat{\theta})\Sigma b(Y_j) + nc'(\hat{\theta}) = 0. \tag{5}$$

It follows from the results of Example 4.11 that

$$E\{b(Y);\theta\} = c'(\theta)/a'(\theta) = \mu(\theta), \quad \text{var}\{b(Y);\theta\} = -\mu'(\theta)/a'(\theta).$$

Now the second derivative of the log likelihood at a solution of (5) is

$$a''(\hat{\theta})\Sigma b(Y_j) + nc''(\hat{\theta}) = -n\{a'(\hat{\theta})\}^2 \text{var}\{b(Y);\hat{\theta}\} \leqslant 0,$$

and so, except in degenerate cases, it follows that the only solutions of (5) are local maxima. If the parameter space is unrestricted, the unique solution of (5) is therefore the m.l.e. Incidentally, one interpretation of the result (5) is that $\Sigma b(Y_j)/n$ is the m.l.e. of its expectation, i.e.

$$\Sigma b(Y_j)/n = E\{b(Y);\hat{\theta}\}.$$

Examples 9.1, 9.2 and 9.4 illustrate the general result that the m.l.e. is a function of the minimal sufficient statistic. The proof is immediate from the factorization theorem of Section 2.2(ii).

A further useful property of m.l.e.'s is their invariance under parameter transformation. That is, if $\chi = \chi(\theta)$ is a function of θ, not
necessarily one-one or differentiable, then the m.l.e. of χ is $\hat{\chi} = \chi(\hat{\theta})$.
This is immediate from the definition (1). In practice, therefore, we
work with the most convenient parameterization and derive the m.l.e.
of any other parameter of interest by transformation. For instance,
in the situation of Example 9.1, if we are interested in the parameter
$\chi = \text{pr}(Y \leqslant y_0)$, then we substitute the m.l.e.'s for μ and σ in
$\chi = \Phi\{(y_0 - \mu)/\sigma\}$.

(ii) Consistency

In the general discussion of estimation with large numbers of
observations a limited, but nevertheless important, role is played by
the notion of *consistency*. Roughly this requires that if the whole
"population" of random variables is observed, then the method of
estimation should give exactly the right answer. Consistency, however,
gives us no idea of the magnitude of the errors of estimation likely
for any given n. There is no single all-purpose definition of consistency,
but three common definitions are as follows.

First, consider estimates T_n based on i.i.d. random variables
$Y_1, \ldots, Y_n$ which are functions of the sample distribution function
$\tilde{F}_n(.)$, where

$$\tilde{F}_n(y) = \frac{1}{n} \Sigma \, \text{hv}(y - Y_{(j)}). \tag{6}$$

If the estimate has the form

$$T_n = t\{\tilde{F}_n(.)\},$$

then we call it *Fisher consistent* if

$$t\{F(.,\theta)\} = \theta. \tag{7}$$

In particular, for discrete random variables this amounts to requiring
that if all sample proportions are equal to the corresponding
probabilities, then the estimate is exactly correct.

A second, and more widely useful, definition is that T_n is *weakly
consistent* for θ if

$$T_n = \theta + o_p(1),$$

where, as explained in Section 9.1, $o_p(1)$ denotes a random variable that is $o(1)$ in probability. Thus weak consistency corresponds to T_n satisfying a weak law of large numbers. Now it can be shown that $\tilde{F}_n(y)$ is consistent for $F(y;\theta)$ uniformly in y (Gnedenko, 1967), and this can be used to show that Fisher consistency implies weak consistency. More will be said about general estimates of the form $t\{\tilde{F}_n(.)\}$ in Section 9.4.

Finally, an estimate T_n is called *strongly consistent* for θ if $T_n = \theta + o(1)$ with probability one, a property corresponding to the strong law of large numbers. Except in connexion with sequential sampling, strong consistency is not statistically a particularly meaningful concept.

For most statistical purposes weak consistency is the most useful of the three definitions; for the rest of this chapter we refer to it simply as *consistency*. Sometimes also we deal with random variables that are consistent for other random variables. Thus the sequence $\{T_n\}$ of random variables is consistent for a proper random variable T if $T_n = T + o_p(1)$; in particular, this implies that the cumulative distribution of T_n tends to that of T.

A key property of maximum likelihood estimates is their consistency under mild conditions. We outline the original elegant proof of this due to Wald (1949). The basis of the argument is the inequality

$$E\left\{\log\frac{f(Y;\theta')}{f(Y;\theta)};\theta\right\} \leqslant \log E\left\{\frac{f(Y;\theta')}{f(Y;\theta)};\theta\right\} = 0 \qquad (8)$$

for all $\theta' \neq \theta$, which is a special case of Jensen's Inequality. If we exclude the trivial situation where two densities can be identical, then strict inequality holds in (8).

For simplicity, suppose that the parameter space Ω is finite dimensional and countable, i.e. $\Omega = \{\theta_1, \theta_2, \dots\}$ and let $\theta = \theta_1$ be the true parameter value. For convenience, let $\theta_2, \theta_3, \dots$ be in order of increasing values of the Euclidean distance $\|\theta' - \theta\| = \|\theta' - \theta_1\|$. Temporarily, we restrict ourselves to i.i.d. random variables $Y_1, \dots, Y_n$. Then the defining property (1) of the m.l.e. can be written

$$\Sigma l(\hat{\theta};Y_j) \geqslant \Sigma l(\theta';Y_j) \quad (\theta' \in \Omega). \qquad (9)$$

The essence of the consistency argument is that (8) and (9) are incompatible unless $\hat{\theta}$ converges to θ, but to show this we have to apply the law of large numbers to the right-hand side of (9) for all

$\theta' \neq \theta$ simultaneously. Now we assume that as $\|\theta' - \theta\|$ tends to infinity, $f(y; \theta')$ tends to zero for almost all y. If the likelihood function is bounded, it then follows quite readily that

$$\lim_{a \to \infty} E\left\{\sup_{\theta' : \|\theta' - \theta\| > a} l(\theta'; Y_j); \theta\right\} = -\infty$$

and consequently, in view of (8), there is a value a_0 such that

$$E\left\{\sup_{\theta' : \|\theta' - \theta\| > a_0} l(\theta'; Y_j) - l(\theta; Y_j); \theta\right\} < 0. \qquad (10)$$

The set $\{\theta' : \|\theta' - \theta\| > a_0\}$ will be equal to $\{\theta_{s+1}, \theta_{s+2}, \ldots\}$ for some fixed s, because of our assumption about the ordering of Ω.

The strong law of large numbers will apply both to the right-hand sum in (9) and to

$$\Sigma \sup_{\theta' : \|\theta' - \theta\| > a_0} l(\theta'; Y_j),$$

for θ the true fixed parameter value. Hence, assuming all expectations to exist, we have from (8) that for $r = 2, \ldots, s$

$$\text{pr}\{\lim n^{-1} \Sigma l(\theta_r; Y_j) < \lim n^{-1} \Sigma l(\theta; Y_j); \theta\} = 1. \qquad (11)$$

In addition, (10) implies that

$$\text{pr}\left\{\lim n^{-1} \sup_{\theta' : \|\theta' - \theta\| > a_0} \Sigma l(\theta'; Y_j) < \lim n^{-1} \Sigma l(\theta; Y_j); \theta\right\} = 1. \qquad (12)$$

The finite set of statements (11) and (12) combine to show that

$$\text{pr}\left\{\lim n^{-1} \sup_{\theta' \neq \theta, \theta' \in \Omega} \Sigma l(\theta'; Y_j) < \lim n^{-1} \Sigma l(\theta; Y_j); \theta\right\} = 1.$$

There is a contradiction with (9) unless

$$\text{pr}\left(\lim_{n \to \infty} \hat{\theta} = \theta; \theta\right) = 1,$$

so that the (strong) consistency of $\hat{\theta}$ is proved. Quite clearly, (12) is not necessary if, in fact, Ω is finite.

In general, if Ω is finite dimensional and compact, then the problem is easily reduced to the countable problem just considered, provided that for $\theta', \theta'' \in \Omega$

$$\lim_{a \to 0} E\{\sup_{\|\theta' - \theta''\| \leqslant a} l(\theta''; Y_j); \theta\} = E\{l(\theta'; Y_j); \theta\}. \qquad (13)$$

This will be satisfied if $f(y;\theta')$ is continuous with respect to θ' in Ω and also if any singularities of $f(y;\theta')$ do not depend on θ'; this last condition is implicit in (10), and is obviously satisfied if the density is bounded.

To see how the conditions of the consistency theorem apply, recall Example 9.1 concerning $N(\mu,\sigma^2)$ variables with log likelihood given by (3). There (8) is easily verified; but for a single random variable the density has a singularity at $y = \mu$ when $\sigma = 0$, and the conditions (10) and (13) are not satisfied. However, if we take pairs of random variables as defining the component terms, then the joint density is bounded and the consistency theorem applies.

For independent but nonidentically distributed random variables the same arguments can be used with the weak law of large numbers, provided that (8) is strict for almost all the densities $f_{Y_j}(y;\theta')$. To achieve this, it is often necessary to group observations, but this is essentially a minor technical problem reflecting the need for identifiability of θ. As we have seen in the simple $N(\mu,\sigma^2)$ problem, grouping may be needed to avoid singularities in the likelihood. Serious difficulties may be encountered with problems where there are infinitely many parameters, and we shall say more about these later. The following example illustrates many of the points made above.

Example 9.5. One-way analysis of variance (ctd). Suppose that Z_{jk} are independently distributed in $N(\mu_j,\sigma^2)$ $(j = 1, \ldots, m; k = 1, \ldots, r)$, all parameters being unknown. Here $\theta = (\mu_1, \ldots, \mu_m, \sigma^2)$ and (8) is an inequality if we take as the notional variable in the consistency theorem $Y_k = (Z_{1k}, \ldots, Z_{mk})$; no fewer of the Z's will do. To satisfy (10) and (13) the Y_k must be taken in pairs to avoid the singularity at $\sigma = 0$. Therefore, if m is fixed, then the m.l.e., namely $(\bar{Z}_{1.}, \ldots, \bar{Z}_{m.}, \text{ss}_w/(rm))$, will be consistent as $r \to \infty$. This is, of course, easy to verify directly from the $N(\mu_j, \sigma^2/r)$ distribution of $\bar{Z}_{j.}$ and the chi-squared distribution with $(r-1)m$ degrees of freedom of ss_w/σ^2.

But now if r is fixed and m tends to infinity, no component of the m.l.e. is consistent. In the context of the consistency theorem, we cannot obtain an infinite number of replications of variables satisfying (8) and the other conditions. This situation occurs because Ω is infinite dimensional. We would, of course, not expect consistent estimates of the means; the point is that σ^2, which is common to all variables, is not consistently estimated either. In fact the variance of $\hat{\sigma}^2$ does converge to zero, but $E(\hat{\sigma}^2) = (r-1)\sigma^2/r$.

At the beginning of the section we discussed the likelihood
equation (2) and its use in deriving the m.l.e. To obtain further general
properties of the m.l.e., we work from the likelihood equation, so
that we need to establish the connexion between it and the m.l.e. in
large samples. To do this we suppose the likelihood to be regular as
defined in Section 9.1.

Now if the likelihood is differentiable around the true parameter
value θ, then (11) implies that for large n the likelihood has a local
maximum in any fixed open interval around θ with probability
tending to one; by differentiability, this maximum satisfies (2).
Further, when the second derivative exists around θ, the consistent
solution of (2) is unique for large n. For suppose that two consistent
solutions exist, both local maxima. Then a consistent local minimum
must exist, at which the second derivative will be positive. However,
since $i.(\theta')$ is assumed positive around $\theta' = \theta$, it follows, at least for
i.i.d. random variables, that the second derivatives around $\theta' = \theta$ will
be negative with probability tending to one. There is thus a contradiction with the occurrence of a local minimum, and hence with the
occurrence of two consistent local maxima.

A more important point in practice is the need to verify that a
solution of (2) used is the m.l.e. In almost all our further discussion
we shall suppose that the likelihood is regular and that the appropriate
solution of (2) is used. To stress, however, that the conditions required
in applying the consistency theorem, and relating it to the likelihood
equation, are not trivial we give an extreme example of Kiefer and
Wolfowitz (1956).

Example 9.6. An inconsistent m.l.e. Let $Y_1, \dots, Y_n$ be i.i.d. with the
density

$$10^{-10} \frac{1}{\sigma\sqrt{(2\pi)}} \exp\left\{-\frac{(y-\mu)^2}{2\sigma^2}\right\} + (1 - 10^{-10})\frac{1}{\sqrt{(2\pi)}}\exp\left\{-\frac{(y-\mu)^2}{2}\right.$$

Now the effect of the first term is for most purposes entirely negligible and the likelihood equation has a solution $\hat{\mu}$ very close to that
for the $N(\mu, 1)$ distribution. Nevertheless the likelihood is unbounded
when μ is set equal to any observed value y_j and $\sigma \to 0$, thus violating
(10) and (13) for any grouping of variables. That is, the overall
suprema of the likelihood do not provide sensible estimates for μ,
whereas the local maximum does. In rough terms the reason for the

anomaly is that, as $\sigma \to 0$, we are in effect allowing a discrete component to the joint distribution of $Y_1, \ldots, Y_n$ and any discrete contribution is "infinitely" greater than a continuous one. At a more practical level, note that if we took account of the inevitable discrete grouped nature of the data, then the anomaly would disappear.

In certain cases it is important to know how the m.l.e. behaves when estimation is restricted to a closed subset Ω^* of Ω. If Ω^* includes the true parameter value θ, it is quite clear that the above discussion applies without change. Suppose, however, that Ω^* does not include θ. Then quite generally the m.l.e. will be weakly consistent for that value θ^* which maximizes $E\{l(\theta'; Y) - l(\theta; Y); \theta\}$ over $\theta' \in \Omega^*$. Often θ^* will be unique for the type of problem arising in practice. The previous consistency proof can be applied with obvious modification. Usually there will not be a permissible solution of the likelihood equation, so that further properties are in general difficult to discuss. An exceptional situation is that of separate families of hypotheses, where Ω^* is equivalent to one family of regular distributions. This situation arises in Section 9.3(iv).

All the previous discussion applies both to the i.i.d. case and to independent non-homogeneous random variables when the dimension of the parameter space is finite. We now describe briefly the other main types of situation that may arise, and how they relate to the earlier discussion.

The simple asymptotic theory of maximum likelihood estimates requires that the dimension of the parameter space is fixed while the number of observations becomes large. The main practical difficulties with the method therefore arise when the dimension of the parameter space is large and comparable with the number of observations. While in some cases the m.l.e. is consistent, very often it is not. The second part of Example 9.5 is an illustration.

One class of problems of this type can be formulated as follows. Let $Y_1, \ldots, Y_n$ be independent vectors with Y_j having density $f_{Y_j}(y_j; \psi, \lambda_j)$, where $\psi, \lambda_1, \ldots, \lambda_n$ are all unknown. Often ψ is called a *structural parameter* and λ_j is called an *incidental parameter*. In many problems of this kind it is possible to obtain a consistent estimate of ψ by working with a conditional likelihood. Specifically, suppose that S_j is sufficient for λ_j ($j = 1, \ldots, n$) for any fixed ψ' and define the conditional likelihood by

$$\text{lik}_{Y \mid S}(\psi'; y|s_1, \ldots, s_n) = \Pi f_{Y_j \mid S_j}(y_j|s_j; \psi'), \qquad (14)$$

as in Section 2.1(iv). This will give a conditional likelihood equation and m.l.e. $\hat{\psi}_c$, say, which can be analysed by the previous arguments. Further discussion of the conditional likelihood is given in part (iii) of this section. Note that with the formulation described here consistent estimation of $\lambda_1, \ldots, \lambda_n$ will not be possible.

For dependent random variables, the same kind of consistency argument as for the i.i.d. case will apply, subject to suitable conditions on the type of dependence. We work with the decomposition

$$f_Y(Y;\theta) = f_{Y_1}(Y_1;\theta) \prod_{j=2}^{n} f_{Y_j|Y^{(j-1)}}(Y_j|Y^{(j-1)};\theta),$$

where $Y^{(j-1)}$ denotes $(Y_1, \ldots, Y_{j-1})$. The log likelihood is again a sum, this time of dependent random variables, and the point to be checked is whether a law of large numbers applies. For some further details, see part (iii) of this section.

Entirely generally, if T_n is a consistent estimate of θ, then for any function $g(.)$ continuous at θ, $g(T_n)$ is consistent for $g(\theta)$.

Thus, in particular, $f(y;\hat{\theta})$ is a consistent estimate of the parametric density $f(y;\theta)$. For arbitrary nonparametric continuous densities, however, consistent estimation is not possible directly by maximum likelihood; indeed formally the maximum likelihood estimate can be taken as the generalised derivative of $\tilde{F}_n(y)$, namely

$$n^{-1} \Sigma \delta(y - Y_j),$$

where $\delta(.)$ is the Dirac delta function. By smoothing over intervals that shrink to zero as $n \to \infty$, but which do so more slowly than $1/n$, it is possible to produce consistent estimates of a probability density subject only to its continuity; see Rosenblatt (1971). In more direct practical terms we can, for any given n, form a good estimate of the amount of probability in a small interval, provided that the interval contains a large number of observations.

The most general finite discrete distribution is the multinomial with cell probabilities $\pi_1, \ldots, \pi_m$, say. If the corresponding observed frequencies are $n_1, \ldots, n_m$, it is easily shown by maximizing the likelihood

$$\frac{n!}{n_1! \ldots n_m!} \pi_1^{n_1} \ldots \pi_m^{n_m}$$

that the maximum likelihood estimate is the set of sample proportions, and this is clearly consistent.

(iii) Asymptotic normality

The basis of large-sample tests and confidence intervals is the general property that the m.l.e. has a limiting normal distribution around the true parameter value as mean and with a variance that is easily calculated. Of course, this is a much stronger result than consistency and requires additional regularity conditions.

We shall assume that $f_Y(y;\theta)$ satisfies the regularity conditions (a) – (d) of Section 9.1. For simplicity, we take first the case of a one-dimensional parameter and suppose that $Y_1, \ldots, Y_n$ are i.i.d. with density $f(y;\theta)$. In particular, then, we have for a single observation that

$$E\{U(\theta);\theta\} = 0, E\{U^2(\theta);\theta\} = E\{-U'(\theta);\theta\} = i(\theta) > 0, \quad (15)$$

where the dash denotes partial differentiation with respect to θ. Also the absolute value of the averaged third derivative of the log likelihood, $u''(Y,\theta')n$, is bounded in the neighbourhood of $\theta' = \theta$ by an integrable function $g(Y)$ whose expectation exists. We shall show that the limiting distribution of $(\hat\theta - \theta)\sqrt{n}$ is $N\{0, 1/i(\theta)\}$.

With the above regularity conditions, $\hat\theta$ is a consistent estimate; expansion of the total score $U_\cdot(\hat\theta)$ in a Taylor series about the true parameter value θ then gives

$$U_\cdot(\hat\theta) = U_\cdot(\theta) + (\hat\theta - \theta) U'_\cdot(\theta) + \tfrac{1}{2}(\hat\theta - \theta)^2 U''_\cdot(\theta^\dagger), \quad (16)$$

where $|\theta^\dagger - \theta| < |\hat\theta - \theta|$. Now the left-hand side of (16) is zero, by the likelihood equation (2). Also $\theta^\dagger$ is necessarily consistent, so that we may rewrite (16) as

$$\frac{1}{\sqrt{n}} \Sigma U_j(\theta)/i(\theta) = \sqrt{n}(\hat\theta - \theta)\left\{\frac{n^{-1} \Sigma U'_j(\theta)}{i(\theta)} + e_n\right\}, \quad (17)$$

where $|e_n| \leqslant |\hat\theta - \theta| g(Y)/i(\theta) = o_p(1)$. The weak law of large numbers applies to the ratio on the right of (17), which is $1 + o_p(1)$. That is,

$$\sqrt{n}(\hat\theta - \theta)\{1 + o_p(1)\} = \frac{1}{\sqrt{n}} \Sigma U_j(\theta)/i(\theta). \quad (18)$$

Application of the central limit theorem to the right-hand side of (18) gives the required limiting result, because of (15).

The regularity conditions used here are unnecessarily restrictive, as can be seen from Example 9.3. There the m.l.e. is the sample median, which is known to be asymptotically normally distributed, even though the likelihood is not differentiable. A full discussion of regularity conditions is given by LeCam(1970).

As an example where asymptotic normality does not hold, consider Example 9.2. Here the m.l.e. is the largest observation. Both exact and limiting distributions are easily calculated and in fact $n(\theta - \hat\theta)$ has a limiting exponential distribution with mean $1/\theta$. The reason that the limiting distribution is not normal is that it is not possible to link $\hat\theta$ approximately with a sum of independent random variables.

An important general point is that $1/i(\theta)$ is the variance of the asymptotic distribution of $(\hat\theta - \theta)\sqrt{n}$ and not, in general, the limit of the exact variance. Fortunately it is the former that is required in using maximum likelihood estimates to construct tests and confidence limits. The distinction between the variance of the asymptotic distribution and the limiting variance is best shown by a simple example.

Example 9.7. Poisson distribution (ctd). Let $Y_1, \ldots, Y_n$ be i.i.d. in the Poisson distribution of mean μ and suppose that we estimate by maximum likelihood the parameter $\theta = 1/\mu$. Then $\hat\theta = 1/\bar{Y}$. Now this has infinite variance for all n; indeed it is an improper random variable in that, for all n, $\bar{Y}$ has a positive probability of being zero. Nevertheless, the limiting distribution of $(\hat\theta - \theta)\sqrt{n}$ is normal with zero mean and variance θ^3. The singularity at $\bar{Y} = 0$ has no effect on the limiting distribution.

The essence of the proof of asymptotic normality is that expansion of the likelihood equation and use of the weak law of large numbers allow us to write approximately

$$i(\theta)\,(\hat\theta - \theta)\sqrt{n} = \frac{1}{\sqrt{n}}\,\Sigma U_j(\theta)\,. \tag{19}$$

Subject to the appropriate regularity conditions an expansion analogous to (19) can be made under much more general conditions and therefore various generalizations of the key result are possible.

In particular, if the observations are i.i.d. but the parameter θ is a vector, then (19) generalizes to

$$\mathbf{i}(\theta)\,(\hat{\theta} - \theta)\sqrt{n} \;=\; \frac{1}{\sqrt{n}}\,\Sigma\,U_j(\theta)\,, \tag{20}$$

where $\mathbf{i}(\theta)$ is the information matrix for a single observation. Equation (20) follows from a multivariate Taylor expansion of the likelihood equation. Thus to first order

$$(\hat{\theta} - \theta)\sqrt{n} \;=\; \frac{1}{\sqrt{n}}\,\mathbf{i}^{-1}(\theta)\,\Sigma\,U_j(\theta)$$

and asymptotic multivariate normality follows from the multivariate central limit theorem. Now the asymptotic mean is zero, because $E\{U(\theta);\theta\} = 0$. To calculate the asymptotic covariance matrix, we have to first order

$$
\begin{aligned}
E\{(\hat{\theta} - \theta)\,(\hat{\theta} - \theta)^{\mathrm{T}}\,n\,;\theta\} \;&=\; n^{-1}\mathbf{i}^{-1}(\theta)E\{U(\theta)U^{\mathrm{T}}(\theta);\theta\}\mathbf{i}^{-1}(\theta) \\
&=\; \mathbf{i}^{-1}(\theta)E\{U(\theta)U^{\mathrm{T}}(\theta);\theta\}\,\mathbf{i}^{-1}(\theta) \\
&=\; \mathbf{i}^{-1}(\theta).
\end{aligned}
$$

An important example is provided by the multinomial distribution.

Example 9.8. Multinomial distribution. In the multinomial distribution with cell probabilities $\pi_1, \ldots, \pi_m$ summing to one, it is convenient to remove the constraint on the parameters by eliminating $\pi_m = 1 - (\pi_1 + \ldots + \pi_{m-1})$, and writing θ for the column vector $(\pi_1, \ldots, \pi_{m-1})$. The log density for a single observation can then be written

$$\log f(Y;\theta) \;=\; \sum_{k=1}^{m-1} \delta_{Yk}\,\log\pi_k \;+\; \delta_{Ym}\,\log(1 - \pi_1 - \ldots - \pi_{m-1})\,,$$

where δ_{st} is the Kronecker delta symbol. Therefore

$$\frac{\partial^2 \log f(Y;\theta)}{\partial\pi_j\partial\pi_k} \;=\; -\frac{\delta_{Yj}\delta_{Yk}}{\pi_j^2} + \frac{\delta_{Ym}}{\pi_m^2}\,,$$

from which, on taking expectations, we have that

$$i_{jk}(\theta) \;=\; \frac{\delta_{jk}}{\pi_j} - \frac{1}{\pi_m}\,. \tag{21}$$

Finally, the asymptotic covariance matrix of $(\hat{\theta} - \theta)\sqrt{n}$ is obtained by inversion as having as its (j, k)th element

$$\delta_{jk} \pi_j - \pi_j \pi_k .$$

If we denote the cell frequencies by $N_1, \ldots, N_m$, then, as noted before, the m.l.e. of π_j is N_j/n. In this case the asymptotic covariance matrix is an exact result.

Provided that θ is finite dimensional, the limiting results generalize directly to situations in which the Y_j, while independent, are not identically distributed. Indeed for (20) to hold in this more general situation we have essentially to check that
(a) a central limit theorem for non-identically distributed random variables applies to the total score

$$U.(\theta) = \Sigma u_j(Y_j, \theta)$$

with a non-singular asymptotic distribution;
(b) a weak law of large numbers applies to the information, i.e. to the convergence in probability of

$$-\frac{1}{n} \frac{\partial}{\partial \theta} \{\Sigma u_j (Y_j ; \theta)\}^{\mathsf{T}}$$

to the limit $i.(\theta)/n$, the average information matrix, assumed here to be non-singular.

Subject to these and a condition on third derivatives, the argument from (20) leads to the result that $(\hat{\theta} - \theta)\sqrt{n}$ is asymptotically multivariate normal with zero mean and covariance matrix $n\, i_{\cdot}^{-1}(\theta)$.

An essential point is that the total information increases indefinitely with n; in certain cases, if, for example, the total information is asymptotically of order n^c, for $c \neq 1$, then a different normalization is required.

Example 9.9. Simple linear regression. Let $Y_1, \ldots, Y_n$ be independently distributed with Y_j having density $g(y_j - \gamma - \beta x_j)$, where the x_j's are known constants and $g(.)$ is a known density. The efficient total score vector for $\theta = (\gamma, \beta)^{\mathsf{T}}$ has components

$$U_{.\gamma}(\theta) = -\sum \frac{g'(Y_j - \gamma - \beta x_j)}{g(Y_j - \gamma - \beta x_j)} , \quad U_{.\beta}(\theta) = -\sum x_j \frac{g'(Y_j - \gamma - \beta x_j)}{g(Y_j - \gamma - \beta x_j)}$$

and the total information matrix is

$$i.(\theta;x) = \begin{bmatrix} n & \Sigma x_j \\ \Sigma x_j & \Sigma x_j^2 \end{bmatrix} i_g ,$$ (22)

where i_g is the intrinsic accuracy of the density $g(.)$ defined in Section 4.8 as

$$\int [\{g'(y)\}^2/g(y)]dy.$$

For sufficiently regular $g(.)$, consistency and asymptotic normality of $\hat{\theta}$ will follow if, according to the general discussion, a central limit theorem with a nonsingular limiting distribution applies to $U.$, and a sufficient condition for this is that $\Sigma(x_j - \bar{x}.)^2 \to \infty$; also, we need the convergence of the average information matrix and the applicability of the weak law of large numbers. For this the same conditions are enough.

The next generalization for consideration is where the number of nuisance parameters increases with the number of observations. As we pointed out in the discussion of consistency, the m.l.e. of the structural parameter need not be consistent in this situation, even if it would be for known values of the nuisance parameters. If consistent, the m.l.e. may have a limiting normal distribution with a variance not given by the above results. A particular example is described in Exercise 9.3. The general position, therefore, is not clear. However, in part (ii) of this section we introduced a particular class of problems in which estimation *via* a conditional likelihood is completely regular. The relevant conditional likelihood is defined in (14), and under some general conditions that we shall not discuss, the previous arguments can be followed to prove consistency and asymptotic normality, with a covariance matrix determined by the inverse of the conditional information matrix. Andersen (1970, 1973) has discussed this in detail including the important question of whether information is lost by the conditioning.

The following simple example illustrates the method.

Example 9.10. Matched pairs for binary data. Suppose that matched pairs of individuals are subject one to a control and the other to a treatment and that a binary observation is made for each, giving for the jth pair the random variables Y_{j1} and Y_{j2}. A possible model for this situation is one that allows arbitrary differences between different pairs but which specifies the same difference between treatment and control in the log odds for "success" versus "failure". That is, if the possible values of each Y are zero and one the density of Y_{j1} and Y_{j2} is

$$\frac{(\psi \lambda_j)^{y_{j1}}}{1 + \psi \lambda_j} \quad \frac{\lambda_j^{y_{j2}}}{1 + \lambda_j} .$$

In fact, inference for ψ is possible by the "exact" methods of Chapter 5, leading to most powerful similar regions. The first step of the argument is the same whether "exact" or asymptotic arguments are used. The statistics $S_j = Y_{j1} + Y_{j2}$ $(j = 1, \ldots, n)$ are sufficient for the nuisance parameters $\lambda_1, \ldots, \lambda_n$ and the conditional distributions are, for an arbitrary pair,

$$f_{Y|S}(0,0|s = 0 ; \psi) = f_{Y|S}(1,1|s = 2 ; \psi) = 1,$$

$$f_{Y|S}(y_1, y_2|s = 1 ; \psi) = \psi^{y_1}/(1 + \psi) \quad (y_1 = 0,1 ; y_2 = 1 - y_1).$$

Thus only pairs with $s = 1$ contribute non-trivially to the conditional likelihood and the conditional maximum likelihood estimate is easily shown to be

$$\hat{\psi}_c(S) = \sum_{\{S_j = 1\}} Y_{1j} / \sum_{\{S_j = 1\}} (1 - Y_{1j}).$$

This is consistent and asymptotically normal if and only if the number of occurrences of $s = 1$ tends to infinity with n ; this imposes weak conditions on the nuisance parameters. The limiting normal distribution given from the conditional information is such that if n_1 is the number of pairs with $s = 1$, then $(\hat{\psi}_c - \psi)$ has mean 0 and variance $\psi(1 + \psi)/n_1$. This is in accord with the binomial distribution arising in the "exact" treatment.

The final generalization is to dependent random variables covering, therefore, inference in time series and stochastic processes. Again it is a case of examining the applicability of the central limit theorem and the weak law of large numbers.

Consider, then, a discrete time stochastic process represented by the random variables $Y_0, Y_1, \ldots, Y_n$, where Y_0 represents the initial conditions. For large n, the influence of Y_0 will usually be negligible. It is convenient to write $y^{(j)}$ for the set of observed values $(y_0, \ldots, y_j)$. Further, for the conditional log likelihood derived from Y_j given $Y^{(j-1)}$, we write

$$l_j(\theta ; Y_j | Y^{(j-1)}) = \log f_{Y_j | Y_{j-1}, \ldots, Y_0}(Y_j | Y_{j-1}, \ldots, Y_0 ; \theta).$$

Then the full log likelihood function is

$$l_{Y_0}(\theta ; Y_0) + \sum_{j=1}^{n} l_j(\theta ; Y_j | Y^{(j-1)}).$$

The argument for consistency depends on a weak law of large numbers for the sum on the right. A necessary condition is that

$$\lim_{|j-k| \to \infty} \text{cov}\{l_j(\theta ; Y_j | Y^{(j-1)}), l_k(\theta ; Y_k | Y^{(k-1)}) ; \theta\} = 0.$$

This implies that dependence is not sustained through a long sequence of conditional likelihoods, and that information accumulates rapidly. The inequality (8), which is central to the proof of consistency, can still be applied, with grouping of observations if $\dim(\theta) > \dim(Y)$.

When $f_Y(y ; \theta)$ is regular, the m.l.e. $\hat{\theta}$ satisfies the likelihood equation $u_.(Y ; \hat{\theta}) = 0$. It is convenient to define

$$u_j(Y_j | Y^{(j-1)}; \theta) = \frac{\partial}{\partial \theta} l_j(\theta ; Y_j | Y^{(j-1)}) \ (j = 1, \ldots, n).$$

These scores have zero expectation under the parameter value θ and we may define the information matrix for the jth observation by

$$i_j(\theta) = E\left\{-\frac{\partial}{\partial \theta} u_j^{T}(Y_j | Y^{(j-1)} ; \theta) ; \theta\right\} \ (j = 1, \ldots, n).$$

The expansion of the m.l.e. can again be made and we have that, to first order,

$$(\hat{\theta} - \theta)\sqrt{n} = n\{i_0(\theta) + i_.(\theta)\}^{-1} \frac{1}{\sqrt{n}} \{u_0(Y_0 ; \theta) + \sum_{j=1}^{n} u_j(Y_j | Y^{(j-1)} ; \theta)\}.$$

Asymptotic normality will follow if the central limit theorem applies and this will be so if the covariances

$$c_{jk}(\theta) = \text{cov}\{u_j(Y_j | Y^{(j-1)}; \theta), u_k(Y_k | Y^{(k-1)}; \theta) ; \theta\}$$

tend to zero as $|j - k| \to \infty$ and the average information matrix tends to a limiting non-singular matrix. Then $(\hat{\theta} - \theta)\sqrt{n}$ has as its limiting distribution $MN_q[0, n i_.^{-1}(\theta)\{\Sigma\Sigma c_{jk}(\theta)\}i_.^{-1}(\theta)]$, because the contribution from Y_0 is negligible asymptotically. Note that this is not so simple as the result for independent random variables. However, for m-dependent Markov processes, i.e. for processes in which

$$f_{Y_j|Y^{(j-1)}}(y_j|y^{(j-1)};\theta) = f_{Y_j|Y_{j-1},\ldots,Y_{j-m}}(y_j|y_{j-1},\ldots,y_{j-m};\theta),$$

the scores are uncorrelated. Then the covariance matrix is again given by inverting the total information matrix.

   In theoretical discussions, although not normally in applications, it is quite often useful to assign particular values to y_0 chosen in order to simplify the likelihood equation.

Example 9.11. First-order autoregressive process. Suppose that the random variables $Y_1,\ldots,Y_n$ are generated by the first-order autoregressive model

$$Y_j = \theta Y_{j-1} + \epsilon_j,$$

where the ϵ_j are i.i.d. random variables with zero mean, variance σ^2 and with the density $g(.)$. Provided that $|\theta|<1$, the process is stationary.

   For this process, $l_j(\theta;y_j|y^{(j-1)}) = \log g(y_j - \theta y_{j-1})$ and consequently

$$i_j(\theta) = E\left\{-\frac{d^2\log g(\epsilon)}{d\epsilon^2}\right\} E(Y_{j-1}^2;\theta),$$

where the first term on the right-hand side is the intrinsic accuracy i_g of the p.d.f. $g(.)$. Now if Y_0 has the stationary distribution of the process, then $E(Y_j^2;\theta) = \sigma^2/(1-\theta^2)$, which is otherwise an approximation for large j. In all cases, we obtain

$$i.(\theta) = \frac{n\sigma^2 i_g}{1-\theta^2} + O(1).$$

It follows that, for regular $g(.)$, the limiting distribution of $\hat\theta$ is asymptotically normal with mean θ and variance

$$n^{-1}(1-\theta^2)(\sigma^2 i_g)^{-1}.$$

For normal $g(.)$, the likelihood equation takes the form

$$\frac{\partial\log f_{Y_0}(Y_0;\hat\theta)}{\partial\theta} + \sum_{j=1}^{n} Y_{j-1}(Y_j - \hat\theta Y_{j-1}) = 0,$$

which is simplified by the assumption $Y_0 = Y_n$ to give

$$\hat{\theta} = \sum_{j=1}^{n} Y_j Y_{j-1} / \sum_{j=1}^{n} Y_j^2 .$$

For this particular case, $i_g \sigma^2 = 1$, so that the large-sample variance of $\hat{\theta}$ is $(1 - \theta^2)/n$.

Stochastic processes in continuous time are more difficult to deal with. Billingsley (1961b) has given a thorough discussion of discrete state Markov processes in continuous time. These will not be discussed here.

We conclude this discussion of the asymptotic distribution of maximum likelihood estimates with some general comments.

First, if we transform from a parameter θ to a parameter $\chi = \chi(\theta)$ by a one-one transformation the maximum likelihood estimates exactly correspond, i.e. $\hat{\chi} = \chi(\hat{\theta})$. Further, provided that the transformation is differentiable at the true parameter point, the correspondence between the asymptotic variances is given by the result of Exercise 4.15 for the transformation of information. In particular, for one-dimensional parameters we have asymptotically that

$$\text{var}(\hat{\chi}) = \left(\frac{\partial \chi}{\partial \theta} \right)^2 \text{var}(\hat{\theta}).$$

This can be obtained also directly from the method of Section 8.4 (iii) for finding the approximate moments of a function of a random variable. Note that, while $\hat{\theta}$ and $\hat{\chi}$ are both asymptotically normal, the rate of approach to the limiting normal form will depend appreciably on the parameterization.

Secondly, because $i_.(\theta)$ will typically depend on the unknown θ, we often need in applications to use alternative forms for the asymptotic variance. Two such are

$$i_.(\hat{\theta}) \quad \text{and} \quad \left[-\frac{\partial^2 l(\theta ; Y)}{\partial \theta^2} \right]_{\theta = \hat{\theta}} ,$$

provided that $i_.(\theta')$ is continuous at $\theta' = \theta$. Quite generally, any consistent estimate of $i_.(\theta)n$ can be used without affecting the limiting distributional result.

Thirdly, note that the conditionality principle applies to the previous discussion. For statistics that are ancillary in the simple sense of Section 2.2(viii) conditioning does not change the maximum likelihood estimate, but does mean that its variance should be found from the conditional information $i_.(\theta|c)$. Now under some, although not all, circumstances C can be taken in a form such that as $n \to \infty$ it converges in probability to a constant. Then the distinction between conditional and unconditional information is not needed in first-order asymptotic theory. When conditioning is used in problems with a large number of nuisance parameters the role of the conditioning statistic is more critical and will usually affect not only the form of the asymptotic variance but the value of the maximum likelihood estimate itself.

Finally our discussion has required the true parameter value to be interior to the parameter space Ω. Without this condition a generally valid Taylor expansion for $\hat{\theta} - \theta$ is clearly not possible. Quite often, when the true parameter value lies on the boundary of the parameter space, a censored normal distribution will result. Rather than give a general discussion we give one example, taken from Hinkley (1973).

Example 9.12. Unsigned normal variables. Let $Y_1, \ldots, Y_n$ be i.i.d. such that $Y_j = |X_j|$, where X_j is $N(\mu, 1)$. This situation arises, for example, if X_j is the difference between a matched pair of random variables whose control and treatment labels are lost. The sign of the parameter μ is clearly unidentifiable, and so we work with $\theta = |\mu|$, with parameter space $\Omega = [0, \infty)$. It will become apparent that if θ were taken to be μ^2 rather than $|\mu|$, then the m.l.e. of θ would behave in a completely regular manner. We have deliberately not done this in order to illustrate a particular phenomenon ; see also Example 4.16.

The log likelihood for a single random variable is

$$l(\theta' ; Y_j) = \text{const} - \tfrac{1}{2} Y_j^2 - \tfrac{1}{2} \theta'^2 + \log \cosh(\theta' Y_j),$$

with efficient score

$$U_j(\theta') = -\theta' + Y_j \tanh(\theta' Y_j) . \tag{23}$$

The conditions for consistency of the m.l.e. are satisfied throughout Ω, and for $\theta > 0$ the complete regularity of the likelihood implies that $(\hat{\theta} - \theta)\sqrt{n}$ has a limiting normal distribution, the information per observation being

$$i(\theta) = 1 - \frac{2e^{-\frac{1}{2}\theta^2}}{\sqrt{(2\pi)}} \int\limits_0^\infty y^2 e^{-\frac{1}{2}y^2} \operatorname{sech}(\theta y) dy.$$

Now if $\theta = 0$, the crucial expansion (16) breaks down. In particular, the efficient score (23) is identically zero at $\theta' = 0$. We can, however, for positive θ', write the Taylor expansion of the log likelihood up to the term in θ'^4, and, on evaluating the derivatives, this becomes for the whole sample

$$l(\theta';Y) = l(0;Y) + \tfrac{1}{2}\theta'^2 (\Sigma Y_j^2 - n) + \tfrac{1}{24}\theta'^4 (-2\Sigma Y_j^4) + o(\theta'^4).$$

Two things follow. First, if $\Sigma Y_j^2 \leqslant n$, then no local maximum of the likelihood occurs near $\theta' = 0$. Thus for large n, $\hat{\theta}$ is zero whenever $\Sigma Y_j^2 \leqslant n$. In other cases, a local maximum does exist and the leading terms of the above expansion give, to first order,

$$\hat{\theta}^2 = \frac{3(\Sigma Y_j^2 - n)}{\Sigma Y_j^4}.$$

Now Y_j^2 has the chi-squared distribution with one degree of freedom when $\theta = 0$, and a simple calculation shows that for large n, $\hat{\theta}^2 \sqrt{n}$ then has approximately a $N(0,2)$ distribution truncated at zero.

Some general results for the m.l.e. when the parameter is on the boundary of Ω are given by Moran (1971).

(iv) Asymptotic efficiency

We have seen in the previous section that, provided some general regularity conditions hold, the limiting distribution of the m.l.e. is normal around the true value as mean and with a variance achieving the Cramér-Rao lower bound of Section 8.3. We call any estimate with this property *asymptotically efficient*. Now this, in effect, pre-supposes that it is impossible to produce estimates with smaller asymptotic variance and, bearing in mind the distinction between expectations and variances and asymptotic expectations and variances, this is highly plausible but not so easily proved. We shall not give a formal proof; in fact there is a minor technical exception which shows the need for additional regularity conditions and this exception we now outline.

Example 9.13. Superefficiency. Let $Y_1, \dots, Y_n$ be i.i.d. in $N(\mu, 1)$, so that $\hat{\mu} = \bar{Y}$. Consider the new estimate

$$T_n = \left\{ \begin{array}{l} \bar{Y}. \ (|\bar{Y}.| \geqslant n^{-1/4}), \\ b\bar{Y}. \ (|\bar{Y}.| < n^{-1/4}). \end{array} \right\}$$

Because $\bar{Y}.$ has the distribution $N(\mu, 1/n)$, the asymptotic distributions of T_n and $\bar{Y}.$ are the same for $\mu \neq 0$. But if $\mu = 0$, T_n has the asymptotic distribution $N(0, b^2/n)$, an improvement on $\bar{Y}.$ if $|b| < 1$. Note that in the definition of T_n for $|\bar{Y}.| < n^{-\frac{1}{4}}$ the mean could be replaced by the median or some other estimate, for suitable b, and produce the same effect.

We call an estimate *superefficient* if for all parameter values the estimate is asymptotically normal around the true value with a variance never exceeding and sometimes less than the Cramér-Rao lower bound. A value of the parameter, zero in the above case, at which the variance is below the Cramér-Rao bound is called a point of superefficiency.

There are three reasons why superefficiency is not a statistically important idea. First, LeCam(1953) has shown that the set of points of superefficiency is countable. Next, it is clear that such estimates give no improvement when regarded as test statistics. Finally, for any fixed n the reduction in mean squared error for parameter points near to the point of superefficiency is balanced by an increase in mean squared error at points a moderate distance away. For these reasons, we shall ignore superefficiency in the rest of the discussion.

Estimates that differ from the m.l.e. by $o_p(n^{-\frac{1}{2}})$ will also be asymptotically efficient. We shall not give a detailed discussion of such alternative methods of estimation. For multinomial distributions with a finite number of cells a fairly general discussion is simple, however. Suppose that there are m cells with probabilities $\pi_1(\theta), \ldots, \pi_m(\theta)$ and that $N_1, \ldots, N_m$ are the corresponding sample frequencies, the total number of trials being n. Write $S = (N_1/n, \ldots, N_m/n)$. Now consider an estimate T_n defined by an estimating equation

$$g_T(T_n, S_1, \ldots, S_m) = 0, \tag{24}$$

where $g_T\{\theta, \pi_1(\theta), \ldots, \pi_m(\theta)\} = 0$. We deal with one-dimensional parameter for simplicity, although there is an immediate generalization.

Provided that the function $g_T(.)$ is continuous, T_n is consistent because S tends in probability to $\pi(\theta)$. We now expand (24) in a

Taylor series, as we did with the likelihood equation, thereby obtaining to first order

$$g_T(T_n,S) = g_T\{\theta,\pi(\theta)\} + (T_n - \theta)\frac{\partial g_T}{\partial x_0} + \sum_{j=1}^{m}\{S_j - \pi_j(\theta)\}\frac{\partial g_T}{\partial x_j}, \quad (25)$$

where $x_0, \ldots, x_m$ are the dummy arguments of $g_T(.)$ and where the derivatives are evaluated at $x_0 = \theta$, $x_j = \pi_j(\theta)$ $(j = 1, \ldots, m)$. The first two terms of (25) vanish and therefore

$$T_n - \theta = -\sum_{j=1}^{m}\{S_j - \pi_j(\theta)\}\frac{\partial g_T}{\partial x_j}\bigg/\frac{\partial g_T}{\partial x_0}. \quad (26)$$

The likelihood equation leads to the special case

$$\hat{\theta} - \theta = \sum \frac{\pi_j'(\theta)}{\pi_j(\theta)}\{S_j - \pi_j(\theta)\}/i(\theta).$$

Thus, for T_n to be asymptotically efficient, we need

$$\frac{\partial g_T}{\partial x_j}\bigg/\frac{\partial g_T}{\partial x_0} = -\pi_j'(\theta)/\{\pi_j(\theta)i(\theta)\} \ (j = 1, \ldots, m). \quad (27)$$

Two particular methods that have been quite widely used as alternatives to maximizing the likelihood are minimum chi-squared and minimum modified chi-squared, corresponding respectively to minimizing

$$\sum\frac{\{N_j - n\pi_j(\theta')\}^2}{n\pi_j(\theta')} \quad \text{and} \quad \sum\frac{\{N_j - n\pi_j(\theta')\}^2}{N_j}. \quad (28)$$

The corresponding estimating equations satisfy (27) and hence the estimates are asymptotically efficient.

A more general family of asymptotically efficient estimates is obtained by applying the method of generalized least squares to transformed observations using an empirical or theoretical estimate of the covariance matrix.

While our initial emphasis is on the construction of asymptotically efficient estimates, it may sometimes be required, for example because of robustness or ease of computation, to consider other estimates. If T_n is asymptotically normal with mean θ and variance σ_T^2/n, we define the *asymptotic efficiency* of T_n to be

$$e(T_n) = \lim n \, (\sigma_T^2 i_{.})^{-1}. \tag{29}$$

For vector parameters we can either make the efficiency calculation component by component in the obvious way or compare the generalized variances. In a similar way, we can define the *asymptotic relative efficiency* of T_n' with respect to T_n'' by $\sigma_{T''}^2 / \sigma_{T'}^2$. All these definitions are made only for asymptotically normal estimates and refer to first-order asymptotic properties.

(v) Asymptotic sufficiency

As explained in Section 9.1, one way of explaining the good asymptotic properties of m.l.e.'s is *via* their asymptotic sufficiency. The proof is implicit in the earlier arguments. Under the usual regularity conditions, an expansion similar to (16) shows that for a one-dimensional parameter

$$l_Y(\theta'\,;Y) = l_Y(\hat\theta\,;Y) + \frac{1}{2}n(\hat\theta - \theta')^2 \left\{ \frac{1}{n} \frac{\partial^2 l_Y(\theta'\,;Y)}{\partial\theta'^2} \right\} + o_p(1),$$

provided that $\theta' - \theta = O(n^{-\frac{1}{2}})$. Alternatively, we can write

$$f_Y(Y\,;\theta') = f_Y(Y\,;\hat\theta) \exp\{-\tfrac{1}{2}(\hat\theta - \theta')^2 i_{.}(\theta) + o_p(1)\}, \tag{30}$$

showing a factorization of the likelihood of the kind required to establish sufficiency. The multiparameter generalization is immediate. Because of the relationship (19) between the m.l.e. and the efficient score vector, the asymptotic sufficiency can also be expressed in terms of $U_{.}(\theta)$.

The form (30) is particularly useful in connexion with the theory of tests, because of the relation it establishes with normal theory. Here, however, we illustrate (30) by establishing the effect on the m.l.e. of a component parameter of setting another component parameter equal to its true value. For simplicity suppose that $\theta = (\psi, \lambda)$ is two dimensional. Thus the unrestricted m.l.e. $\hat\theta = (\hat\psi, \hat\lambda)$ has the distribution

$$MN_2\{\theta, i_{.}^{-1}(\theta)\},$$

where

$$i_{.}(\theta) = \begin{bmatrix} i_{.\psi\psi}(\theta) & i_{.\psi\lambda}(\theta) \\ i_{.\lambda\psi}(\theta) & i_{.\lambda\lambda}(\theta) \end{bmatrix}.$$

Suppose now that we want the m.l.e of λ when ψ is given. If we approximate to the likelihood by the joint normal density of $(\hat{\psi}, \hat{\lambda})$, the restricted m.l.e. is, by the usual properties of the bivariate normal density,

$$\tilde{\lambda} = \hat{\lambda} + i_{.\lambda\lambda}^{-1}(\theta) \, i_{.\lambda\psi}(\theta) \, (\hat{\psi} - \psi), \tag{31}$$

and the associated limiting variance is, of course, $i_{.\lambda\lambda}^{-1}(\theta)$. This agrees to first order with the result of expanding both $U_.(\hat{\psi}, \hat{\lambda})$ and $U_.(\psi, \tilde{\lambda})$ and solving for $\tilde{\lambda}$ to first order. The expression (31) is valid in the general multiparameter case.

(vi) Computation of maximum likelihood estimates

Iterative numerical techniques will very often be necessary to compute m.l.e.'s and their asymptotic variances. Detailed discussion of computational procedures is quite outside the scope of this book, but the following general points should be noted. Use of good numerical procedures becomes, of course, of increasing importance as dim(θ) increases.

Choice of starting point for iteration is important. Preliminaryy transformation of the parameter so that the information matrix is approximately diagonal can be very helpful. If the estimates for some components can be expressed explicitly in terms of the others it may be best to use this fact, unless some all-purpose program is being employed.

If an iterative procedure of Newton–Raphson type is used, involving analytical rather than numerical second derivatives, there is a choice between using either observed second derivatives or expected second derivatives, i.e. information. That is, the iterative schemes can be defined by

$$\hat{\theta}^{(r+1)} = \hat{\theta}^{(r)} - \left\{ \frac{\partial^2 l_Y(\theta; Y)}{\partial \theta^2} \right\}^{-1}_{\theta = \hat{\theta}^{(r)}} U_.(\hat{\theta}^{(r)}),$$

or by

$$\hat{\theta}^{(r+1)} = \hat{\theta}^{(r)} + i_.^{-1}(\hat{\theta}^{(r)}) \, U_.(\hat{\theta}^{(r)}).$$

There seems to be some empirical evidence that the second, where feasible, is to be preferred.

One theoretical point connected with both iteration schemes is that, if started with a consistent estimate, they lead in one step to an asymptotically efficient estimate. This is easily seen by comparison with the basic approximation (19).

In practice, it may be clear on general grounds that there is a unique maximum to the likelihood function. If not, some check on the global form of the likelihood function is desirable ; for example, the occurrence of several maxima of about the same magnitude would mean that the likelihood-based confidence regions are formed from a number of disjoint regions and summarization of data *via* a maximum likelihood estimate and its asymptotic variance could be very misleading.

(vii) Higher-order properties

We now examine further aspects of the large-sample behaviour of maximum likelihood estimates, by computing higher-order approximations to moments of the limiting normal distribution. To do this rigorously requires further differentiability assumptions about the likelihood, but these are self-evident and will not be discussed. The results are useful in the limited sense that they give correction terms for mean and variance, etc. Also, in principle, they provide a basis for comparing different asymptotically efficient estimates.

To start, consider the second-order expression for the mean of the limiting distribution of $\hat{\theta}$. Taking the one-parameter i.i.d. variable case for simplicity, we have by an extension of (16) that

$$0 = U_.(\hat{\theta}) = U_.(\theta) + (\hat{\theta} - \theta) U_.'(\theta) + \tfrac{1}{2}(\hat{\theta} - \theta)^2 U_.''(\theta) + O_p(n^{-\frac{1}{2}}).$$
(32)

The first-order solution to (32) is given by (19), from which the limiting normal distribution is derived. Taking expectations through (32), we obtain

$$E(\hat{\theta} - \theta) E\{U_.'(\theta)\} + \text{cov}\{\hat{\theta}, U_.'(\theta)\} + \tfrac{1}{2}E(\hat{\theta} - \theta)^2 E\{U_.''(\theta)\}$$
$$+ \tfrac{1}{2}\text{cov}\{(\hat{\theta} - \theta)^2, U_.''(\theta)\} = O(n^{-\frac{1}{2}}).$$
(33)

At this point it is convenient to introduce the notation that for a single observation

$$\kappa_{r.s}(\theta) = E[\{U(\theta)\}^r \{U'(\theta)\}^s]$$
(34)

and to note that

$$E\{U''(\theta)\} = -3\kappa_{11}(\theta) - \kappa_{30}(\theta).$$

Now the first covariance in (33) is obtained to order n^{-1} by substitution from (19), whereas the second covariance is $o(n^{-1})$. Some manipulation then gives

$$E(\hat{\theta} - \theta) = -\frac{\kappa_{11}(\theta) + \kappa_{30}(\theta)}{2ni^2(\theta)} + o(n^{-1}) = \frac{b(\theta)}{n} + o(n^{-1}), (35)$$

say. The corresponding result for multidimensional θ follows by applying (32) to each component. The bias term in (35), which is of order n^{-1}, can be estimated by $n^{-1}b(\hat{\theta})$, so defining a corrected m.l.e.

To obtain a similar second-order approximation to the variance and higher moments of $\hat{\theta}$, more terms are required in (32), but the calculation runs parallel to that for the bias. In particular, for the variance we have

$$\text{var}(\hat{\theta}) = \frac{1}{ni(\theta)} + \frac{2b'(\theta)}{n^2 i(\theta)} + \left[\frac{2\{\kappa_{20}(\theta)\kappa_{02}(\theta) - \kappa_{11}^2(\theta)\} + \{\kappa_{11}(\theta) + \kappa_{30}(\theta)\}^2}{2n^2 i^4(\theta)}\right]$$
$$+ o(n^{-2}). \tag{36}$$

The second term on the right-hand side vanishes if $\hat{\theta}$ is replaced by $\hat{\theta} - b(\hat{\theta})/n$, the bias-corrected estimate. The third term on the right-hand side can be shown to depend particularly on the choice of parameter and the proximity of $f_Y(y;\theta)$ to an exponential family form.

The same methods can be applied to any estimating equation, in particular to the multinomial estimating equation (24). There it is possible to show that the bias-corrected versions of efficient estimates have larger second-order variances than the m.l.e. Some details of these properties are described in Exercise 9.8. For a general discussion of these results, and of their relation to information loss arising from the use of a single estimate, see Fisher (1925), Rao (1963) and Efron (1974).

An important application of this type of argument is to improve distributional approximations for test statistics related to m.l.e.'s; see Section 9.3. The following example illustrates a use in estimation of residuals (Cox and Snell, 1968, 1971).

Example 9.14. Modified residuals. Suppose that $y_1, \ldots, y_n$ are observations such that each random variable $Y_j = g_j(\theta, \epsilon_j)$, where $\epsilon_1, \ldots, \epsilon_n$ are i.i.d. random variables. Then the residual r_j may be defined implicitly by $y_j = g_j(\hat{\theta}, r_j)$, where $\hat{\theta}$ is the m.l.e. of θ based on y. Assuming that we can express r_j explicitly as $h_j(y_j, \hat{\theta})$, a simple expansion gives

$$E(R_j \,;\theta) \;=\; E(\epsilon_j) + \frac{b(\theta)}{n} E\left\{\frac{\partial h_j(Y_j,\theta)}{\partial \theta}\right\}$$

$$+ \frac{1}{ni(\theta)}\left[E\left\{\frac{\partial h_j(Y_j,\,\theta)}{\partial \theta}\,U_j(\theta)\right\} + \tfrac{1}{2}E\left\{\frac{\partial^2 h_j(Y_j\,;\theta)}{\partial \theta^2}\right\}\right] + o(n^{-1}).$$

A similar calculation gives the difference between the variances of R_j and ϵ_j. These first approximations to discrepancies between R_j and ϵ_j are useful in defining modified residuals whose means and variances agree with those of the errors to order n^{-1} ; alternatively the properties of test statistics based on the R_j can be evaluated.

An alternative way of reducing the bias of m.l.e.'s is by the jackknife ; see Section 8.4(iv). Also, because of the approximate linearity of the m.l.e. as expressed in (19), the pseudo-values of (8.31) could be used to estimate variance. While, so far as we know, this has not been investigated in detail, it is hard to see what advantage the jackknife approach could have in this context, over the consistent approximation to variance based on the second derivative of the log likelihood. The connexion between the jackknife and the calculation of asymptotic variance for location parameter problems is discussed in Section 9.4.

9.3   Large-sample parametric significance tests

(i) General remarks
In Chapters 3–5 we developed the exact theory of significance tests for parametric hypotheses. Many problems are such that no satisfactory exact procedure exists, either because sufficiency and invariance do not produce simple reduction of the data, or because the distribution theory is intractable, or because optimal properties are not uniform in the set of alternative parameter values. In the last case, the arguments of Section 4.8, used also in Section 6.3, apply a criterion of local optimality to develop tests for single parameters. Section 5.4 briefly mentioned some methods for dealing with more complicated problems, where sufficiency and invariance fail to produce unique test statistics. One such method is that of the maximum likelihood ratio, which is the subject of this section.

The general maximum likelihood (m.l.) ratio test extends the likelihood ratio test, optimal for simple hypotheses, to situations where one or both hypotheses are composite. This has some intuitive appeal, but the results have no guaranteed optimality properties for fixed sample size n. Of course, the m.l. ratio will always be a function of the sufficient statistic, and will be invariant in an invariant testing problem ; see Section 5.3. Thus, in particular cases, it will recover procedures previously derived from other stronger considerations. With increasingly large sample size, however, the m.l. ratio test becomes optimal under the assumed regularity conditions, essentially because only local alternatives are relevant for very large n, and because of the asymptotic sufficiency of the unrestricted m.l.e. in problems involving hypotheses that are not disjoint or separate. In such cases, the normal structure of the asymptotically sufficient statistic leads to asymptotically similar and invariant tests.

Large-sample confidence intervals based on the likelihood function are derived directly from significance tests according to the procedure of Chapter 7. Our subsequent treatment of significance tests is, therefore, intended to apply also to interval estimation. Special points of detail and interpretation are discussed in Chapter 7. Some brief comments on large-sample interval estimates are, however, given in the final part of this section.

(ii) Simple null hypotheses

Although the theory of this section is of most value for composite null hypotheses, it is convenient to begin with simple null hypotheses. Suppose then that Y has p.d.f. $f(y ; \theta)$, $\theta \in \Omega$. As in Chapter 4, we suppose that the simple null hypothesis H_0 asserts that $\theta = \theta_0$, where θ_0 is an interior point of Ω, and that the alternative hypothesis is $H_A : \theta \in \Omega_A \subset \Omega$. Usually $\Omega_A = \Omega - \{\theta_0\}$, but this is not always the case.

For a specific alternative $\theta = \theta_1$, the best test of H_0 has critical regions of the form

$$\left\{ y ; \frac{f(y ; \theta_1)}{f(y ; \theta_0)} \geq c \right\},$$
(37)

which satisfy the consistent nesting requirement (4.8). Sometimes (37) defines a uniformly most powerful test for θ_1 in some specific direction away from θ_0, but this is not generally so and, even when it is, there may be other directions against which we would like to test. Nevertheless, we can generally define a m.l. ratio critical region which, in effect, simultaneously tests against all alternatives. This is the region containing large values of

$$e^{\frac{1}{2}W'} = \frac{\sup\limits_{\theta' \in \Omega_A} \text{lik}(\theta'; Y)}{\text{lik}(\theta_0; Y)}.$$

If a uniformly most powerful test exists, this criterion defines it. Otherwise W' achieves a simple invariant data reduction, whose properties are to be investigated.

When $\Omega_A = \Omega - \{\theta_0\}$, a more convenient expression is the equivalent statistic $W = \max(0, W')$ defined by

$$e^{\frac{1}{2}W} = \text{lik}(\hat{\theta}; Y)/\text{lik}(\theta_0; Y). \tag{38}$$

Example 9.15. Normal-theory linear model (ctd). Suppose that Y_j are distributed independently in $N(x_j^T\beta, \sigma_0^2)$ for $j = 1, \ldots, n$ and that σ_0^2 is known. Then $\hat{\beta} = (\Sigma x_j x_j^T)^{-1} \Sigma x_j Y_j$ and substitution into

$$\text{lik}(\beta; Y) \propto \exp\left\{-\frac{1}{2\sigma_0^2} \Sigma (Y_j - x_j^T\beta)^2\right\}$$

gives the m.l. ratio

$$e^{\frac{1}{2}W} = \exp\left\{\frac{1}{2\sigma_0^2} (\hat{\beta} - \beta_0)^T (\Sigma x_j x_j^T) (\hat{\beta} - \beta_0)\right\}.$$

In this particular case, $\hat{\beta}$ has exactly the $MN\{\beta, \sigma_0^2(\Sigma x_j x_j^T)^{-1}\}$ distribution, and so, under H_0, W has exactly a chi-squared distribution, by a standard property of the multivariate normal distribution. Notice that the m.l. ratio here is uniformly most powerful invariant with respect to orthogonal transformations.

For finite n, the null distribution of W will generally depend on n, and on the form of the p.d.f. of Y. However, there is for regular problems a uniform limiting result as $n \to \infty$. To see this, consider for simplicity the single parameter case. Then, expanding W in a Taylor series, we get, as in (16),

$$W = 2\{l(\theta_0; Y) - l(\theta_0; Y) - (\hat{\theta} - \theta_0) U_.(\hat{\theta}) - \tfrac{1}{2}(\hat{\theta} - \theta_0)^2 U_.'(\theta^\dagger)\},$$

where $|\theta^\dagger - \theta_0| \leqslant |\hat{\theta} - \theta_0|$. Then, since $U_.(\hat{\theta}) = 0$, and because $\theta^\dagger$ is consistent,

$$W = -n(\hat{\theta} - \theta_0)^2 \left\{ \frac{1}{n} \frac{\partial}{\partial \theta_0} U.(\theta_0) \right\} + o_p (1).$$ (39)

By the asymptotic normality of $\hat{\theta}$ and the convergence of $U.'(\theta_0)/n$ to $-i.(\theta_0)/n$, described in Section 9.2(iii), W has, under H_0, a limiting chi-squared distribution on one degree of freedom.

For multiparameter situations, (39) is immediately generalized to

$$W = -n(\hat{\theta} - \theta_0)^T \left\{ \frac{1}{n} \frac{\partial}{\partial \theta_0} U.^T(\theta_0) \right\} (\hat{\theta} - \theta_0) + o_p(1),$$ (40)

and the asymptotic normality of $\hat{\theta}$ indicates that, under H_0, W has a limiting chi-squared distribution with degrees of freedom equal to $\dim(\theta)$.

Notice that the limiting distribution of W follows also immediately from the asymptotic sufficiency of $\hat{\theta}$ described by (30), because the m.l. ratio based on the approximate normal distribtuion of $\hat{\theta}$ is equal to

$$\exp \left\{ \tfrac{1}{2} (\hat{\theta} - \theta_0)^T i.(\theta_0) (\hat{\theta} - \theta_0) \right\},$$

which gives directly the chi-squared distribution for the exponent W. This calculation is legitimate provided that we use (30) only for $|\theta' - \theta_0| = o(1)$, but we can restrict ourselves to such values knowing that the likelihood maximum occurs within $O_p(n^{-\frac{1}{2}})$ of the true parameter value θ. This point is discussed in more detail later.

In the light of these remarks, we notice also that the limiting chi-squared distribution of W under H_0 applies when Ω_A is *any* fixed parameter set surrounding θ_0, i.e. a set for which θ_0 is not a boundary point. This is because the consistency of $\hat{\theta}$ implies that for any such Ω_A

$$\sup_{\theta' \in \Omega_A} l(\theta' ; Y) = \sup_{\theta' \in \Omega - \theta_0} l(\theta' ; Y) + o_p(1).$$

Further valuable information is uncovered by examining (40) more closely. First, the quadratic form in (40) is itself asymptotically equivalent to the m.l. ratio. If we replace the matrix $-n^{-1} \partial U.^T(\theta_0)/\partial \theta_0$ by $n^{-1} i.(\hat{\theta})$, to which it is asymptotically equivalent, then we get the *m.l. test statistic*

$$W_e = (\hat{\theta} - \theta_0)^T i.(\hat{\theta}) (\hat{\theta} - \theta_0).$$ (41)

In an alternative and asymptotically equivalent version, $i.(\hat{\theta})$ is replaced by $i.(\theta_0)$.

A second large-sample equivalent to W is obtained from (40) by recalling from (20) that, when $\theta = \theta_0$,

$$\hat{\theta} - \theta_0 = \mathbf{i}_{.}^{-1}(\theta_0)\, U_{.}(\theta_0) + O_p(n^{-1}).$$

Substitution into (40) gives as asymptotically equivalent to W the *score statistic*

$$W_u = U_{.}^{\mathrm{T}}(\theta_0)\mathbf{i}_{.}^{-1}(\theta_0)\, U_{.}(\theta_0). \tag{42}$$

An advantage of this version of the likelihood ratio is that the m.l.e. need not be calculated, although, of course, in practice the m.l.e. may be required in its own right. In fact, W_u is the multiparameter invariant version of the local likelihood ratio test statistic as discussed in Section 4.8, using the large-sample normal distribution under H_0 of the efficient score vector. We shall return to this in more detail later.

In practice, in tests involving one degree of freedom it will be rather more meaningful to use $\sqrt{W_e}$ or $\sqrt{W_u}$, with an appropriate sign, rather than W_e or W_u, but in the discussion here we use W_e or W_u for uniformity of presentation.

Before discussing further properties of the likelihood statistics, we give some examples.

Example 9.16. Non-linear regression. Suppose that $Y_j = g_j(\beta) + \epsilon_j$ ($j = 1, \ldots, n$), where $\epsilon_1, \ldots, \epsilon_n$ are i.i.d. in $N(0, \sigma_0^2)$, and σ_0^2 is known. Let $g_j(\beta)$ be a non-linear function of the scalar parameter β, such as $g_j(\beta) = \exp(\beta x_j)$. Then, with rare exceptions, solution for $\hat{\beta}$ has to be iterative. Provided that $g_j(\beta)$ is differentiable at $\beta = \beta_0$, the W_u statistic (42) for testing $H_0 : \beta = \beta_0$ against $H_A : \beta \neq \beta_0$ is calculated from

$$U_{.}(\beta_0) = \Sigma g_j'(\beta_0)\{ Y_j - g_j(\beta_0)\},$$

$$i_{.}(\beta_0) = \sigma_0^2 \Sigma \{g_j'(\beta_0)\}^2.$$

When $g_j(\beta)$ is linear in β, the statistics W, W_e and W_u all reduce to the standardized version of $(\hat{\beta} - \beta_0)^2$, where $\hat{\beta}$ is the least squares estimate of β. Notice that near $\beta = \beta_0$ we have approximately $g_j(\beta) = g_j(\beta_0) + (\beta - \beta_0)g_j'(\beta_0)$. Thus $U_{.}(\beta_0)$ is proportional to the least squares regression statistic for the approximating local linear model.

Example 9.17. Multinomial distribution (ctd). Consider n observations from the m-cell multinomial distribution with cell probabilities

$\pi_1, \ldots, \pi_m$. The maximum likelihood estimate of $\theta = (\pi_1, \ldots, \pi_{m-1})$ was discussed in Example 9.8. We have, in particular, that for cell frequencies $N_1, \ldots, N_m$, $\hat{\theta} = (N_1/n, \ldots, N_{m-1}/n)$,

$$l(\theta; Y) = \sum_{j=1}^{m} N_j \log \pi_j$$

and

$$n^{-1} i_.(\theta) = \frac{1}{\pi_m} 11^T + \text{diag}\left(\frac{1}{\pi_1}, \ldots, \frac{1}{\pi_{m-1}}\right).$$

The three statistics W, W_e and W_u for testing $H_0: \pi = \pi^0$ against $H_A: \pi \neq \pi^0$ are easily seen to be

$$W = 2 \sum_{j=1}^{m} N_j \log\left(\frac{N_j}{n\pi_j^0}\right),$$

$$W_e = \sum_{j=1}^{m} \frac{(N_j - n\pi_j^0)^2}{N_j},$$

$$W_u = \sum_{j=1}^{m} \frac{(N_j - n\pi_j^0)^2}{n\pi_j^0}.$$

The limiting chi-squared distribution of these statistics has $m-1$ degrees of freedom. Both W_e and W_u are referred to as chi-squared goodness of fit statistics; the latter, often called the Pearson chi-squared statistic, was mentioned in Example 3.2. The large-sample chi-squared distribution of W_u was first derived by Pearson (1900). Of course, W_u has no computational advantage over W_e in this case.

Recall from Section 9.2 that the efficient estimation of β in the parametric multinomial distribution $\pi(\beta)$ is often most conveniently done *via* the corresponding parametric version of W_u. The composite hypothesis that π has a particular parametric form $\pi(\beta)$ will be discussed in Example 9.21.

Example 9.18. Poisson regression model. Let $Y_1, \ldots, Y_n$ be independent, Y_j having a Poisson density with mean $a + \theta x_j$, for fixed constants $a, x_1, \ldots, x_n$ such that $\Sigma x_j = 0$; thus a is the known overall mean of the Y_j. This formulation is not the technically natural one for the Poisson distribution and does not give uniformly most powerful tests for θ, but it is a permissible model if $a + \theta x_j > 0$ for all x_j. Consider the null hypothesis $H_0: \theta = 0$ with alternative $H_A: \theta \neq 0$. The log likelihood function is

$$l(\theta \; ; Y) \; = \; -na + \Sigma \, Y_j \log(a + \theta x_j) - \Sigma \log Y_j!,$$

from which it is clear that no explicit exact solution is available for $\hat{\theta}$. However, the W_u statistic (42) has a particularly simple form, since $U_.(0) = \Sigma x_j Y_j/a$ and $i_.(0) = \Sigma x_j^2/a$. Thus

$$W_u \; = \; (\Sigma x_j Y_j)^2/(a \Sigma x_j^2).$$

Notice that, as in the non-linear model of Example 9.16, W_u is the same as the linear least squares test statistic and the corresponding uniformly most powerful test statistic for regression in the natural parameter.

That the m.l. ratio statistic W is equivalent to the score statistic W_u, which arose in the context of locally most powerful tests in Section 4.8, suggests the large-sample optimality of the m.l. ratio method. We now develop this notion in some detail. First consider the properties of W, W_e and W_u under the alternative hypothesis with $\theta \neq \theta_0$. For simplicity, we deal with the case of i.i.d. variables and a one-dimensional parameter θ. The arguments apply to the non-homogeneous multiparameter case.

Suppose that $\theta = \theta_A \neq \theta_0$. Then the expansion for W generalizing (40) can be written

$$W \; = \; 2\{l(\theta_A \; ; Y) - l(\theta_0; Y)\} - (\hat{\theta} - \theta_A)^2 U'_.(\theta_A) + o_p(1), \quad (43)$$

wherein the second term has the limiting chi-squared distribution. By the assumption (8) on the density $f(y \; ; \theta)$, the first term in (43) is unbounded ; this must be so quite generally in order for $\hat{\theta}$ to be consistent. The result is that, for a fixed $\theta_A \neq \theta_0$, and for any finite value d,

$$\mathrm{pr}(W > d \; ; \theta_A) \to 1 \, . \qquad (44)$$

That is, for any fixed level α, the m.l. ratio test has power approaching one, for any fixed alternative. We call a test with this property a *consistent test*. The same argument applies to the statistics W_e and W_u by expanding $\hat{\theta} - \theta_0$ and $U_.(\theta_0)$ about θ_A.

Now consider alternatives that are local, specifically sequences $\{\theta_n\}$ such that θ_n converges to θ_0. With regularity of the likelihood in some interval surrounding $\theta' = \theta_0$, it is easily verified from (18) that $(\hat{\theta} - \theta_n)\sqrt{n}$ has the limiting $N\{0, i^{-1}(\theta_0)\}$ distribution and that (43) becomes

$$W = 2\sqrt{n}(\theta_n - \theta_0)\frac{1}{\sqrt{n}}U.(\theta_n) + n(\theta_n - \theta_0)^2 i(\theta_0) + n(\hat{\theta} - \theta_n)^2 i(\theta_0) + o_p(1).$$

(45)

Therefore, if $(\theta_n - \theta_0)\sqrt{n}$ diverges, then (44) holds with $\theta_A = \theta_n$. But if $(\theta_n - \theta_0)\sqrt{n}$ converges to δ, say, then (45) implies that

$$W = n(\hat{\theta} - \theta_0)^2 i(\theta_0) + o_p(1),$$

which is approximately chi-squared with non-centrality parameter $\rho^2 = \delta^2 i(\theta_0)$. That is, for local alternatives $\theta_n = \theta_0 + \delta/\sqrt{n}$, the m.l. ratio statistic has non-trivial power properties and W is asymptotically equivalent to the square of a $N(\rho,1)$ variable. The same result applies to both W_e and W_u.

In the multiparameter case for independent variables generally, this discussion extends immediately to show that W is a consistent test statistic for alternatives θ_n such that $(\theta_n - \theta_0)\sqrt{n}$ diverges, whereas for $\theta_n = \theta_0 + \delta/\sqrt{n}$, the statistic W has a limiting chi-squared distribution with non-centrality parameter $n^{-1}\delta^T i.(\theta_0)\delta$ and degrees of freedom equal to $\dim(\theta)$. The same result holds for W_e and W_u.

Consideration of the sequence of alternatives $\theta_n = \theta_0 + \delta/\sqrt{n}$ is dictated solely by the desire to approximate to the power in a region of modest power, and has no direct physical significance.

The connexion between the m.l. ratio statistic, W, and the locally most powerful tests of Section 4.8 is thus that only local alternatives are important for very large n. Now, for the single parameter case, critical regions determined by W_u, and hence those for W_e and W, are asymptotically equivalent to locally most powerful unbiased critical regions of the form (4.74). This is because the symmetric limiting distribution of $U.(\theta_0)/\sqrt{n}$ and the unbiasedness condition together imply that the term proportional to $U.(\theta_0)$ in (4.74) goes to zero. Still considering the single parameter case, and restricting attention to alternatives of the form $\theta_n = \theta_0 + \delta/\sqrt{n}$, we have, by the argument used in Section 4.8, that the most powerful one-sided tests have critical regions asymptotically equivalent to

$$\{y \,;\, U.(\theta_0)/\sqrt{n} \geqslant c_1\} \quad \text{and} \quad \{y \,;\, U.(\theta_0)/\sqrt{n} \leqslant -c_2\} \quad (46)$$

for $\delta < 0$ and $\delta > 0$ respectively, where $c_1, c_2 > 0$ are independent of δ for any fixed significance levels. If the levels of the two regions approximated in (46) are equal, so that the union of the two regions is unbiased, no other unbiased test has asymptotically superior power.

But the symmetry of the limiting distribution of $n^{-\frac{1}{2}} U_{.}(\theta_0)$ shows that this asymptotically optimal unbiased test is based solely on W_u. Thus the m.l. ratio and its equivalents are asymptotically most powerful unbiased, since they are equivalent to W_u.

For the multiparameter case, no uniformly most powerful unbiased test exists for the reasons discussed in Section 4.9, essentially because different most powerful tests will be appropriate for different directions in the parameter space. In the present context, the asymptotically most powerful unbiased test for alternatives $\theta_n = \theta_0 + \delta b/\sqrt{n}$, with b a fixed vector and δ a scalar, has critical regions determined by

$$W_u(b) = \{b^T U_{.}(\theta_0)\}^2/\{b^T i_{.}(\theta_0)b\},$$

which has a limiting chi-squared distribution, and $W_u(b)$ will depend on b ; $W_e(b)$ is defined similarly in terms of $\hat{\theta} - \theta_0$. It is simple to verify that

$$\sup_b W_u(b) = U_{.}^T(\theta_0)\, i_{.}^{-1}(\theta_0)\, U_{.}(\theta_0) = W + o_p(1). \quad (47)$$

Thus the m.l. ratio test is equivalent to the most significant value among all one-directional asymptotically most powerful test statistics. Now notice from (23) that W and the quadratic form in (47) are invariant under arbitrary transformations of the parameter θ, in particular one-one transformations. The m.l. ratio is therefore the asymptotically most powerful unbiased test of H_0 that is invariant under parameter transformations leaving θ_0 distinct from Ω_A.

For a detailed general mathematical treatment of the properties of W, W_e and W_u, see Wald (1943).

As we have said before, the power function of a significance test is used in theoretical discussion, not directly in analysing data. However, the parameter defining the power function may be useful for description and estimation. A more special point arises for the chi-squared goodness of fit statistic of Example 9.17, where the non-centrality parameter ρ^2 for local alternatives $\pi_j - \pi_j^0 = O(n^{-\frac{1}{2}})$ is expressible as

$$\left\{ \sum_{j=1}^m \pi_j^0 \log \pi_j - \sum_{j=1}^m \pi_j^0 \log \pi_j^0 \right\} \sqrt{n} + o(1). \quad (48)$$

Suppose that the test statistic is to be used for grouped data from a

continuous hypothetical p.d.f. $f_0(y)$, where the grouping into m
intervals with probabilities $\pi_1^0, \ldots, \pi_m^0$ is open to choice. If no
particular direction in the alternative parameter space is to be considered, then a sensible choice of grouping, on the basis of maximizing
large-sample power, is that which minimizes the term $\Sigma\pi_j^0\log\pi_j^0$ in
(48), often called the entropy. A simple calculation shows that this
leads to the choice $\pi_j^0 = 1/m$ for each j, so that the range of the continuous variable is broken up into intervals with equal probability
under the null density $f_0(y)$. Note that for a fixed alternative p.d.f.
$f_A(y)$, change in the discrete null distribution π^0 effects a change in
the discrete alternative distribution. Where some knowledge of the
likely form of $f_A(y)$ is available, the choice of uniform multinomial
probabilities may be very inappropriate, but in such a case a more
refined analysis, not usually *via* the chi-squared test, should be made.

Our discussion of the m.l. ratio applies to non-homogeneous
independent random variables, as in Examples 9.16 and 9.18, and to
Markov dependent random variables. This is easily checked by
referring to the appropriate asymptotic properties of $\hat{\theta}$ in Section
9.2 (iii) for such cases. For non-Markovian dependent random variables,
however, the m.l. ratio may not have the same limiting chi-squared
properties.

So far we have considered the regular parametric hypothesis where
θ_0 is interior to $\Omega_A \cup \{\theta_0\}$ and where this parameter space is continuous. When this is not so, $\hat{\theta}$ does not have a limiting normal distribution
and some modification of the previous arguments is required. Suppose,
first, that the likelihood is completely regular in some parameter
space Ω with θ_0 interior to Ω, but that θ_0 is on the boundary of
$\Omega_A \cup \{\theta_0\}$. Then the asymptotic normal property of the unrestricted
m.l.e. is still available, and for the problem restricted to $\Omega_A \cup \{\theta_0\}$
we just need to take account of the truncation of the limiting normal
distribution. A more detailed discussion of this will be given for
composite hypotheses in part (iv) of this section, but a simple example
will illustrate the point.

Example 9.19. One-sided test for single parameter. Suppose that θ is
one-dimensional and that $\Omega_A = \{\theta \; ; \theta > 0\}$, $\theta_0 = 0$. In practice we
would use the locally most powerful test with critical region
$\{y \; ; u.(\theta_0) > c\}$, and apply the limiting normal property of $U.(\theta_0)$
to calculate significance probabilities. This is a simple equivalent of

the m.l. ratio method which we are about to examine. Temporarily, write $\Omega^* = \{\theta \; ; \theta \geqslant 0\}$, in which the m.l.e. $\hat{\theta}^*$ is given by

$$\hat{\theta}^* = \max(0, \hat{\theta})$$

with probability tending to one, by (30). It follows that the m.l. ratio has the representation

$$W = \begin{cases} \hat{\theta}^2 i_.(0) + o_p(1) & (\hat{\theta} \geqslant 0), \\ o_p(1) & (\hat{\theta} < 0), \end{cases}$$

from which the asymptotic normality of $\hat{\theta}$ gives

$$\text{pr}(W > d \; ; \theta = 0) = \tfrac{1}{2}\,\text{pr}(X_1^2 < d) + o(1),$$

where X_1^2 is a random variable with the chi-squared distribution on one degree of freedom.

If, however, θ_0 is actually a boundary point of the full parameter space, Ω, then the unrestricted m.l.e. does not have a limiting normal distribution and we often require a higher-order expansion for W than (39). This situation is illustrated in the model of Example 9.12 when $H_0: \theta = 0$, in which the appropriate expansion is easily seen to lead to

$$\text{pr}(W > d \; ; \theta = 0) = \tfrac{1}{2}\text{pr}(X_1^2 > d) + o(1).$$

When Ω_A is separate from θ_0, i.e. when for some $\epsilon > 0$ we have $\|\theta' - \theta_0\| \geqslant \epsilon, \theta' \in \Omega_A$, the previous discussion of the m.l. ratio does not apply at all. Problems of this type are treated in part (iv) of this section.

It is important in practice to know how well the m.l. ratio, and its large-sample equivalents, behave for moderately large sample sizes. For example, is W actually superior to W_e and W_u when a finer analysis to $O(n^{-1})$ is carried out? Is some modification of the score statistic possible which gives a more sensitive test against detectable alternatives? We postpone these important questions to part (vi) of this section after we have examined the m.l. ratio test for composite null hypotheses.

(iii) Composite null hypotheses involving subspaces
We now extend the treatment of the m.l. ratio to problems of the type considered in Chapter 5, where the null hypothesis H_0 specifies

not a single parameter value θ_0 but a general subset Ω_0 of Ω. With
alternative hypothesis $H_A : \theta \in \Omega_A$, the m.l. ratio statistic is defined as

$$e^{\frac{1}{2}W'} = \frac{\sup\limits_{\theta' \in \Omega_A} \text{lik}(\theta' ; Y)}{\sup\limits_{\theta' \in \Omega_A} \text{lik}(\theta' ; Y)}. \tag{49}$$

Often, especially when $\Omega_0 \cup \Omega_A = \Omega$, it is convenient to work with

$$e^{\frac{1}{2}W} = \frac{\sup\limits_{\theta' \in \Omega_0 \cup \Omega_A} \text{lik}(\theta' ; Y)}{\sup\limits_{\theta' \in \Omega_0} \text{lik}(\theta' ; Y)}, \tag{50}$$

with the relationship $W = \max(0, W')$.

In Chapter 5 we distinguished between several types of composite
hypothesis problems. The most familiar problem, with which we deal
in this part, has Ω_0 as a subset of Ω where θ is expressible as a
function of a lower dimensional parameter β; that is, for $\theta \in \Omega_0$,

$$\theta_j = g_j(\beta_1, \ldots, \beta_{q'}) \quad (j = 1, \ldots, q ; q' < q). \tag{51}$$

The most straightforward situation is when $\theta = (\psi, \lambda)$ and $\Omega_0 =$
$\{\theta ; \psi = \psi_0\}$, so that λ is a separate nuisance parameter; as in Chapter
5, it is convenient to write $\theta = (\psi, \lambda)$ in general discussion, rather
than use full vector notation. Because the m.l. ratio is invariant under
parameter transformation, the general parametric restriction (51) can
be put in the form $\theta = (\psi_0, \lambda)$, which is most convenient for our
discussion, although not necessarily for numerical work.

Suppose, then, that the vector Y has p.d.f. $f(y ; \psi, \lambda)$, where
$H_0: \psi = \psi_0$ and the alternative hypothesis is $H_A : \psi \neq \psi_0$, with λ
as a nuisance parameter. As in the simple hypothesis problem, the
limiting distribution of the m.l. ratio W is determined by the
limiting normal distribution of $\hat{\theta} = (\hat{\psi}, \hat{\lambda})$. Under the null hypothesis,
λ is estimated by maximum likelihood with $\psi = \psi_0$ fixed, so that
the discussion of constrained maximum likelihood in Section 9.2(v)
will be needed here. We use the notation $\hat{\lambda}_0$ to denote the m.l.e. of
λ subject to $\psi = \psi_0$. Then, by Taylor series expansion about the
parameter point (ψ_0, λ), we obtain

$$W = 2\{l(\hat{\psi},\hat{\lambda};Y) - l(\psi_0,\lambda;Y)\} - 2\{l(\psi_0,\hat{\lambda}_0;Y) - l(\psi_0,\lambda;Y)\}$$

$$= \begin{bmatrix} \hat{\psi} - \psi_0 \\ \hat{\lambda} - \lambda \end{bmatrix}^{\mathrm{T}} i.(\psi_0,\lambda_0) \begin{bmatrix} \hat{\psi} - \psi_0 \\ \hat{\lambda} - \lambda \end{bmatrix} - \begin{bmatrix} 0 \\ \hat{\lambda}_0 - \lambda \end{bmatrix}^{\mathrm{T}} i.(\psi_0,\lambda_0) \begin{bmatrix} 0 \\ \hat{\lambda}_0 - \lambda \end{bmatrix} + o_p(1).$$

$$(52)$$

This expansion is valid provided that $\psi - \psi_0 = O(n^{-\frac{1}{2}})$, λ being the irrelevant true value of the nuisance parameter. By the result (31) connecting $\hat{\lambda}_0$ to $(\hat{\psi},\hat{\lambda})$, (52) reduces to

$$W = (\hat{\psi} - \psi_0)^{\mathrm{T}} \{i._{\psi\psi}(\psi_0,\lambda) - i._{\psi\lambda}^{\mathrm{T}}(\psi_0,\lambda)i._{\lambda\lambda}^{-1}(\psi_0,\lambda)i._{\psi\lambda}(\psi_0,\lambda)\}(\hat{\psi} - \psi_0) + o_p(1).$$

$$(53)$$

The matrix $i.(\psi_0,\lambda)$ has been partitioned according to the partition (ψ,λ) of θ, and the matrix of the quadratic form in (53) is, in fact, $\{i.^{\psi\psi}(\psi_0,\lambda)\}^{-1}$, the inverse of the covariance matrix in the limiting marginal normal distribution of $\hat{\psi}$. Therefore, the limiting distribution of W is chi-squared with degrees of freedom equal to $\dim(\psi)$, or, in more general notation, $\dim(\Omega) - \dim(\Omega_0)$. The distribution is central under H_0.

One important feature of this result is that the m.l. ratio test is shown to be asymptotically similar, because the limiting chi-squared distribution does not involve λ. This is to be anticipated from the normal asymptotically sufficient statistic $\hat{\theta} = (\hat{\psi},\hat{\lambda})$, because exact similar tests are available for components of the multivariate normal mean.

As with simple hypotheses, the expansion of W leads to two asymptotically equivalent test statistics. The first, immediate from the quadratic form in (53), is the m.l. statistic.

$$W_e = (\hat{\psi} - \psi_0)^{\mathrm{T}} i.(\hat{\psi} : \hat{\lambda})(\hat{\psi} - \psi_0), \qquad (54)$$

where $i.(\psi : \lambda)$ is the inverse of the covariance matrix $i.^{\psi\psi}(\psi,\lambda)$ of the limiting normal distribution of $\hat{\psi}$, given in full by

$$i.(\psi : \lambda) = i._{\psi\psi}(\psi,\lambda) - i._{\psi\lambda}^{\mathrm{T}}(\psi,\lambda)i._{\lambda\lambda}^{-1}(\psi,\lambda)i._{\psi\lambda}(\psi,\lambda).$$

A second equivalent to W is derived from the representation of $\hat{\psi} - \psi_0$ in terms of $U.(\psi_0,\lambda)$. Equation (20) gives such a representation for $\hat{\theta} = (\hat{\psi},\hat{\lambda})$, so that partitioning $U.(\psi_0,\lambda)$ into $U._{\psi}(\psi_0,\lambda)$ and $U._{\lambda}(\psi_0,\lambda)$ according to the partition (ψ,λ) of θ, we obtain

$$\hat{\psi} - \psi_0 = i.^{-1}(\psi_0 : \lambda)\{U._{\psi}(\psi_0,\lambda) + i.(\psi_0 : \lambda)i.^{\psi\lambda}(\psi_0,\lambda)U._{\lambda}(\psi_0,\lambda)\} + o_p(1).$$

Substitution into (53) and some simplification gives the score statistic

$$W_u(\lambda) = \{U_{.\psi}(\psi_0,\lambda) - b.(\psi_0,\lambda)U_{.\lambda}(\psi_0,\lambda)\}^T i_.^{\psi\psi}(\psi_0,\lambda)$$

$$\{U_{.\psi}(\psi_0,\lambda) - b.(\psi_0,\lambda)U_{.\lambda}(\psi_0,\lambda)\} \tag{55}$$

as asymptotically equivalent to W, where $b.(\psi_0,\lambda) = i_{.\lambda\lambda}^{-1}(\psi_0,\lambda)$
$i_{.\psi\lambda}(\psi_0,\lambda)$ is the matrix of regression of $U_{.\psi}(\psi_0,\lambda)$ on $U_{.\lambda}(\psi_0,\lambda)$. Of
course, $W_u(\lambda)$ involves the unknown value λ of the nuisance para-
meter, but the substitution of $\hat\lambda_0$ for λ in (55) adds only a $o_p(1)$ term,
since $\hat\lambda_0 - \lambda = O_p(n^{-\frac{1}{2}})$, so that a usable form of the score statistic is
obtained from $W_u(\lambda)$ by replacing λ by $\hat\lambda_0$:

$$W_u = W_u(\hat\lambda_0). \tag{56}$$

In fact any estimate $\tilde\lambda$ such that $\tilde\lambda - \lambda = O_p(n^{-\frac{1}{2}})$ may be substituted
for λ in (55), but $\hat\lambda_0$ is usually the obvious choice, because then only
the parametric model in Ω_0 needs to be fitted.

A further important general point about the W_u statistic is that
often the parameter λ can be defined in such a way that $b.(\psi_0,\lambda) = 0$,
making W_u a quadratic form in $U_{.\psi}(\psi_0,\hat\lambda_0)$ alone. This is a particular
advantage when the model with $\psi = \psi_0$ is simpler than that for
general ψ, although often the calculation of $(\hat\psi,\hat\lambda)$ will be of intrinsic
interest.

By the asymptotic equivalence of W, W_e and W_u, the latter two
statistics also have approximate central chi-squared distributions
under H_0 with degrees of freedom equal to $\dim(\psi) = \dim(\Omega) - \dim(\Omega_0)$.
This also follows directly from the asymptotic normality of $\hat\psi$ and
the efficient score.

The previous discussion of non-null behaviour extends to the
present situation ; in particular, for local alternatives $\psi = \psi_0 + \delta_\psi/\sqrt{n}$
the non-centrality parameter is $n\delta_\psi^T i.(\psi_0 : \lambda)\delta_\psi$.

We now give some examples to illustrate the use of W, W_e and W_u.

Example 9.20. Two-sample test. Let $Y_1, \ldots, Y_{n_1}$ be i.i.d. with p.d.f.
$f(y;\beta_1,\beta_2,\ldots,\beta_q)$ and let $Y_{n_1+1}, \ldots, Y_{n_1+n_2}$ be i.i.d. with p.d.f.
$f(y;\beta_1^*,\beta_2,\ldots,\beta_q)$ and suppose that $H_0 : \beta_1 = \beta_1^*$ with alternative
$H_A : \beta_1 \neq \beta_1^*$, $\beta_2, \ldots, \beta_q$ being additional nuisance parameters. This
is a quite general two-sample testing problem involving one compenent
parameter ; the analysis is easily extended to deal with hypotheses
about q' components ($q' \leq q$). Bearing in mind the advantage of

making $\mathbf{b}.(\psi_0,\lambda)$ zero by suitable choice of parameter λ, we define the indicator I_j to be equal to 0 for $j = 1, \ldots, n_1$, and equal to 1 for $j = n_1 + 1, \ldots, n_1 + n_2$ and use the following orthogonalized form of the p.d.f.

$$f_{Y_j}(y\,;\psi,\lambda) = f\left\{y\,;\lambda_1 + \left(I_j - \frac{n_2}{n_1 + n_2}\right)\psi,\lambda_2, \ldots, \lambda_q\right\}$$

$$= \begin{cases} f(y\,;\beta_1, \ldots, \beta_q) & (j = 1, \ldots, n_1), \\ f(y\,;\beta_1^*, \ldots, \beta_q) & (j = n_1 + 1, \ldots, n_1 + n_2). \end{cases} \tag{57}$$

The null hypothesis is now $H_0: \psi = 0$, with $\lambda = (\lambda_1, \ldots, \lambda_q)$ as nuisance parameter, and we shall derive the form of the W_u statistic defined in (56). To do this, we use the temporary notation

$$l_1(\psi,\lambda\,;Y^{(1)}) = l(\psi,\lambda\,;Y_1, \ldots, Y_{n_1})$$

and

$$l_2(\psi,\lambda\,;Y^{(2)}) = l(\psi,\lambda\,;Y_{n_1+1}, \ldots, Y_{n_1+n_2}).$$

Then

$$U_{.\psi}(0,\lambda) = \left[\frac{\partial l(\psi,\lambda\,;Y)}{\partial \psi}\right]_{\psi=0}$$

$$= \frac{n_1}{n_1 + n_2}\left[\frac{\partial l_2(\psi,\lambda\,;Y^{(1)})}{\partial \lambda_1}\right]_{\psi=0} - \frac{n_2}{n_1+n_2}\left[\frac{\partial l_2(\psi,\lambda\,;Y^{(2)})}{\partial \lambda_1}\right]_{\psi=0} \tag{58}$$

This implies the required orthogonality, $\mathbf{b}.(0,\lambda) = 0$, because

$$E\left\{\left[\frac{\partial l(\psi,\lambda\,;Y)}{\partial \psi}\right]_{\psi=0}^{\mathrm{T}}\left[\frac{\partial l(\psi,\lambda\,;Y)}{\partial \psi}\right]_{\psi=0}\,;\psi = 0\right\} = 0,$$

so that (56) becomes

$$W_u = U_{.\psi}^2(0,\hat{\lambda}_0)\,i_{.\psi\psi}^{-1}(0,\hat{\lambda}_0),$$

with $\hat{\lambda}_0$ the m.l.e. in $l(0,\lambda\,;Y)$. The limiting chi-squared distribution of W_u applies here when n_1 and n_2 both tend to infinity, a necessary condition for unbounded growth of $i_{.\psi\psi}(0,\lambda)$. This is clear on general grounds from (58).

Example 9.21. Multinomial distribution (ctd). In Example 9.17 we discussed testing goodness of fit of a specified multinomial distribution $\pi = (\pi_1, \ldots, \pi_m)$. Now we suppose that π is defined under H_0 to be of the parametric form $\pi(\beta)$, where $\dim(\beta) = q < m - 1$, while the union of H_0 and H_A, the alternative hypothesis, is the general one that π is an arbitrary multinomial distribution. As before, we use the sufficient sample cell frequencies $N_1, \ldots, N_m$.

The null hypothesis $H_0 : \pi = \pi(\beta)$ is in the general form (51), which to carry over to our general discussion should be put in the form $\pi = (\psi, \lambda)$, with $H_0 : \psi = \psi_0$, λ being unspecified. This can be done by choosing λ to be those π_j forming a one-one function of β. Supposing these to be $(\pi_1, \ldots, \pi_q)$ for simplicity, ψ is then the independent set $\{\pi_{q+1} - \pi_{q+1}(\lambda), \ldots, \pi_{m-1} - \pi_{m-1}(\lambda)\}$, so that H_0 becomes $\psi = 0$. The preceding theory then applies to show that W, W_e and W_u have limiting chi-squared distributions with $m - 1 - q$ degrees of freedom.

The form of the W_u statistic is most easily derived by direct expansion of W. We have

$$W = 2 \sum_{j=1}^{m} N_j \log\left\{\frac{N_j}{n\pi_j(\hat{\beta})}\right\} = 2 \sum_{j=1}^{m} N_j \log\left\{1 + \frac{N_j - n\pi_j(\hat{\beta})}{n\pi_j(\hat{\beta})}\right\}.$$

(59)

Collecting the first two terms in the expansion of the logarithm, and using the fact that $N_j/\{n\pi_j(\hat{\beta})\} = 1 + o_p(1)$ by consistency of $\hat{\beta}$ and N_j/n, we obtain, apart from a term $o_p(1)$,

$$W = \sum_{j=1}^{m} \frac{\{N_j - n\pi_j(\hat{\beta})\}^2}{n\pi_j(\hat{\beta})} = \sum_{j=1}^{m} \frac{\{N_j - n\pi_j(\hat{\beta})\}^2}{N_j}.$$

(60)

The first quadratic form is the W_u statistic, and the second is the W_e statistic.

Notice that W_u and W_e correspond closely to the chi-squared statistics for the simple hypothesis in Example 9.17. More importantly, W_u and W_e are the minimum values of the chi-squared and modified chi-squared criteria for efficient estimation of β as described in Section 9.2(iv).

Quite clearly, although we have not said so explicitly, any asymptotically efficient estimate $\tilde{\beta}$ may be used in place of the m.l.e. in W_e, W_u or W. This can easily be verified by expansion about the m.l.e. However, the same limiting chi-squared property does not hold if we substitute an estimate not asymptotically equivalent to the m.l.e. This point is particularly apposite if the multinomial distribution $\pi(\beta)$ is a discretized version of a continuous distribution obtained by grouping variables $Y_1, \ldots, Y_n$. In this situation, the m.l.e. $\hat{\beta}^{\dagger}$ based on the original variables will usually be more efficient than the multinomial m.l. estimate $\hat{\beta}$ based on $N_1, \ldots, N_m$ for large n. It is then reasonably clear that substitution of $\hat{\beta}^{\dagger}$ for $\hat{\beta}$ in either W_e or W_u will give statistics that tend to have larger values, at least for large n ; for example, we know that $\hat{\beta}$ is asymptotically equivalent to the value of θ' which minimizes either expression in (28); see Exercise 9.16.

Example 9.22. Choice between models. Suppose that $f_1(y;\lambda_1)$ and $f_2(y;\lambda_2)$ are two p.d.f.'s that are candidates for the p.d.f. of the i.i.d. variables $Y_1, \ldots, Y_n$, such that $\{f_k(y;\lambda_k), \lambda_k \in \Omega_k\}$ $(k = 1,2)$ are distinct, or separate, families. If these two p.d.f.'s represent two extremes, it might be appropriate to consider the family of distributions with p.d.f.

$$f_Y(y;\psi,\lambda) \propto \{f_1(y;\lambda_1)\}^{\psi} \{f_2(y;\lambda_2)\}^{1-\psi}. \tag{61}$$

Then to test whether or not $f_1(y;\lambda_1)$ is the correct p.d.f. for some $\lambda_1 \in \Omega_1$, we can test $H_0 \colon \psi = \psi_0 = 1$.

For simplicity, suppose λ_1 and λ_2 to be one-dimensional. Define $l_k(\lambda_k;Y_j) = \log f_k(Y_j;\lambda_k)$ and $l'_k(\lambda_k;Y_j) = \partial l_k(\lambda_k;Y_j)/\partial\lambda_k$ $(k=1,2;$ $j=1,\ldots,n)$. Then, carrying through the calculation of the scores used in the W_u statistic (56), we find that

$$U_{.\psi}(1,\lambda) = \Sigma\left(l_1(\lambda_1;Y_j) - l_2(\lambda_2;Y_j) - E\left[\{l_1(\lambda_1;Y_j)-l_2(\lambda_2;Y_j)\};\lambda_1\right]\right).$$

$$U_{.\lambda}(1,\lambda) = \begin{bmatrix} \Sigma l'_1(\lambda_1;Y_j) \\ 0 \end{bmatrix}.$$

From these relations we find that

$$b_.(1,\lambda) = \left(\frac{\mathrm{cov}\{l_1(\lambda_1;Y)-l_2(\lambda_2;Y), l'_1(\lambda_1;Y);\lambda_1\}}{\mathrm{var}\{l'_1(\lambda_1;Y);\lambda_1\}}, 0\right),$$

and

$$i_.^{\psi\psi}(1,\lambda)=n^{-1}\left[\mathrm{var}\{l_1(\lambda_1;Y)-l_2(\lambda_2;Y);\lambda_1\}- b_.^2(1,\lambda)\mathrm{var}\{l'_1(\lambda_1;Y);\lambda_1\}\right]^{-1}$$

The W_u statistic (56) is calculated by substitution of $\hat{\lambda}_0$, the m.l.e. of λ under H_0. In this problem H_0 asserts that $f_1(y;\lambda_1)$ is the p.d.f., so that the first component of the vector $\hat{\lambda}_0$ is $\hat{\lambda}_1$. But λ_2 does not appear in the p.d.f. under H_0, so that the substituted value can be any value in Ω_2. One possible choice of course is $\hat{\lambda}_2$. The asymptotic chi-squared result will apply to W_u, which has one degree of freedom.

Notice that in this large-sample context, under H_0, $\hat{\lambda}_1$ converges to λ_1 and $\hat{\lambda}_2$ converges to $\lambda_2(\lambda_1)$, say, depending on λ_1. The W_u test for large n is effectively testing $H_0 : \psi = 1$ in the family (61) with $\lambda_2 = \lambda_2(\lambda_1)$. It might, therefore, be more appropriate to use $\lambda_2(\hat{\lambda}_1)$ instead of $\hat{\lambda}_2$ in the W_u test. However, it is not clear which estimate for λ_2 is best in general. In the special case of two normal-theory linear regression models with distinct regressor variables, the W_u test using $\lambda_2(\hat{\lambda}_1)$ reproduces the standard exact regression test for adequacy of the single variable regression.

For further treatment of this problem, see Cox (1961, 1962) and Atkinson (1970). We shall consider the direct likelihood ratio test between $\psi = 0$ and $\psi = 1$ in part (iv) of this section.

Example 9.23. Multivariate one-way analysis of variance (ctd). In Example 5.24, we described the m.l. ratio test of the hypothesis $H_0 : \mu_1 = \ldots = \mu_m$ against $H_A : \mu_j$ not all equal, with independent samples from the multivariate normal distributions $MN_p(\mu_1, \Sigma), \ldots, MN_p(\mu_m, \Sigma)$, the covariance matrix Σ being part of the nuisance parameter.

If the observations are $Y_{jk} (k = 1, \ldots, r_j ; j = 1, \ldots, m)$ and $n = \Sigma r_j$, we find quite easily that

$$\hat{\mu}_j = \bar{Y}_{j.} \ (j = 1, \ldots, m), \ n \, \hat{\Sigma} = SS_w = \Sigma\Sigma \, (Y_{jk} - \bar{Y}_{j.})(Y_{jk} - \bar{Y}_{j.})^T .$$

Then with $SS_b = \Sigma r_j (\bar{Y}_{j.} - \bar{Y}_{..})(\bar{Y}_{j.} - \bar{Y}_{..})^T$, where $n\bar{Y}_{..} = \Sigma r_j \bar{Y}_{j.}$, the m.l. ratio statistic is

$$e^{\frac{1}{2}W} = \left(\frac{|SS_w + SS_b|}{|SS_w|} \right)^{\frac{1}{2}n}$$

It is of interest to examine the corresponding W_e and W_u statistics given by (54) and (56), it being more convenient here to discuss the m.l.e. statistic W_e. For this problem, we may write

$$\psi = \text{col}(\mu_2 - \mu_1, \ldots, \mu_m - \mu_1), \ \lambda = \text{col}(\mu_1, \sigma_1, \ldots, \sigma_p),$$

$$\tag{62}$$

where $\Sigma = (\sigma_1, \ldots, \sigma_p)$, and ψ and λ are both column vectors, as usual. The null hypothesis is that $\psi = 0$ with alternative $\psi \neq 0$. Now the m.l.e. of ψ is

$$\hat{\psi} = \text{col}(\bar{Y}_{2.} - \bar{Y}_{1.}, \ldots, \bar{Y}_{m.} - \bar{Y}_{1.}).$$

Because

$$i.(\mu_1, \ldots, \mu_m, \Sigma) = \text{diag}\{i.(\mu_1), \ldots, i.(\mu_m), i.(\Sigma)\},$$

with $i.(\mu_j) = r_j \Sigma^{-1}$ $(j = 1, \ldots, m)$, it is fairly straightforward to calculate $i.(\psi : \lambda)$ and thence to show that

$$W_e = \Sigma r_j (\bar{Y}_{j.} - \bar{Y}_{..})^{\mathsf{T}} \hat{\Sigma}^{-1} (\bar{Y}_{j.} - \bar{Y}_{..}) = n\text{tr}(\text{SS}_b \text{SS}_w^{-1}) ; \quad (63)$$

this is equal to n times the sum of the eigenvalues of $\text{SS}_b \text{SS}_w^{-1}$. This statistic can be derived directly by expansion of the m.l. ratio statistic W in terms of the eigenvalues of $\text{SS}_b \text{SS}_w^{-1}$.

An important extension of these results is to testing the null hypothesis that the vector means lie in a q dimensional subspace. Details of this are given in Exercise 9.14.

One of the situations where the usual likelihood approach falls down is when there are infinitely many nuisance parameters. Recall that in Sections 9.2 (ii) and (iii) we considered problems where Y_j are independent with p.d.f. $f_Y(y; \psi, \lambda_j)$ $(j = 1, \ldots, n)$. We examined an alternative method of likelihood estimation that can be used when a statistic $T = t(Y)$ exists that is sufficient for λ in the p.d.f. $f_Y(y; \psi, \lambda)$ when ψ is fixed, and such that T does not depend on λ. The approach is to apply likelihood methods to the conditional likelihood

$$\text{lik}(\psi; y|t_1, \ldots, t_n) = \frac{\Pi f_Y(y_j; \psi, \lambda_j)}{\Pi f_T(t_j; \psi, \lambda_j)}. \quad (64)$$

Given regularity conditions on this conditional likelihood, a consistent asymptotically normal m.l.e. is derived from it, when one usually does not exist from the unconditional likelihood. To test a composite hypothesis involving ψ in this situation, our earlier discussion of the m.l. ratio test can be applied *in toto* to the conditional likelihood. Of course, the incidental nuisance parameter is already eliminated, but part of ψ may also be a nuisance parameter. An alternative to explicit conditioning is sometimes the use of invariance arguments, implicitly reducing the problem to the conditional likelihood framework.

Example 9.24. One-way analysis of variance (ctd). Let Y_{jk} be independent $N(\lambda_j, \psi)$ $(j = 1, \dots, m\,; k = 1, \dots, r)$ with r fixed, and consider $H_0: \psi = \psi_0$, with alternative $H_A: \psi \neq \psi_0$. The m.l.e. of λ_j is $\bar{Y}_{j.}$ for each j, and $\hat{\psi} = \Sigma\Sigma(Y_{jk} - \bar{Y}_{j.})^2/(mr) = \mathrm{SS}_w/(mr)$; also $\hat{\lambda}_0 = \hat{\lambda}$. Consequently the m.l. ratio statistic is

$$W = mr\left\{ -\log\left(\frac{\hat{\psi}}{\psi_0}\right) - 1 + \frac{\hat{\psi}}{\psi_0}\right\}.$$

But under H_0, $mr\hat{\psi}/\psi_0$ has the chi-squared distribution with $m(r-1)$ degrees of freedom, so that $\hat{\psi} = \psi_0(r-1)/r + O_p(n^{-\frac{1}{2}})$ and W diverges with probability one as $m \to \infty$.

In this case, $\bar{Y}_{j.}$ is sufficient for λ_j given ψ, and the conditional likelihood (64) is proportional to

$$\psi^{-\frac{1}{2}m(r-1)}\exp\{-\mathrm{SS}_w/(2\psi)\},$$

which leads to the conditional m.l. ratio statistic

$$e^{\frac{1}{2}W_c} = \left\{\frac{\mathrm{SS}_w}{m(r-1)\psi_0}\right\}^{-\frac{1}{2}m(r-1)}\exp\left\{-\frac{1}{2}m(r-1) + \frac{\mathrm{SS}_w}{2\psi_0}\right\},$$

to which the standard asymptotic theory applies as $m \to \infty$.

The point of this example is not that W_c is preferable to W as a statistic; indeed, both are equivalent to the chi-squared statistic SS_w/ψ_0 and provided that we use its exact chi-squared property no problems arise. Critical regions are determined by small and large values of the chi-squared statistic in accordance with the null chi-squared distribution. It is when we appeal to the standard asymptotic distribution of W that things go wrong. In more complicated problems it would often be difficult or impossible to pick out from W an equivalent statistic whose distribution under H_0 is exactly known.

Another device that is sometimes used for problems with incidental nuisance parameters is to assume that these parameters are random effects. For example, suppose that we wish to test the homogeneity of $\theta_1, \dots, \theta_n$ for variables $Y_1, \dots, Y_n$ with densities $f(y; \theta_j)$ $(j = 1, \dots, n)$. One approach is to suppose that $Y_1, \dots, Y_n$ have some identifiable mixed density $\int f(y; \theta)\, dG(\theta; \xi)$, where the hypothesis $\theta_1 = \dots = \theta_n$ is translated into a simple hypothesis about ξ. If $G(\theta; \xi)$ were the $N(\xi_1, \xi_2)$ distribution, the hypothesis would be that $\xi_2 = 0$; in this particular case, the induced hypothesis is that the parameter ξ lies on the boundary of the parameter space, and non-standard expansions might be required to discover limiting properties of the m.l. ratio. Some discussion of this approach is given by Moran (1971).

To some extent, the random effects approach is effectively an empirical Bayes approach to the problem, a subject which is taken up with a different emphasis in Section 10.7.

(iv) Other composite hypotheses
So far we have dealt with composite hypotheses of the form $H_0: \theta \in \Omega_0$, where Ω_0 is a Euclidean subspace of Ω, equivalent to a set of the form $\{\theta ; \theta = (\psi_0, \lambda), \lambda \in \Omega_\lambda\}$. As we indicated in Section 5.1, many interesting problems are not of this type. Two possibilities are (a) hypotheses involving q dimensional subsets of the q dimensional parameter space, such as that θ lies in some specified spherical region; and (b) hypotheses involving separate families of distributions, e.g. the problems underlying Examples 5.9 and 9.22.

In general, we are here interested in the problem of testing $H_0: \theta \in \Omega_0$ versus $H_A: \theta \in \Omega_A$, where Ω_0 is not a subspace of Ω as in the previous section, and where Ω_A is not necessarily $\Omega - \Omega_0$. Two distinct situations exist, according as Ω_0 and Ω_A do or do not touch; technically speaking, to touch means to have limit points in common. Before discussing general behaviour of the m.l. ratio in such situations, it is instructive to consider the following simple example.

Example 9.25. Bivariate normal mean (ctd). Suppose that the single two-dimensional random variable $Y = (Y_1, Y_2)$ is $MN_2(\theta, I)$ and consider the m.l. ratio test of the null hypothesis that $\|\theta\| \leqslant 1$ with alternative that $\|\theta\| > 1$; thus the null hypothesis asserts that θ lies in the unit circle with centre at the origin. Then we have immediately that the m.l. ratio statistic W', defined in (49), is

$$W' = \inf_{\|\theta\| \leqslant 1} \{(Y_1 - \theta_1)^2 + (Y_2 - \theta_2)^2\} - \inf_{\|\theta\| > 1} \{(Y_1 - \theta_1)^2 + (Y_2 - \theta_2)^2\}.$$

$$W = \begin{cases} \{(Y_1^2 + Y_2^2)^{\frac{1}{2}} - 1\}^2 & (Y_1^2 + Y_2^2 > 1), \\ 0 & \text{otherwise}, \end{cases}$$

from which it is easy to see that

$$\text{pr}(W \leqslant d \,; \theta) \;=\; \text{pr}\{X_2^2(\|\theta\|^2) \leqslant (1 + \sqrt{d})^2 \}, \tag{65}$$

where $X_2^2(\|\theta\|^2)$ is a chi-squared variable with two degrees of freedom and with non-centrality $\|\theta\|^2$. Under the null hypothesis, (65) has a minimum on the boundary of the unit circle, so that the unbiased m.l. ratio test reduces to a test of $H_0': \|\theta\| = 1$, with alternative that $\|\theta\| > 1$; this test was obtained in Example 5.19 as an unbiased invariant test. Notice that, in terms of polar coordinates, Ω_0 is on one side of a "hyperplane" defined by the constancy of a component parameter.

This example brings out two points. First, obviously, the simple chi-squared results of the previous subsections no longer apply. Secondly, the significance probability has an upper bound which is attained on the boundary between Ω_0 and Ω_A; equivalently, the size of a fixed size test is attained on the boundary. In fact, for large samples, the result of the example will apply to any density $f(y\,;\theta)$ for which the unrestricted m.l.e. $\hat\theta$ is asymptotically $MN(\theta, n^{-1}\mathbf{I})$, because of the asymptotic sufficiency property characterized in (30). We shall continue to make use of this property in what follows.

In general, suppose that the true parameter value θ is an interior point of the null hypothesis space Ω_0, and hence a non-zero distance away from the alternative space Ω_A. Then the m.l. ratio statistic W' tends to $-\infty$ as n increases. More precisely,

$$W'/n \;=\; 2 \inf_{\Omega_A} E\{l(\theta'\,;Y) - l(\theta\,;Y)\,;\theta\} + o_p(1). \tag{66}$$

This can be verified by Taylor series expansion of W', using the consistency of the m.l.e's restricted in turn to Ω_0 and to Ω_A. We shall denote these m.l.e.'s by $\hat\theta_0$ and $\hat\theta_A$. Under the conditions of the consistency theorem of Section 9.2(ii), $\hat\theta_0$ and $\hat\theta_A$ are consistent for θ and θ_A respectively, at least for i.i.d. variables, where θ_A is the value of θ' which gives the minimum on the right-hand side of (66). We shall suppose θ_A unique, although this is not essential to the discussion of W'. One consequence of (66) is that if $Y_1, \ldots, Y_n$ are i.i.d. and if we suppose that

$$\mu(\theta, \theta') \;=\; E\{l(\theta, Y_j) - l(\theta'\,; Y_j)\,;\theta\} > 0 \quad (\theta \in \Omega_0, \theta' \in \Omega_A), \tag{67}$$

then, for all positive ϵ, $\text{pr}(W \leqslant \epsilon\,;\theta)$ tends to zero. Note that the inequality (67) is the strict case of Jensen's Inequality, used in (8).

Thus, only when θ is a boundary point of Ω_0 and Ω_A is there a non-degenerate limiting distribution for W'. In other cases, some standardization of W' is required to obtain a non-degenerate limiting distribution.

Let us first examine the situation where θ is a boundary point of both Ω_0 and Ω_A. Then both $\hat{\theta}_0$ and $\hat{\theta}_A$ are consistent for θ if the unrestricted m.l.e. $\hat{\theta}$ is consistent, which we shall assume. We shall show that W' has a limiting distribution determined by an equivalent m.l. ratio test based on the asymptotically sufficient statistic $\hat{\theta}$, under the regularity conditions of our previous discussion. The consistency of $\hat{\theta}_0$ and $\hat{\theta}_A$ ensures that, for large n, W' depends only on the behaviour of the likelihood around θ, when the approximation (30) is valid, so that local Taylor series expansions of W' are justified, as in earlier parts of this Section. Some further simplification is possible when we consider the likelihood locally at θ, namely that we can approximate to Ω_0 and Ω_A without affecting limiting distributions. Specifically, we can approximate to Ω_0 and Ω_A by "linear" sets Ω_0^* and Ω_A^* with the property that if θ' is a member of one of the sets, so are all parameter values $\theta' + a(\theta' - \theta)$ for $a > 0$. The sets Ω_0^* and Ω_A^* must be such that distances between them and the sets they approximate are small relative to distances from θ. For example, if Ω_0 and Ω_A are circles touching at θ, then Ω_0^* and Ω_A^* are half-planes separated by the tangent at θ.

To carry out the required Taylor series expansions, we note that, for regular problems, $\hat{\theta}_0$ and $\hat{\theta}_A$ are both of the form $\theta + O_p(n^{-\frac{1}{2}})$. Then we have, using (30), that

$$W' = 2\{l(\hat{\theta}_A ; Y) - l(\hat{\theta}_0 ; Y)\}$$
$$= \inf_{\Omega_0} (\hat{\theta} - \theta')^{\mathrm{T}} \mathbf{i}_.(\theta) (\hat{\theta} - \theta') - \inf_{\Omega_A} (\hat{\theta} - \theta')^{\mathrm{T}} \mathbf{i}_.(\theta)(\hat{\theta} - \theta') + o_p(1).$$

$$(68)$$

The values of θ' involved here are within $O(n^{-\frac{1}{2}})$ of θ. Therefore, if Ω_0^* and Ω_A^* are approximating sets of the type described earlier, (68) may be expressed in terms of minimizations over Ω_0^* and Ω_A^*, with an error of $o_p(1)$. The limiting distribution of W' is then obtained from the fact that $\hat{\theta}$ is approximately $MN \{\theta, \mathbf{i}_.^{-1}(\theta)\}$. Further details are given by Chernoff (1954). Notice the similarity between (68) and the m.l. ratio of Example 9.25.

From the discussion preceding the derivation of (68), it is quite clear that for large n, with θ still on the boundary of Ω_0 and Ω_A,

$$\text{pr}\,(W' \leqslant d\,;\theta') \leqslant \text{pr}\,(W' \leqslant d\,;\theta)\ (\theta' \in \Omega_0).$$

For fixed values of θ' in Ω_0 or Ω_A, not equal to θ, the m.l. ratio test is consistent, by (67). A more detailed treatment of the expansion for W' is given by Feder (1968), who derives the local power function. If the true value θ is not in either Ω_0 or Ω_A, then (66) generalizes in an obvious way; for example, for i.i.d. variables, we obtain

$$W'/n = \underset{\theta' \in \Omega_0}{2\inf}\ \mu(\theta,\theta') - \underset{\theta' \in \Omega_A}{2\inf}\ \mu(\theta,\theta') + o_p(1)\,, \qquad (69)$$

with $\mu(\theta,\theta')$ defined by (67). Thus W' tends to $+\infty$ or $-\infty$ depending on which of Ω_0 and Ω_A is "closer" to θ.

Now consider the situation where Ω_0 and Ω_A are disjoint parameter sets, i.e. with no boundary point in common. For simplicity, we suppose that $Y_1, \ldots, Y_n$ are i.i.d. variables. We still assume the likelihood to be regular in the usual sense, and we continue to assume that (8) holds, which now may be expressed in the form

$$\underset{\theta' \in \Omega_A}{\inf}\ \mu(\theta,\theta') > 0 \quad (\theta \in \Omega_0), \qquad (70)$$

with a corresponding inequality if $\theta \in \Omega_A$. As usual, we discuss the limiting behaviour of W' by developing the Taylor series expansion of each factor in the m.l. ratio. If the m.l.e.'s of θ restricted to Ω_0 and Ω_A are again $\hat{\theta}_0$ and $\hat{\theta}_A$, respectively, then

$$W' = 2\{l(\hat{\theta}_A\,;Y) - l(\hat{\theta}_0\,;Y)\}. \qquad (71)$$

Without loss of generality, we shall take the true value of θ to be in Ω_0 and to emphasise this we shall here denote it by θ_0. This implies that $\hat{\theta}_0$ and $\hat{\theta}$ are asymptotically equivalent and that $\hat{\theta}_0$ may be treated as a solution of the likelihood equation $U_{.}(\theta') = 0$ for large n. The other m.l.e. $\hat{\theta}_A$ can behave in one of two ways. Either it converges to a point on the boundary of Ω_A, or to an interior point. In the latter case, $\hat{\theta}_A$ is asymptotically equivalent to a solution of $U_{.}(\theta') = 0$, and hence has a limiting normal distribution. In both cases $\hat{\theta}_A$ converges to θ_{0A}, say, which minimizes $\mu(\theta_0, \theta')$ over $\theta' \in \Omega_A$. Because of (70), the law of large numbers implies that W' diverges to $-\infty$, reflecting the fact that for separate hypotheses consistent discrimination is possible. A more detailed analysis of W' is therefore required to obtain non-degenerate large-sample approximations.

We deal first with the case of bimodal likelihoods, where both $\hat{\theta}_0$ and $\hat{\theta}_A$ may be treated as solutions of the likelihood equation. That is, in particular, we assume

$$E\{U_.(\theta_0);\theta_0\} = E\{U_.(\theta_{0A});\theta_0\} = 0 . \tag{72}$$

Expanding the two terms in the m.l. ratio statistic (71) in Taylor series, and using (72), we obtain

$$\frac{W'}{2n} = \mu(\hat{\theta}_0,\hat{\theta}_{0A}) + \frac{1}{n}\sum \{l(\theta_{0A};Y_j) - l(\theta_0;Y_j) - \mu(\theta_0,\theta_{0A})$$

$$- c^T(\theta_0;Y)\,i^{-1}(\theta_0)U_j(\theta_0)\} + O_p(n^{-1}), \tag{73}$$

where, for a single observation,

$$c(\theta_0;Y) = \text{cov}\{l(\theta_{0A};Y) - l(\theta_0;Y),U(\theta_0);\theta_0\}.$$

Here $\mu(\hat{\theta}_0,\hat{\theta}_{0A})$ is the value of $\mu(\theta_0,\theta_{0A})$ when $\hat{\theta}_0$ is substituted for $\theta_0;\theta_{0A}$ is a function of θ_0. In fact, the limiting distribution derived from (73) would be unchanged if $\mu(\hat{\theta}_0,\hat{\theta}_A)$ were used to centralize $W'/(2n)$. Applying the central limit theorem to (73), we find that

$$T' = \frac{W'}{2n} - \mu(\hat{\theta}_0,\hat{\theta}_{0A}) \tag{74}$$

has a limiting normal distribution with zero mean and variance

$$\sigma_n^2 = n^{-1}[\text{var}\{l(\theta_{0A};Y) - l(\theta_0;Y);\theta_0\} - c^T(\theta_0,Y)i^{-1}(\theta_0)c(\theta_0;Y)]. \tag{75}$$

Because $\mu(\theta',\theta_A)$ is zero when θ' is in Ω_A, T' will tend to be negative if the true parameter value is in Ω_A, and so the approximate significance probability of t'_{obs} will be the lower tail, $\Phi(t'_{obs}/\sigma_n)$.

The above results generalize in an obvious way to the case of non-homogeneous variables.

Example 9.26. Geometric distribution versus Poisson distribution (ctd). Consider again the problem discussed in Example 9.22, that of testing between two alternative families of distribtuions for i.i.d. variables $Y_1, \ldots, Y_n$. As a specific example, suppose that Ω_0 corresponds to the family of Poisson distributions with general p.d.f.

$$\frac{e^{-\lambda}\lambda^y}{y!} \quad (y = 0,1,\ldots) , \tag{76}$$

and that Ω_A is the family of geometric distributions with general p.d.f.

$$\frac{\lambda^y}{(1+\lambda)^{y+1}} \quad (y = 0,1, \ldots). \tag{77}$$

An exact similar region test of Ω_0 with alternative Ω_A is available as discussed in Example 5.9, but the exact null distribution is intractable.

The two densities (76) and (77) are regular with respect to the nuisance parameter λ. Strictly speaking, to fit the general discussion we should have $\theta = (\psi, \lambda)$, where ψ is a two-valued parameter equal to ψ_0 for $\theta \in \Omega_0$ and ψ_A for $\theta \in \Omega_A$. Then, obviously,

$$\theta_0 = (\psi_0, \bar{Y}_.) \quad \text{and} \quad \hat{\theta}_A = (\psi_A, \bar{Y}_.),$$

and $\theta_{0A} = (\psi_A, \lambda_0)$ if $\theta_0 = (\psi_0, \lambda_0)$. Since ψ is constant and known within Ω_0 and Ω_A, the general discussion from (72) to (75) applies with the modification that ψ is ignored; that is, the score $U.(\theta)$ is the derivative of $l(\theta ; Y)$ with respect to λ and $i(\theta_0)$ is equal to $i(\lambda_0)$ for the Poisson distribution. Then we find that the centred m.l. ratio statistic T' is

$$T' = -\Sigma \log Y_j! + n \sum_{k=0}^{\infty} \log k! \, (e^{-\bar{Y}} \cdot \bar{Y}^k)/k! \, ,$$

and the variance σ_n^2 of the normal approximation is consistently estimated by

$$\frac{1}{n} \left[\mathrm{var}\{\log Y! ; \theta = (\psi_0, \bar{y}_.)\} - \frac{\mathrm{cov}^2\{Y, \log Y! ; \theta = (\psi_0, \bar{y}_.)\}}{\bar{y}_.} \right]. \tag{78}$$

That is, moments are evaluated according to the Poisson distribution with λ set equal to the sample m.l.e., $\bar{y}_.$. This example is treated in detail, with relevant numerical calculations, by Cox (1962).

So far as we know, there has been no general investigation into the relative merits of using $\mu(\hat{\theta}_0, \hat{\theta}_{0A})$ and $\mu(\hat{\theta}_0, \hat{\theta}_A)$ to standardize W' as in (74).

Finally, suppose that the likelihood function is unimodal for large n, so that the second equality in (72) does not hold. Then the Taylor series expansion of W' from (71) gives, again for i.i.d. variables,

$$\frac{W'}{2n} = \frac{1}{n} \Sigma \{l(\theta_{0A} ; Y_j) - l(\theta_0; Y_j)\} - (\hat{\theta}_A - \theta_{0A}) E\{U(\theta_{0A}); \theta_0\} + O_p(n^{-1}).$$

$$\tag{79}$$

In this situation θ_{0A} will be a boundary point of Ω_A. The general limiting properties of W' are difficult to discuss because the asymptotic sufficiency of $\hat{\theta}$ cannot be used for the large deviation $\theta_{0A} - \theta_0$.

(v) Asymptotic efficiency

In part (ii) of this section, we outlined the asymptotic efficiency of the m.l. ratio test, and its large-sample equivalents, showing, in particular, the correspondence to locally most powerful tests based' on the score $U(\theta)$, introduced earlier in Section 4.8. This asymptotic efficiency is directly related to that of the m.l.e. discussed in Section 9.2(iv), because the m.l. ratio test has critical regions asymptotically equivalent to those determined by large values of $\pm\sqrt{n}(\hat{\theta} - \theta_0)$.

Consider any statistic T which has a limiting $N\{\theta, n^{-1}\sigma_T^2(\theta)\}$ distribution, θ being one-dimensional. Then the m.l. ratio test of $H_0: \theta = \theta_0$ based on T has critical region asymptotically equivalent to

$$\{y \; ; |t - \theta_0|\sqrt{n} \geqslant k_{\frac{1}{2}\alpha}^* \sigma_T(\theta_0)\},$$

with significance level approximately α. The large-sample power function, to first order, for $\theta - \theta_0 = \delta n^{-\frac{1}{2}}$ is

$$\Phi\left\{-k_{\frac{1}{2}\alpha}^* - \frac{\delta}{\sigma_T(\theta_0)}\right\} + 1 - \Phi\left\{k_{\frac{1}{2}\alpha}^* - \frac{\delta}{\sigma_T(\theta_0)}\right\}, \qquad (80)$$

which is symmetric about $\delta = 0$. The local power is determined, to first order, by the second derivative of (80), which is

$$2k_{\frac{1}{2}\alpha}^* \; \Phi'(k_{\frac{1}{2}\alpha}^*) \, / \, \{\sigma_T^2(\theta_0)\} . \qquad (81)$$

The upper bound on this derivative is attained for the m.l. ratio test based on the original data, or equivalently for the m.l. test, since $\sigma_T^2(\theta_0) \geqslant n \, i^{-1}(\theta_0)$.

If we compare two consistent asymptotically normal statistics T_1 and T_2 with limiting variances $n^{-1}\sigma_{T_1}^2(\theta)$ and $n^{-1}\sigma_{T_2}^2(\theta)$, (80) shows that they have asymptotically the same power for the same local values of θ if the sample sizes n_1 and n_2 are related by

$$n_1^{-1}\sigma_{T_1}^2(\theta_0) = n_2^{-1}\sigma_{T_2}^2(\theta_0) .$$

Thus, if efficiency is measured by relative sample sizes, then the *asymptotic relative efficiency* of T_1 to T_2 is

$$e(T_1 : T_2) = \frac{\sigma_{T_2}^2(\theta_0)}{\sigma_{T_1}^2(\theta_0)} , \qquad (82)$$

or, equivalently, the ratio of power function curvatures at θ_0. This is the efficiency measure introduced in Section 9.2(iv). Notice that the measure $e(T_1 ; T_2)$ is independent of $k^*_{\frac{1}{2}\alpha}$, and hence of the size of the critical regions. The same measure is appropriate for local alternatives when derived from the significance probability point of view.

Of course, it is unnecessary to restrict test statistics to be consistent estimates of θ. More generally, we might suppose that T is consistent for some function $\mu(\theta)$ which is a monotone function of θ, at least near to $\theta = \theta_0$. A corresponding general definition of asymptotic relative efficiency, often called *Pitman efficiency,* relates to statistics T that are asymptotically $N\{\mu_T(\theta), \sigma_T^2(\theta)\}$. The simple calculation leading to (80) generalizes immediately to give the asymptotic relative efficiency of two such statistics T_1 and T_2 as

$$e(T_1 : T_2) = \left\{ \frac{\mu'_{T_1}(\theta_0)}{\mu'_{T_2}(\theta_0)} \right\}^2 \left(\frac{\sigma_{T_2}^2(\theta_0)}{\sigma_{T_1}^2(\theta_0)} \right) . \tag{83}$$

Note that the sensitivity measure $\mu'_T(\theta_0)/\sigma_T(\theta_0)$ is invariant under monotonic transformations of T, as is to be expected.

Section 6.3(iv) discussed asymptotic relative efficiency of one-sided rank tests. Generally, the measure (83) is obtained as the square of the ratio of power slopes at $\theta = \theta_0$ for one-sided tests based on asymptotically normal statistics. One illustration is Example 6.6. We omit further details.

For multidimensional parameters, where the m.l. ratio test is asymptotically equivalent to a chi-squared test with the quadratic form $(\hat{\theta} - \theta_0)^T i.(\theta_0) (\hat{\theta} - \theta_0)$, power is measured by the non-centrality index $(\theta - \theta_0)^T i.(\theta_0) (\theta - \theta_0)$. The relative efficiency analogous to $e(T_1 : T_2)$ in (83) will then be the ratio of non-centralities, which reduces to (83) for alternatives in any single direction away from $\theta = \theta_0$. Similar results apply for composite hypotheses.

(vi) Higher-order properties
In Section 9.2(vii), we briefly discussed some of the higher-order properties of the m.l.e., in particular those which give more accurate

approximations to the moments. A parallel analysis can be carried out for the test statistics discussed in this Section. Several objectives need to be distinguished. First, there is the practically important matter of checking the adequacy of, and possibly improving, the chi-squared approximation to the distribution under the null hypotheses. Secondly, there is the more detailed comparison of power properties for tests that to the first order have the same power. In making such comparisons it is, of course, essential that the more refined version of the distribution under the null hypothesis is used. Finally, and in some ways most fundamentally, there is the objective of obtaining new test statistics having higher-order optimality properties.

The detailed calculations are very complicated if carried through in any generality. Therefore we concentrate here on explaining the method and outlining one or two results.

We start, for simplicity, with a one-dimensional parameter θ, H_0 being the simple null hypothesis $\theta = \theta_0$, with two-sided alternatives $H_A : \theta \neq \theta_0$. To first order, the statistics

$$W = 2\{l(\hat{\theta} ; Y) - l(\theta_0 ; Y)\}, W_e = (\hat{\theta} - \theta_0)^2 i.(\hat{\theta}), W_u = U^2.(\theta_0) i.^{-1}(\theta_0)$$

are equivalent, but, in general, their second-order properties under H_0 will not be the same, an initial important distinction being that W and W_u, but not W_e, are invariant under transformations of the parameter. To examine the higher-order properties of, say, W, we take the Taylor series expansion up to terms in $\hat{\theta} - \theta_0$ of order n^{-1}, involving third derivatives of $U.(\theta_0)$. Extensive algebra leads to the result that

$$E(W;\theta_0) = 1 + \frac{\kappa_{02.}(\theta_0) - \kappa_{40.}(\theta_0) - 2\kappa_{21.}(\theta)}{4i^2(\theta_0)}$$

$$+ \frac{5\{\kappa_{30.}(\theta_0) + 3\kappa_{11.}(\theta_0)\}^2 - 24\kappa_{11.}(\theta_0)\{\kappa_{30.}(\theta_0) + 2\kappa_{11.}(\theta_0)\}}{12i.^3(\theta_0)} + O(n^{-2})$$

$$= 1 + r(\theta_0)/n + O(n^{-2}), \tag{84}$$

say. This shows that the modified statistic

$$T^* = W/\{1 + r(\theta_0)/n\}$$

has expectation under H_0 agreeing with that of the approximating chi-squared distribution with one degree of freedom to order n^{-1}. In fact, an extremely complicated calculation (Lawley, 1956) shows that all moments of T^* give corresponding agreement, and, moreover, that this holds for multiparameter problems including those with nuisance parameters. The general discussion is best carried through with orthogonalized parameters, i.e. with a diagonal information matrix.

The corresponding results for the test statistics W_e and W_u are more complicated in that different correction factors are required for the moments of different orders, so that if improved approximations to significance levels are required, then expansion in terms of Laguerre polynomials, analogous to the Edgeworth series, is in principle necessary.

Thus we have the important practical conclusion that if a higher-order approximation to the first moment of the m.l. ratio statistic can be obtained, then a more refined approximation to the significance level is possible. This is particularly valuable in multivariate analysis and Bartlett (1938, 1947) obtained a considerable number of important tests and, by direct calculation, their correction factors. The correction factor is often best obtained from first principles ; we give a simple example.

Example 9.27. Normal mean with unknown variance (ctd). Let $Y_1, \ldots, Y_n$ be i.i.d. in $N(\mu, \sigma^2)$ with $H_0: \mu = 0$ and $H_A: \mu \neq 0$, the variance being unknown. Here

$$W = n \log \left(1 + \frac{T^2}{n-1} \right),$$

where T is the standard Student t statistic. Direct expansion shows that

$$W = n \left\{ \frac{T^2}{n-1} - \frac{T^4}{2(n-1)^2} \right\} + O_p(n^{-2}).$$

We now need for the T statistic the second moment correct to order n^{-1} and the fourth moment correct to order 1. Note that these can be found by elementary expansions without knowledge of the explicit form of the Student t distribution. The modified statistic is

$$T^* = \frac{W}{1 + 3/(2n)}.$$

As a numerical example, if $n = 6$ and $t_{obs} = 4$ the exact two-sided significance level is 0.102. The chi-squared approximations without and with correction factor are 0.060 and 0.093, respectively.

The m.l. ratio statistic has its simple distributional properties only for two-sided alternatives. For one-sided tests for a single parameter, it is easier to work directly with the efficient score and with a limiting normal approximation rather than with a chi-squared approximation. This is particularly direct if there are no nuisance parameters, for then we can write down simply the null hypothesis moments or cumulants of $U.(\theta_0)$. This typically introduces a correction of order $n^{-\frac{1}{2}}$ in the calculation of significance levels. An alternative is to transform the test statistic to approximate symmetry, and in fact

$$\left[U.(\theta_0) - \frac{\kappa_{30.}(\theta_0)}{6i^2(\theta_0)} \{U^2(\theta_0) - i.(\theta_0)\} \right] \{i.(\theta_0)\}^{-\frac{1}{2}} \qquad (85)$$

has zero mean, variance $1 + O(n^{-1})$ and skewness zero to order $n^{-\frac{1}{2}}$. Similar corrections can be carried out when nuisance parameters are present, although the bias terms arising from the elimination of nuisance parameters need careful attention (Bartlett, 1953a, b, 1955).

The second-order analysis of power for the statistics W, W_u and W_e requires the setting up of an Edgeworth expansion for the joint distribution of

$$U.(\theta_0)/\sqrt{n} \quad \text{and} \quad U'(\theta_0)/n.$$

Peers (1971) has given a detailed analysis and has shown that no one statistic is uniformly superior. Choice between the test statistics on the basis of higher-order power would, in general, require an unrealistic degree of precision in the specification of alternatives. The m.l. ratio statistic is unbiased to order $n^{-\frac{1}{2}}$.

Other statistics, for example those in which $i.(\theta_0)$ is replaced by $-U'(\hat{\theta})/n$, can be handled by similar methods.

Next we consider briefly the derivation of improved test statistics *via* second-order theory. In a single parameter problem the use of the test statistic $U.(\theta_0)$ can be regarded as approximating to the likelihood function close to the null value by the exponential family density

$$f(y; \theta) = f(y; \theta_0) \exp\{(\theta - \theta_0) U.(\theta_0)\}.$$

This suggests taking the approximation to a second term and using, therefore, a test statistic which is a combination of $U.(\theta_0)$ and $U'.(\theta_0)$. The correction to W_u will be of order $n^{-\frac{1}{2}}$ and corresponds closely in spirit to (4.66), which was derived by maximizing power at an alternative not quite local to θ_0. Higher-order inference based explicitly on the two-term exponential family approximation has been considered by Efron (1974).

A final general point is that it is possible to base higher-order comparisons on the behaviour in the extreme tail of the null distribution, this behaviour being assessed by a law of large deviations rather than by a refinement of the central limit theorem. Behaviour in regions of extremely small significance levels seems, however, of little direct statistical interest.

(vii) Asymptotic interval estimation

The role of significance tests in the construction of interval estimates was discussed at length in Chapter 7. The same relationship holds in asymptotic theory and therefore the asymptotically efficient tests described above can be used for interval estimation. We therefore restrict ourselves here to some brief comments on asymptotic interval estimation.

Much the simplest and widely used such procedure for a single parameter, with or without nuisance parameters, is to give the m.l.e. together with an asymptotic standard error derived either from the information matrix at the m.l.e. point or from the matrix of observed second derivatives of the log likelihood. Thus, in the single parameter case, the asymptotic normal distribution leads to the asymptotic $1 - \alpha$ confidence interval

$$[\hat{\theta} - k_{\frac{1}{2}\alpha}^* \{i.(\hat{\theta})\}^{-\frac{1}{2}}, \hat{\theta} + k_{\frac{1}{2}\alpha}^* \{i.(\hat{\theta})\}^{-\frac{1}{2}}]. \tag{87}$$

In terms of the m.l. test statistic W_e this is exactly equivalent to

$$\{\theta ; (\hat{\theta} - \theta)^2 i.(\hat{\theta}) \leqslant k_{\frac{1}{2}\alpha}^{*2} = c_{1,\alpha}^*\}, \tag{88}$$

showing precisely the connexion of (87) with the procedure for constructing interval estimates from significance tests.

There are three main reasons for sometimes considering procedures other than (87) and its obvious multiparameter generalization. First, (87) is not invariant under transformations of the parameter, and while, for sufficiently large n, all "reasonable" functions of the parameter will give practically the same answer, there can be advantages in having exact invariance. Secondly, it may be required to use more refined distributional results than the normal approximation for the m.l.e. Finally, (87) will be seriously misleading if the situation really calls for very unsymmetrical intervals or for several disjoint intervals. While, of course, in the limit these situations arise with negligible probability and can be ignored in asymptotic theory, nevertheless other methods, in particular those based more explicitly on the m.l. ratio, can be expected to behave more sensibly than (87).

Thus if $\theta = (\psi, \lambda)$ and $\Omega_0 = \{\psi_0, \lambda \in \Omega_\lambda\}$, then the set

$$\{\Omega_0 ; \sup_\Omega l(\theta' ; Y) - \sup_{\Omega_0} l(\theta' ; Y) \leqslant \tfrac{1}{2} c^*_{d,\alpha}\} \qquad (89)$$

based on the m.l. ratio test gives an asymptotic confidence region for ψ of size α, where $c^*_{d,\alpha}$ is the upper α point of the chi-squared distribution with degrees of freedom $d = \dim(\psi)$. This is particularly suitable for graphical presentation if $\dim(\psi) = 1$ or 2. Note that (89) is invariant under transformation of the parameter of interest and that, if there are no nuisance parameters, then the region is likelihood-based. This last property means that in cases where the asymptotic distribution theory is a poor approximation, regions of the right general shape are obtained, even though the confidence coefficient attached to any one region may depart from its nominal value.

Regions based directly on the efficient scores, i.e. using the W_u test statistic, have the advantage that, at least when there are no nuisance parameters, moments of the test statistic are obtained directly. Thus, for a one-dimensional parameter θ, confidence regions can be obtained from

$$\{\theta ; |U_.(\theta)/\{i_.(\theta)\}^{\frac{1}{2}}| \leqslant k^*_{\frac{1}{2}\alpha}\}. \qquad (90)$$

These will usually be intervals. Again the procedure is invariant under transformation of the parameter. The test statistic has exactly mean zero and unit variance and corrections based on its higher moments are easily introduced. Bartlett (1953a,b, 1955) has given a detailed discussion including the case with nuisance parameters.

Finally, if the simple method (87) based directly on the m.l.e. and its standard error is used, then it is worth considering transforming the parameter. One possibility is to make the information function constant, i.e. in the single parameter case to define a new parameter θ^* by

$$\frac{d\theta^*}{d\theta} \propto \{i.(\theta)\}^{-\frac{1}{2}} .$$

A second possibility is to make the log likelihood function nearly symmetrical about its maximum; both these have as their objective the removal of discrepancies from likelihood-based regions based on higher-order theory.

9.4   Robust inference for location parameters

In the previous sections of this Chapter we have concentrated on procedures quite directly based on the likelihood function and which are asymptotically optimal. There are, however, a number of reasons for considering also procedures that are not fully optimal. One is ease of computation. Another is width of validity, in the sense of having some relevant properties under a wide range of circumstances. Related to this is a third aspect, that of robustness, especially in the presence of contamination by extreme observations. This has already been discussed to some extent in Section 8.6 in connexion with the point estimation of location parameters and we now take up this discussion more systematically.

Suppose that $Y_1, \dots, Y_n$ are ostensibly i.i.d. with density $f_0(y\,;\theta)$. We suppose further that either (a) $f_0(y\,;\theta)$ represents a particularly tentative distributional assumption, or (b) atypical observations are likely to occur, usually in the form of outliers. The unknown true density $f(y\,;\theta)$ depends on the parameter of interest θ, and, bearing in mind the above background, we want an estimate or representative statistic T_n for θ that will be reasonably efficient if $f_0(y\,;\theta)$ is correct, but which is not too sensitive to the likely departures from $f_0(y\,;\theta)$.

An important family of estimates may be represented in the form $T_n = t_n(\tilde{F}_n)$. That is, T_n is derived from the sample distribution function $\tilde{F}_n$, or equivalently from the order statistics, by a functional transformation, possibly depending on n. Two simple examples are the mean and median given respectively by

$$\overline{Y}. = \int x d\tilde{F}_n(x) ,$$

$$\tilde{Y} = \begin{cases} Y_{(\frac{1}{2}n+\frac{1}{2})} = \tilde{F}_n^{-1}\left(\frac{n+1}{2n}\right) & (n \text{ odd}), \\[2mm] \frac{1}{2}Y_{(\frac{1}{2}n)} + \frac{1}{2}Y_{(\frac{1}{2}n+1)} = \frac{1}{2}\tilde{F}_n^{-1}(\frac{1}{2}) + \frac{1}{2}\tilde{F}_n^{-1}(\frac{1}{2} + \frac{1}{2}n^{-1}) & (n \text{ even}), \end{cases}$$

where in each case $\widetilde{F}_n^{-1}(v) = \inf\{y : \widetilde{F}_n(y) = v\}$. Now for the mean, the functional form does not depend on n, i.e. we can write $\overline{Y} = t(\widetilde{F}_n)$. The general theoretical discussion is much simplified by restricting attention to statistics of this form; in other cases we can usually choose a suitable $t(.)$ such that $t_n(\widetilde{F}_n) - t(\widetilde{F}_n)$ is negligible for large n. Thus for the median the appropriate and obvious choice is $t(\widetilde{F}_n) = \widetilde{F}_n^{-1}(\frac{1}{2})$.

It can be shown (Gnedenko, 1967) that $\widetilde{F}_n(y)$ is uniformly consistent for the true distribution $F(y; \theta)$ and that therefore, provided that $t(.)$ is continuous, $T_n = t(\widetilde{F}_n)$ is consistent for $t\{F(.; \theta)\}$, which for simplicity we shall denote by $t(F)$. Note that T_n is by definition Fisher consistent for $t(F)$ in the sense of Section 9.2(ii). If the underlying distribution departs from the assumed form $F_0(y; \theta)$, and if $t(F_0) = \theta$, then clearly $t(F)$ need not in general equal θ. In this situation, we may be able to take $t(F)$ as defining the parameter of the distribution that is of real interest. This is straightforward when dealing with the location of symmetrical distributions, because, by imposing suitable symmetry on T_n, we can ensure that our estimate is consistent for the centre of symmetry. All our examples concern the location of symmetrical distributions, although the theoretical discussion is quite general.

The approximate behaviour of T_n relative to θ depends, then, on the behaviour of $t(\widetilde{F}_n) - t(F)$, where $\widetilde{F}_n(y)$ is a consistent and asymptotically normal estimate of $F(y; \theta)$. When $F(y; \theta)$ is discrete, we can make a Taylor expansion of $t(\widetilde{F}_n) - t(F)$ in terms of derivatives of $t(.)$ with respect to discrete probabilities. We did this for the multinomial estimating equation (24) in Section 9.2(iv).

When $F(y; \theta)$ is allowed to be continuous, a more general Taylor expansion is required, in terms of von Mises derivatives. Here the basic notion is that $t\{(1 - \epsilon)F + \epsilon K\}$ is a differentiable function of ϵ, and the first derivative of $t(.)$ with respect to F at y, denoted by $d_{t,F}(y)$, is defined by

$$\lim_{\epsilon \to 0} \frac{t\{(1 - \epsilon)F + \epsilon K\} - t(F)}{\epsilon} = \int d_{t,F}(z) dK(z). \qquad (91)$$

Equivalently, $d_{t,F}(y)$ is given explicitly from (91) by choosing $K(z) = \text{hv}(z - y)$, the unit Heaviside function. The second derivative $d_{t,F}(y_1, y_2)$ is defined in a completely analogous way, as are higher derivatives, but is not needed here.

We can now express the difference $t(\widetilde{F}_n) - t(F)$ in a Taylor series using the expansion

$$t\{(1 - \epsilon)F + \epsilon K\} = t(F) + \epsilon \int d_{t,F}(z)dK(z) + O(\epsilon^2).$$

Thus, with $\epsilon = n^{-1}$ and $K(z) = n\widetilde{F}_n(z) - (n - 1)F(z;\theta)$, we obtain

$$t(\widetilde{F}_n) = t(F) + \int d_{t,F}(z)d\widetilde{F}_n(z) + O_p(n^{-1}), \qquad (92)$$

because $\int d_{t,F}(z)dF(z;\theta) = 0$ by the definition (91), and $\widetilde{F}_n(y) - F(y;\theta)$ is $O_p(n^{-\frac{1}{2}})$. Now, using the definition (6) for $\widetilde{F}_n(y)$, (92) becomes

$$t(\widetilde{F}_n) - t(F) = n^{-1}\Sigma d_{t,F}(Y_j) + O_p(n^{-1}). \qquad (93)$$

The random variables $d_{t,F}(Y_j)$ are i.i.d. with zero mean and variance

$$\sigma_{t,F}^2 = \int d_{t,F}^2(y)dF(y;\theta), \qquad (94)$$

so that an application of the standard central limit theorem to (93) gives the result that

$$\frac{\{t(\widetilde{F}_n) - t(F)\}\sqrt{n}}{\sigma_{t,F}}$$

is asymptotically $N(0,1)$. The derivation, of course, implicitly assumes that $d_{t,F}(y)$ is not identically zero, which is a valid assumption for the linear estimates that we shall discuss in the sequel.

We now concentrate on estimates of the location parameter in location families with densities of the form $g(y - \mu)$, where μ is the centre of symmetry. One simplification arising from location invariance is that the variance (94) is independent of $\theta = \mu$. All the estimates that we consider satisfy $t(F) = \mu$. Note that we are effectively assuming the scale of the variables to be known, and set arbitrarily equal to unity. Some comments on the effect of unknown scale parameter will be made later.

The types of estimates $t(\widetilde{F}_n)$ that have been proposed for the location parameter μ fall into several categories. One is the class of linear combinations of order statistics

$$T_n = n^{-1}\Sigma c_{nj}Y_{(j)},$$

whose asymptotic properties are described in Section 4 of Appendix 2. The appropriate form for $t(\widetilde{F}_n)$ is obtained when we choose $c_{nj} = c\{j/(n + 1)\}$, so that

$$t(\widetilde{F}_n) = \int_0^1 c(v)\,\widetilde{F}_n^{-1}(v)\,dv, \tag{95}$$

subject to

$$\int c(v)dv = 1 \quad \text{and} \quad \int c(v)dv = 0 \,;$$

these restrictions are simply equivalent to $t(F) = \theta$ in the general formulation. The limiting normal distribution of (95) for this case is identical to that described in Appendix 2, with the approximation (A2.25) for the variance $\sigma_{t,G}^2$. When a finite number of order statistics are used, their joint asymptotic normality can be used directly to obtain the appropriate standardized version of (95).

Example 9.28. Linear combination of three order statistics. As a simple illustration of the linear robust location estimates, consider, for $p < \frac{1}{2}$,

$$T_n = c_1 Y_{([np])} + c_2 Y_{([\frac{1}{2}n])} + c_1 Y_{(n-[np]+1)},$$

which is essentially a combination of the median and symmetrically placed sample quantiles. Here the asymptotically equivalent statistic is

$$t(\widetilde{F}_n) = c_1 \widetilde{F}_n^{-1}(p) + c_2 \widetilde{F}_n^{-1}(\tfrac{1}{2}) + c_1 \widetilde{F}_n^{-1}(1-p). \tag{96}$$

To calculate the derivative $d_{t,F}(y)$, we need the derivative of $F^{-1}(v)$ according to the definition (91), where now $F(y) = G(y-\mu)$. Without loss of generality suppose that $\mu = 0$. Then defining $\xi_v = G^{-1}(v)$, the von Mises derivative of $G^{-1}(v)$ is easily seen to be

$$d_{\xi_v,G}(y) = \frac{v\,\mathrm{sgn}(y-\xi_v)}{g(\xi_v)} \,.$$

Hence, for the linear combination (96), we have

$$d_{t,G}(y) = \frac{c_1 p\,\mathrm{sgn}(y-\xi_p)}{g(\xi_p)} + \frac{c_2\,\mathrm{sgn}(y)}{2g(0)} + \frac{c_1(1-p)\,\mathrm{sgn}(y-\xi_{1-p})}{g(\xi_{1-p})},$$

so that $d_{t,G}(Y)$ is a trinomial variable. In general $d_{t,F}(y) = d_{t,G}(y-\mu)$. The variance in (94) is easily calculated, and we omit the details. A numerical illustration will be given later.

Example 9.29. Trimmed mean. A location estimate particularly favoured for dealing with symmetric outliers is the trimmed mean, obtained by deleting a certain proportion of extreme order statistics and averaging the remainder. Thus, with $r = [np]$,

$$T_n = \frac{1}{n-2r} \sum_{j=r+1}^{n-r} Y_{(j)} , \qquad (98)$$

and

$$t(\widetilde{F}_n) = \frac{1}{1-2p} \int_p^{1-p} \widetilde{F}_n^{-1}(v)dv = \frac{1}{1-2p} \int_{\xi_p}^{\xi_{1-p}} xd\widetilde{F}_n(x) .$$

Since integration is a linear operation, the von Mises derivative of $G^{-1}(v)$ used in Example 9.28 can be applied here to give

$$d_{t,G}(y) = \begin{cases} \xi_p/(1-2p) & (y < \xi_p) , \\ y/(1-2p) & (\xi_p \leqslant y \leqslant \xi_{1-p}) , \\ \xi_{1-p}/(1-2p) & (\xi_{1-p} < y), \end{cases}$$

where again $\xi_v = G^{-1}(v)$. Consequently the variance of the large-sample normal approximation for $(T_n - \mu)\sqrt{n}$ is, for all μ,

$$\sigma_{t,G}^2 = \frac{1}{(1-2p)^2} \left\{ \int_{\xi_p}^{\xi_{1-p}} y^2 dG(y) + 2p\xi_p^2 \right\} . \qquad (99)$$

A different type of estimate is defined, by analogy with the likelihood equation (2) and the multinomial estimating equation (24), as the solution to

$$n^{-1}\Sigma\zeta(Y_j - T_n) = 0 , \qquad (100)$$

where $\zeta(.)$ has a shape similar to the likelihood score function ; the particular choice $\zeta(v) = g'(v)/g(v)$ makes T_n a solution to the likelihood equation. The functional form $t(\widetilde{F}_n)$ is defined implicitly by expressing (100) as

$$\int \zeta\{y - t(\widetilde{F}_n)\}d\widetilde{F}_n(y) = 0 . \qquad (101)$$

Estimates defined in this way are called M-estimates. Assuming that $\zeta(v)$ is differentiable, we find that for arbitrary location μ

$$d_{t,F}(y) = \zeta(y - \mu)/\int \zeta'(v)g(v)dv. \qquad (102)$$

The motivation for (100) is that the weight attached to an observation is measured by $\zeta(.)$; thus, for example, when $g(.)$ is the standard normal density, $g'(v)/g(v) = v$. Any outlying observations can be deemphasised by reducing $|\zeta(v)|$ in the range of unlikely values. Particular forms of $\zeta(v)$ have been developed by Huber (1964) and others to produce low values of $\sigma_{t,G}^2$ simultaneously for several densities $g(y)$.

An important feature of the expression $t(\widetilde{F}_n)$ for a statistic is pointed out by (102), namely that $d_{t,F}(y)$ measures the influence on T_n of an "observation error" $y - \mu$. This is also apparent from the definition of the von Mises derivative in (91) if we set $K(z) = $ hv$(z - y)$. Thus, $\epsilon d_{t,F}(y)$ is the approximate change in the value of T_n if an additional $n\epsilon$ observations with value y are obtained. This makes $d_{t,F}(y)$ a useful tool for examining the sensitivity of the estimate to particular types of observations such as outliers.

Now consider the variance (94) of the limiting normal distribution for $(T_n - \mu)\sqrt{n}$. If $d_{t,F}(y)$ is a continuous function, then it is consistently estimated by $d_{t,\widetilde{F}_n}(y)$, since T_n is consistent for μ and $\widetilde{F}_n(y)$ is consistent for $F(y) = G(y - \mu)$. Therefore, we have as a consistent estimate for $\sigma_{t,F}^2$ the expression

$$S_t^2 = \int d_{t,\widetilde{F}_n}^2(y)\, d\widetilde{F}_n(y) = n^{-1} \Sigma\, d_{t,\widetilde{F}_n}^2(Y_j). \tag{103}$$

We omit the technical details. Thus for large-sample inference about μ we can treat

$$\frac{(T_n - \mu)\sqrt{n}}{\{n^{-1} \Sigma\, d_{t,\widetilde{F}_n}^2(Y_j)\}^{\frac{1}{2}}}.$$

as approximately $N(0,1)$. This applies to the trimmed mean in Example 9.29, for which we estimate $G(y)$ in (99) by $\widetilde{F}_n(y + T_n)$ to obtain

$$S_t^2 = \frac{1}{(1-2p)^2}\left\{\frac{1}{n}\sum_{j=[np]+1}^{n-[np]} (Y_{(j)}-T_n)^2 + p(Y_{([np])}-T_n)^2 + p(Y_{(n-[np]+1)}-T_n)^2\right\}$$

$$\tag{104}$$

The estimate (103) does not apply to the linear combination (96), because $d_{t,G}(y)$ cannot be consistently estimated in that case; with some restriction on the density $g(.)$, the individual density values arising in $\sigma_{t,G}^2$ for that case can be consistently estimated, as outlined in Section A2.3.

The estimate S_t^2 in (103) has a strong connexion with the jackknife variance estimate described in Section 8.4(iv). To see this, let $\widetilde{F}_{n-1,j}(y)$ be the sample distribution function with Y_j omitted, i.e.

$$\widetilde{F}_{n-1,j} = \frac{1}{n-1} \sum_{k \neq j} \mathrm{hv}(y - Y_k).$$

Then, applying (91) with $F(y) = \widetilde{F}_{n-1,j}(y)$, $K(y) = \mathrm{hv}(y - Y_j)$ and $\epsilon = n^{-1}$, we obtain

$$t(\widetilde{F}_n) - t(\widetilde{F}_{n-1,j}) = n^{-1} d_{t,\widetilde{F}_{n-1,j}}(Y_j) + o_p(n^{-1}).$$

Noting that, for continuous $d_{t,G}(y)$, the right-hand side is $n^{-1} d_{t,G}(Y_j - \mu) + o_p(n^{-1})$, we see that

$$n(T_n - T_{n-1,j}) = d_{t,G}(Y_j - \mu) + o_p(1),$$

whereas the jackknife pseudo-value defined in (8.31) is

$$T_j^{\mathrm{P}} = nT_n - (n-1)T_{n-1,j}.$$

The difference between the two is negligible to first order. It follows that the jackknife estimate (8.32) is equivalent to (103) for large n. Bearing in mind our reference to $d_{t,F}(y) = d_{t,G}(y - \mu)$ as a sensitivity index for $t(\widetilde{F}_n)$, we find that the jackknife pseudo-values, T_j^{P}, represent sample approximations to this index at each sample value. The jackknife method of estimating standard errors is known to break down when $d_{t,G}(y)$ has discontinuities.

The variance estimate S_t^2 in (103) also has a direct use in deriving simple adaptive estimates of μ. For example, suppose that we consider two possible estimates $t_1(\widetilde{F}_n)$ and $t_2(\widetilde{F}_n)$ with associated limiting variances $\sigma_{t_1,G}^2/n$ and $\sigma_{t_2,G}^2/n$. Then, because $S_{t_1}^2$ and $S_{t_2}^2$ are consistent, choice of the location estimate with lower estimated variance will, for large n, be approximately equivalent to use of the better estimate. This basic procedure can be generalized in several ways. One is to consider a family of estimates, say the trimmed means of Example 9.29, where p is arbitrary, subject to $0 \leqslant p < \frac{1}{2}$. Then, for a given value of p, let us temporarily denote the variance estimate (104) by $S_t^2(p)$, with the true variance $\sigma_t^2(p)$ given by (99). Suppose that $\sigma_t^2(p)$ has a minimum at $p = p^*$ such that $0 < p^* < \frac{1}{2}$. Because $S_t^2(p)$ is consistent for $\sigma_t^2(p)$, it is quite easy to show that the value $\widetilde{p}$ which minimizes $S_t^2(p)$ is consistent for p^*. This is enough to prove that the trimmed mean using $p = \tilde{p}$ is a consistent approximation to the optimal trimmed mean, whatever the underlying density $g(.)$. Some numerical illustrations for the trimmed mean will be given later.

Adaptive estimates can be based also on general linear combinations of order statistics. A relatively simple case is the estimate (96) based on three order statistics, where, for fixed p, the optimal weights c_1 and c_2 which minimize the limiting variance depend on $g(0)$ and $g(\xi_p) = g(\xi_{1-p})$. As we show in Section A2.3, the density function may, under some smoothness assumptions, be consistently estimated. Consequently the optimal weights c_1 and c_2 and optimal p value may be consistently estimated for arbitrary $g(.)$. A much more ambitious approach is to estimate the weight function $c(.)$ which minimizes the limiting variance (A2.25) for the general linear estimate. One method is to group order statistics symmetrically and use weighted averages of the group sums, with weights determined empirically. This method is used by Takeuchi (1971) and Johns (1974). A full account of adaptive estimation is beyond the scope of this book.

So far, we have ignored the problem of the scale parameter. Location estimates defined by the likelihood-type equation (100) depend on the scale, inasmuch as the function $\zeta(u)$ is used to diminish the effect of extreme values or outliers. For an arbitrary known scale parameter σ, (100) is generalized to

$$\frac{1}{\sigma} \int \psi \left\{ \frac{y - t(\tilde{F}_n)}{\sigma} \right\} d\tilde{F}_n(y) = 0. \qquad (105)$$

When σ is unknown, a corresponding equation can be used to define an estimate $s(\tilde{F}_n)$. By analogy with the likelihood equation, this equation is usually taken to be

$$\int \left\{ \frac{y - t(\tilde{F}_n)}{s(\tilde{F}_n)} \right\} \psi \left\{ \frac{y - t(\tilde{F}_n)}{s(\tilde{F}_n)} \right\} d\tilde{F}_n(y) = 1, \qquad (106)$$

which is to be solved in conjunction with (105), in which $s(\tilde{F}_n)$ replaces σ. The limiting distribution of $t(\tilde{F}_n)$ and $s(\tilde{F}_n)$ is derived from the bivariate generalization of the Taylor expansion (92), one result being that, if $\psi(v) = -\psi(-v)$, then the limiting distribution of $t(\tilde{F}_n)$ is independent of that of $s(\tilde{F}_n)$ and is unaffected by σ being unknown.

Other methods can be used for estimating the scale parameter, and the large-sample theory goes through in an obvious way. Notice that the estimate of precision derived from (103) does not involve σ.

To illustrate the types of location estimate that we have been considering and their robustness relative to more standard estimates, we now give some numerical results derived both from the asymptotic theory and from small sample simulations. We compare the following five estimates:

(a) the sample median, approximately $\widetilde{F}_n^{-1}(\frac{1}{2})$;

(b) the sample mean, $\bar{Y}_{\cdot}$,

(c) the linear combination of tertiles and median given approximately by $0.3\,\widetilde{F}_n^{-1}(\frac{1}{3}) + 0.4\,\widetilde{F}_n^{-1}(\frac{1}{2}) + 0.3\,\widetilde{F}_n^{-1}(\frac{2}{3})$;

(d) the trimmed mean (98) with estimated optimal trimming proportion p;

(e) the M-estimate (100) with

$$\xi(y) = \begin{cases} k\,\mathrm{sgn}(y) & (|y| > k), \\ y & (|y| \leqslant k). \end{cases}$$

The first two estimates are maximum likelihood ones for the double exponential and normal distribution respectively, and are conventional simple estimates of location.

The numerical results pertain to four different underlying distributions, namely

(i) standard normal, $N(0,1)$;

(ii) a mixture of $N(0,1)$ and $N(0,9)$ with mixing proportions 0.95 and 0.05, respectively;

(iii) double exponential with density $f(y) = \frac{1}{2}e^{-|y|}$;

(iv) Cauchy with density $f(y) = \{\pi(1 + y^2)\}^{-1}$.

Although these distributions are all in "standard" form, note that they have different variances, namely 1.00, 1.40, 2.00 and ∞, respectively. Comparative measures of scale may be based on quantiles, such as $F^{-1}(0.95)$ which has values 1.00, 1.07, 1.40 and 3.84 relative to the standard normal distribution (i).

For each combination of estimate and underlying distribution, Table 9.1 gives the variance $\sigma_{t,F}^2$ of the limiting normal distribution of $\sqrt{n}(T_n - \mu)$ calculated from (94) and the definition of $d_{t,F}(y)$ in (91); case (b) (iv) is of course an exception.

The absolute rates of convergence to the limiting normal distribution differ from case to case, and in particular we might expect the adaptive estimate (d) to have slow convergence. It is therefore useful to examine corresponding numerical results for a typical sample size.

For this Table 9.1 also gives, in italics, Monte Carlo estimates of the variances for $n = 20$, which are abstracted from the extensive numerical study by Andrews *et al* (1972). These results confirm that asymptotics are too optimistic especially for estimate (d).

Comparisons of the variances in Table 9.1 should strictly only be made between estimates for a fixed distribution, unless proper allowance is made for the differences in effective scale for the distributions. For example, if we decide that the mixture of normal distributions (ii) is comparable to $N(0, \sigma_0^2)$, then for estimate (c) we might justifiably compare the large-sample variances $1.27 \sigma_0^2$ and 1.35.

Even though numerical comparisons such as those derived from Table 9.1 are useful in indicating the more robust location estimates, for inference purposes one also needs information on the accuracy with which the variance $\sigma_{t,F}^2$ can be estimated, by jackknife methods or otherwise. This is a subject beyond the scope of our present discussion.

TABLE 9.1

Asymptotic variances of $\sqrt{n}(T_n - \mu)$ for five estimates T_n defined in (a) − (e) with four underlying distributions. Monte Carlo estimates of variance for $n = 20$ in italics

	Estimate				
	(a)	(b)	(c)	(d)	(e)
Distribution					
(i)	1.57	1.00	1.27	1.00	1.11
	1.50	*1.00*	*1.23*	*1.11*	*1.11*
(ii)	1.68	1.40	1.35	1.15	1.20
	1.56	*1.43*	*1.32*	*1.22*	*1.22*
(iii)	1.00	2.00	1.42	1.00	1.32
	1.37	*2.10*	*1.35*	*1.48*	*1.43*
(iv)	2.47	∞	2.50	2.28	3.02
	2.90	*—*	*3.1*	*3.5*	*3.7*

In principle, the theory we have outlined for robust location estimation can be extended to cover much more general problems.

For example, the likelihood-type equation (100) can be used for linear regression problems, replacing $Y_j - \mu$ by the general residual $Y_j - E(Y_j)$, with $E(Y_j)$ expressed parametrically ; see Huber (1973). For multivariate location data some difficulty arises because different observations may be deemphasised or omitted on different axes, if estimation is carried out one coordinate at a time. This may not be appropriate, depending on the type of departure from standard conditions anticipated. Difficulties arise in the asymptotic theory for multivariate estimates, because the multivariate sample distribution function is complicated.

For estimating the parameter θ in an arbitrary univariate p.d.f. $f(y ; \theta)$, the methods applied for location estimation can be used, in principle, by operating on the individual scores $u_j(Y_j, \theta)$. Thus, for example, we can trim the extreme score values. More generally, the likelihood equation

$$n^{-1} \Sigma\, u_j(Y_j, T_n) = 0$$

can be replaced by the weighted form

$$n^{-1} \Sigma\, c_{nj} u_{(j)} (Y_j, T_n) = 0 , \qquad (107)$$

where $u_{(j)}(Y_j, \theta)$ are the ordered scores for a given value of θ. The asymptotic theory of linear combinations will apply here to the Taylor expansion of (107) about θ. Of course, for arbitrary parametric p.d.f.'s, there is the question of whether θ still has a meaning when the density is not of an assumed form. This question does not arise in the symmetric location case considered earlier, where θ is a unique physical index for symmetric p.d.f.'s, but for unsymmetric situations the different estimates such as mean and median are estimating different physical constants.

Bibliographic notes

The first systematic account of the maximum likelihood method of estimation was given by Fisher (1922, 1925), who derived the property of asymptotic efficiency and related it to sufficiency. Subsequently there were several discussions of consistency and asymptotic efficiency, notably those of Wilks (1938) and Cramér (1946) ; Cramér's conditions were analysed in detail by Kulldorf (1957). A general treatment of consistency under mild conditions was given by Wald (1949), with extensions by Kiefer and Wolfowitz (1956).

LeCam (1953) surveys much of the post–1925 literature on maximum likelihood and more recently (LeCam, 1970) has given a detailed technical account of the asymptotic normality of the m.l.e. ; see also Walker (1969) and Dawid (1970). Norden (1972) provides an extensive bibliography on m.l.e.

Second-order properties of m.l.e. are considered by Fisher (1925), Bartlett (1953a,b, 1955), Rao (1961, 1962, 1963), Shenton and Bowman (1963) and Efron (1974).

Other efficient methods of estimation such as those based on chi-squared criteria are studied by Fisher (1925), Neyman (1949), Barankin and Gurland (1951) and Ferguson (1958).

Problems involving in the limit infinitely many parameters are discussed by Neyman and Scott (1948) and Kiefer and Wolfowitz (1956). The use of conditional likelihoods in such problems (Bartlett, 1936a,b, 1937) has been studied systematically by Andersen (1970, 1973) and from a different point of view by Kalbfleisch and Sprott (1970).

The asymptotic theory of the m.l. ratio test statistic in regular problems was given in general terms by Wilks (1938) and Wald (1941, 1943), the original formulation being due to Neyman and Pearson (1928). Wald established the asymptotic equivalence with other forms of test statistic. The use of the efficient score for testing and for confidence limits is due to Bartlett (1953a, b, 1955) and to Neyman (1959) ; for a recent account of the relations between the different test statistics, see Moran (1970). Another asymptotically equivalent form, the Lagrange multiplier statistic, is due to Aitchison and Silvey (1958, 1960) and Silvey (1959) and is particularly suitable when the null hypothesis is specified by constraints.

Problems with composite hypotheses with general boundaries were analysed by Chernoff (1954) and in more detail by Feder (1968). Cox (1961, 1962) discussed tests involving separate families.

Inference for Markov processes is described by Billingsley (1961a, 1961b).

Second-order results for asymptotically efficient tests, including the development of correction factors, were given by Bartlett (1937, 1953a,b, 1955) and Lawley (1956) and more recently by Peers (1971). Higher-order approximations for multivariate tests are discussed in detail by Box (1949). For a different approach *via* the laws of large deviations, see, in particular, Hoeffding (1965) and Brown (1971). A discussion of asymptotic relative efficiency alternative to that of Section 9.3(v) is due to Bahadur ; for a summary of his work on this see his monograph (Bahadur, 1971).

Anscombe (1953) discusses the asymptotic theory of sequential estimation.

A modern approach to distributional problems connected with test statistics that depend on the sample distribution function is by weak convergence (Billingsley, 1968). For a discussion of tests of goodness of fit, see Durbin (1973) and for earlier work on the chi-squared goodness of fit test with grouped continuous data, see Watson (1958) and Chernoff and Lehmann (1954).

The early history of robust estimation is not clear, although the general idea of calculating estimates after possibly deleting aberrant observations is old. One mid-19th century reference is to the work of Mendeleev. Recent developments stem from Tukey (1960) and Hodges and Lehmann (1963). A thorough survey is given by Huber (1972) and the book of Andrews *et al* (1972) describes a major sampling study as well as much other information.

The general theory of differentiable statistical functions is discussed by von Mises (1947) and Fillipova (1962).

Further results and exercises

1. Suppose that the independent pairs of random variables $(Y_1, Z_1), \ldots, (Y_n, Z_n)$ are such that Y_j and Z_j are independent in $N(\xi_j, \sigma^2)$ and $N(\beta\xi_j, \tau^2)$, respectively. That is, the means of Y_j and Z_j are functionally related by $E(Z_j) = \beta E(Y_j)$. Show that the two solutions of the likelihood equation (2) are

$$\tilde{\beta} = \pm \left(\frac{\Sigma Z_j^2}{\Sigma Y_j^2}\right)^{\frac{1}{2}}, \tilde{\sigma}^2 = \frac{1}{2n}\left(\Sigma Y_j^2 - \frac{\Sigma Y_j Z_j}{\tilde{\beta}}\right),$$

$$\tilde{\tau}^2 = \frac{1}{2n}(\Sigma Z_j^2 - \tilde{\beta} \Sigma Y_j Z_j), \tilde{\xi}_j = \frac{1}{2}(Y_j + Z_j/\tilde{\beta}),$$

with the larger likelihood value attained at the point with $\tilde{\beta} \Sigma Y_j Z_j > 0$. By considering the log likelihood maximized with respect to σ and τ for fixed β and ξ, with ξ on the line joining the vectors y and z/β, show that the stationary value with larger likelihood is, in fact, a saddle-point of the likelihood function, i.e. that no local maximum exists.

A more detailed examination of the likelihood function when account is taken of rounding errors shows that there are two local maxima, the values of β corresponding closely to the regression coefficient of Z on Y and the reciprocal of the regression coefficient of Y on Z.

[Section 9.2(i) ; Solari, 1969 ; Copas, 1972]

2. Assuming the regularity conditions of Section 9.1, prove the consistency of a solution of the likelihood equation using the properties of $U.(\theta')$ for θ' in $(\theta - \delta, \theta + \delta)$, where δ is arbitrarily small. Give an example where this proof cannot be applied but where Wald's more general argument does apply.

[Section 9.2 (ii) ; Cramér, 1946]

3. Let $Y_{jk} (k = 1, \ldots, r_j ; j = 1, \ldots, m)$ be independent with Y_{jk} distributed in $N(\mu, \sigma_j^2)$, all parameters being unknown. The minimal sufficient statistic is $(\bar{Y}_{1.}, \ldots, \bar{Y}_{m.}, SS_1, \ldots, SS_m)$, where, as usual, $SS_j = \Sigma(Y_{jk} - \bar{Y}_{j.})^2$. For estimating μ, verify that the likelihood equation is a special case of the equation

$$\sum_{j=1}^{m} \frac{a_j(\bar{Y}_{j.} - \mu')}{SS_j + r_j(\bar{Y}_{j.} - \mu')^2} = 0.$$

Notice that each term on the left-hand side has zero expectation when $\mu' = \mu$, the true value. Now assume $r_1, r_2, \ldots$ to be fixed with m becoming large. Using a generalization of the result of Exercise 2, show that the solution $\tilde{\mu}_a$ of the estimating equation is consistent under appropriate conditions on the a_j's and the r_j's. Hence, show that under these conditions $\tilde{\mu}_a$ has a limiting normal distribution whose variance can be made uniformly lower than that for the m.l.e., by a suitable choice of the a_j's.

Comment on the result and its connexion with Example 5.8.

[Section 9.2 (iii) ; Neyman and Scott, 1948]

4. In a linear model under the usual assumptions, the regression parameter β is estimated with covariance matrix $(x^T x)^{-1} \sigma^2$; show that the response to be observed at an arbitrary point $\tilde{x}$ in the factor space can be predicted with mean squared error $\sigma^2 \{1 + \tilde{x}^T (x^T x)^{-1} \tilde{x} \}$, where the second term represents the uncertainty arising from not knowing β.

Consider, now, a more general situation in which at a point x in factor space the corresponding response Y_x has p.d.f. $f_{Y_x}(y;x,\theta)$. From some data, θ is estimated by maximum likelihood with asymptotic covariance matrix $i_.^{-1}(\theta)$. It is required to measure the uncertainty arising in predicting the response at an arbitrary point $\tilde{x}$ in factor space and attributable to lack of knowledge of θ. Justify the measurement of uncertainty with θ known by $E\{-\log f_{Y_x}(Y;\tilde{x},\theta);\theta\}$. Show further that when θ is estimated a reasonable measure of total uncertainty is $E\{-\log f_{Y_x}(Y;\tilde{x},\hat{\theta});\theta\}$. Define the component of uncertainty attributable to lack of knowledge of θ by the difference of these quantities, and show by the usual large-sample approximations that this is $\text{tr}\{i(\theta;\tilde{x})i_.^{-1}(\theta)\}$, where $i(\theta;\tilde{x})$ is the information matrix calculated from one observation at $\tilde{x}$. Verify that this gives the previous results for the linear model.

[Section 9.2 ; White, 1973]

5. Let $Y_1,\ldots,Y_n$ be independent with Y_j having the p.d.f. $f(y;\psi,\lambda_j)$, and suppose that S_j is minimal sufficient for λ_j with ψ fixed, with the property that S_j is functionally independent of ψ. This is the incidental parameter problem. Show that, if ψ and λ_j are scalars, then the Cramér–Rao lower bound on the variance of an unbiased estimator of ψ is, with $\theta = (\psi,\lambda_1,\ldots,\lambda_n)$,

$$\{i_{.00}(\theta) - \Sigma i_{.0j}^2(\theta)i_{.jj}^{-1}(\theta)\}^{-1},$$

where

$$i_{.00}(\theta) = E\left\{-\frac{\partial^2 l(\theta;Y)}{\partial\psi^2};\theta\right\}, \quad i_{.0j}(\theta) = E\left\{-\frac{\partial^2 l(\theta;Y)}{\partial\psi\partial\lambda_j};\theta\right\}$$

and

$$i_{.jj}(\theta) = E\left\{-\frac{\partial^2 l(\theta;Y)}{\partial\lambda_j^2};\theta\right\} \quad (j=1,\ldots,n).$$

It can be shown that this lower bound is attained asymptotically by the conditional m.l.e. of ψ if the p.d.f. is a member of the exponential family with S_j satisfying the extended definition of ancillarity given in Section 2.2 (viii).

Investigate the asymptotic efficiency of the conditional m.l.e. of ψ for the cases

(a) Y_j bivariate normal with means λ_j and $\lambda_j + \psi$ and identity covariance matrix ;

(b) Y_j bivariate exponential with means $1/\lambda_j$ and $1/(\psi + \lambda_j)$, the components being independent.

[Sections 9.2 (iii), (iv) ; Andersen, 1970]

6. Let $Y_1, \dots, Y_n$ be successive variables in a two-state stationary Markov chain with transition probability $\mathrm{pr}(Y_j = 1 | Y_{j-1} = 0) = \lambda$ and stationary probability $\psi = \mathrm{pr}(Y_j = 1)$. Show that the variance of the limiting normal distribution of the m.l.e. $\hat{\psi}$ is $\psi(1 - \psi)(1 - 2\psi + \lambda)/\{n(1 - \lambda)\}$, which is also the variance of $\bar{Y}$. Hence verify that the explicit m.l.e. of λ with fixed ψ replaced by $\bar{Y}$ is asymptotically efficient.

[Sections 9.2 (iii), (iv) ; Klotz, 1973]

7. Let $Y_1, \dots, Y_n$ be independent, each Y_j having a Poisson distribution with mean $\theta_1 + \theta_2 x_j$ for fixed x_j ; the parameter (θ_1, θ_2) is restricted to make each mean positive. Show that the m.l.e. of (θ_1, θ_2) is asymptotically equivalent to a weighted least squares estimate. Make a comparison with the unweighted least squares estimate, including the special case $\theta_2 = 0$.

[Section 9.2 (iv)]

8. The asymptotic sufficiency of the m.l.e. $\hat{\theta}$ in regular problems implies only that the information lost in summarizing Y by $\hat{\theta}$ is asymptotically negligible relative to $i.(\theta)$. Show that, for one-dimensional θ, the loss of information is $E_{\hat{\theta}} \mathrm{var}\{U.(\theta)|\hat{\theta} ; \theta\}$, which to first order can be derived from the variance of $U'.(\theta)$ conditionally on $U.(\hat{\theta}) = 0$. A similar calculation would be appropriate for a general consistent estimating equation.

Consider specifically the regular multinomial situation where $N_1, \dots, N_m$ are cell frequencies in a sample of size n corresponding to cell probabilities $\pi_1(\theta), \dots, \pi_m(\theta)$. Note that the N_j can be treated as independent variables with Poisson distributions of means $n\pi_j(\theta)$ constrained by $\Sigma N_j = n$, and that $U.(\hat{\theta}) = 0$ is approximately equivalent to another linear restriction on the N_j. Hence derive an explicit expression for $i.(\theta) - i_{\hat{\theta}}(\theta)$.

Use the same arguments to derive the information loss in using a general asymptotically efficient estimating equation. Discuss these results and their relation to second-order approximations for variances of the limiting normal distributions of efficient estimates.

[Section 9.2 (vii) ; Fisher, 1925 ; Rao, 1961 ; Efron, 1974]

9. Suppose that $Y_1, \ldots, Y_n$ are i.i.d. with continuous p.d.f.

$$f(y;\theta) = \begin{cases} c(\theta)d(y) & (a \leqslant y \leqslant b(\theta)), \\ 0 & \text{elsewhere} \end{cases},$$

where $b(\theta)$ is a monotone function of the single parameter θ. Show that the m.l.e. of θ is $b^{-1}(Y_{(n)})$, and hence that the m.l. ratio criterion for testing $H_0 : \theta = \theta_0$ against the two-sided alternative $\theta \neq \theta_0$ is, under H_0, given by

$$W = -2n \log \int_a^{Y_{(n)}} c(\theta_0)d(y)dy,$$

which has exactly the chi-squared distribution with two degrees of freedom.

Investigate the multi-sample extension in which m samples are used to test $\theta_1 = \ldots = \theta_m$.

[Section 9.3 (ii) ; Hogg, 1956]

10. Let $Y_1, \ldots, Y_n$ be i.i.d. with regular p.d.f. $f(y;\theta)$, where $\dim(\theta) > 1$. Suppose that θ is restricted to satisfy $\zeta(\theta) = 0$, where $\zeta(.)$ is differentiable. Use the method of Lagrange multipliers to derive the constrained likelihood equation and derive the limiting normal distribution of the constrained m.l.e. $\tilde{\theta}$.

If the restriction $\zeta(\theta) = 0$ is a null hypothesis with alternative that $\zeta(\theta) \neq 0$, show that the m.l. ratio statistic is asymptotically equivalent to the standardized form of $\zeta^2(\hat{\theta})$, where $\hat{\theta}$ is the usual unconstrained m.l.e. Show that the corresponding version of the score statistic W_u is based on $U(\tilde{\theta})$.

[Sections 9.2 (iii), 9.3 (iii) ; Aitchison and Silvey, 1958]

11. Consider a two-dimensional contingency table with cell frequencies N_{jk} ($j = 1, \ldots, r$; $k = 1, \ldots, c$) and corresponding probabilities π_{jk}. For the null hypothesis of row-column independence with general alternative derive the forms of the W_u and W_e statistics.

[Section 9.3 (iii)]

12. Suppose that $Y_1, \ldots, Y_n$ are independent, Y_j being distributed in $N(\mu_j, \sigma^2)$ where either $\mu_j = \beta w_j$ or $\mu_j = \gamma z_j$ ($j = 1, \ldots, n$), all parameters being unknown and the w_j's and z_j's being known constants. Derive the explicit form of the W_u statistic of Example 9.22 for

testing between these separate models, paying particular attention to the alternative forms arising from use of the general estimates $\hat{\lambda}_2$ and $\lambda_2(\hat{\lambda}_1)$. Compare the results with the usual exact analysis of this problem.

[Section 9.3 (iii) ; Walker, 1970]

13. A bivariate random variable (X, Y) of the form $X = U + V$, $Y = U + W$, where U, V and W have independent Poisson distributions, is said to have a bivariate Poisson distribution. Note that the marginal distributions of X and Y are Poisson and that X and Y are independent if and only if $E(U) = 0$. Consider the construction of an asymptotic test of independence from independent pairs $(X_1, Y_1), \ldots, (X_n, Y_n)$. Show that the m.l. ratio statistic is computationally difficult to find, but that the efficient score statistic W_u, with the nuisance parameters $E(X)$ and $E(Y)$ estimated by $\bar{X}$ and $\bar{Y}$, leads to consideration of the test statistic $\Sigma (X_j - \bar{X})(Y_j - \bar{Y}) \sqrt{(\bar{X}.\bar{Y})}/\sqrt{n}$, having asymptotically a standard normal distribution under the null hypothesis.

[Section 9.3 (iii) ; Neyman, 1959]

14. Let $Y_1, \ldots, Y_n$ be i.i.d. in the density $f(y ; \psi, \lambda)$ and let H_0 be the composite hypothesis that $(\psi, \lambda) \in \Omega_{\psi, 0} \times \Omega_\lambda$ with alternative that $(\psi, \lambda) \in (\Omega_\psi - \Omega_{\psi, 0}) \times \Omega_\lambda$. Show that the m.l. ratio criterion for testing H_0 is asymptotically equivalent to that for testing $H_{0\lambda}$: $(\psi, \lambda) \in \Omega_{\psi, 0} \times \{\lambda\}$ with the substitution $\lambda = \hat{\lambda}$ if and only if $i_{.\psi\lambda} = 0$.

Use this result in the context of Example 9.23 to show that the homogeneity of m means can be tested using $\hat{\Sigma}$ as a fixed covariance matrix. Using the same argument, find an approximation to the m.l. ratio criterion for testing the hypothesis that, for $j = 1, \ldots, m$,

$$\mu_j = \nu_0 + \sum_{k=1}^{q} \beta_{jk} \nu_k ,$$

where $\nu_0, \ldots, \nu_q$ are unknown vectors normalized to $\nu_k^T \nu_k = 1$ and spanning a q dimensional subspace, $q < m$. This can be done most conveniently for fixed Σ by minimizing the likelihood exponent with respect to the β_{jk} first, then ν_0 and finally $\nu_1, \ldots, \nu_q$.

[Section 9.3 (iii)]

15. Consider a discrete stationary stochastic process with r states, with M_{jk} the number of one-step transitions from state j to state k

in a series of n consecutive observations. Show that the m.l. ratio statistic for testing the null hypothesis of serial independence against the alternative of first-order Markov dependence is asymptotically equivalent to the chi-squared statistic

$$T = \sum \sum \frac{(M_{jk} - \widetilde{M}_{jk})^2}{\widetilde{M}_{jk}},$$

where $\widetilde{M}_{jk} = M_{j.} M_{.k}/(n-1)$. Derive the corresponding W_u statistic directly.

[Section 9.3 (iii) ; Anderson and Goodman, 1957]

16. In a regular parametric problem, a consistent estimate $T = t(Y)$ has a limiting $MN_q \{\theta, \mathbf{v}(\theta)\}$ distribution. Show that T and the m.l.e. $\hat{\theta}$ have a joint limiting normal distribution with cross-covariance matrix $\mathbf{i}_.^{-1}(\theta)$. Now for testing the null hypothesis $H_0: \theta = \theta_0$ against the alternative $\theta \neq \theta_0$ consider the statistic W_T defined by

$$\exp(\tfrac{1}{2} W_T) = \text{lik}(T ; Y)/\text{lik}(\theta_0 ; Y).$$

By comparing W_T with the m.l. ratio, show that the limiting distribution of W_T under H_0 is the same as that of

$$\sum_{j=1}^{q} Z_j^2 + \sum_{j=1}^{q} \{1 - \gamma_j(\theta)\} Z_{q+j}^2 ,$$

where $Z_1, \ldots, Z_{2q}$ are i.i.d. in $N(0, 1)$ and the $\gamma_j(\theta)$ are eigenvalues of $\mathbf{i}_.(\theta) \mathbf{v}(\theta)$.

Use a similar argument to examine the limiting distribution of the chi-squared statistic of Example 9.21 when the frequencies N_j are obtained by grouping continuous observations. Here T is the usual m.l.e. based on the N_j and $\hat{\theta}$ is the m.l.e. based on the original continuous data.

[Section 9.3 (iii) ; Chernoff and Lehmann, 1954 ; Watson, 1958]

17. Two random variables X and Y taking non-negative integer values x and y such that $x + y \leqslant n$ are called *F-independent* if for some functions $\alpha(.), \beta(.)$ and $\gamma(.)$

$$\text{pr}(X = x, Y = y) = \alpha(x) \beta(y) \gamma(n - x - y).$$

Two random variables U and V taking non-negative integer values are called *quasi-independent* over a set $\mathcal{S}$ of values if for functions $\alpha'(.)$ and $\beta'(.)$

$$\text{pr}(U = u, V = v) = \alpha'(u)\beta'(v) \ (u,v \in \mathcal{S}).$$

Discuss the relation between these definitions and give examples of practical situations where one is applicable and not the other.

Show how to obtain m.l. ratio tests of the null hypotheses of F-independence and quasi-independence.
[Section 9.3 (iii) ; Darroch, 1971 ; Darroch and Ratcliff, 1973 ;
Goodman, 1968]

18. Let $Z_1, \ldots, Z_n$ be i.i.d. in $N(\mu, \sigma^2)$, and let $Y_j = |Z_j|$. Show that using Y the m.l. ratio test of $H_0 : \mu = 0$ against the alternatives $H_A : \mu \neq 0$, with σ^2 unknown, is asymptotically equivalent to that based on large values of $T = \Sigma \, Y_j^4/(\Sigma \, Y_j^2)^2$. What is the limiting distribution of T ?

[Section 9.3 (iv) ; Hinkley, 1973]

19. For symmetric location densities $h(y - \theta)$, the distance measure $\mu(\theta, \theta_A)$ is independent of θ for any pair of densities. Calculate the measure for all pairs among the densities : normal, Cauchy, double exponential, and logistic, with a suitable choice of scale. Following the discussion of separate composite hypotheses in Section 9.3 (iv), investigate the m.l. ratio tests between each pair of densities, paying particular attention to power properties.

[Section 9.3 (iv)]

20. For the problem of testing $H_0 : \mu = 0$ versus $H_A : \mu \neq 0$ in the $N(\mu, \sigma^2)$ distribution with σ^2 unknown, compare numerically the exact distributional properties of the W_u and W_e statistics with the limiting chi-squared distribution. Derive second-order corrections for W_u and W_e based on first moments and determine numerically whether any improvement has been made over the first-order results.

[Sections 9.3 (iii), 9.3 (vi)]

21. Use asymptotic methods to obtain confidence limits for both parameters in the model of Exercise 4.13 and to test the adequacy of the model.

[Section 9.3 (vii)]

10 BAYESIAN METHODS

10.1 Introduction

Throughout the previous discussion our approach has been as follows. The parameter θ, regarded as the central objective of the statistical analysis, is an unknown constant. So long as we work on the basis that the model is correct, we must be prepared for the possibility that the parameter may take any value within the specified parameter space. The data are our guide to the true value. Therefore, we use methods of analysis that have sensible properties whatever the true value of θ. We have seen, especially in Chapters 4—9, various ways of formalizing this approach for particular problems, all based on a repeated sampling principle that, whatever the true value of θ, some desirable and physically interpretable properties should hold in hypothetical repetitions.

We now consider situations in which a different attitude is taken towards θ ; namely, it is regarded as the realized value of a random variable Θ, with p.d.f. $f_\Theta(\theta)$, called the *prior distribution*. Moreover this p.d.f. is, for the most part, regarded as known independently of the data under analysis, although in Section 10.7 we deal with so-called empirical Bayes procedures, where $f_\Theta(\theta)$ is regarded as unknown and to be estimated from relevant data.

The introduction of the prior distribution has two consequences. First, $f_\Theta(\theta)$ will often in some sense contain information about the value θ and, to the extent that this is correct information, it is likely often to sharpen the inference about θ. Secondly, the fact that Θ is a random variable greatly clarifies the methods of analysis to be used. For it is clear that information about the realized value of any random variable, based on observation of related random variables, is summarized in the conditional distribution of the first random variable, given all relevant information. That is, if we observe $Y = y$, we need to compute the conditional p.d.f. of Θ, given $Y = y$; this is called the *posterior distribution*. Further, if we are interested only in certain components of a vector parameter Θ, then we have just to integrate out the nuisance parameters from the posterior p.d.f. in order to find the conditional p.d.f. of the parameters of interest. Finally, if we are interested not directly in the parameter but in some independent future observation $Y^\dagger$ depending on θ, it will be possible to obtain the conditional distribution of the value of $Y^\dagger$ given the data from the posterior distribution of Θ.

That is, the gain achieved by introducing the prior p.d.f. is partly that it provides a way of injecting additional information into the analysis and partly that there is a gain in logical clarity. All this is entirely uncontroversial, given the formulation described above ; the arguments that have surrounded this subject for a long time centre entirely on the legitimacy and desirability of introducing the random variable Θ and of specifying its prior distribution.

The theorem that combines the prior distribution and the data to form the posterior distribution is a simple result in probability theory first given by Thomas Bayes in 1763. The approach to statistical inference based on the systematic use of this theorem is also named after Bayes, although the adjective Bayesian is often used also for a particular approach in which subjective probabilities are heavily emphasized.

Before dealing in Section 10.4 with the central problem of the meaning to be attached to prior distributions we describe Bayes's theorem and the mechanics of its use.

10.2   Bayes's theorem

Bayes's theorem is concerned with reversing the order of the statements in a conditional probability, i.e. with relating $\mathrm{pr}(A|B)$ and $\mathrm{pr}(B|A)$. In its simplest form there is an event A and a collection of events $B_1, B_2, \ldots$, the latter forming a partition of the sample space. That is, one and only one of the events $B_j(j = 1,2, \ldots)$ occurs. Then Bayes's theorem is used in any one of the following three equivalent forms :

$$\mathrm{pr}(B_j|A) = \frac{\mathrm{pr}(A|B_j)\,\mathrm{pr}(B_j)}{\mathrm{pr}(A)} \tag{1}$$

$$= \frac{\mathrm{pr}(A|B_j)\,\mathrm{pr}(B_j)}{\Sigma\,\mathrm{pr}(A|B_k)\,\mathrm{pr}(B_k)} \tag{2}$$

$$\propto \; \mathrm{pr}(A|B_j)\,\mathrm{pr}(B_j), \tag{3}$$

where the proportionality in (3) refers to variation with j for fixed A.

Form (1) is an immediate consequence of the definition of conditional probability, provided that $\mathrm{pr}(A) \neq 0$. Version (2) follows on using the law of total probability for $\mathrm{pr}(A)$.

In the particular kind of application with which we shall be concerned, A will represent the data and $B_1, B_2, \ldots$ alternative explanations or hypotheses about the probability mechanism generating the data. In the great majority of the applications that we shall consider, the events A, $B_1, B_2, \ldots$ concern random variables and therefore it is convenient to rephrase (1)–(3) in terms of the p.d.f.'s of random variables and, moreover, to use a notation in line with our particular application. Then for random variables Θ and Y, we have that in the usual notation

$$f_{\Theta|Y}(\theta|y) = \frac{f_{Y|\Theta}(y|\theta)f_\Theta(\theta)}{f_Y(y)} \tag{4}$$

$$= \frac{f_{Y|\Theta}(y|\theta)f_\Theta(\theta)}{\int f_{Y|\Theta}(y|\theta')f_\Theta(\theta')d\theta'} \tag{5}$$

$$\propto f_{Y|\Theta}(y|\theta)f_\Theta(\theta). \tag{6}$$

Equations (4)–(6) correspond respectively to (1)–(3). For continuous random variables, we make the usual assumption that the "regular" forms of the conditional density are taken whenever the conditioning events have zero probability. Note that $f_{Y|\Theta}(y|\theta)$ is the likelihood; when θ is not a random variable, it was previously denoted by $f_Y(y;\theta)$. An important feature of (5) and (6) is that $f_\Theta(\theta)$ need not be a proper density, i.e. need not have integral equal to unity, in order for $f_{\Theta|Y}(\theta|y)$ to be a proper density. We shall say more about such improper prior distributions later.

Some simple illustrations of Bayes's theorem will now be given ; in the following sections some more thorough discussion of its formal consequences will be developed.

Example 10.1 Industrial inspection. As a very simplified model of a special problem in industrial inspection, suppose that a process can be in one of two states : B_1, good and B_2, bad. In B_1 items are defective with small known probability p_1 and in B_2 the corresponding probability is a larger value p_2, again assumed known. Suppose that $\text{pr}(B_1) = \pi_1$, $\text{pr}(B_2) = \pi_2$, these probabilities being based, for example, on the performance of the process over a long time. Finally, let the event A be that an item is observed to be defective.

We use Bayes's theorem in the form (3) :

$$\text{pr}(B_1|A) \propto \text{pr}(A|B_1)\,\text{pr}(B_1) = p_1\pi_1,$$

$$\text{pr}(B_2|A) \propto \text{pr}(A|B_2)\,\text{pr}(B_2) = p_2\pi_2.$$

Because one or other of B_1 and B_2 must occur, the normalizing constant is the reciprocal of the sum of the right-hand sides, so that

$$\text{pr}(B_j|A) = \frac{p_j\pi_j}{p_1\pi_1 + p_2\pi_2}\ (j = 1,2). \tag{7}$$

For example, if $\pi_1 = 0.9$, $\pi_2 = 0.1$ and $p_1 = 0.05$, $p_2 = 0.5$, then the posterior probability that the system is in the good state B_1 given a defective is 9/19 ; if the observation is that the item is non-defective, then the corresponding posterior probability for state B_1 is 171/172. These values show how the observation modifies the prior probability 0.9.

In some ways a more meaningful way of writing (7) is in the form

$$\frac{\text{pr}(B_1|A)}{\text{pr}(B_2|A)} = \frac{\text{pr}(A|B_1)}{\text{pr}(A|B_2)} \times \frac{\text{pr}(B_1)}{\text{pr}(B_2)}, \tag{8}$$

according to which the posterior odds for B_1 versus B_2 is the likelihood ratio multiplied by the prior odds. By (2) the same result holds when there are more than two possible states B, provided that we work with the odds of B_i relative to B_j, i.e. $\text{pr}(B_i)/\text{pr}(B_j)$.

This example can clearly be generalized in many ways.

Example 10.2. Parameter of binomial distribution. We now deal with an example in which the version of Bayes's theorem for random variables is used. Suppose that Y is binomially distributed with index n and with parameter θ and that the prior distribution of Θ is of the beta form with density

$$\theta^{a-1}(1-\theta)^{b-1}/B(a,b) \qquad (9)$$

for known constants a and b. By (6), the posterior density of Θ given $Y = y$ is proportional to

$$\binom{n}{y}\theta^{y}(1-\theta)^{n-y} \times \theta^{a-1}(1-\theta)^{b-1}/B(a,b) . \qquad (10)$$

Note that, in (10), we are concerned with a function of θ for fixed y. On inserting a normalizing constant to make the integral over $(0,1)$ equal to one, we have the posterior density

$$\theta^{y+a-1}(1-\theta)^{n-y+b-1}/B(y+a,n-y+b) . \qquad (11)$$

This is another beta density, the indexing quantities (a,b) in the prior density having been changed in the light of the data to $(y+a, n-y+b)$.

Now, given any distribution for Θ, the probability that the next observation will be a "success" is the expectation of Θ over its distribution. Thus, given the observation y and the prior density, the probability that the next observation is a success is the mean of (11) and is therefore

$$\frac{B(y+a+1, n-y+b)}{B(y+a, n-y+b)} = \frac{y+a}{n+a+b} . \qquad (12)$$

In particular, if all observations so far have been successes, so that $y = n$, then the probability that the next trial is a success is $(n+a)/(n+a+b)$. The special case $a = 1$ and $b = 1$, corresponding to a uniform prior distribution, leads to the probability $(n+1)/(n+2)$; this result is known as Laplace's law of succession. It gives the probability that an outcome that has always been observed in the past will be observed in the next trial. The particular prior distribution that leads to this result will, however, not often be very sensible.

One important point to note is that the probability that the next two observations are successes is *not* the square of (12), as is clear from calculation of $E(\Theta^2| Y = y)$ according to (11).

One important general consequence of Bayes's theorem is that, because the posterior density involves the data only through the likelihood, the sufficiency principle is automatically satisfied. More explicitly, if the likelihood factorizes in the form $m_1(s,\theta)m_2(y)$, then the factor $m_2(y)$ can be absorbed into the normalizing constant of proportionality in Bayes's theorem. Example 10.2 illustrates this result.

Further, because the posterior density is the distribution of Θ given the whole data, any ancillary statistic for the problem is automatically held constant and a special argument about possible ancillarity is unnecessary.

Finally, the strong likelihood principle is automatically satisfied. If we have two random systems generating proportional likelihood functions for a common parameter, and if the prior density is the same in the two cases, then so is the posterior density. Thus, in particular, suppose that we observe r successes in n trials, the trials being independent given the constant probability of success θ. Then the posterior density of Θ is the same whether r is the observed value of a binomial random variable for fixed n or whether n is the observed value of a negative binomial random variable for fixed r or whether some more complicated data-dependent sequential stopping rule is used. It is, of course, necessary that the stopping rule depend only on the data and not on supplementary information about θ ; see Example 2.34.

10.3 Conjugate prior distributions

Although Bayes's theorem can be used to combine any prior distribution with any likelihood, it is convenient, especially in theoretical discussions, to take special forms for the prior distribution that lead to simple answers. Example 10.2 is an instance where the binomial likelihood and the beta prior combine to give a posterior distribution of simple form. In general, for a given model $f_Y(y;\theta)$, we can look for a family of prior distributions such that the posterior distributions also are in the same family, only the indexing quantities being changed. Such a family is called *closed under sampling* or *conjugate* with respect to $f_Y(y;\theta)$. Of course, the reasonableness of using such a distribution needs careful consideration in each application.

One special kind of conjugate family is that with support at a finite or enumerable number of points, the indexing quantities being the relevant probabilities ; Example 10.1 is an instance with a two-point prior distribution. While very simple prior distributions such as these are sometimes useful in particular problems, their value is severely limited by the fact that, no matter how much data are obtained, only the initial isolated parameter values are considered possible.

To look for continuous conjugate prior densities, we consider i.i.d. observations having the exponential family distribution with density

$$f_Y(y;\theta) = \exp\{a(\theta)b(y) + c(\theta) + d(y)\}. \tag{13}$$

Quite clearly, if we combine this with the prior density

$$f_\Theta(\theta;k_1,k_2) \propto \exp\{k_1 a(\theta) + k_2 c(\theta)\}, \tag{14}$$

then a simple conclusion emerges. In fact, if the data consist of independent random variables $Y_1, \ldots, Y_n$ having the density (13), the posterior density is

$$f_\Theta\{\theta;k_1 + \Sigma\,b(y_j), k_2 + n\}. \tag{15}$$

A few special examples of (14) are summarized in Table 10.1.

TABLE 10.1

Some conjugate prior distributions

Distribution of Y	Parameter	Conjugate prior distribution
Binomial	Prob. of success	Beta
Poisson	Mean	Gamma
Exponential	Reciprocal of mean	Gamma
Normal	Mean (variance known)	Normal
Normal	Variance (mean known)	Inverse Gamma

The only one of these conjugate distributions calling for special mention is that for the variance of a normal distribution. The density of a normal distribution of zero mean and variance τ is

$$\frac{1}{\sqrt{(2\pi)}} \exp\left(-\frac{y^2}{2\tau} - \frac{1}{2}\log\tau\right),$$

so that, in line with the general discussion around (14), the conjugate prior density for T is proportional to $\tau^{-b}\exp(-a/\tau)$ and can be written on reparameterization as

$$f_T(\tau;b,\tau_0) = \frac{\{(b-2)\tau_0\}^{b-1}\tau^{-b}\exp\{-(b-2)\tau_0/\tau\}}{\Gamma(b-1)}, \quad (16)$$

where τ_0 is the expectation of T. This is the density of the reciprocal of a variate having a gamma distribution; $d_b = 2(b-1)$ plays the role of "degrees of freedom".

Example 10.3. Ratio of Poisson means. As a simple example of a problem involving more than one parameter, suppose that the random variables Y_1 and Y_2 are independently distributed in Poisson distributions of means θ_1 and θ_2 and that the prior distributions are independent gamma distributions of means m_1 and m_2 and indices b_1 and b_2, respectively. Then the posterior joint density of Θ_1 and Θ_2 is proportional to

$$(e^{-\theta_1}\theta_1^{y_1})(b_1\theta_1/m_1)^{b_1-1}\exp(-b_1\theta_1/m_1) \times (e^{-\theta_2}\theta_2^{y_2})(b_2\theta_2/m_2)^{b_2-1}\exp(-b_2\theta_2/m_2)$$
$$(17)$$

That is, the means are independently distributed in gamma distributions; a statement equivalent to (17) is that the posterior distribution of Θ_i is such that $2(m_i + b_i)\Theta_i/m_i$ is distributed as chi-squared with $2(y_i + b_i)$ degrees of freedom.

Now suppose that we are interested in the ratio $\rho = \theta_1/\theta_2$ of the means. This is represented by the random variable $P = \Theta_1/\Theta_2$. The marginal density of P can be obtained from (17) by the usual device of a change of variables, followed by an integration with respect to the second variable. In fact, however, this is exactly the same calculation as is involved in obtaining the F distribution from the ratio of chi-squared variables. It follows, therefore, that the posterior distribution of P is such that

$$\frac{(m_1 + b_1)\,m_2\,(y_2 + b_2)}{(m_2 + b_2)\,m_1\,(y_1 + b_1)}\,P$$

has an F distribution with degrees of freedom $(2y_1 + 2b_1, 2y_2 + 2b_2)$.

Example 10.4. Multinomial distribution (ctd). Suppose that $Y_1, \ldots, Y_n$ are independent random variables with the multinomial distribution $\mathrm{pr}(Y_j = l) = \theta_l$ over categories $l = 1, \ldots, m$, such that $\Sigma\theta_l = 1$. Then the likelihood function is proportional to

$$\prod_{j=1}^{m} \theta_j^{n_j},$$

where n_j is the sample frequency in the jth category. This generalizes the situation of Example 10.2, and it is clear from (14) that the conjugate prior distribution for $\Theta = (\Theta_1, \ldots, \Theta_m)$ is the Dirichlet density

$$\frac{(\Sigma a_j)! \, \Pi \theta_j^{a_j}}{\Pi a_j!} \quad (\Sigma \theta_j = 1), \tag{18}$$

and that the posterior density for Θ is then

$$\frac{(n + \Sigma a_j)! \, \Pi \theta^{a_j + n_j}}{\Pi(a_j + n_j)!}.$$

If the number of categories is large, then the conjugate prior distribution here implies that $\Theta_1, \ldots, \Theta_m$ are pairwise approximately independent. This will often be inappropriate. For example, if $\theta_1, \ldots, \theta_m$ are derived from a continuous density by grouping, then it is usually reasonable to suppose that the values $\theta_1, \ldots, \theta_m$ are, in some sense, smooth. We shall discuss more elaborate methods of handling this type of situation in Section 10.4(iv).

Notice that if in (18) the a_j are all equal, then we have uniformity of prior knowledge for all θ_j, e.g. the prior expected value of each component probability would be $1/m$.

Example 10.5. Normal distribution of unknown mean and variance. Let the model for observed random variables $Y_1, \ldots, Y_n$ be that they are i.i.d. in $N(\mu, \tau)$. Suppose the prior distribution is that μ and τ are independent, the first being normal with mean ξ_0 and variance v_0 and the second having the form (16).

Then the joint posterior density of the mean and variance, being proportional to the product of the likelihood and the prior density, is in turn proportional to

$$\tau^{-\frac{1}{2}n} \exp\left\{-\frac{\Sigma(y_j - \mu)^2}{2\tau}\right\} \exp\left\{-\frac{(\mu - \xi_0)^2}{2v_0}\right\} \tau^{-b} \exp\left\{-\frac{(b-2)\tau_0}{\tau}\right\},$$

$$\tag{19}$$

where later we shall write $d_b = 2(b - 1)$.

If we want the marginal posterior density of μ, then we integrate with respect to τ, using the fact that

$$\int_0^\infty \tau^{-c} \exp\left(-\frac{k}{2\tau}\right) d\tau = \left(\frac{2}{k}\right)^{c-1} \Gamma(c-1).$$

There results a density which, on writing $\bar{y}_. = \Sigma y_j/n$ and $\mathrm{MS} = \Sigma(y_j - \bar{y}_.)^2/(n-1) = \mathrm{SS}/(n-1)$, is proportional to

$$\{n(\bar{y}_. - \mu)^2 + \mathrm{SS} + (d_b - 2)\tau_0\}^{-\frac{1}{2}(d_b + n)} \exp\{-(\mu - \xi_0)^2/(2\nu_0)\}.$$

$$(20)$$

The first factor on its own corresponds to μ having a posterior Student t distribution around $\bar{y}_.$ with degrees of freedom $n - 1 + d_b$ and estimated variance

$$\frac{\mathrm{SS} + (d_b - 2)\tau_0}{(n - 1 + d_b)n}.$$

$$(21)$$

This is modified by the second factor in (20), i.e. the one corresponding to the prior normal density of μ.

A special case of (20), easily treated from first principles, is obtained by letting b tend to infinity. This gives the posterior density of μ when the variance is known to be τ_0, say. In fact, the posterior distribution of μ is then normal with mean and variance respectively

$$\mu_W(\bar{y}_., \sigma_0^2/n ; \xi_0, \nu_0) = \frac{\dfrac{\bar{y}_.}{\sigma_0^2/n} + \dfrac{\xi_0}{\nu_0}}{\dfrac{1}{\sigma_0^2/n} + \dfrac{1}{\nu_0}},$$

$$(22)$$

$$\sigma_W^2(\sigma_0^2/n, \nu_0) = \frac{1}{\dfrac{1}{\sigma_0^2/n} + \dfrac{1}{\nu_0}}.$$

These correspond to the optimum weighted combination of the two estimates of μ, one with variance σ_0^2/n and the other with variance ν_0. The moments in (22) will be used again in the sequel.

Another special case of (20) that is of theoretical interest corresponds to the improper prior probability element proportional to $d\mu d\sigma/\sigma \propto d\mu d\tau/\tau$. This can be obtained here by taking $\tau_0 = 0$, $d_b = 0$, and letting $\nu_0 \to \infty$. It then follows directly from (20) that the posterior density of μ is proportional to

$$\left\{ \frac{1}{n(\bar{y}_. - \mu)^2 + (n-1)\,\mathrm{MS}} \right\}^{\frac{1}{2}n},$$

i.e. that conditionally on the data, the unknown mean has the form

$$\bar{y}_. + T_{n-1} \left(\frac{\mathrm{MS}}{n} \right)^{\frac{1}{2}},$$

where T_{n-1} is a random variable having the Student t distribution with $n-1$ degrees of freedom.

An immediate extension of this result gives the distribution of the difference of two means of normal distributions when the variances of the two distributions are treated as separate unknown parameters with independent prior distributions; for alternative discussion of this problem, see Section 5.2 (iv). Assuming, then, that the means and variances μ_1, μ_2 and σ_1^2, σ_2^2 have the improper prior probability element proportional to $d\mu_1 d\mu_2 (d\sigma_1/\sigma_1)(d\sigma_2/\sigma_2)$, and that two independent sets of normally distributed observations are available, it follows that, conditionally on the data, the difference of the two means has the form

$$\bar{y}_{1.} - \bar{y}_{2.} + T_{n_1 - 1} \left(\frac{\mathrm{MS}_1}{n_1} \right)^{\frac{1}{2}} - T_{n_2 - 1} \left(\frac{\mathrm{MS}_2}{n_2} \right)^{\frac{1}{2}}, \qquad (23)$$

where $T_{n_1 - 1}$ and $T_{n_2 - 1}$ are independently distributed in Student t distributions. In this, MS_1 and MS_2 are, of course, to be treated as constants. The expression (23) can conveniently be written

$$\bar{y}_{1.} - \bar{y}_{2.} + \left(\frac{\mathrm{MS}_1}{n_1} + \frac{\mathrm{MS}_2}{n_2} \right)^{\frac{1}{2}} (T_{n_1 - 1} \sin a - T_{n_2 - 1} \cos a),$$

with

$$\tan a = \{(\mathrm{MS}_1/n_1)/(\mathrm{MS}_2/n_2)\}^{\frac{1}{2}}.$$

That is, for numerical purposes, the distribution of a weighted combination of Student t distributions has to be tabulated as a function of the two degrees of freedom and of the angle a (Fisher and Yates, 1963). The distribution is called the Behrens-Fisher distribution; it was obtained by these two authors for this problem using a non-Bayesian argument, the correctness or justification of which has been the subject of much controversy.

10.4 Interpretation of Bayesian probability statements

(i) General remarks

In the previous section we examined some formal manipulations involving prior and posterior distributions. We now discuss the interpretation of such statements.

Bayes's theorem combines a prior density for the parameter with the likelihood to produce a posterior distribution for the parameter. Availability of the likelihood has been assumed in most of the preceding chapters of this book and needs no special discussion at this point.

Essentially three interpretations can be given to prior distributions,

(a) as frequency distributions;

(b) as normative and objective representations of what it is rational to believe about a parameter, usually in a situation of ignorance;

(c) as a subjective measure of what a particular individual, "you", actually believes.

We now discuss these in turn. There is deliberately some overlap with the arguments rehearsed in Chapter 2.

(ii) Prior frequency distributions

Sometimes the parameter value may be generated by a stable physical random mechanism, whose properties are either known or can be inferred by the analysis of suitable data. An instance is where the parameter is a measure of the properties of a batch of material in an industrial inspection problem. Here observations on previous batches allow the estimation of the prior frequency distribution, provided, of course, that the mechanisms underlying the process are sufficiently stable to be a guide to the interpretation of the current set of data. In particular cases of nonstationarity, a special model could be fitted that would still allow the incorporation of the historical data into the conclusion to be reached from the current observations.

Such an application is straightforward except for the estimation of the prior distribution, which will normally involve disentangling from previous data the variation in the parameter, i.e. correcting for the presence of random error. If the amount of previous data is large, then we can regard the prior frequency distribution as known, but in other cases some allowance may be necessary for errors of estimation in the prior distribution; we return to this briefly in the discussion of empirical Bayes procedures in Section 10.7.

Because in this kind of application all probabilities involved have a physical interpretation in terms of frequencies, no special new problems of interpretation arise.

(iii) Rational degrees of belief

If the application of Bayes's theorem in statistics is restricted to situations where there is a prior frequency distribution, the method is important but of essentially limited usefulness. In most situations arising in science and technology the parameters are unique constants, representing in idealized form underlying "true" properties of the system under investigation. Partly because of the mathematical attractiveness of arguments based on Bayes's theorem, there have been many attempts to give Bayesian arguments a much greater range of applicability, by a suitably extended definition and interpretation of probability in terms of degrees of belief.

The most notable account of an approach in which probability represents rational degree of belief is in the book by Jeffreys (1961 ; 1st ed., 1939). For most purposes, the prior distributions used there represent degrees of belief in a state of initial ignorance about the parameter, the posterior distributions therefore representing evidence contributed by the data. To this extent, the objective of the techniques is the same as that of the non-Bayesian methods set out in the earlier part of the book, rather than that of trying to incorporate other forms of information.

A central problem is therefore that of specifying a prior distribution for a parameter about which nothing is known.

If there are only a finite number of possibilities, then it is natural to give them equal prior probability ; not to do so would indicate prior knowledge that one possibility is more likely than another. Even this case, however, is open to the difficulty that if one of the component possibilities is later subdivided into two and the new set of possibilities treated as equally likely, then all the prior probabilities change. This, while disturbing, is not a fatal objection, since one can argue that in the second situation new information has been introduced and the whole problem changed.

When we are dealing with a continuously varying parameter there are similar but much more serious difficulties. Suppose, first, that there is a scalar parameter θ taking values in an interval, often the whole real line. If we wish to express a state of initial ignorance about θ, it is natural to take a uniform prior density, at least if the parameter space is of finite extent. Unfortunately, however, if Θ is uniform, an arbitrary non-linear function $g(\Theta)$ is not and, therefore, some criterion is necessary to indicate which particular function of Θ is to be taken as uniform; in any case, the very fact that some functions of Θ are nonuniformly distributed and hence do not represent "ignorance" may be taken as an indication that the whole idea of a prior distribution uniquely and exactly representing ignorance is untenable.

Two arguments have been put forward to choose an ignorance prior. One is that if the problem has some invariance properties, then the prior should have corresponding structure. This leads to a link with Haar Measures (Brillinger, 1963; Stone and von Randow, 1968). In particular, for a location parameter, all intervals $(a, a + h)$ should have the same prior probability for any given h and all a. This implies that the location parameter itself must have a uniform prior density; the fact that this is an *improper density*, i.e. has infinite integral over the whole line, does not matter, at least as far as the formal theory goes. Similarly, for a scale parameter, we may require that all intervals (a, ka) have the same prior probability for any given $k > 1$ and for all $a > 0$. This leads to the parameter σ, say, having density proportional to $1/\sigma$, so that the probability of (a, ka) is proportional to $\log k$, for all a. Again the density is improper.

These and other similar densities amount to a special choice of indexing constants in the conjugate families discussed in Section 10.3. Thus in the binomial situation of Example 10.2, we are led to the beta posterior distribution of (11) with $a = b = \frac{1}{2}$. In Example 10.3 concerning two Poisson distributions, we take $b_1 = b_2 = \frac{1}{2}$, and m_1 and m_2 arbitrarily large. Thus

$$\frac{y_2 + \frac{1}{2}}{y_1 + \frac{1}{2}}\, P \tag{24}$$

has a posterior F distribution with degrees of freedom $(2y_1 + 1, 2y_2 + 1)$.

Again, in the normal distribution with both parameters unknown, we take an element of prior probability proportional to

$$d\mu\, d\sigma/\sigma. \tag{25}$$

That is, in addition to taking the special prior densities above, we take μ and σ independent. This leads in (20) to the conclusion that

μ has a posterior Student t distribution with $n - 1$ degrees of freedom centred on $\bar{y}_.$, and with a scale constant $(\text{MS}/n)^{\frac{1}{2}}$, this implying posterior interval statements that are formally identical to confidence intervals based on the Student t statistic.

For a general situation involving a vector parameter, the argument outlined above for a general scalar parameter suggests taking the prior density to be inversely proportional to the square root of the information determinant.

As in Example 10.5, the posterior distribution corresponding to an ignorance prior distribution can often be obtained formally as the limit of the posterior distribution for a very diffuse proper distribution. The latter is referred to as a vague prior distribution, the case in Example 10.5 being that of the $N(\xi_0, \nu_0)$ distribution with very large ν_0. Notice, however, that such a vague prior does give information about the parameter of interest. Thus, in the example referred to, we know that *a priori* $\text{pr}(M > \xi_0) = \frac{1}{2}$, whereas no such statement can be deduced from an ignorance prior. In some situations it is useful to distinguish between vague and ignorance priors, particularly for multiparameter problems, where ignorance priors lead to some difficulties.

For general single parameter cases, Jeffreys suggests that we take as uniformly distributed that function of the parameter for which the information function $i(\theta)$ of Section 4.8 is constant. The justification is that, at least locally, the problem is a location parameter one in terms of the transformed parameter. Thus if θ is a binomial proportion, the information function for $\sin^{-1}\sqrt{\theta}$ is constant and if we take this function to have a uniform prior density, then the density for Θ itself is proportional to $\theta^{-\frac{1}{2}}(1 - \theta)^{-\frac{1}{2}}$. Similarly, if μ is the mean of a Poisson distribution, then this argument leads to a prior distribution with density proportional to $\mu^{-\frac{1}{2}}$. Generally, the prior density is proportional to $\{i(\theta)\}^{\frac{1}{2}}$. Incidentally, this implies that if we contemplate two different kinds of random variable both having distributions depending on the same parameter θ, then the prior density of Θ will in general be different in the two cases, which is rather unsatisfactory.

In some cases, the use of prior distributions expressing ignorance in the manner outlined above leads to results agreeing exactly or approximately with non-Bayesian sampling theory work. For example, confidence intervals as described in Chapter 7 sometimes have the property that their posterior probability content is equal to the confidence coefficient. The single-parameter case has been investigated by Lindley (1958). Example 10.5 is a particularly simple illustration. However in other situations, particularly those involving more than two parameters, ignorance priors lead to different and entirely unacceptable answers, as illustrated by Example 2.39.

This example is concerned with random variables $Y_1, \ldots, Y_n$ independently normally distributed with unit variance and with means $\mu_1, \ldots, \mu_n$, the parameter of interest being $\delta^2 = \mu_1^2 + \ldots + \mu_n^2$. If the μ_j are given independent uniform prior distributions, then the posterior distributions are independent $N(y_j, 1)$. Briefly, this implies that δ^2 is $d^2 + n$ with an error of order $\sqrt{n}$, where $d^2 = y_1^2 + \ldots + y_n^2$. This contrasts with the confidence interval solution according to which δ^2 is $d^2 - n$ with an error of order $\sqrt{n}$. That is, the Bayesian interval is calculated by a rule that is very likely to give misleading answers in hypothetical repetitions.

The explanation lies in the prior distribution that has been taken, which in effect asserts that it is highly likely that the μ_j are extremely dispersed. This is a very unreasonable expression of initial ignorance for practical situations. We return to this point in Example 10.8.

The difficulty in this case can be avoided by agreeing to base the inference on the statistic D^2, for example because of a reduction based on invariance under orthogonal transformations. The distribution of D^2 depends only on δ^2. However, this does not destroy the essential point of the example.

An example which is, in some ways, even more striking than that just discussed has been given by Mitchell (1967).

Example 10.6. Exponential regression. Suppose that $Y_1, \ldots, Y_n$ are independently normally distributed with constant variance σ^2 and with

$$E(Y_j) \; = \; \gamma + \beta \rho^{x_0 + ja} \quad (j = 1, \ldots, n). \tag{26}$$

That is, observations on an exponential regression function are made at equally spaced values of x, with spacing a, where $x_0 \geqslant 0$ and $a > 0$. Suppose also that it is assumed known that $0 \leqslant \rho \leqslant 1$, so that the asymptote is approached in the direction of increasing x.

A seemingly plausible prior distribution is to take γ, β, ρ and σ as independently distributed, with γ, β and σ having the improper prior distributions discussed earlier and ρ being uniformly distributed on $[0,1]$.

That is the prior probability element is proportional to $d\gamma d\beta d\rho d\sigma/\sigma$.

For any observations y, it turns out that the marginal posterior density of ρ is proportional to

$$\{\rho^{x_0}(1-\rho^a)\,h(\rho,y)\}^{-1}, \tag{27}$$

where $h(.,.)$ is bounded and has no zeros in $[0,1]$. This posterior is an improper distribution on $[0,1]$ and can only be interpreted as meaning that zero and one are the possible values of ρ, whatever the data. This is entirely unacceptable.

Incidentally, note that if (26) is rewritten

$$E(Y_j) = \gamma + \beta e^{-\kappa(x_0+ja)}, \tag{28}$$

the prior density for κ is exponential with unit mean. This, while a legitimate density, is far from uniform, illustrating again the difficulty of representing ignorance in a prior density.

We now attempt a brief assessment of the approach based on rational degrees of belief, in particular the taking of prior distributions to represent ignorance. The approach is in many ways an attractive one, in that it enables the calculation of probability distributions for unknown parameters that appear to summarize what can be learned from the data. Yet the objections seem fatal. They are

(a) in some situations, especially with many parameters, what appear to be prior distributions representing ignorance in fact contain much information, usually quite misleading (*cf.* Examples 10.5 and 10.6) ;

(b) improper priors, such as that used for the mean of a normal distribution, attach arbitrarily smaller probability to, say, the interval $(-10^{10}, 10^{10})$ than to the complement of this interval. This is, however, defensible if we were in fact totally ignorant of the parameter, a situation which in practice would not arise ;

(c) a flat prior density for a parameter Θ may represent a quite peaked density for $g(\Theta)$, a non-linear function of possible interest, and the criteria for selecting ignorance priors are rather arbitrary (Example 10.6).

In the subjective theory to be dealt with in the next subsection the prior distributions considered above do sometimes appear, always, however, as approximations to what we have termed vague prior distributions.

(iv) Subjective degrees of belief

We now turn to the approach to the use of Bayes's theorem that has received most attention in recent years. This is that a probability represents a subjective degree of belief held at a particular time by a particular person, "you", given certain information. A number of detailed accounts are available and a thorough development will not be attempted here ; see Good (1950), Savage (1954), Raiffa and Schlaifer (1961), Savage *et al* (1962), Cornfield (1969), DeGroot (1970) and Lindley (1971a).

It is assumed that for any uncertain proposition B, "you" have an attitude to its truth measured by a number $\mathrm{pr}(B)$; this is conditional on all available information. It is a crucial assumption that all such probabilities are comparable.

To determine your value of $\mathrm{pr}(B)$, hypothetical or real betting games are considered, involving modest stakes. These can take various forms, but one of the simplest is to suppose that there is available a set of uncertain events $\{C_p\}$, the probability of occurrence of C_p being known and agreed equal to p. In principle, one such event is required for each p, $0 \leqslant p \leqslant 1$. Suppose that "you" have the option of

(a) 1 money unit if C_p occurs,
     0 money units if C_p does not occur,   (29)

or

(b) 1 money unit if B occurs,
     0 money units if B does not occur,   (30)

and that "you" must choose one and only one of these options. If p is adjusted so that "you" are indifferent between options (a) and (b), then your value of $\mathrm{pr}(B)$ is p. If it is known that B is an outcome of a physical random system with known frequency or physical probability p_f, and nothing else relevant is known, then there is clearly an expected loss of money units unless $p = p_f$.

The availability of the set $\{C_p\}$ is useful for an elementary introduction to subjective probability, but is not required for a development from minimal axioms.

The next step is to show that in determining the probabilities of more complex events by the above procedure, either

(I) the ordinary rules of probability theory, including Bayes's
    theorem, are obeyed,

or

(II) strategies are followed that are internally inconsistent, in that

they imply decisions with clearly less than optimal performance in the betting context.

The proof of this will not be given here.

The conclusion is not that betting behaviour as actually observed
satisfies (I), but only that it is reasonable to require that, in a study
of how subjective probabilities ought to be combined, the usual rules
of probability should be followed ; in particular, prior information
should be combined with data in accordance with Bayes's theorem.
All prior distributions determined by applying the hypothetical
betting procedure are proper, so that the difficulties of Example 10.6
cannot arise.

In principle, the argument connecting (29) and (30) applied
repeatedly can be used to determine the prior density of a scalar or
vector parameter. At a practical level, the determination will be very
difficult when there are many parameters, although it may sometimes
be simplified by considerations of symmetry. Mathematically the
application of Bayes's theorem is much simplified if a reasonable
approximation can be found within the relevant conjugate family.
When the prior density is judged to be practically constant over the
range of parameter values for which the likelihood is appreciable, we
may be able to use one of the flat priors of part (iii) as an approximation. In particular, the results of Examples 10.2–10.5 can be used,
the precise choice of prior distribution being unimportant if the
amount of data is large.

One possible way to construct subjective probabilities is by the
device of "imaginary results", which is recommended by Good (1950).
The idea is most easily introduced by the following simple example,
described by Good (1965).

Example 10.7. Parameter of binomial distribution (ctd). Consider
the binomial situation of Example 10.2, where we suppose that the
prior distribution for the binomial proportion Θ has the conjugate
beta form (9). Then (12) gives the posterior probability that the
next observation is a success. Suppose that "success" and "failure"
correspond to the occurrence of head and tail on the tossing of a
coin, which prior to any experimentation seems to be fair. Then we
would take $a = b$ in (9).

Now imagine that the sampling experiment yields 1 tail (failure)
and $n - 1$ heads (successes). Then how large should n be in order
that we would just give odds of 2 to 1 in favour of a head occurring

at the next toss of the coin? For the value of n given in answer to this question, we have the probability identity, deduced from (12), that

$$\frac{n - 1 + a}{n + 2a} = \frac{2}{3} . \tag{31}$$

This gives the appropriate value of a in the beta approximation to the subjective prior.

For example, the choice $a = b = 1$ giving Laplace's law of succession would derive from the answer $n = 4$, which for most coin-tossing problems would seem very small. Thus, in this case, the use of imaginary results might highlight the inadequacy of a fairly conventional prior distribution.

In more complicated problems, the use of imaginary results is correspondingly more difficult. The basic idea is to find a readily-appraisable characteristic of the problem, and then to equate the direct subjective appraisal with the appraisal derived from use of Bayes's theorem with a suitable form of prior. In the multinomial situation of Example 10.4, such a characteristic is the probability of obtaining two successive observations in the same category. In general, it would be advantageous to use more than one set of imaginary results as a check on internal consistency.

In all this discussion it should not be taken for granted that a prior density from the conjugate family will be adequate.

We now use the problem of Example 2.39 again, both to illustrate the above discussion and to bring out some new points.

Example 10.8. Sum of squared normal means (ctd). The same notation is used as in Example 2.39. The form of the subjective prior distribution is, in principle, to be settled by consideration of each special case and there are no general arguments to indicate the form it will take. It is, nevertheless, reasonable to hope that, if all the means are initially on the same footing, one can use a common normal prior distribution with mean ξ_0 and variance ν_0 for all the μ_j's, independently.

Then the posterior distribution of μ_j is normal with mean $(y_j + \xi_0/\nu_0)/(1 + 1/\nu_0)$ and variance $1/(1 + 1/\nu_0)$, as shown in Example 10.5. Therefore, again because of posterior independence of the means, a non-central chi-squared distribution gives the posterior distribution of the random variable $\Delta^2 = \Sigma \mu_j^2$. To obtain results in a form comparable to those of Example 2.39, we work with a normal approximation according to which, for large n,

$$E(\Delta^2 | Y = y) = \frac{d^2 + 2n\bar{y}.\xi_0/\nu_0 + n\xi_0^2/\nu_0^2}{(1 + 1/\nu_0)^2} + \frac{n}{1 + 1/\nu_0} . \tag{32}$$

If $\nu_0 \to \infty$, then we recover the result for the improper prior density considered previously, but it is unlikely that we would use this result knowing that it implies the individual means are highly dispersed.

There now arises a general difficulty. What if the prior information and that contributed by the data are inconsistent ; in this particular case, what if $\bar{y}$. and ξ_0 are appreciably different? If the prior distribution has been determined correctly by the betting procedure outlined above, it cannot be wrong in the sense that it does represent "your" prior attitude. On the other hand, this attitude may be based on a misconception ; another possible explanation of discrepancy is inadequacy of the probability model, or bias in the data. If there is strong inconsistency between the data and prior knowledge, then the combination of information *via* Bayes's theorem is irrelevant and misleading. As with other problems involving the combination of information, the homogeneity of such information should be considered.

One possible cause for difficulty in this problem is that we may have over-specified our prior knowledge, for example by fixing the values of ξ_0 and ν_0 when we are unsure as to the values that these should take. It is worth considering this in some detail.

One approach to the choice of prior, given that the μ_j are assumed independent $N(\xi_0, \nu_0)$, is to use preliminary subjective ideas about characteristics of the prior distribution. For example, we might consider upper and lower limits for any one of the means. Thus, if we can obtain an interval for which one would allow a subjective probability of, say, 0.90 that this interval includes μ_1, then we can equate the limits of the interval to $\xi_0 \pm 1.64\nu_0^{\frac{1}{2}}$. A similar approach would be to consider imaginary results, as in Example 10.7. In the present case it might be possible, given imaginary values of sample mean and variance from $N(\mu_1, 1)$, to determine the sample size necessary to make the subjective probabilities $\mathrm{pr}(Y > a_1)$ and $\mathrm{pr}(Y > a_2)$ equal to p_1 and p_2. This would give two equations in ξ_0 and ν_0, whose solutions determine the specific form of the subjective prior distribution.

With both of these approaches, it may not be possible in practice to give specific subjective values to the appropriate quantities. For example, we may arrive at the statement that the lower 0.05 prior probability limit for μ_j is somewhere between b_1 and b_2. This would imply that we have imprecise knowledge about ξ_0 and ν_0, and a natural thing to do is to assign a prior density to these quantities. We examine this possibility in the next example.

One formal possibility when ξ_0 and ν_0 are unknown might be to estimate them from the data according to some sensible rule. This, however, is contrary to the spirit of the subjective Bayesian approach, and falls in the category of empirical Bayes methods discussed in Section 10.7.

In discussions of the prior distribution so far, we have supposed that the subjective prior can be arrived at in a single step, so to speak. Thus, in Example 10.7 we envisaged a unique answer emerging from the imaginary results argument, corresponding to a single choice of n. It could happen, however, that only upper and lower limits for n can be obtained. One general way to handle this apparent difficulty is to assess one's subjective prior knowledge in two or more stages. Thus, in the case of Example 10.7, we might decide first that on the basis of imaginary results the prior distribution for the binomial proportion has the symmetric conjugate beta form, with index a somewhere between a_1 and a_2. Then, in the second stage, we might represent our ignorance about a by saying that it is represented by a random variable A with some flat prior distribution concentrated between the limits a_1 and a_2.

The same situation sometimes arises in multiparameter problems, such as those of Examples 10.4 and 10.8. The following example illustrates the use of two-stage prior distributions in the situation of Example 10.8.

Example 10.9. Several normal means. Suppose that $Y_1, \ldots, Y_n$ are independent, with Y_j having the $N(\mu_j, 1)$ density. As before, we suppose that the prior distributions of $\mu_1, \ldots, \mu_n$ are independent normal, with unknown common mean ξ and variance ν. This represents the first stage of our subjective prior information, and in the second stage we bring in the information about ξ and ν.

To be completely general, we should treat both ξ and ν as random variables, possibly with the conjugate normal and gamma distributions as in Example 10.5. However, let us assume that ν is known, equal to ν_0. If ξ does have a conjugate normal prior distribution with mean ξ_0 and variance χ_0, then we can combine all the prior information about $\mu_1, \ldots, \mu_n$ and easily deduce that the vector μ has a joint normal prior with mean $\xi_0 \mathbf{1}$ and covariance matrix

$$\nu_0 \mathbf{I} + \chi_0 \mathbf{J} = \begin{bmatrix} \nu_0 + \chi_0 & \chi_0 & \cdots & \chi_0 \\ \chi_0 & \nu_0 + \chi_0 & \cdots & \chi_0 \\ \vdots & \vdots & & \vdots \\ \chi_0 & \chi_0 & \cdots & \nu_0 + \chi_0 \end{bmatrix}. \quad (33)$$

Here $\mathbf{1}^T = (1, \ldots, 1)$ and $\mathbf{J} = \mathbf{11}^T$. Notice that in this prior distribution the means are correlated.

We are now in a position to apply Bayes's theorem, which takes the form

$$f_{M\mid Y}(\mu\mid y ; \xi_0, \nu_0, \chi_0) \propto f_{Y\mid M}(y\mid \mu)\, f_M(\mu ; \xi_0, \nu_0, \chi_0). \quad (34)$$

Here the random variable corresponding to μ is denoted by M. Substituting the two joint normal densities on the right-hand side, and using the identity

$$(\mathbf{I} + a\mathbf{J})^{-1} = \mathbf{I} - \frac{a}{1 + na}\, \mathbf{J},$$

we find that

$$f_{M\mid Y}(\mu\mid y ; \xi_0, \nu_0, \chi_0) \propto \exp\{-\tfrac{1}{2}(\mu - \mathbf{b}z)^T\, \mathbf{b}^{-1}\, (\mu - \mathbf{b}z)\},$$

where

$$\mathbf{b}^{-1} = (1 + 1/\nu_0)\mathbf{I} - \{\chi_0/(\nu_0^2 + n\chi_0\nu_0)\}\mathbf{J},$$

$$z = y + \xi_0(\nu_0\mathbf{I} + \chi_0\mathbf{J})^{-1}\mathbf{1}; \quad (35)$$

that is, the means have a joint normal distribution with mean vector $\mathbf{b}z$ and covariance matrix $\mathbf{b}$, which is not diagonal for finite values of ν_0.

A special case of particular interest is $\chi_0 \to \infty$, corresponding to vague prior knowledge of the common mean ξ. Here we obtain from (35) the result that $\mu_1, \ldots, \mu_n$ have a joint normal posterior distribution with

$$E(M_j|y) = \frac{y_j + \bar{y}./\nu_0}{1 + 1/\nu_0} \tag{36}$$

and covariance matrix

$$\frac{\nu_0}{\nu_0 + 1} \; \mathbf{I} - \frac{1}{n\nu_0} \mathbf{J} \; . \tag{37}$$

Comparing (36) with equation (22) of Example 10.5, we see that $\bar{y}.$ is substituted for the prior mean ξ; the individual posterior means are shrunk from y_j, the m.l.e., towards their average $\bar{y}.$.

If a non-degenerate prior distribution for ν is required, the resulting calculation of the posterior distribution of μ is very complicated, as one might anticipate from (33). A detailed discussion is given by Lindley and Smith (1972), who treat the much more general case of an arbitrary linear model. In fact, their advice is to use estimates for variance parameters, which is not a strictly Bayesian procedure; we shall say something more about this problem in Section 10.7.

In principle, we could apply the results (36) and (37) to the problem of Example 10.8, namely the calculation of a posterior distribution for $\delta^2 = \mu_1^2 + \ldots + \mu_n^2$. This is no longer straightforward because the means are correlated, but for reasonably large n it is not difficult to derive the corresponding normal approximation. We omit the details.

We now give some general comments on the approach to statistical inference based on subjective prior distributions.

(a) The determination of prior distributions, especially when there are many parameters, is likely to be difficult. Considerations of symmetry can be useful, as can the use of imaginary results. Distributions representing vague knowledge may often be relevant, but care is necessary in their choice. In multiparameter problems it may be useful to consider the prior information in stages.

(b) In some situations, for example in some business decision problems, there may be little or no relevant "hard" data, and assessment of information from other sources is important. The determination of prior distributions by considering hypothetical bets may then be a very useful device for clarifying an individual's opinion, and it may lead to greater understanding among individuals. If a generally agreed prior distribution can be obtained, and the rational basis for it understood, it may be possible to use it in subsequent

probability calculations. However, uncritical acceptance of a subjective distribution, whilst possibly in order in private determination
of consistent behaviour, does not seem reasonable in rational discussion.

(c) In the statistical analysis of scientific and technological data,
there will nearly always be some external information, such as that
based on previous similar investigations, fully or half-formed
theoretical predictions, and arguments by vague analogy. Such
information is important in reaching conclusions about the data at
hand, including consistency of the data with external information,
and in planning future investigations. Nevertheless, it is not at all
clear that this is usually best done quantitatively, expressing vague
prior knowledge by a probability distribution, rather than qualitatively.
Further, the initial assumption of subjective probability is that all
uncertainties are comparable, so that a probability based on some
entirely unspecified feeling is to be treated in the same way as one
based on careful investigation and measurement under controlled
conditions. For narrow purposes of betting behaviour this may be
so, but it does not seem reasonable for careful discussion of the
interpretation of data.

(d) It is important not to be misled by somewhat emotive words
like subjective and objective. There are appreciable personal elements
entering into all phases of scientific investigations. So far as statistical
analysis is concerned, formulation of a model and of a question for
analysis are two crucial elements in which judgement, personal
experience, etc. play an important role. Yet they are open to rational
discussion : for example, if there is a disagreement about the model
appropriate for the data, this can, to some extent, be investigated
by further analysis of the data and any previous relevant data, as
well as by finding whether the disagreement has any appreciable
effect on the inference drawn from the data. Given the model and a
question about it, it is, however, reasonable to expect an objective
assessment of the contribution of the data, and to a large extent this
is provided by the methods of the earlier chapters.

It is sometimes argued that the deduction of the properties of
subjective probability, and in particular of their combination *via*
Bayes's theorem, is so compelling that any methods inconsistent
with these properties should be discarded. In particular, the strong
likelihood principle is obeyed by subjective probabilities. Hence, it is
argued, all methods that do not obey the strong likelihood principle,

and that includes most sampling theory methods, should be rejected as equivalent to internally inconsistent behaviour. A particular example, mentioned a number of times already, concerns independent binary trials, where the posterior density for the binomial proportion puts the number of successes and number of trials on an equal logical footing, irrespective of which is random and which is fixed.

The arguments for not accepting a solely Bayesian viewpoint have already been implicitly outlined. The assumption that all aspects of uncertainty are directly comparable is often unacceptable. More importantly, the primary objective of statistical inference is usually in some sense that of ending up near the "true" model, rather than of being internally consistent. The notion of comparing long-run performance with the "true" values of unknown parameters serves to link the analysis of data with the underlying "true" situation. Despite the rather arbitrary nature of some of the criteria based on this concept, they seem to us in the last analysis more compelling than those based purely on internal consistency.

(v) Summing up

Of the three ways of interpreting prior distributions, the first, as a frequency distribution is uncontroversial but of limited applicability ; it is, however, very important to adopt a Bayesian analysis when a frequency prior distribution is available. The second interpretation, as an objective distribution representing ignorance, seems to us attractive in principle but untenable in the forms proposed so far. The third approach, *via* subjective probabilities based on betting behaviour, is of quite wide potential applicability. For the initial stages of the careful interpretation of data, the approach is, in our opinion, inapplicable because it treats information derived from data as on exactly equal footing with probabilities derived from vague and unspecified sources. For the later stages of an analysis, and, more importantly, in situations where the amount of satisfactory data is small, the approach is likely to be sometimes valuable. It does, however, seem essential that the basis of any prior distribution used needs explicit formulation in each application and the mutual consistency of different kinds of information that are merged needs examination.

At a more theoretical level it is always instructive to compare solutions of related problems obtained by different approaches or under somewhat different assumptions. From this point of view, Bayesian theory is both interesting and valuable.

10.5   Bayesian versions of some previous types of procedure

We now discuss briefly the Bayesian analogues of the main types of procedure developed earlier in the book.

(i) Interval estimation

The simplest problem, and the one about which little need be said, is interval estimation. The posterior distribution of a parameter is a complete account of what can be learned from the data and the prior information, assuming the correctness of the specification. The presence of nuisance parameters causes no difficulty, in principle, because the posterior distribution of those of interest can always be formed by integration, once the full posterior distribution of all parameters has been found. The posterior distribution may be summarized for convenient presentation in one of several ways :

(a) If the distribution is exactly or approximately of a standard functional form, then a specification of the numerical values of the parameters will be a very convenient way of describing the posterior distribution. In particular, for posterior distributions that are nearly normal the mean and standard deviation can be used. A transformation of the parameter, if sufficiently simple, may be useful. When the parameters are multidimensional, the mean and covariance matrix can be used if the distribution is approximately normal.

(b) If the parameter is one-dimensional and the posterior distribution, while not of any simple mathematical form, is unimodal, it will often be convenient to give a few upper and lower percentage points. These serve the same purpose as upper and lower confidence limits. In some applications it may be, say, only the upper limits that are of any practical interest.

(c) If the parameter is multidimensional, or if the density is multimodal, then the simplest description of the posterior distribution may be *via* a series of regions of high posterior density. That is, if $p(\theta)$ is the posterior density, we give the α level region as

$$\mathcal{R}_\alpha = \{\theta ; p(\theta) \geqslant c_\alpha\}, \tag{38}$$

where c_α is chosen so that

$$\int_{\mathcal{R}_\alpha} p(\theta)d\theta = 1 - \alpha. \tag{39}$$

We might, for example, take $\alpha = 0.10, 0.05, 0.01$, etc.

All these are however, just devices for the concise description of probability distributions.

Very similar remarks apply when the object is the prediction of future observations rather than parameter values.

Example 10.10. Future observation from normal distribution.
Suppose that $Y_1, \ldots, Y_n$ are i.i.d. in the $N(\mu, \sigma_0^2)$ distribution, the mean being the only unknown parameter. Suppose also that the prior distribution of μ is normal with mean ξ_0 and variance ν_0. Let $Y^\dagger$ be a single future observation from the same distribution which is, given μ, independent of $Y_1, \ldots, Y_n$. Thus we can write

$$f_{Y^\dagger|M,Y}(y^\dagger|\mu, y) = f_{Y^\dagger|M}(y^\dagger|\mu) \propto \exp\left\{-\frac{(y^\dagger - \mu)^2}{2\sigma_0^2}\right\} ; \quad (40)$$

also, by the previous work in Example 10.5, the conditional density of μ given $Y = y$, i.e. the posterior density, is proportional to

$$\exp\left[-\frac{\{\mu - \mu_W(\bar{y}_., \sigma_0^2/n ; \xi_0, \nu_0)\}^2}{2\sigma_W^2(\sigma_0^2/n, \nu_0)}\right], \quad (41)$$

in the notation of (22). The posterior density of $Y^\dagger$ is

$$f_{Y^\dagger|Y}(y^\dagger|y) = \int_{-\infty}^{\infty} f_{Y^\dagger|M,Y}(y^\dagger|\mu, y) f_{M|Y}(\mu|y) \, d\mu ,$$

and it follows from (40) and (41) that the future observation $Y^\dagger$ is normally distributed with mean and variance

$$\mu_W(\bar{y}_., \sigma_0^2/n ; \xi_0, \nu_0) \quad \text{and} \quad \sigma_0^2 + \sigma_W^2(\sigma_0^2/n, \nu_0) , \quad (42)$$

respectively. Essentially, we have expressed $Y^\dagger$ as the sum of two independent normal random variables M and $Y^\dagger - M$ whose sum is normally distributed with the usual combination of means and variances.

A similar argument would apply if we were interested in the mean or other quantity calculated from a set of future observations. Again the information is summarized in a probability distribution and previous remarks about summarizing distributions apply.

(ii) Point estimation
The above remarks deal with the Bayesian analogue of confidence

intervals and tolerance intervals for future observations. The discussion of point estimation will largely be deferred to Chapter 11,
where a loss function is introduced. From several points of view,
however, if it is required to summarize the posterior distribution in
a single quantity, then the mean is often the most sensible. In
particular, if the prior density is exactly or approximately constant,
then the use of the mean of the likelihood with respect to the parameter is indicated ; this is identical with the maximum likelihood
estimate only in special cases. This is discussed more fully in Section
11.7.

In certain circumstances the difficulty of calculation of the posterior
mean naturally leads one to consider the mode as a point estimate,
which corresponds more closely to the maximum likelihood estimate.
Some applications of modal estimation arise in the discussion of
empirical Bayes methods in Section 10.7.

(iii) Significance test analogues

Finally, we discuss the analogues of significance tests in the Bayesian
approach. First, the pure significance tests of Chapter 3 have no
analogue, in that a situation in which only the null hypothesis is
explicitly formulated, although other possible models are in the
background, cannot be handled within the Bayesian framework. This
is a weakness of the Bayesian approach, considered in isolation.

When alternative hypotheses are entered explicitly into the
formulation, the null hypothesis can arise in at least two ways; see
Section 4.1. The first is when a null hypothesis value, say θ_0,
separates a continuous interval of possible values of a scalar parameter θ into two qualitatively different regions. Here, from the
Bayesian point of view, the natural question to consider from the
posterior density is "what is the probability, given the data, that say
$\theta > \theta_0$?" In cases where the prior distribution is very flat, it
commonly happens that the above posterior probability is equal to,
or nearly equal to, the one-sided significance level considered in the
corresponding sampling-theory procedures. This being the case, it is
sometimes useful to consider what prior distribution leads to the
same result as the sampling-theory significance test. If there is a sharp
discrepancy between this and the prior distribution determined
subjectively, then we may feel inclined against the usual significance
test. There is, however a difference of emphasis as well as of interpretation between the two approaches. For, in the development of

Chapter 4, the alternative values have a different logical status from the null value, in that the former should serve only to mark out the types of departure from the null hypothesis which it is desired to detect. These ideas are illustrated by the following example due to Altham (1969).

Example 10.11. Two-by-two contingency table. Consider a two-dimensional contingency table, such as we have discussed before in Example 5.3, but now with two categories in each dimension. Then let θ_{jk} be the probability that an individual belongs to the (j,k)th cell, i.e. in the jth row and kth column, for $j = 1,2$ and $k = 1,2$. The sufficient statistics for a sample of size n are the cell frequencies N_{jk}, with likelihood

$$\prod_{j=1}^{2} \prod_{k=1}^{2} \theta_{jk}^{n_{jk}}. \tag{43}$$

This is a special multinomial situation, and the conjugate prior distribution deduced from (18) in Example 10.4 is

$$f_{\Theta}(\theta) \propto \prod_{j=1}^{2} \prod_{k=1}^{2} \theta_{jk}^{a_{jk}-1}. \tag{44}$$

Now suppose that we are interested in the interaction, or association, between rows and columns, which is the cross-ratio

$$\psi = \frac{\theta_{11}\theta_{22}}{\theta_{12}\theta_{21}}.$$

This is equal to one if and only if $\theta_{jk} = \theta_{j.}\theta_{.k}$ for all j and k, where $\theta_{j.} = \theta_{j1} + \theta_{j2}$, etc., as in our standard notation. To distinguish between positive and negative association, we calculate the posterior probability that Ψ is less than one, say. Here the usual null hypothesis $H_0\colon \psi = 1$ of no interaction separates the two types of association. In view of the above remarks, it is easy to see that

$$\mathrm{pr}(\Psi < 1 \mid n_{11}, \dots, n_{22}) = \mathrm{pr}\left(\frac{\Theta_{11}}{\Theta_{1.}} < \frac{\Theta_{21}}{\Theta_{2.}} \,\middle|\, n_{11}, \dots, n_{22}\right). \tag{45}$$

The posterior density for all the parameters is

$$\frac{(n_{..} + \Sigma\Sigma a_{jk})! \, \Pi\Pi\theta^{n_{jk}+a_{jk}-1}}{\Pi\Pi(n_{jk} + a_{jk})!}, \tag{46}$$

which follows directly from (43) and (44).

To calculate (45), we notice first from (46) that the two variables $\Theta_{11}/\Theta_{1.}$ and $\Theta_{21}/\Theta_{2.}$ are independent with posterior densities proportional to

$$z^{n_{11}+a_{11}-1}(1-z)^{n_{12}+a_{12}-1} \quad \text{and} \quad z^{n_{21}+a_{21}-1}(1-z)^{n_{22}+a_{22}-1} \; ,$$

respectively. Then it is not difficult to show that

$$\mathrm{pr}(\Psi<1|n_{11},\ldots,n_{22}) = \sum_{k=k_{\min}}^{n_{12}+a_{12}-1} \binom{n_{.1}+a_{.1}-1}{k}\binom{n_{.2}+a_{.2}-1}{n_{2.}+a_{2.}-k}\binom{n_{..}+a_{..}-2}{n_{1.}+a_{1.}-1}$$

(47)

where $k_{\min} = \max(n_{21}+a_{21}-n_{12}-a_{12}, 0)$. It is interesting to compare this result with the usual exact one-sided significance test for no interaction, $\psi = 1$, with alternative that $\psi < 1$. The uniformly most powerful similar test, known as Fisher's exact test, is described in Example 5.3. For cell frequencies n'_{jk}, the exact significance probability is given by (47) with the identities

$$n'_{jj} = n_{jj} + a_{jj} \qquad (j=1,2) \, ,$$

$$n'_{jk} = n_{jk} + a_{jk} - 1 \qquad (1 \leqslant j \neq k \leqslant 2) \, .$$

Consequently, the posterior probability $\mathrm{pr}(\Psi < 1 \mid n_{11}, \ldots, n_{22})$ is equal to the significance probability derived from $n_{11}, \ldots, n_{22}$ if the prior density is the special case

$$f_{\Theta}(\theta) \propto \theta_{12}^{-1}\theta_{21}^{-1}.$$

This is an improper prior, but its form nevertheless seems to suggest that the significance test is equivalent to using a prior assumption of association. However, as we pointed out above, there is a difference in interpretations between the Bayesian and sampling theory results.

A second kind of null hypothesis is the one considered because it is thought quite likely to be true or nearly true. For example, this might be appropriate in a comparison of control and treatment groups of individuals. This needs a different analysis from the Bayesian viewpoint. In the simplest case where there are only two simple hypotheses H_0 and H_A we have immediately from Bayes's theorem that, in an obvious notation,

$$\frac{\text{pr}(H_0 \mid Y = y)}{\text{pr}(H_A \mid Y = y)} = \frac{\text{pr}(H_0)}{\text{pr}(H_A)} \times \frac{f_0(y)}{f_A(y)}. \tag{48}$$

That is, the posterior odds for H_0 versus H_A are the prior odds multiplied by the likelihood ratio. Note that, although the likelihood ratio again enters in an essential way, the solution here is simpler than that associated with the Neyman–Pearson lemma in that we do not have to examine the distribution of the likelihood ratio considered as a random variable.

With several possible distinct alternatives $H_{A_1}, \ldots, H_{A_k}$ we can proceed similarly and in particular calculate the posterior odds for H_0 being true versus its being false as

$$\frac{\text{pr}(H_0 \mid Y = y)}{\text{pr}(\bar{H}_0 \mid Y = y)} = \frac{\text{pr}(H_0) f_0(y)}{\sum\limits_{j=1}^{k} \text{pr}(H_{A_j}) f_{A_j}(y)}, \tag{49}$$

where $\bar{H}_0$ represents the event that H_0 is false, here the union of the H_{A_j}.

One case likely to arise in applications is where there is a scalar parameter θ and the null hypothesis asserts that $\theta = \theta_0$. Often this is called a *sharp hypothesis* because it specifies a single parameter value. Then (49) generalizes to

$$\frac{\text{pr}(H_0 \mid Y = y)}{\text{pr}(\bar{H}_0 \mid Y = y)} = \frac{p_0 f_{Y \mid \Theta}(y \mid \theta_0)}{(1 - p_0) \int\limits_{\theta \neq \theta_0} p_A(\theta) f_{Y \mid \Theta}(y \mid \theta) \, d\theta}, \tag{50}$$

where $p_A(\theta)$ is the conditional prior p.d.f. of Θ given that $\Theta \neq \theta_0$ and p_0 is the prior probability of H_0 being true. That is, the prior distribution of Θ consists of an atom of probability p_0 at $\theta = \theta_0$ and a density $(1 - p_0) p_A(\theta)$ over the remaining values. Note that the prior density is assumed proper. An instructive comparison of this procedure with an ordinary sampling-theory significance test is obtained by examining a special case in more detail (Lindley, 1971a).

Example 10.12. Bayesian significance test for a normal mean.
Suppose that $Y_1, \ldots, Y_n$ are i.i.d. in $N(\mu, \sigma_0^2)$ with known variance σ_0^2. Let the null hypothesis be that $\mu = \mu_0$ and have prior probability p_0, and, under the alternative, let μ have a normal density with mean μ_0 and variance v_A. Of course, it would be possible to gain

some extra generality by having a different mean under the altern-
tive. Under our special assumption we may, however, without loss of
generality take $\mu_0 = 0$.

Then we have for the posterior odds of H_0 versus $\bar{H}_0$, from (50),

$$
\frac{\text{pr}(M=0|Y=y)}{\text{pr}(M\neq 0|Y=y)} = \frac{p_0}{1-p_0} \times \frac{\exp\left(-\dfrac{\Sigma y_j^2}{2\sigma_0^2}\right)}{\dfrac{1}{\sqrt{(2\pi v_A)}} \displaystyle\int_{-\infty}^{\infty} \exp\left\{-\dfrac{\Sigma(y_j-\mu)^2}{2\sigma_0^2}\right\} \exp\left(-\dfrac{\mu^2}{2v_A}\right) d\mu}
$$

$$
= \frac{p_0}{1-p_0} \times \left(1 + \frac{n v_A}{\sigma^2}\right)^{\frac{1}{2}} \exp\left\{-\frac{n\bar{y}_.^2}{2\sigma_0^2(1 + n^{-1}\sigma_0^2/v_A)}\right\}. \tag{51}
$$

Thus, if $\bar{y}_. = 0$, then the posterior odds in favour of H_0 increase with
n, the dependence on sample size being approximately proportional
to $\sqrt{n}$. Further, suppose that $|\bar{y}_.| = k^*_{\frac{1}{2}\alpha} \sigma_0/\sqrt{n}$, so that in an
ordinary two-sided significance test the deviation from H_0 has
significance level α. Then the right-hand side of (51) is, for large n
and fixed σ_0^2 and v_A, approximately equal to

$$
\left(\frac{p_0}{1-p_0}\right)\left(\frac{n v_A}{\sigma_0^2}\right)^{\frac{1}{2}} \exp\left(-\frac{1}{2} k^{*2}_{\frac{1}{2}\alpha}\right). \tag{52}
$$

That is, for sufficiently large n, a sample just significant at, say, the
1% level in an ordinary significance test contributes evidence in
favour of H_0. The dependence of posterior odds on n for smaller
values of n is shown, from (51), in Table 10.2 for the special case
$p_0 = \frac{1}{2}, v_A = \sigma_0^2$.

Comparison of (51) and (52) with the result of an ordinary significance test provides a very good illustration of the difference between
the present Bayesian approach and the significance tests of Chapters
3 and 4. First, if appropriate numerical values of p_0 and v_A can be
found for a particular application, clearly (51) gives a much stronger
conclusion than the significance tests of Chapter 3 ; see, especially,
the comment (vii) of Section 3.4 on the limitations of pure significance tests. The two approaches can, however, be formally reconciled
in two ways.

TABLE 10.2

Relation between sampling theory and Bayesian
significance tests

Sample size n	Posterior odds for H_0 versus $\bar{H}_0$ at significance level	
	0.1	0.01
1	0.365	0.0513
10	0.283	0.00797
10^2	0.690	0.0141
10^4	6.70	0.132
10^6	66.8	1.31
10^8	668	13.1

One way is as follows. Consider a fixed value of n. If $|\bar{y}.| \gg \sigma_0/\sqrt{n}$, then the answer is clear cut; if $|\bar{y}.| \ll \sigma_0/\sqrt{n}$, then there can be no effective information in the data against H_0. The interesting case, where the significance test is required, is when $|\bar{y}.|$ and $\sigma_0/\sqrt{n}$ are comparable. Hence the effective range of alternatives is a range of μ values of order $\sigma_0/\sqrt{n}$ on each side of $\mu = 0$. We can represent this conventionally by taking $\nu_A = c\sigma_0^2/n$, where c is a universal constant, a reasonable value for which might to 10. If $|\bar{y}.| = k^*_{\frac{1}{2}\alpha} \sigma_0/\sqrt{n}$, then we find that the posterior odds (51) in favour of H_0 is a universal function of the significance level, namely

$$\left(\frac{p_0}{1 - p_0}\right) (1 + c)^{\frac{1}{2}} \exp\left\{-\frac{k^*_{\frac{1}{2}\alpha}}{2(1 + 1/c)}\right\}.$$

Thus, in a formal way, use of a pure significance test is equivalent to the use of the above conventional prior density. Of course, it is of the essence of the Bayesian argument that one should use not conventional prior distributions but ones chosen to represent the particular circumstances of the application; in particular, such prior distributions would be independent of n.

Another way to reconcile the two approaches is to say that the interpretation of the significance level in the pure significance test should depend on n. Notice from (52) that the posterior odds in favour of H_0 will be approximately constant if $k^{*2}_{\frac{1}{2}\alpha}$ is proportional to $\log n$. Hence it can easily be shown that if

$$\alpha = \text{const } n^{-\frac{1}{2}} \log n, \tag{53}$$

then the posterior odds are approximately independent of n. That is, the two approaches are compatible if a fixed interpretation of the significance test vis-à-vis the truth of H_0 demands that the significance probability be smaller the larger n is, as in (53). Of course, in the significance test approach, we do not measure numerically the degree of truth of a hypothesis.

From the discussion in Section 9.3, we know that, for very large n, the significance test will reject H_0 with high probability unless H_0 is almost exactly true. It may well be that differences from H_0 can arise that are so small as to be of no practical importance. This changes the character of the problem and requires a different solution in both the sampling theory and Bayesian approaches. For example, in the Bayesian approach, we might have a non-degenerate distribution for μ under the null hypothesis, such as a rectangular distribution or even, as an approximation, $N(0,\nu_0)$ with $\nu_0 \ll \nu_A$.

An extension of the central formula (50) is possible when nuisance parameters are present. Thus if $\theta = (\psi,\lambda)$ and the null hypothesis is that $\psi = \psi_0$, the natural extension of (50) is

$$\frac{\mathrm{pr}(H_0|Y=y)}{\mathrm{pr}(\bar{H}_0|Y=y)} = \frac{p_0}{1-p_0} \times \frac{\int p_0(\lambda)f_{Y|\Psi,\Lambda}(y|\psi_0,\lambda)d\lambda}{\underset{\psi \neq \psi_0}{\iint} p_A(\psi,\lambda)f_{Y|\Psi,\Lambda}(y|\psi,\lambda)d\psi d\lambda}, \quad (54)$$

in an obvious notation. Notice that we have allowed the conditional prior distribution of Λ to be different in H_0 and in $\bar{H}_0$. This degree of generality is not usually required, and, if we can assume the conditional prior density of Λ to be continuous as $\psi \to \psi_0$, then we obtain a much simpler result for the posterior odds (Dickey and Lientz, 1970). The continuity assumption amounts to saying that the prior density of Λ when H_0 is true is

$$p_0(\lambda) = \frac{p_A(\psi_0,\lambda)}{\int p_A(\psi_0,\lambda')d\lambda'}, \quad (55)$$

where $p_A(\psi,\lambda)$ is continuous everywhere. Now substitution into the general result (54) gives

$$\frac{\mathrm{pr}(H_0|Y=y)}{\mathrm{pr}(\bar{H}_0|Y=y)} = \frac{p_0}{1-p_0} \times \frac{\int p_A(\psi_0,\lambda)f_{Y|\Psi,\Lambda}(y|\psi_0,\lambda)d\lambda/\int p_A(\psi_0,\lambda)d\lambda}{\iint p_A(\psi,\lambda)f_{Y|\Psi,\Lambda}(y|\psi,\lambda)d\psi d\lambda}$$

$$= \frac{p_0}{1-p_0} \times \frac{f_{\Psi|Y,\bar{H}_0}(\psi_0|y,\bar{H}_0)}{f_{\Psi|\bar{H}_0}(\psi_0|\bar{H}_0)}, \quad (56)$$

on applying Bayes's theorem. Of course, the density of Ψ at ψ_0 under $\bar{H}_0$ is determined as a limit. Thus the effective likelihood ratio, the second factor on the right-hand side, is the ratio of posterior and prior densities of Ψ at ψ_0.

This result can be applied quite readily in several of the earlier examples, such as Examples 10.3 and 10.5. A relatively simple problem arising from Example 10.4 would be to test the hypothesis of equal multinomial probabilities.

10.6   Asymptotic Bayesian theory

Examples have already arisen of the general result that with a large amount of data the posterior distribution depends primarily on the likelihood. This is most easily seen when the observations have a density within the exponential family and the conjugate prior is taken. It then follows from (15) that the quantities that index the posterior density are, as the number of observations n increases, dominated by the contributions from the data, which are proportional to n in probability. Moreover, the exponential family density itself is asymptotically normal, because it can be regarded as a convolution to which the central limit theorem applies. Hence the posterior density is, with probability one, asymptotically normal with mean and variance depending on the likelihood and not on the prior.

In fact a much more general result, which we now sketch, holds. Remember, however, that if a set of possibilities is given zero prior probability then that set has zero posterior probability whatever the data.

Suppose for simplicity that $Y_1, \dots, Y_n$ are i.i.d. each with the density $f(y|\theta)$, where the vector parameter Θ has prior density $f_\Theta(\theta)$. We shall assume $f(y|\theta)$ to be regular, as in Chapter 9, and suppose that the prior density is continuous at the "true" parameter value. Then the posterior density of Θ is proportional to

$$\exp\left\{ \sum_{j=1}^{n} \log f(y_j|\theta) + \log f_\Theta(\theta) \right\},$$

and if $l(\theta; y)$ denotes the log likelihood function as previously, then we can expand this posterior density about the maximum likelihood point and take it as proportional to

$$\exp\{l(\hat{\theta};y)\}\exp\{-\tfrac{1}{2}(\theta-\hat{\theta})^{\mathrm{T}}\left(-\frac{\partial^2 l(\theta;y)}{\partial\theta^2}\right)_{\theta=\hat{\theta}}(\theta-\hat{\theta})+\log f_{\Theta}(\theta)\}+o(\|\theta-\hat{\theta}\|^2);$$

$$(57)$$

cf. the expansion in (9.30). The matrix of observed second derivatives is the sum of n independent terms and is of order n. Therefore, provided that $\log f_{\Theta}(\theta)$ is $O(1)$ and is smooth in the neighbourhood of $\theta = \hat{\theta}$, its contribution is, to first order, constant over the range of interest, i.e. for $\theta = \hat{\theta} + o(1)$, and can be absorbed into the normalizing constant. That is, the posterior density is asymptotically normal with mean $\hat{\theta}$ and covariance matrix

$$\left[-\frac{\partial^2 l(\theta;y)}{\partial\theta^2}\right]_{\theta=\hat{\theta}}. \tag{58}$$

This is very closely connected with the corresponding result for maximum likelihood estimates derived in Section 9.2. Note, however, the occurrence in (58) of the matrix of observed second derivatives rather than the information matrix $i.(\theta)$ of expected second derivatives.

The argument leading to (58) requires only that the likelihood should become increasingly concentrated around its maximum, and hence is much more widely applicable than just to i.i.d. random variables. Walker (1969) and Dawid (1970) have given careful discussion of the regularity conditions under which, with probability one, the posterior distribution is asymptotically normal. These are very similar to the regularity conditions required for asymptotic normality of the m.l.e. By taking further terms in the expansion (57), a series representation for the posterior density can be obtained (Lindley, 1961 ; Johnson, 1970).

10.7   Empirical Bayes procedures

We now return to the situation where the prior distribution is a frequency distribution, i.e. where the parameter values are generated by a stable physical random mechanism. Industrial acceptance sampling and process control are examples of the type of practical situation to which the results may be applied.

One fairly general formulation is as follows. Suppose that $Y_1, \ldots, Y_n$ are independently distributed, Y_j having the p.d.f.

$f_{Y_j}(y;\psi_j,\lambda)$, where λ is a constant nuisance parameter. Let the Ψ_j's be independent and identically distributed with the prior density $f_\Psi(\psi;\zeta)$, where ζ is a further unknown parameter. In a fully Bayesian treatment $\psi_1,\ldots,\psi_n,\lambda$, and possibly also ζ, would derive from a joint prior density. We shall assume here that λ and ζ are not generated randomly, and so are fixed unknowns.

Suppose that interest is concentrated on one of the ψ_j's and, without loss of generality, let it be ψ_n. The following discussion applies equally well to a group of ψ_j's considered simultaneously.

Now the posterior density of Ψ_n for fixed λ,ζ and data y involves the data only through y_n and is

$$f_{\Psi_n|Y}(\psi_n|y;\lambda,\zeta) \propto f_{Y_n}(y_n;\psi_n,\lambda)f_{\Psi_n}(\psi_n;\zeta); \qquad (59)$$

the other observations $y_1,\ldots,y_{n-1}$ do not enter because of the strong independence assumptions. Within the present formulation, the posterior distribution (59) is not available because it involves the unknowns λ and ζ. The empirical Bayes approach is to estimate these quantities, and to do this we examine the joint density of the observations with the ψ_j's integrated out. For this, the density of Y_j is

$$f_{Y_j}(y_j;\lambda,\zeta) = \int f_{Y_j}(y_j;\psi,\lambda)f_\Psi(\psi;\zeta)d\psi. \qquad (60)$$

Then the posterior likelihood of (λ,ζ) using all the data is

$$\prod_{j=1}^{n} f_{Y_j}(y_j;\lambda,\zeta), \qquad (61)$$

and hence "good" estimates $\tilde{\lambda}$ and $\tilde{\zeta}$, for example maximum likelihood estimates, can be found. This suggests obtaining from (59) the estimated posterior density of Ψ_n as

$$\tilde{f}_{\Psi_n|Y}(\psi_n|y) = \frac{f_{Y_n}(y_n;\psi_n,\tilde{\lambda})f_{\Psi_n}(\psi_n;\tilde{\zeta})}{\int f_{Y_n}(y_n;\psi,\tilde{\lambda})f_{\Psi_n}(\psi;\tilde{\zeta})d\psi}. \qquad (62)$$

If we are interested in a point estimate for ψ_n, then the mean of this estimated posterior distribution is a natural candidate, although in certain cases the mode may be much easier to calculate.

If n is large, then the difference between (59) and (62) will be small, provided that the errors in the estimates $\tilde{\lambda}$ and $\tilde{\zeta}$ are relatively small, e.g. of order $n^{-1/2}$. This simply means that the parametric densities involved can be consistently estimated. A more refined theory has much in common with the discussion in Section 9.2 and will not be given here.

Note that in the fully Bayesian treatment, the posterior density of (Ψ, Λ, Z) would be found and then integrated with respect to $\psi_1, \ldots, \psi_{n-1}, \lambda$ and ζ.

We illustrate these ideas with two examples.

Example 10.13. Several normal means (ctd). We take up again the problem discussed in Example 10.9, but now in a more general framework. Thus we suppose that Y_{jk} $(j = 1, \ldots, m; k = 1, \ldots, r)$ are independent normally distributed variables with variance σ_w^2, Y_{jk} having mean μ_j. The prior frequency distribution of μ_j is normal with mean ξ and variance σ_b^2. In the general notation $\psi_j = \mu_j$, $\lambda = \sigma_w^2$ and $\zeta = (\xi, \sigma_b^2)$ and interest is assumed concentrated on μ_m.

Now for fixed $(\xi, \sigma_b^2, \sigma_w^2)$, the posterior density of μ_m is normal with mean and variance respectively

$$\mu_W (\bar{y}_m ., \sigma_w^2 / r ; \xi, \sigma_b^2) \quad \text{and} \quad \sigma_W^2 (\sigma_w^2 / r, \sigma_b^2), \qquad (63)$$

where $\bar{y}_m . = \Sigma y_{mk}/r$. This is (59). If the distribution of the Y_{jk}'s is considered in terms of $(\xi, \sigma_b^2, \sigma_w^2)$, integrating out the μ_j's, then we have the one-way component of variance model in which

$$Y_{jk} = \xi + \eta_j + \epsilon_{jk}, \qquad (64)$$

where $E(\eta_j) = E(\epsilon_{jk}) = 0$, $\text{var}(\eta_j) = \sigma_b^2$, $\text{var}(\epsilon_{jk}) = \sigma_w^2$ and all η_j's and ϵ_{jk}'s are independently normally distributed. Efficient estimates, not quite the maximum likelihood ones, are thus, provided that $m > 1$ and $r > 1$,

$$\tilde{\xi} = \bar{y}.. , \tilde{\sigma}_w^2 = \frac{\Sigma\Sigma (y_{jk} - \bar{y}_j.)^2}{m(r - 1)} , \tilde{\sigma}_b^2 = \max\left\{\frac{r\Sigma(\bar{y}_j. - \bar{y}..)^2}{m - 1} - \tilde{\sigma}_w^2, 0\right\}.$$

$$(65)$$

Therefore the estimated distribution (62) for μ_m is obtained by replacing $(\xi, \sigma_b^2, \sigma_w^2)$ in (63) by the estimates (65). In particular, a point estimate of μ_m obtained as the mean of the estimated prior distribution is

$$\mu_W (\bar{y}_m ., \tilde{\sigma}_w^2 / r ; \bar{y}.., \tilde{\sigma}_b^2) . \qquad (66)$$

Notice that this estimate lies between $\bar{y}_m .$ and $\bar{y}..$, being closer to $\bar{y}..$ when the data support the hypothesis that σ_b^2 is zero. Some further aspects of this particular estimate will be examined in Section 11.8.

Of course, because of the symmetry of the whole model, the corresponding estimate of an arbitrary mean, μ_j, is obtained from (66) by replacing $\bar{y}_m$. with $\bar{y}_j$.

The posterior distribution cannot be estimated with a single observation on each mean, $r = 1$, because the model (64) is then unidentifiable.

A result similar to (66) is obtained by Lindley and Smith (1972), who follow a subjective approach in which σ_w^2 has a conjugate prior distribution. Their estimate for μ_m amounts to substitution of the posterior m.l.e. for σ_w^2 in place of $\tilde{\sigma}_w^2$, which they use because the mean of the marginal posterior distribution for μ_m cannot be calculated explicitly.

Note, finally, that the model of this example is the same as that for the random-effects model of analysis of variance, although the emphasis here is on the estimation of μ_m rather than on the estimation of the components of variance.

Example 10.14. Compound Poisson distribution. Suppose that $Y_1, \ldots, Y_n$ have independent Poisson distributions with means $\mu_1, \ldots, \mu_n$. Let the prior distribution of the μ_j's be the gamma distribution, symbolically $\Gamma(v\,;\beta)$, with density

$$\frac{\beta(\beta x/v)^{\beta-1}\,e^{-\beta x/v}}{\Gamma(\beta)} \qquad (x \geqslant 0) \tag{67}$$

of mean v and index β. Here the parameter λ in the general formulation is not required and $\zeta = (v,\beta)$. For given ζ, the posterior distribution of μ_n given $Y_n = y_n$ is the $\Gamma\{v(y_n + \beta)/(v + \beta)\,;y_n + \beta\}$ distribution. In particular the posterior mean is

$$\frac{v(y_n + \beta)}{v + \beta}. \tag{68}$$

The marginal distribution of the Y_j's, when the Poisson distribution is compounded with the gamma distribution (67), is the negative binomial distribution

$$f_{Y_j}(y\,;v,\beta) = \left(\frac{\beta}{\beta + v}\right)^\beta \frac{\Gamma(\beta + y)}{\Gamma(\beta)y!} \left(\frac{v}{\beta + v}\right)^y.$$

From this, the maximum likelihood estimates $(\hat{v}, \hat{\beta})$ can be found, and, in particular, $\hat{v} = \bar{y}_{..}$. The estimate of μ_n is then the estimate of the posterior mean (68), that is

$$\frac{\bar{y}_{.}(y_n + \hat{\beta})}{\bar{y}_{.} + \hat{\beta}} . \tag{69}$$

Note the general similarity between this and the normal-theory result (66), in that the estimate is a combination of the overall mean $\bar{y}_{.}$ and the individual component estimate with weights depending on the estimated dispersion of the means.

Although we have discussed estimation of the posterior distribution for a single component Ψ_n, the method clearly applies to estimation of the joint posterior distribution of any or all of the parameters $\Psi_1, \ldots, \Psi_n$. This joint distribution may be of particular interest in tests of homogeneity of the parameter values $\psi_1, \ldots, \psi_n$, and empirical Bayes versions of Bayesian sharp hypothesis tests (Section 10.5 (iii)) could be constructed.

One application concerns the hypothesis H_0 of equality of multinomial proportions for the model of Example 10.4. If the conjugate distribution (18) with all a_j equal to a constant a were appropriate, given $\bar{H}_0$, then, in the general formulation above, we have $\psi_j = \theta_j, \zeta = a$, and λ is not required. Several problems involving multinomial distributions are discussed from the empirical Bayes point of view by Good (1965).

The formulation adopted so far for empirical versions of Bayesian procedures is completely parametric. That is, both the prior distribution and the distribution of the observations given the parameter values are specified in parametric form. Appreciable work has, however, been done on procedures for use when the prior distribution is arbitrary. Such procedures are often called non-parametric empirical Bayes. We illustrate these by describing estimation of the posterior mean for discrete exponential families of observations.

Example 10.15. Discrete exponential family with arbitrary prior distribution. Suppose that $Y_1, \ldots, Y_n$ are independently distributed, each Y_j being discrete with an exponential family density of the special form

$$f_{Y_j}(y; \theta_j) = \exp\{y \log\theta_j + c(\theta_j) + d(y)\} \ (y = 0, 1, \ldots) , \tag{70}$$

$\Theta_1, \dots, \Theta_n$ being i.i.d. with continuous prior density $p(\theta)$. Then the posterior distribution of Θ_n, say, is proportional to

$$\theta_n^{y_n} \exp\{c(\theta_n)\}p(\theta_n).$$

The posterior mean of Θ_n is thus

$$\frac{\int \theta^{y_n+1} \exp\{c(\theta)\}p(\theta)d\theta}{\int \theta^{y_n} \exp\{c(\theta)\}p(\theta)\,d\theta}$$

$$= \exp\{d(y_n) - d(y_n+1)\}\, \frac{f_Y(y_n+1)}{f_Y(y_n)}, \qquad (71)$$

where $f_Y(y)$ is the marginal density of each Y_j, given by

$$f_Y(y) = \int \theta^y \exp\{c(\theta) + d(y)\}p(\theta)d\theta.$$

Notice that in (71) the first factor is determined by the conditional distribution of the observations. The ratio in the second factor is estimated by

$$\frac{\text{no. of } y_1, \dots, y_n \text{ equal to } y_n+1}{\text{no. of } y_1, \dots, y_n \text{ equal to } y_n}, \qquad (72)$$

i.e. the ratio of sample proportions at the values $y_n + 1$ and y_n. Hence the posterior mean is estimated without use of the prior distribution, and requires only the knowledge that the parameter values are generated by a stable random mechanism.

These results can be applied in particular to the case of Poisson random variables described in Example 10.14. The Poisson distribution is of the exponential form (70) with $\theta_j = \mu_j$ and $d(y) = -\log y!$, so that from (71) and (72) the non-parametric empirical Bayes posterior mean for μ_n is

$$(y_n+1)\, \frac{\text{frequency of value } y_n+1}{\text{frequency of value } y_n}. \qquad (73)$$

It can be shown that the estimate (73), and its general counterpart, have good asymptotic properties, but, even in what one would consider large samples, these estimates can be very poor; the ratio (72) in general has quite appreciable scatter unless the mean is very small. Much work has been done on smoothing the estimates, essentially making the assumption that the prior distribution has certain natural smooth properties.

An important point to note in this example is that a special feature, equation (71), of the posterior mean is being used.

Bibliographic notes

Bayesian arguments were widely used in the 19th century, notably by Laplace ; uncritical use of uniform prior distributions was particularly condemned by J.A.Venn. Following Fisher's further strong criticisms in the 1920's and 1930's, especially of the use of prior distributions to represent ignorance (Fisher, 1959, 1st. ed., 1956), interest in the approach was for a time at a relatively low level, except for problems where the prior distribution has a frequency interpretation.

The highly original book of Jeffreys (1961, 1st ed., 1939) gave both a general account of probability as measuring rational degree of belief and also a considerable number of special results.

The relatively recent resurgence of interest in the Bayesian approach dates from the books of Good (1950) and Savage (1954) ; see Savage (1962) for a brief general statement of the Bayesian approach together with comments and criticisms from other points of view. Important work on the foundations of subjective probability was done by Ramsey (1931) and de Finetti (1937, 1972). DeGroot (1970) has given a careful account of the axiomatic basis of subjective probability as well as general development, especially linked to decision-making. For a review of measurement of subjective probabilities see Hampton, Moore and Thomas (1973). Lindley (1971a) gives a general review of Bayesian statistics.

Empirical Bayes methods not involving parametric specification of the prior distribution are due to Robbins (1956) ; see Maritz (1970) and Copas (1969) for general discussion.

Good (1965) gives a detailed account of parametric empirical Bayes methods associated with multinomial distributions.

Further results and exercises

1. Consider $n_1 + n_2$ experimental units, the jth being characterized by an unknown control value y_j and an unknown treatment value $z_j = y_j + \theta$. It is possible to observe either the control or the treatment value on a particular unit, but not both. Suppose that n_1 units are assigned to the control at random. Derive the likelihood function

and describe its information content for θ. Now suppose that the y_j are a random sample from the p.d.f. $f(y)$, and that Θ has a prior density $g(\theta)$. Show that the posterior distribution of Θ is not the same as the prior distribution, even though no information about θ can be derived from the likelihood alone without appeal to the repeated sampling principle.

[Section 10.2 ; Cornfield, 1969]

2. Let Y be the number of "ones" in n observations corresponding to random sampling without replacement of a finite population of size m containing θ "ones". Thus Y has a hypergeometric distribution. Suppose further that the prior distribution of Θ is itself hypergeometric. Prove that, given that $Y = y$, the posterior distribution of $\Theta - y$, the number of "ones" remaining in the population, does not involve y. Explain the physical basis of the result and comment on its implications for acceptance sampling.

[Section 10.2 ; Mood, 1943]

3. Consider a sequence of three independent Bernoulli trials with constant probability θ of success, and suppose that *a priori* Θ is equally likely to be 1/4 or 3/4. If the result of the first trial is a success, show that the posterior probability that the next two trials will be failures is the same as that for a success followed by a failure. Comment on this.

[Section 10.2]

4. Let $Y_{jk} (j = 1, \ldots, m ; k = 1, \ldots, r)$ follow a normal-theory components of variance model, i.e. $Y_{jk} = \mu + \eta_j + \epsilon_{jk}$, where the η_j's and ϵ_{jk}'s are independently normally distributed with zero means and variances respectively σ_b^2 and σ_w^2. Define $\sigma^2 = r\sigma_b^2 + \sigma_w^2$. The minimal sufficient statistic is, in the usual notation, $(\overline{Y}_{..}, \mathrm{SS}_b, \mathrm{SS}_w)$. Two Bayesian analyses are contemplated for μ, in the first of which only $(\overline{Y}_{..}, \mathrm{SS}_b)$ is to be used. Explain why the exclusion of SS_w might be thought to be sensible, and why this is incorrect.

In the first analysis using $(\overline{Y}_{..}, \mathrm{SS}_b)$, the prior density is taken to be proportional to $d\mu d\sigma/\sigma$, σ_w not being involved. Deduce from the results of Example 10.5 that the posterior distribution of μ is proportional to the Student t distribution with $m - 1$ degrees of freedom. The second analysis uses the full minimal sufficient statistic with prior density proportional to $d\mu d\sigma_w d\sigma/(\sigma_w \sigma)$ over the region $\sigma \geqslant \sigma_w$. Prove that this leads to a more dispersed posterior distribution for μ than the first analysis.

Comment on the surprising aspects of this last result and its implications for the consistent choice of improper prior distributions.
[Sections 10.3 and 10.4 (iii) ; Stone and Springer, 1965]

5. Let Y be a $N(\mu, \sigma_0^2)$ variable. Contrast the posterior expected values of μ given $Y = y$ for large y when the priors are the standard normal and standard Student t distributions.
[Section 10.4 (iv); Lindley, 1968; Dawid, 1973]

6. Comment critically on the following discussion of the connexion between the subjective and the frequency interpretations of probability. Let $A_1, \ldots, A_n$ be uncertain events from a large number n of quite separate situations, e.g. from separate fields of study. Suppose that they all have for "you" approximately the same subjective probability p. Further let F be the event that the proportion of $A_1, \ldots, A_n$ that are true is between $p - \epsilon$ and $p + \epsilon$, for some fixed small ϵ, e.g. $\epsilon = 10^{-2}$. Then

(i) because the A_j refer to different situations, it is reasonable for "you" to treat them as independent ;

(ii) because subjective probability obeys the ordinary laws of probability, the weak law of large numbers implies that for large enough n the subjective probability of F is close to one ;

(iii) therefore "you" should, to be consistent, be prepared to bet heavily that F is true, i.e. that your subjective probabilities have a hypothetical frequency interpretation.
[Section 10.4 (iv)]

7. An observation y is associated with a constant θ such that y is represented by a random variable Y with density $f(y ; \theta)$. In repeated experimentation, the successive values of θ have a known frequency distribution, so that a particular θ may be represented by the random variable Θ with this distribution. A $1 - \alpha$ confidence region $\mathcal{R}_\alpha(y)$ for the value θ associated with y is to be such that the doubly unconditional probability $\text{pr}\{\Theta \in \mathcal{R}_\alpha(Y)\}$ is equal to $1 - \alpha$. The optimal confidence region is said to be that with smallest expected area. Apply the general Neyman-Pearson lemma of Exercise 4.2 to show that the optimal confidence region is

$$\{\theta \; ; f_{\Theta\mid Y}(\theta\mid y) \geqslant k_\alpha\},$$

where k_α corresponds to the confidence level $1 - \alpha$.

Compare this method of constructing confidence regions with the classical Bayes and sampling theory methods in the case of three independent observations from the uniform distribution on $(0, \theta)$, when Θ has prior p.d.f. 2θ on the interval $(0, 1)$.

[Section 10.5 (i) ; Neyman, 1952]

8. A Poisson process of unknown rate ρ is observed in continuous time. The prior p.d.f. of P is

$$\frac{t_0(t_0\rho)^{n_0-1}e^{-t_0\rho}}{\Gamma(n_0)}.$$

It is observed that the nth event occurs at time t. Obtain the posterior p.d.f. of P. Prove that the probability that the subsequent interval $(t, t + h)$ contains no events is $(t_0 + t)^{n_0+n}/(t_0 + t + h)^{n_0+n}$. For what values of (t_0, n_0) does the posterior distribution of P give answers agreeing formally with those obtained by confidence interval arguments?

[Section 10.5]

9. Suppose that H_0 and H_1 are simple hypotheses, and that the statistic $T = t(Y)$ is a monotone increasing function of the likelihood ratio $\mathrm{lik}(H_1 ; Y)/\mathrm{lik}(H_0 ; Y)$, so that large values of T are evidence against H_0. What relationships are there between the observed posterior likelihood ratio and the significance probability $\mathrm{pr}(T \geqslant t_{\mathrm{obs}} ; H_0)$?
[Sections 10.5 (iii) and 4.3 ; Barnard, 1947 ; Pratt, 1965 ; Efron,
1971]

10. Let y_1 and y_2 be the numbers of successes in sets of n_1 and n_2 Bernoulli trials, the oorresponding probabilities of success being θ_1 and θ_2. Suppose that the sharp hypothesis $H_0: \theta_1 = \theta_2$ has positive prior probability and that if H_0 is not true then Θ_1 and Θ_2 have independent beta prior distributions. Using a suitable prior distribution for the nuisance parameter when H_0 is true, calculate the ratio of the posterior and prior odds of H_0.

[Section 10.5 (iii)]

11. Suppose that there are two alternative models for the vector random variable Y, both models containing nuisance parameters. Denote the corresponding p.d.f.'s by $g(y ; \phi)$ and $h(y ; \chi)$. Let the prior probabilities of the models be π_g and π_h and, conditionally on the correctness of the relevant model, let the prior p.d.f.'s of Φ and X be $p_g(\phi)$ and $p_h(\chi)$. Adapt the asymptotic Bayesian argument of Section 10.6 to show that the ratio of the posterior probabilities of the two models is approximately

$$\frac{\pi_g}{\pi_h} \times \frac{g(y ; \hat{\phi})}{h(y ; \hat{\chi})} \times \frac{(2\pi)^{\frac{1}{2}q_g}}{(2\pi)^{\frac{1}{2}q_h}} \times \frac{p_g(\hat{\phi})}{p_h(\hat{\chi})} \times \frac{\Delta_\phi^{-\frac{1}{2}}}{\Delta_\chi^{-\frac{1}{2}}} ,$$

where Δ_ϕ and Δ_χ are the information determinants associated with the estimation of ϕ and χ, and q_g and q_h are the dimensions of the parameters.

Examine, in particular, the special case where both models are normal-theory linear models with known variance.

Comment on the difficulties of assigning a numerical value to $p_g(\hat{\phi})/p_h(\hat{\chi})$. Produce some justifications for this ratio being approximately proportional to $(\Delta_\phi/\Delta_\chi)^{\frac{1}{2}}$, at least when the degrees of freedom in the two models are equal.

[Sections 10.5 (iii) and 10.6 ; Cox, 1961 ; Box and Hill, 1967 ; Box and Henson, 1970 ; Box and Kanemasu, 1973]

12. Obtain an empirical Bayes solution to the problem of estimating a normal mean from a number of sets of data with different variances. In the notation of Example 5.8, suppose that the number m of sets of data is large and that the variances $\sigma_1^2, \ldots, \sigma_m^2$ have a prior density of the conjugate inverse gamma form. Show that the mode of the m.l.e. of the posterior distribution of the mean μ is given by a statistic of the same general form as (5.28).

[Section 10.7]

13. Let $Z_1, \ldots, Z_{n+r-1}$ be i.i.d. in $N(\mu, \sigma^2)$, and suppose that

$$Y_j = \sum_{k=j}^{r+j-1} Z_k (j = 1, \ldots, n).$$ Thus observations $y_1, \ldots, y_n$ are

unweighted moving averages of r successive values, i.e. $z_1 + \ldots + z_r, \ldots, z_n + \ldots + z_{n+r-1}$. Calculate the general form of the posterior mean of each Z_k when μ and σ^2 are known, and suggest an empirical Bayes estimate for Z_k when μ and σ^2 are unknown constants. Derive explicit results for the case $r + 1 > n \geqslant 2$, dealing separately with the two situations $k = 1, \ldots, n - 1, r + 1, \ldots, n+r-1$ and $k = n, \ldots, r$. Comment on the results.

[Section 10.7 ; Cox and Herzberg, 1972]

11 DECISION THEORY

11.1 Examples of statistical decision problems

In the previous Chapter we studied the effect of introducing a prior probability distribution for the unknown parameters in a statistical model. In the light of the data, a posterior distribution can be calculated for the unknown parameter which puts together the information in the prior distribution and in the data. We now introduce a further element into the problem. It is supposed that one of a number of courses of action has to be chosen, a list of the available actions or decisions being available. Further, a numerical measure, called the *utility*, is available measuring the "gain" accruing from any combination of true parameter value and decision. Utility is measured on such a scale that the objective can be taken to be to maximize expected utility.

Thus two new features are being introduced. One is the idea that the analysis has the aim not of, in some sense, assessing the information that is available about the unknown parameter but rather that of choosing between a number of clearly formulated alternative courses of action. The second feature is the availability of the utility function as a quantitative measure of the consequence of a given course of action.

The type of decision mentioned above, which we call a *terminal decision*, may well be preceded by subsidiary decisions concerning how many and what type of observations should be taken. We return to this in Section 11.9, but for the moment concentrate on the case where a single terminal decision is to be made. We deal later also with the case, common for example in control theory, where a number of related terminal decisions are to be made in sequence.

We give first a few examples of statistical decision problems. As models of real problems they are inevitably very idealized.

412

Example 11.1. Discrimination between two groups. The following is the most primitive form of decision problem. The observable variable Y follows one of two possible known p.d.f.'s $f_1(y)$ and $f_2(y)$; the corresponding prior probabilities are π_1 and π_2. Suppose that there are two possible decisions, d_1 and d_2, and that the utility of reaching decision d_k when the true distribution is $f_j(.)$ is u_{jk}.

One interpretation is in terms of the industrial inspection problem of Example 10.1, where Y is the number of defectives having a binomial distribution with one of two possible values for the true proportion defective.

If one decision is preferred whichever is the true state, then the decision problem is trivial. We exclude this case and assume the decisions labelled so that $u_{11} > u_{12}$ and $u_{22} > u_{21}$. It is then sometimes convenient to interpret the decision d_1 as meaning "act as if the true distribution is $f_1(.)$", although this is purely for convenience of exposition. The central emphasis is on the possible courses of practical action and their consequences and not on any pseudo-inferential interpretation.

In the industrial inspection problem d_1 might be the decision to accept a batch, whereas d_2 is the decision to reject it.

Another interpretation of the problem is in terms of discriminant analysis ; here the two densities might, for example, be multivariate normal distributions.

In this example both the number of possible true distributions and the number of possible decisions take their smallest non-trivial values, namely two. We shall see that the solution determining the optimum decision for any given y is very simple. Much of that simplicity is retained when the numbers of states and decisions remain finite.

Example 11.2. One-shot control problem. Suppose that the scalar random variable Y is normally distributed with mean μ and known variance σ_0^2 and that the prior distribution of the unknown parameter μ is normal wtih mean ξ_0 and variance ν_0. On the basis of the observation, y, it is required to choose a real number t, say ; we can interpret this as a decision to adjust the mean of the process by t, thus converting a mean of μ into one of $\mu - t$. Thus we can denote the possible decisions by d_t, where t takes on real values.

To complete the specification we have to specify the consequences of a particular combination (μ, t). For this, suppose that the objective is to adjust the process mean to a target value, say zero. Provided that the cost of making the adjustment is fixed independently of t and μ, the utility of the combination will thus be

$$u(\mu,t) = a - w(\mu - t),\tag{1}$$

where $w(x)$ measures the loss arising from being off target by an amount x. It will often be reasonable, at least as a first approximation, to take squared error for the loss, so that with, $b > 0$,

$$u(\mu,t) = a - b(\mu - t)^2.\tag{2}$$

Other forms of utility function are possible. For example, there may be a cost, c, associated with making any non-zero adjustment to the process, no matter how small, and in that case a possible utility is

$$u(\mu,t) = \begin{cases} a - b\mu^2 & (t = 0), \\ a - b(\mu - t)^2 - c & (t \neq 0). \end{cases}\tag{3}$$

The specification is now complete and the objective is to give a rule for choosing t once y is observed.

While this set-up can be given other interpretations, we have described here one in terms of a single control decision; practical control problems usually involve chains of interrelated decisions.

Example 11.3. Choice between two treatments. As a third example of a simple decision problem, suppose that there are two treatments and that it is required to choose between them on the basis of relevant data. Suppose that the two treatments correspond to normal distributions of observations of the same known variance σ_0^2 and unknown means μ_1 and μ_2, the observations representing, for example, yields. It will turn out to be convenient for this particular problem to give the prior density of $\mu_2 - \mu_1$ and we take this to be normal with mean ξ_0 and variance ν_0; it will be assumed that the prior distribution of $\mu_1 + \mu_2$ is independent of that of $\mu_1 - \mu_2$.

Suppose that there are two possible decisions, d_1 and d_2, corresponding respectively to the choice of treatment 1 and 2 as the "better" treatment, i.e. as the one giving the higher response. We take the utility corresponding to decision d_j to depend on μ_j, and for simplicity suppose the relation to be linear. However, because of the possibly differing set-up costs the two linear functions are taken to be different, although with the same slope, so that

$$u(\mu_1,\mu_2;d_1) = a_1 + b\mu_1, \quad u(\mu_1,\mu_2;d_2) = a_2 + b\mu_2.\tag{4}$$

In the simplest form of the problem the data consist of independent sets of observations from the two treatments with means $\overline{Y}_{1.}$ and $\overline{Y}_{2.}$, which have variances σ_0^2/n, where n is the sample size, assumed for simplicity to be the same for the two treatments.

The object is to specify which decision is to be reached for any given $\overline{y}_{1.}$ and $\overline{y}_{2.}$.

11.2   Formulation of terminal decision problems

The most general formulation of a one-step terminal decision problem requires the following elements to be specified :

(i) an observable random variable Y with sample space $\mathcal{Y}$, with a family of possible p.d.f.'s for Y, namely $f_Y(y\;;\theta)$;

(ii) the parameter space Ω for θ ;

(iii) a prior p.d.f. $f_\Theta(\theta)$ over the parameter space Ω ;

(iv) a set of possible decisions $d \in \mathcal{D}$;

(v) a utility function $u(\theta,d)$ defined for all possible pairs (θ,d), i.e. available on the product space $\Omega \times \mathcal{D}$.

A *decision function* is a rule specifying for any possible $Y = y$ which decision is to be taken, i.e. it is a function from $\mathcal{Y}$ into $\mathcal{D}$; we denote such a decision function by $\delta(y)$. Sometimes it is convenient to contemplate decision functions which allow us to choose at random among two or more decisions; then $\delta(y)$ has to be interpreted as a probability distribution on $\mathcal{D}$. We call such a decision function a *randomized decision function.* Such decision functions are normally of interest only in very theoretical discussions.

TABLE 11.1

Components of decision problems for Examples 11.1 − 3

	11.1	*11.2*	*11.3*
Distribution of random variable	$f_\theta(y)$	$N(\mu,\sigma_0^2)$	$\overline{Y}_{1.}:N(\mu_1,\sigma_0^2/n)$ $\overline{Y}_{2.}:N(\mu_2,\sigma_0^2/n)$
Parameter space	Two point $\{1,2\}$	$-\infty<\mu<\infty$	$-\infty<\mu_1,\ \mu_2<\infty$
Prior distribution	π_1,π_2	$N(\xi_0,\nu_0)$	$\mu_1-\mu_2:N(\xi_0,\nu_0)$ $\mu_1+\mu_2$ indep.
Decision space	Two point $\{d_1,d_2\}$	$-\infty<t<\infty$	Two point $\{d_1,d_2\}$
Utility function	$u(\theta_j,d_k)=u_{jk}$	$u(\mu,t)=a-w(\mu-t)$	$u(\mu_1,\mu_2;d_k)=a_k+b\mu_k$

Table 11.1 lists the five elements set out above for the three examples of the previous section. The objective is to find the decision function that maximizes the expected utility.

Before giving, in the next section, the solution to this problem, we discuss briefly some of the implications of this formulation.

First, to what extent in practice is the idea of decision-making important as the objective of analysis? The qualitative idea that we should clarify alternative courses of action and their consequences is undoubtedly important and widely helpful. It is arguable that all statistical analyses have an objective that has at least some decision-making element to it, and to this extent the qualitative ideas of decision theory seem very important. However, in a fundamental research problem the alternative decisions might be (i) close the investigation and publish, (ii) do such and such a further experiment, (iii) abandon the line of work, etc. While it might indeed be fruitful to list the possibilities and their possible consequences, it seems unlikely that quantitative assessment of consequences would often be useful, or indeed that the decision-making element is more than one facet of this kind of problem.

In more immediately "practical" investigations the decision element is often crucial. Even here, however, it is unlikely that single major decisions will or should be made by mechanical application of a decision rule. The better approach will be to isolate for critical discussion the separate aspects entering into the final decision ; what is contributed by the data? what are the assumed utilities and what is the basis for their calculation? Most importantly, have all the possible decisions been looked at? The contribution of statistical ideas to major decision making is more likely to be in the clarification of these separate elements than in the provision of a final optimum decision rule.

Note that an optimum decision rule will give the decision appropriate under given conditions. It will not indicate whether the evidence for that decision is overwhelming or whether, say, several decisions have about the same expected utility. At the least, therefore, we need some comparison of the expected utilities of the main competing decisions and examination of sensitivity to the main assumptions. In smaller scale decisions, and especially in decisions that have to be made repeatedly under similar conditions, the use of a mechanical decision rule is much more likely to be valuable. For this reason, the most immediate applications of decision theory are likely to be in

such fields as control theory, industrial and biological acceptance
testing and some kinds of medical diagnosis; in all these, considerable
numbers of similar decision-making situations are likely to arise and
the usefulness of a formal decision rule is much greater.

In addition to the immediate data on which the decision is to be
based, a prior distribution and a utility are needed. The considerations affecting the specification of a prior distribution were discussed
in Section 10.4. There is the following additional consideration,
however. When the objective is the interpretation of data, the need
to keep separate what is reasonably firmly based on the data and
what is subjective impression from some unspecified source seems
compelling. In decision making, however, prior information must be
taken account of and if the decision function is to be completely
specified the prior information has to be formulated quantitatively.
The case for using subjective probabilities is thus much stronger than
in the inferential situation, although the need to scrutinize their basis
remains. The only alternative to the use of subjective probabilities
will often be to abandon the idea of a wholly quantitative decision-
making procedure, to formulate as clearly as possible what can be
inferred from the data and to merge this with the prior information
qualitatively. We shall in Sections 11.5 and 11.6 consider what can
be done more formally about incompletely specified decision-making
problems.

The possible difficulties of obtaining an appropriate utility function
will be examined in Section 11.4.

11.3   Solution of the fully specified terminal decision problem

The general solution of the terminal decision problem can be given
quite simply. The objective is to maximize expected utility. By a
basic property of bounded conditional expectations, we have for any
given decision function $\delta(.)$ that, in an obvious notation,

$$E_{Y,\Theta} u\{\Theta, \delta(Y)\} = E_Y (E_{\Theta \mid Y} [u\{\Theta, \delta(Y)\} \mid Y]). \qquad (5)$$

That is, the overall expected utility is the expected utility for given
data $Y = y$, in turn expected over the marginal distribution of y. Now
the maximum of this is achieved if and only if, for each y, we choose
that decision which maximizes the expected utility over the distribution given $Y = y$, i.e. over the posterior distribution. For any
decision rule that did not have this property at, say $Y = y'$, could

immediately be improved by changing it at $Y = y'$ and leaving it unchanged for all other y. Of course, when the distribution of Y is continuous, we could, in principle, change the decision on a set of y-values of measure zero without changing the expected utility. This is, however, for reasons that have been discussed in other contexts, not a matter of practical import.

Thus the procedure for finding the optimum decision is as follows.
(a) For $Y = y$, find the posterior distribution over the parameter space. (b) Then compute for each possible decision the expected utility with respect to the posterior distribution. (c) Finally choose that decision with greatest expected utility.
Such a decision rule is called the *Bayes decision rule* for the problem.

Before giving some examples of the calculation of such decision rules, there are three general points to be made. First, because the probability calculations are *via* Bayes's theorem, the strong likelihood principle is satisfied. Observations giving proportional likelihood functions, whether or not from the same random system, lead to the same optimal decision. Secondly, the role of randomized decision rules in this formulation is clear. If, for given y, there is a unique decision yielding largest expected utility, then any randomized rule is disadvantageous. If there are several decisions giving the same expected utility, then, given the formulation, any of these decisions will do and, in particular, any randomized mixture of them. Here randomized decision functions are acceptable but do not have any positive advantages, always within the particular formulation. Thirdly, at a more immediately practical level, if we are applying this procedure to a particular complex problem, then we need to carry out step (b), the comparison of expected utilities for competing decisions, only for the observations actually obtained. There is no need to examine the decision that would have been reached for observations that might have been obtained, but which, in fact, were not.

Of course, if we were involved in a choice between alternative experiments, with each one of which is associated an optimal decision rule, then a comparison of the two maximum unconditional expected utilities would be appropriate. Here, however, we are involved with a given experiment and there is nothing to be gained by a so-called preposterior analysis.

Example 11.4. Discrimination between two groups (ctd). We take up
the problem formulated in Example 11.1. Given $Y = y$, the posterior
probabilities of the two possible true distributions 1 and 2 are
respectively

$$\frac{\pi_1 f_1(y)}{\pi_1 f_1(y) + \pi_2 f_2(y)} \quad \text{and} \quad \frac{\pi_2 f_2(y)}{\pi_1 f_1(y) + \pi_2 f_2(y)}. \tag{6}$$

Thus, if we take decision d_1, then the expected utility is

$$\frac{\pi_1 f_1(y) u_{11} + \pi_2 f_2(y) u_{21}}{\pi_1 f_1(y) + \pi_2 f_2(y)}, \tag{7}$$

with the corresponding expression

$$\frac{\pi_1 f_1(y) u_{12} + \pi_2 f_2(y) u_{22}}{\pi_1 f_1(y) + \pi_2 f_2(y)} \tag{8}$$

if decision d_2 is taken. Therefore, d_1 is the better decision if and only
if

$$\pi_1 f_1(y)(u_{11} - u_{12}) > \pi_2 f_2(y)(u_{22} - u_{21}). \tag{9}$$

If the relation is reversed, with $>$ replaced by $<$, then d_2 is the
better decision, whereas if the two sides are equal it is immaterial
which decision is taken.

Note that, as would be expected on general grounds, the decision
rule depends only on the difference between the alternative utilities
for different decisions for a given state. That is, we could, without
loss of generality, have taken $u_{11} = u_{22} = 0$ for the purpose of calculating the optimum decision, although not, of course, for calculating
the expected utility achieved. If we write $w_{12} = u_{11} - u_{12}$ and
$w_{21} = u_{22} - u_{21}$, then we call the w's *decision losses* or *regrets* ; for
they give the loss of utility for a given state as compared with the
best utility achievable if we were certain of that state. Then the
optimum decision rule is

$$\frac{\pi_1 f_1(y)}{\pi_2 f_2(y)} \cdot \frac{w_{12}}{w_{21}} \quad \begin{cases} > 1, \text{ take } d_1, \\ = 1, \text{ take } d_1 \text{ or } d_2, \\ < 1, \text{ take } d_2. \end{cases} \tag{10}$$

We have had two previous discussions, in Sections 4.3 and 10.5 (iii),
of a situation in which there are only two possible true distributions,
or states. In the first, testing a simple null hypothesis against a simple

alternative, the two distributions entered the problem in quite different ways. The solution involved not only the observed value of the likelihood ratio but also its distribution under one of the hypotheses. In the second version of the problem, the two distributions are treated on the same logical basis, and we summarize the information contained in the data and the prior distribution by computing the posterior odds for one state versus the other. The data enter only through the observed value of the likelihood ratio. In our present and final version of the problem, the solution is essentially an immediate deduction from the solution of the second problem.

Example 11.5. One-shot control problem (ctd). For the situation of Example 11.2, the posterior distribution of μ given $Y = y$ is normal with mean and variance respectively, in the notation of Example 10.5,

$$\mu_W(y, \sigma_0^2; \xi_0, \nu_0) \quad \text{and} \quad \sigma_W^2(\sigma_0^2, \nu_0). \tag{11}$$

Under the squared error utility function (2) the conditional expected utility for an arbitrary decision d_t depends on the expected squared deviation of μ from t. It is thus a minimum when t is chosen to be the mean of the posterior distribution. Therefore the optimum decision rule is to take

$$t = \mu_W(y, \sigma_0^2; \xi_0, \nu_0). \tag{12}$$

The corresponding expected utility is

$$a - \frac{b}{1/\sigma_0^2 + 1/\nu_0}. \tag{13}$$

Under the more complicated utility (3) which incorporates a cost c associated with any non-zero adjustment, the calculation is essentially similar. If a non-zero adjustment is made it should be that given by (12) and the corresponding expected utility is (13) minus c, the cost of adjustment. If a zero adjustment is made, then the expected utility is

$$a - b\mu_W^2(y, \sigma_0^2; \xi_0, \nu_0) - \frac{b}{1/\sigma_0^2 + 1/\nu_0}.$$

Thus it is better not to adjust if and only if

$$|y/\sigma_0^2 + \xi_0/\nu_0| < (1/\sigma_0^2 + 1/\nu_0)(c/b)^{\frac{1}{2}}. \tag{14}$$

It is instructive to compare this with the result of a significance test of the null hypothesis $\mu = 0$. This is designed to measure how consistent the data are with $\mu = 0$. It takes no account of prior information or costs. Depending on what these are, it is perfectly possible to have cases where there is overwhelming evidence that the true mean is not zero and yet it is best to make no adjustment, and conversely to have data that are quite consistent with $\mu = 0$ and yet for which an adjustment to the process is advisable. In particular, if the prior information is negligible, i.e. $\nu_0 \gg \sigma_0^2$, then the critical value of $| y |$ is $(c/b)^{\frac{1}{2}}$ and is independent of σ_0^2.

Thus, although it is tempting to interpret the possible decisions as

$$d_0: \text{act as if } \mu = 0,$$

$$d_t: \text{act as if } \mu = t,$$

it is essential to appreciate that two quite different problems are involved in the significance test and in the decision problem.

Example 11.6. Choice between two treatments (ctd). With the set-up of Example 11.3 it is simplest to examine directly the difference between the expected utility for decision d_2 and that for decision d_1. In fact, it follows directly from the utility function (4) that the difference is

$$a_2 - a_1 + b \left\{ \frac{\bar{y}_{2.} - \bar{y}_{1.}}{2\sigma_0^2/n} + \frac{\xi_0}{\nu_0} \right\} \bigg/ \left(\frac{1}{2\sigma_0^2/n} + \frac{1}{\nu_0} \right).$$

Thus d_2 is the better decision if and only if

$$\frac{\bar{y}_{2.} - \bar{y}_{1.}}{2\sigma_0^2/n} > \left(\frac{1}{2\sigma_0^2/n} + \frac{1}{\nu_0} \right) \left(\frac{a_1 - a_2}{b} - \frac{\xi_0}{\nu_0} \right).$$

To interpret this, note that the utilities of the two treatments are equal when $\mu_2 - \mu_1$ equals $(a_1 - a_2)/b$. Thus the optimum decision rule is to take the second treatment if and only if the posterior mean difference exceeds this break-even level. The remarks of the previous example about the relation between decision-making and significance tests are again apposite.

The significance test is largely irrelevant if attention is concentrated on the decision, whereas the decision rule alone gives no hint of the strength of evidence pointing to the decision taken.

11.4   Utility

In previous discussion we have assumed that a prior distribution is known over the parameter space, expressing quantitatively all relevant information other than the data, and, in addition, that the utility of any given combination of parameter and decision is known. There is an extensive literature on utility and its properties and in the present section we merely discuss some of the points most likely to arise in applications. For a full treatment see DeGroot (1970) and Fishburn (1969).

In the present context, utility is a number defined when we have specified an eventuality and the probability that the eventuality arises; more specifically, an eventuality is a combination of a parameter value and a decision. If eventuality $\&$ occurs with probability p, then we write the utility as $u(\&,p)$. If $p = 1$ we can simplify the notation to $u(\&)$.

Utility is required to have two key properties. First, the ordering of utilities gives qualitative preferences:

if $\&_1$ is preferred to $\&_2$, then $u(\&_1) > u(\&_2)$;

if $\&_1$ and $\&_2$ are equally desirable, then $u(\&_1) = u(\&_2)$;   (15)

if $\&_2$ is preferred to $\&_1$, then $u(\&_1) < u(\&_2)$.

Similar inequalities hold if the eventualities are uncertain. Secondly, there is a property of combination which ensures that expected utility is the relevant quantity for discussion when complicated possibilities are considered. Let $\& = (\&_1, \&_2; p_1, p_2)$ denote that eventuality $\&_1$ occurs with probability p_1 and $\&_2$ with probability $p_2 = 1 - p_1$. Then it is required that

$$u(\&) = p_1 u(\&_1) + p_2 u(\&_2).$$ (16)

In general, when we are faced with a probability distribution over a set of eventualities, the utility is given by the expected value of the utilities of the separate eventualities.

If all the eventualities can be assessed in terms of a single quantity, for example money or time, then the rank ordering of preference will cause no difficulty. In other cases, however, even the ranking of eventualities may be difficult. For example, in a clinical problem one might have to compare two treatments, one leading to a quick but painful cure, the other to a slow but painless cure. Which of these

possibilities has the higher utility, i.e. is preferred? If the probabilities
of cure are different in the two cases this adds a further complication.
This is rather an extreme example, although quite often apparently
incommensurable elements are involved. Another example is where
the possibility of reaching a wrong conclusion in a purely scientific
investigation has to be matched against expense in the economic
sense and the consequences of delay in reporting the conclusions. It
may often be useful to score the component aspects in some individually meaningful way, i.e. to set up a vector utility function. Decision
calculations can then be repeated with several different weightings of
the components, although it is of the essence of the problem that in
the last analysis some weighting does have to be chosen either
explicitly or by implication.

Assuming now that the eventualities have been ranked in order of
preference, we have to satisfy the expectation property (16). Clearly
utility is undetermined to within a linear transformation of positive
slope, i.e. we can choose the origin and unit of utility arbitrarily. If
the eventualities are characterized in terms of money and the amounts
involved do not vary too greatly, then it will often be a sensible
approximation to take utility and money as equivalent. However, if
the money values vary over a wide range, then this will not be
adequate.

For example, for most individuals the possibilities
  (a) gain $£10^4$ with certainty,
  (b) gain $£2 \times 10^4$ with probability $\frac{1}{2}$ and gain zero with probability
     $\frac{1}{2}$,
will not be regarded as equally preferable, i.e. will not have equal
utility. By (16), however, the utilities would be equal if utility is
equivalent to money.

In fact, the relation between the utilities of (a) and (b) depends on
the circumstances. For a large company, the utilities might indeed be
almost equal, whereas for most private individuals the utility of (a)
would be appreciably greater than that of (b), and a person who
needs $£1.5 \times 10^4$ immediately to avoid total ruin and disgrace would
prefer (b). Even when all eventualities can be characterized in terms
of money, there is no universal procedure for translating money into
utility.

It was shown by von Neumann and Morgenstern (1953) that, subject to some reasonable axioms, a utility function does exist for any
particular individual, "you"; it expresses the relations holding between

your preferences if inconsistencies are to be avoided. Your utilities can, in principle, be determined by arguments very similar to those used in Chapter 10 to find subjective probabilities. Indeed, a careful treatment would deal simultaneously with the existence of subjective probabilities and utilities. For convenience of exposition we have here separated the two topics and now suppose that clear-cut probabilities are available. Take the origin of utility to be the unchanged present state and take the unit of utility as some definite eventuality $\mathcal{E}_1$, e.g. the certain gain of £1. To find the utility of some eventuality, $\mathcal{E}$, with utility greater than one, we find a probability p such that the two possibilities

(a) $\mathcal{E}_1$ is certain to occur; (b) with probability p, $\mathcal{E}$ occurs and with probability $1 - p$ zero utility is achieved,

are regarded as equally desirable. Then the utility of $\mathcal{E}$ is $1/p$. A similar kind of argument can be used if the utility of $\mathcal{E}$ is less than one.

This argument serves both to establish a utility function when all possibilities are expressed in terms of, say, money and to establish a weighting of the different components of utility. Many of the comments made in connexion with subjective probability are relevant. Utilities determined in this way lead to internally consistent decision-making, but there is no guarantee that "your" utilities are justified in any external sense, and for this reason critical analysis of the basis of any determination of utilities may be very important. Difficult problems arise when decisions have to be taken collectively by people with inconsistent utility functions ; these are outside the present discussion.

In some applications in which utility is a non-linear function of money, values below a certain level have a special interpretation as "ruin". An alternative to converting money into utility is to maximize expected money gain subject to a constraint requiring the probability of "ruin" to be small. In practice, this is quite often a convenient evasion of the difficult task of assigning a utility to "ruin".

In the above discussion, utility is defined for each combination of parameter value and decision. In some cases, it may be helpful to derive this utility by considering first the utility corresponding to an as yet unobserved future state, represented by a random variable Z. Let the p.d.f. of Z, the prognosis distribution, be the known function $f_Z(z\,;\theta,d)$ of the parameter θ and the decision d and let $u(z)$ be the utility of a final state z. Then, with given parameter θ and decision d, we can associate the utility

$$\int u(z) f_Z(z; \theta, d) dz . \tag{17}$$

In cases where the distribution of Z can be estimated from relevant data and the consequences of the decision are best assessed from the final state z,(17) provides a convenient way of deriving a utility function. It was first suggested by Aitchison (1970b) in a medical context where z is the final state of the patient and the parameter θ gives the correct diagnosis. In Aitchison's example there were a small number of possible disease types, so that θ was restricted to a small number of values.

In some applications, particularly those concerned with economic problems, inputs of utility may be expected at various times in the future. Especially when utility is in effect money, it is reasonable to discount utility that will not be realized immediately by assigning utility u achieved at time t a present value of $ue^{-\gamma t}$, where γ is a rate of compound interest. The justification for this is that the present value invested now at compound interest would grow to u at time t. Of course, the choice of an appropriate value of γ in any particular context needs care. More generally, if the rate of input of utility at time t is $\rho_u(t)$, then the present value is

$$\int_0^\infty \rho_u(t) e^{-\gamma t} dt . \tag{18}$$

To summarize, the essential difficulties of decision-making may lie at one or more of several different points. In the treatment of the previous sections, it is assumed that the essential problem is that of the efficient use of data in the light of the fact that the true parameter value, representing the true state of affairs, is unknown. In other problems, the main difficulty may lie in assessing the consequences of a particular combination of parameter value and decision, i.e. in assessing the prognosis distribution. In yet other cases, there may be little "hard" data and the main part of the task may be in assembling the prior information in reliable and useful form. In all cases the possibility of having overlooked types of decision that are open is likely to be important. Thus, while the analysis given earlier is superficially a very general one it does not necessarily capture the essence of the practical difficulties of decision-making.

11.5   Incompletely specified decision problems

(i) General

In the previous sections, we have discussed the terminal decision problem in which all the components entering the problem are specified quantitatively. We now turn to problems where one or more of the components previously assumed known quantitatively is not available. These are most likely to be the prior distribution and the utility. It can be argued that in such cases we should insist that the best possible quantitative assessment is used and that it is dodging the real issue to discuss theoretical approaches based on less than the full specificattion, e.g. on a specification that excludes a prior distribution.

While we do not accept this view, it is certainly true that prior knowledge and utility are essential elements of the situation and if they are not introduced quantitatively, then consideration of them should come in qualitatively at some stage, at least by implication. Any decision analysis that appears to avoid all reference to the ideas has to be viewed suspiciously.

(ii) Neither utility nor prior distribution available

If there is quantitative information about neither utility nor prior information, it may be possible to get some useful results by consideration of the probabilities of decision errors of various kinds. While some subjective choices of appropriate error rates will have to be made, this approach is valuable in giving the decision maker some idea of the consequences of any rules proposed and of ensuring that the procedures adopted in different situations are interrelated in a roughly reasonable way. This approach is most likely to be useful when a series of broadly similar problems are to be faced.

Example 11.7. Acceptance sampling for a binomial parameter. As a specific example, consider a two-decision problem concerning a binomial parameter θ ; we can interpret the two decisions d_1 and d_2 as corresponding, respectively, to accepting and rejecting a batch whose true fraction defective is θ. If we had a utility function and a prior density for θ of the two-point type, then the analysis of Example 11.1 would be immediately relevant ; if we had a continuous prior density, then the optimum procedure could again be found fairly easily, the form of the answer being particularly simple if the density is of the conjugate beta form. Suppose that, in the absence of these pieces of information, we choose two values $\theta_1 < \theta_2$ such that if $\theta = \theta_1$ then we quite clearly prefer d_1, and if $\theta = \theta_2$, then we quite clearly prefer d_2. More specifically, we choose acceptable error rates α_1 and α_2 and require that, if possible,

$$\mathrm{pr}(d_2;\theta_1) \leqslant \alpha_1 , \quad \mathrm{pr}(d_1;\theta_2) \leqslant \alpha_2. \tag{19}$$

We assume that the data on which the decision is to be based lead to a binomially distributed random variable Y with index n and parameter θ. The sensible decision rules will be of the form

$$\text{take decision } d_1 \text{ if and only if } y \leqslant c. \tag{20}$$

It follows that if we use the decision rule (20), then the probability of reaching decision d_1, as a function of θ, is

$$\mathrm{pr}_c(d_1;\theta) = \sum_{y=0}^{c} \binom{n}{y} \theta^y (1-\theta)^{n-y} . \tag{21}$$

For any given sample size n and value of c, (21) can be plotted against θ and forms what is called the *operating characteristic* curve of the decision rule. To satisfy both the conditions (19) the operating characteristic curve must pass through or above the point $(\theta_1, 1-\alpha_1)$ and through or below the point (θ_2, α_2). We can by trial and error find the smallest n and the appropriate c so that the conditions are satisfied. If n is preassigned, then it may be impossible to satisfy (19) and then the requirements will have to be relaxed, either by increasing one or both error probabilities, or by moving the values of θ_1 and θ_2 further apart. The type of argument sketched here has been extensively elaborated in connection with industrial acceptance sampling.

Prior probabilities and utilities enter qualitatively in the specification of the scheme. Clearly there is no point in seeking protection against situations that are highly unlikely to arise or in which a decision error has unimportant consequences.

One objection to this formulation, even accepting the absence of utilities and prior probabilities, is that it is unnecessarily complicated ; only two constants, n and c, have to be chosen and to do this, specification of four things is required. It might be simpler, say, to equate c/n to the approximate break-even point and to specify one further requirement to fix n. The specification is redundant in that the same decision rule corresponds to many different specifications of the type

(19). All this would have considerable force if one were concerned with simple sampling procedures only of the present type ; when it comes to comparing schemes of quite different types, including those requiring multiple sampling, then the specification taken here seems useful.

Formally and historically there is a direct relation between the operating characteristic, as defined here, and the power function of a test of the null hypothesis $\theta = \theta_2$ against alternatives $\theta < \theta_2$, working at significance level α_2. There is, however, a major difference from the treatment of power given in Chapters 4 and 5, where tests are used to assess the evidence in data and no question of a pre-assigned particular level of significance arises.

While a considerable element of subjective choice enters into the selection of a decision rule by this method, the approach does seem clearly preferable to an entirely arbitrary choice of procedure. The consequences of applying the rule can be studied and their implications understood ; different situations can be treated in a relatively appropriate way.

Other decision problems can be tackled in terms of a consideration of possible error probabilities, at least when the number of possible decisions is small. The remarks made in the discussion of Example 11.7 will apply fairly generally.

(iii) Prior distribution available but not utility

Sometimes, previous similar data can be used to specify a prior distribution, but no quantitatively useful utility function can be found. No elaborate discussion seems necessary. The posterior distribution over the parameter space is obtained. In addition, one might attempt to mark out regions of the parameter space in which the various decisions are (a) optimum, (b) not far short of optimum, (c) bad. Study of the posterior distribution in the light of these regions would then indicate the appropriate decision.

(iv) Utility available but not prior distribution

The final case to be considered is where a utility function is known, but where no prior distribution over the parameter space can be used. This situation has been extensively studied and the ideas involved are of general interest. We therefore give a more detailed discussuion in a separate section.

11.6   Decision theory without prior distributions

(i) Regret and the risk function

For some purposes, it is convenient to work not directly with utilities but rather with the loss of utility arising from having made a decision other than the optimum one for the parameter value in question. This leads us to define quite generally a *regret function* or *decision loss*, a special case of which has already been introduced in the discussion of Example 11.4, discrimination between two groups. This function serves primarily to simplify some formulae, although in the discussion of the so-called minimax method the use of regret rather than, say, minus utility has an essential effect on the answer.

For a given parameter value θ and decision d, we therefore define the regret or loss to be

$$w(\theta,d) \;=\; \sup_{d^* \in D} u(\theta,d^*) - u(\theta,d) . \tag{21}$$

The supremum is the utility that would be achieved if we knew the true value θ. The regret is non-negative. For a given decision rule $\delta(Y)$ we can calculate, for each fixed parameter value θ, the expected regret and we call this the *risk function* of the decision rule and denote it by $r(\theta,\delta)$. Thus

$$r(\theta,\delta) = E_Y[w\{\theta,\delta(Y)\};\theta] , \tag{22}$$

where the expectation is over the density $f_Y(y;\theta)$ of Y for fixed θ. This characterizes any decision rule by a function of θ; the definition is easily adapted to randomized rules.

If we have a prior distribution for Θ, then we can average (22) over θ, obtaining what is called the *Bayes risk*. This associates a single number with each decision rule, and comparison of alternative rules is then straightforward. In fact, the Bayes average of (22) is a constant minus the expected utility, so that this approach is equivalent to that of Section 11.3. The difficulty of proceding without a prior distribution is that we have to compare alternative decision rules on the basis of functions of the parameter rather than through single measurements.

Example 11.8. Normal mean with known variance (ctd). As an example of the calculation of risk functions, suppose that we observe $Y_1, \ldots, Y_n$ i.i.d. in $N(\mu,\sigma_0^2)$. The possible decisions d_t are indexed by real numbers t, the utility function being

$$u(\mu,d_t) = a - b(\mu - t)^2,$$

where $b > 0$; one application is the control problem of Example 11.2.

If we know the value of μ, we take $t = \mu$ and gain utility a; therefore, the regret or loss function is

$$w(\mu,d_t) = b(\mu - t)^2. \qquad (23)$$

The following are some possible decision rules δ, i.e. some procedures for choosing a value t given y. In the general notation, $\delta(y) = t$.

(i) δ_1: Take t to be the sample mean $\bar{Y}_.$, i.e. $\delta_1(Y) = \bar{Y}_.$;

(ii) δ_2: Take t to be the sample median, $\tilde{Y}$, say ;

(iii) δ_3: Take t to be the fixed value μ_0, whatever the data ;

(iv) δ_4: Take t to be the Bayes estimate that would be used if μ had a normal prior distribution of mean ξ_0 and variance ν_0, i.e. take

$$\delta_4(Y) = \mu_W(\bar{Y}_., \sigma_0^2/n ; \xi_0,\nu_0).$$

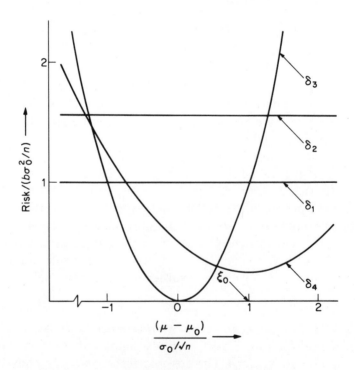

Fig. 11.1. Graphical representation of risk functions for Example 11.8. Special case : $\nu_0 = \sigma_0^2/n$; $\xi_0 - \mu_0 = \sigma_0/\sqrt{n}$.

These are four among many possible decision rules. To calculate the risk function for each, we fix μ and take the expected value of (23). This is easily done ; for convenience we use the result that the variance of the sample median is approximately $\frac{1}{2}\pi\sigma_0^2/n$. The four risk functions are in fact as follows :

$$r(\mu,\delta_1) \ = \ b\sigma_0^2/n \ ,$$

$$r(\mu,\delta_2) \ \simeq \ \tfrac{1}{2}\pi b\sigma_0^2/n \ ,$$

$$r(\mu,\delta_3) \ = \ b(\mu-\mu_0)^2 \ ,$$

$$r(\mu,\delta_4) \ = \ b\left(\frac{1}{n/\sigma_0^2 + 1/\nu_0}\right)^2\left\{\frac{n}{\sigma_0^2} + \frac{(\mu-\xi_0)^2}{\nu_0}\right\}.$$

These are shown graphically in Fig. 11.1. The precise relations between the four functions are determined by the dimensionless parameters $(\sigma_0^2/n)/\nu_0$ and $(\xi_0-\mu_0)/(\sigma_0\sqrt{}/n)$, and in the figure these are both taken to be one. Qualitatively similar relations hold entirely generally.

Some general conclusions can be drawn from this.

(a) Whatever the value of μ, δ_2 has a larger risk than δ_1. This follows directly from the sufficiency of $\bar{Y}_{.}$.

(b) The best rule for any given μ depends markedly on μ, even if we exclude rules such as δ_3 which are, of course, very good near the fixed point and very bad well away from it.

(c) The rules δ_1 and δ_2 have constant risks. This is a special feature closely connected with the invariance of the problem and decision rules under translation.

(d) The Bayes rule δ_4 is good when μ is close to the assumed prior mean ξ_0. It has, as noted above, the property of minimizing the risk averaged over the assumed prior distribution.

The comments on this example provide motivation for some of the following general definitions.

(ii) Admissibility

The decision rule δ_2 of Example 11.8 has the bad property that there is another decision rule, δ_1, better than it whatever the value of the unknown parameter. We call a decision rule such as δ_2 *inadmissible.* The formal definition is that a decision rule δ is inadmissible if there exists another decision rule δ' such that for all $\theta \in \Omega$

$$r(\theta,\delta) \geqslant r(\theta,\delta') , \qquad (24)$$

with strict inequality for some θ. If a decision rule is not inadmissible it is called *admissible*. Thus any admissible rule is one that cannot be dominated.

It can be argued that, to the extent that we accept the formulation of a decision problem and do not consider further things like robustness and ease of computation, inadmissible rules can be discarded. For if a rule δ is inadmissible, then we can, in principle, find a different decision rule whose performance as measured by the risk function is nowhere worse and somewhere better than than of δ. Note, however, that if δ is inadmissible and if δ'' is a *particular* admissible rule, then it does not follow that in a particular instance δ'' is preferable to δ. For example, if we had to choose in Example 11.8 simply between the use of the median δ_2 and the use of the fixed estimate rule δ_3, then we might very well prefer the median.

A general prescription, therefore, is to find all the admissible decision rules for a problem and then from them to choose a rule which has good performance over the range of parameter values which qualitative prior knowledge and the data indicate to be relevant.

A useful fact is that any Bayes decision rule obtained by taking a proper prior distribution over the whole parameter space must be admissible. To see this, suppose that the Bayes decision rule for a prior density $f_\Theta(\theta)$ is δ_B and that δ^* is a decision rule satisfying

$$r(\theta,\delta^*) \leqslant r(\theta,\delta_B) , \qquad (25)$$

with strict inequality on a parameter set of positive probability under the prior density. Then clearly the corresponding Bayes risks satisfy

$$E_\Theta\{r(\Theta,\delta^*)\} < E_\Theta\{r(\Theta,\delta_B)\}, \qquad (26)$$

which contradicts the fact that the Bayes rule minimises Bayes risk.

Notice that if $f_\Theta(\theta)$ is improper, and if (25) is satisfied with strict inequality on some bounded set, then the strict inequality in (26) does not necessarily follow. For example, if the improper prior is the limit of a sequence of proper prior distributions, then the probability of any bounded set in Ω tends to zero. Delicate mathematical arguments are involved in determining whether a limit of proper Bayes rules is admissible, the answer depending largely on the dimensionality of the parameter space. For a large class of problems, Bayes decision rules with respect to improper priors are admissible in one and two dimensions, but not in higher dimensions. The admissibility is closely related to the group of transformations under which the problem is invariant, and a full mathematical treatment of the question has been given by Brown (1966).

For the estimation problem of Example 11.8, all the decision rules δ_4 are admissible because they are proper Bayes rules. The decision rule δ_1 is the limit of δ_4 as $\nu_0 \to \infty$ and so corresponds to an improper prior uniform over the whole real line. In this case the improper rule is admissible, as can be deduced from the Cramér-Rao inequality (8.9). Some details are given in Exercise 11.5.

For Example 11.4, in which there are just two decisions and two possible parameter values, all admissible rules are of the form (10) in which d_1 is taken if $f_1(y)/f_2(y) > c$, d_2 is taken if the ratio is less than c, and it is immaterial what is done if equality holds. Here c may be zero or infinite.

The idea of admissibility is important in a negative sense in that it enables certain rules to be excluded. Understandably, an important area of investigation in the development of admissibility ideas has been that of conventional sampling-theory procedures, and many interesting results have been obtained. We discuss the most remarkable one in Section 11.8, involving the parameters of the linear model, and Exercise 11.7 deals with the inadmissibility of the normal-theory estimate of variance.

An obvious drawback to the idea of admissibility is that, in many cases, there are so many admissible rules that the restriction to them is only a very small step towards the selection of a rule for actual use.

(iii) Minimax rule

We now introduce one way of choosing a particular rule from among the set of admissible rules. While mathematically interesting, the method is unfortunately statistically rather unappealing : in principle it evades all reference to prior knowledge, quantitative or qualitative.

For any particular decision rule δ with risk function $r(\theta,\delta)$, there is a maximum risk, possibly infinite. We characterise δ by this maximum, i.e. by

$$m(\delta) = \sup_{\theta \in \Omega} r(\theta,\delta) . \qquad (27)$$

For the rules of Example 11.8,

$$m(\delta_1) = \frac{b\sigma_0^2}{n}, \; m(\delta_2) \simeq \frac{\pi b\sigma_0^2}{2n}, \; m(\delta_3) = m(\delta_4) = \infty.$$

The minimax rule δ^* is the one that minimizes $m(\delta)$, i.e. for all δ

$$m(\delta^*) \leqslant m(\delta).$$

That is, the minimax rule is such that the worst expected regret is kept as small as possible. The artificiality of this can be seen quite easily by noting that δ^* could have worse properties than δ for nearly all values of θ, but have a slightly lower maximum risk.

The minimax rule is appropriate in two-person zero-sum games, where there are two adversaries with conflicting goals and on a symmetrical footing. The notion, however, that the decision maker is faced with a malevolent opponent who chooses the true parameter values is usually far-fetched.

One case in which the minimax rule is defensible is where the risk associated with it is small. It can then be argued that use of δ^* ensures that, whatever the true parameter value, the expected loss of utility is small ; while there may be an apparently better rule, any improvement can only be small and may carry with it the danger of seriously bad performance in some region of the parameter space.

The minimax rule could be applied directly to negative utility rather than to regret. This would lead to a different answer in general, as can be seen by following through the general development, or from the following extreme example.

Example 11.9. Distinction between minimax regret and minimax loss. Suppose that there are just two parameter values θ_1 and θ_2, the first being a "bad" state where all utilities are low and the second a "good" state where all utilities are high. To simplify matters we will omit intermediate stages and suppose that just two decision rules are under discussion and that the expected utilities are as shown in the following array :

	δ_1	δ_2
θ_1	5	0
θ_2	900	1000

If we work with the expectation of minus utility, then we can write the array

$$
\begin{array}{ccc}
 & \delta_1 & \delta_2 \\
\theta_1 & -5^* & 0^* \\
\theta_2 & -900 & -1000
\end{array}
\tag{28}
$$

For each decision the maximum has been marked with an asterisk, and the smaller of these values corresponds to decision rule δ_1, which is therefore the minimax rule. If, however, as in the initial discussion, we work with regrets, the array is

$$
\begin{array}{ccc}
 & \delta_1 & \delta_2 \\
\theta_1 & 0 & 5^* \\
\theta_2 & 100^* & 0
\end{array}
\tag{29}
$$

and the minimax regret rule is δ_2.

It is sometimes thought that the minimax rule is one of extreme pessimism, in that it assumes that a particularly unfavourable parameter value is likely to arise. This is so for (28), where the decision rule is, in effect, chosen to do as well as possible under the "bad" parameter value. It is not so for minimax regret, based on (29); the emphasis here is on doing as well as possible in situations where the expected decision loss is high and this may correspond to any level of actual achieved utility.

We re-emphasize that the justification for working with expected regret and expected utility throughout this and the subsequent discussion is the basic linear property of utilities.

(iv) Geometric interpretation of Bayes and minimax rules
The relation between the Bayes and the minimax rules can best be seen by a geometrical representation of the risk function. This can be used directly in any case in which there are only a finite number of possible values of the parameter, and hence by extension in other cases also. To be as explicit as possible, we illustrate the idea by a simple special case.

Example 11.10. Two-decision problem for the Poisson distribution.
Suppose that Y has a Poisson distribution of mean μ and that μ has

one of the two values 1 and 2. Let there be two possible decisions d_1 and d_2, which can be interpreted, for example, in terms of accepting and rejecting a batch in industrial acceptance sampling. Let the regret function be

$$w(1, d_1) = 0, w(1, d_2) = 50, w(2, d_1) = 100, w(2, d_2) = 0 .$$

For any decision rule, the risk function is defined at just the two points $\mu = 1$ and $\mu = 2$. Table 11.2 gives the risk functions for a selection of decision rules. The first few rules are "sensible" ones in which d_2 is taken if the value of y is sufficiently large. Later decision rules in the table are mathematically well-defined but are physically silly.

TABLE 11.2.

Risk function for a simple problem about
Poisson distributions

Decision rule : Take d_1 if and only if	Risk at $\mu = 1$	Risk at $\mu = 2$
δ_1 : never	50	0
δ_2 : $y = 0$	31.6	13.5
δ_3 : $y = 0,1$	13.2	40.6
δ_4 : $y = 0,1,2$	4.02	67.7
δ_5 : $y = 0,1,2,3$	0.95	85.7
δ_6 : $y = 0, \dots , 4$	0.18	94.7
δ_7 : always	0	100
δ_8 : y even	21.6	50.9
δ_9 : $y = 1,2, \dots$	18.4	86.5
δ_{10} : $y = 2,3, \dots$	36.8	67.7

Each decision rule is described by two risks, the expected losses calculated one when $\mu = 1$, the other when $\mu = 2$. The results are represented graphically in Fig. 11.2. There are two axes, one for each value of μ, and each decision rule leads to one point.

Quite generally, in any problem with a finite number of possible parameter values, a graphical representation of the risk function is possible with one axis for each parameter value; by extension, in smooth problems with an infinite number of parameter values, the conclusions can be obtained as a limit.

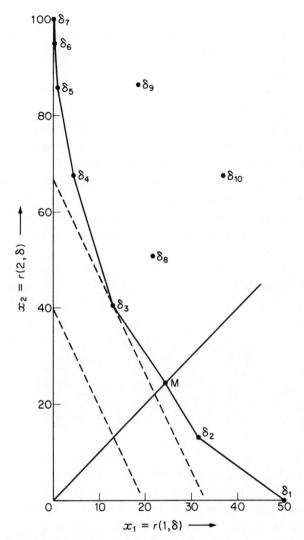

Fig. 11.2. Risks of decision rules in two-decision Poisson problem, Example 11.10. Decision rules $\delta_1, \ldots, \delta_{10}$ defined in Table 11.2 ; minimax rule, M ; $----$: lines of constant Bayes risk for $\pi_1 = \frac{2}{3}, \pi_2 = \frac{1}{3}$.

The following properties of the graphical representation hold quite generally although, in order to be as specific as possible, they will be explained in terms of the special case of Example 11.10.

(a) Provided that we allow randomized decision functions, the set of achievable points in the diagram is convex. For let P' and P'' be any two points in the diagram achieved by decision functions δ' and δ'', respectively. Now suppose that δ' is followed with probability p and δ'' is followed with probability $1 - p$. For each value of the unknown parameter the risks under the new strategy are weighted averages of those under the components δ' and δ''. Therefore the randomized strategy corresponds to the point $pP' + (1 - p)P''$, i.e. to a point on the line joining P' and P''. Thus, for any two points in the achievable set, the set also contains the whole of the line joining them, and this is the requirement for the set to be convex.

(b) Provided that the risks are calculated from regrets and that decision rules always leading to the same decision are allowable, the set of achievable points is bounded below by the coordinate planes and intersects these planes. This is clear in our particular example. By always following d_1, the optimum decision if in fact $\mu = 2$, we ensure that the risk at $\mu = 2$ is zero. All risks are non-negative. Quite generally, by always taking the decision optimum for a particular parameter value, we ensure that the regret is identically zero at that parameter value.

(c) The bounded convex set of achievable points has an achievable "lower", i.e. "south-west", boundary. Any point not on that boundary corresponds to an inadmissible decision rule. For, starting from such a point not on the boundary we can, by moving "south-west" towards the boundary, find a new point for which no coordinate is increased and for which some are decreased. This proves the inadmissibility of the strategy corresponding to the original point. In our example, applications of the results of Example 11.4 shows that $\delta_1, \ldots, \delta_7$ are Bayes rules and hence admissible rules, and more generally that any decision rule of the form "choose d_1 for $y \leqslant k$ and d_2 otherwise" is admissible. Therefore, the risks of each of these rules gives a point on the boundary of the achievable region. Furthermore, the boundary is made up of the lines joining successive pairs of these admissible points. For example, any point P on the line joining the points corresponding to δ_1 and δ_2 does itself correspond to one of the Bayes rules (10) in which randomization is used when $y = 0$, and so P corresponds to an admissible rule. Finally, the rules δ_8, δ_9 and δ_{10} are clearly inadmissible.

(d) Suppose that we want to find on the diagram the Bayes strategy corresponding to a given prior distribution, in the particular example to the prior probabilities (π_1, π_2) with $\pi_1 + \pi_2 = 1$. That is, we are now interested not in the components of the risk function but in their weighted average. In Fig. 11.2 any point on the line

$$\pi_1 x_1 + \pi_2 x_2 = a \qquad (30)$$

has Bayes risk a. For very small a the line (30) does not intersect the achievable region. Imagine a gradually increased. There will be a smallest a for which (30) contains at least one point on the boundary of the achievable region; the existence of such a line, called a supporting line or plane to the convex set, follows from the general properties of such sets.

Figure 11.2 illustrates this for the special problem taking $\pi_1 = \frac{2}{3}, \pi_2 = \frac{1}{3}$.

For the supporting plane corresponding to a particular prior distribution, one of two things may happen. The plane may intersect the boundary in a unique point. This gives the unique Bayes decision rule for the prior distribution in question. However, it is clear from Fig. 11.2 that for certain very special values of π_1 and π_2 the supporting line coincides with one of the straight-line sections of the lower boundary. Thus if $\pi_1 = 0.857$, the supporting line coincides with the section joining the points corresponding to the rules "take d_1 if $y = 0, 1, 2$" and "take d_1 if $y = 0, 1, 2, 3$". It follows that, for this very special prior distribution, the Bayes decision rule is not unique but that the above two rules, or any random mixture of them, are equivalent. In other words, in this particular case, the optimum decision for $y \neq 3$ is unique, but it is immaterial what is done if $y = 3$.

Quite generally, therefore, the Bayes rule is either unique or there may be several rules with the same Bayes risk; the geometrical characterization is that the latter case holds if and only if the supporting plane coincides with a planar section of the lower boundary of the achievable region.

(e) Finally, we give a geometrical interpretation of the minimax rule. For this, note that any decision rule is for this purpose characterized by the largest of its coordinates according to (27); in Fig. 11.2 the unit line has been drawn and points below this line have maximum risk given by the x_1 coordinate and those above the line have x_2 as the relevant coordinate. It is clear that the minimax rule is given by the intersection of the lower boundary and the unit line. This is shown by the point M in the diagram. For any other achievable point has its relevant coordinate larger than that of M. Note that the point M will always exist, provided that the conditions of point (b) are applicable. Further, the point M is unique and, unless it happens to coincide with a pure Bayes strategy, it will be a particular random mixture of such strategies. In the particular example the minimax rule is to take d_1 if $y = 0$ and to take d_1 with probability 0.195 if $y = 1$, but otherwise to take d_2. It is a very particular Bayes rule corresponding to the prior distribution $\pi_1 = 0.4$, $\pi_2 = 0.6$, called the *least favourable prior distribution* for the problem.

We shall not discuss the mathematical niceties of the existence and form of the minimax rule in continuous cases, but it is useful to indicate some properties of minimax rules which are suggested by the above arguments. As is indicated in the discrete case, the minimax rule usually makes the risk function constant. We can easily see that if δ is an admissible rule with constant risk, then it must be a minimax rule. For if not, then there exists a rule δ' such that

$$r(\theta,\delta') \leqslant \sup r(\theta,\delta') < \sup r(\theta,\delta) = r(\theta,\delta),$$

which contradicts the admissibility of δ. Since Bayes rules are admissible, we can often find a minimax rule by examing a particular class of Bayes rules, such as that defined by conjugate prior distributions. The prior distribution which yields the minimax rule, i.e. the least favourable prior distribution, may be improper. Whether or not the minimax rule seems appropriate may be judged by examining the behaviour of the least favourable prior. Finally, the minimax rule may be a randomized rule and need not be unique.

In the problem of Example 11.8, the estimation of a normal mean, the admissibility of the sample mean (Exercise 11.5) and the fact that the mean squared error is equal to the constant value σ_0^2/n show that the sample mean is the minimax estimation rule for squared error regret. The following example illustrates the construction of a minimax rule from a class of admissible estimates.

Example 11.11. Minimax estimate of binomial proportion. Let Y be the number of successes in n independent Bernoulli trials with constant probability of success θ. If we take squared error regret, i.e. $w(\theta,d) = (d - \theta)^2$, and assume a conjugate Beta prior for Θ as in Example 10.2, then the Bayes estimate for θ is the posterior mean

$$\delta(Y) = \frac{Y + a}{n + a + b}.$$

The risk function here is given by (bias)^2 plus variance and is

$$r(\theta,\delta) = \frac{\{(a + b)\theta - a\}^2 + n\theta(1 - \theta)}{(n + a + b)^2},$$

which must be constant if $\delta(Y)$ is to be a minimax estimate. Equating the coefficients of θ^2 and θ to zero gives immediately that $a = b = \frac{1}{2}\sqrt{n}$, so that the minimax estimate of θ is

$$\frac{Y + \frac{1}{2}\sqrt{n}}{n + \sqrt{n}},$$

with risk equal to $\frac{1}{4}n^{-1}(1 + 1/\sqrt{n})^{-2}$. This compares with the risk $\theta(1 - \theta)/n$ of the unbiased maximum likelihood estimate Y/n, whose maximum risk is $\frac{1}{4}n^{-1}$. Note that the least favourable prior distribution in this case is symmetrically concentrated around $\theta = \frac{1}{2}$, and depends on the sample size.

11.7   Decision theory representation of common problems

In the first part of this book we concentrated on three main types of statistical procedure, significance tests, interval estimates and point estimates. Other problems that came up occasionally concerned future observations rather than parameter values. It is useful to see that there are rough decision-theory parallels for these problems. Some of these have already been used in previous sections to illustrate the basic concepts of decision theory.

The simplest analogue of a significance testing problem is a two-decision problem with a two-valued regret function. Suppose that there is a single parameter θ, with θ_0 a particular value ; if $\theta \leqslant \theta_0$, then d_1 is the preferred decision, whereas if $\theta > \theta_0$, then d_2 is the preferred decision. Further, let the utility be

$$u(\theta,d_1) = \begin{cases} 1 & (\theta \leqslant \theta_0), \\ 0 & (\theta > \theta_0), \end{cases} \qquad u(\theta,d_2) = \begin{cases} 0 & (\theta \leqslant \theta_0), \\ 1 & (\theta > \theta_0), \end{cases} \qquad (31)$$

so that there is unit regret if a wrong decision is taken. For any decision rule δ, the risk function is thus

$$r(\theta,\delta) = \begin{cases} \text{pr}(d_2 ; \theta) & (\theta \leqslant \theta_0), \\ \text{pr}(d_1 ; \theta) & (\theta > \theta_0), \end{cases} \qquad (32)$$

and is essentially equivalent to the power function of a test of $\theta \leqslant \theta_0$ against alternatives $\theta > \theta_0$ as this was described in Chapter 4. Without a prior distribution, some further criterion is required to define a test in the decision-theoretic sense, e.g. it is usual to place an upper bound on the first of the two risk components in (32), and then the size of the test corresponds to $r(\theta_0, \delta)$.

For composite hypotheses involving nuisance parameters, the regret function may be taken to be invariant under the group of transformations leaving the testing problem invariant, as described in Section 5.3. This reduces the decision problem to one involving maximal invariants.

The relation of the decision problem with significance testing is no more than a crude resemblance, the objective of significance testing being to assess evidence of inconsistency with a null hypothesis, rather than to reach a decision. This point is illustrated in Example 11.5. In a genuine two-decision problem it is unlikely that the simple loss function (31) will be appropriate.

A rather similar decision-theoretic representation can be given for interval estimation. The allowable decisions correspond to intervals on the θ axis for one-dimensional parameters, or more generally to sets in Ω. Let d_A denote the decision to adopt set A. The simple utility function analogous to (31) is

$$u(\theta, d_A) = \begin{cases} 1 & (\theta \notin A), \\ 0 & (\theta \in A), \end{cases} \tag{33}$$

and the risk function for any decision rule δ is

$$r(\theta, \delta) = \text{pr}\{\theta \notin A_\delta(Y); \theta\}, \tag{34}$$

where $A_\delta(Y)$ is the set associated with Y under rule δ. Thus a $1 - \alpha$ confidence region can be interpreted as a decision rule with constant risk α. Again the parallel is only a very formal one. A genuinely decision-theoretic formulation of the problem would probably involve a more complicated utility function than (33) in which the loss of utility when $\theta \notin A$ depends on the distance of θ from A and in which there is a penalty depending on the size of A; see Exercise 11.3. Such a version would be applicable to some problems of search.

Both significance testing and interval estimation are concerned with explicit assessment of uncertainty and hence are not properly regarded as decision problems. Point estimation in the narrow sense, however, when regarded as the end-point of an analysis, does not involve an explicit statement of uncertainty and hence can reasonably be treated as a decision problem. If d_t is the decision to adopt the estimated value t, then a non-randomized decision rule assigns a value $t(Y)$ to the parameter. The regret function $w(\theta, d_t)$ will typically be an increasing function of the distance between t and θ and will vanish if and only if $t = \theta$. If, in particular, in suitable units

$$w(\theta, d_t) = (t - \theta)^2, \tag{35}$$

then the risk is the mean squared error considered as a function of θ. We have used this already in Examples 11.2, 11.8 and 11.11. Of course, it is typical of non-Bayesian theory that no estimate minimizes risk uniformly in θ. Some additional *ad hoc* criterion or restriction of the estimate is then necessary. One criterion is the minimax one described in the previous section. A more conventional restriction is to the class of unbiased estimates, but this often contradicts the basic decision theory criterion of admissibility. An outstanding and important example of this is discussed in the next section.

Another criterion that may be considered is that of invariance, according to which, for example, an estimate $t(Y)$ of an unrestricted one-dimensional location parameter θ should satisfy

$$t(Y + a1) = t(Y) + a \; (-\infty < a < \infty) \tag{36}$$

if the regret function depends only on $t - \theta$. There is a strong connexion between invariance and unbiasedness. The following example introduces a quite general invariant estimate.

Example 11.12. Pitman estimate of location. Suppose that $Y_1, \ldots, Y_n$ are i.i.d. with continuous p.d.f. $h(y - \theta)$, with $\Omega = (-\infty, \infty)$. Let $\mathcal{G}$ be the group of translations, i.e. with typical element g_a transforming y into $y + a1$. The corresponding group of transformations on Ω consists of the same translations, and the squared error regret function (35) is invariant.

The estimate $t(Y)$ is invariant if it satisfies (36). We know from Example 5.14 that a maximal invariant under $\mathcal{G}$ is the set of differences $Z_j = Y_{(j)} - Y_{(1)}(j = 2, \ldots, n)$. Then, on transforming from Y to $(Y_{(1)}, Z_2, \ldots, Z_n)$, the risk function of $t(Y)$ can be written

$$r(\theta, t) = n! \int_{-\infty}^{\infty} \int_0^{\infty} \ldots \int_0^{\infty} \{t(y_{(1)}, y_{(1)} + z_1, \ldots, y_{(1)} + z_n) - \theta\}^2 h(y_{(1)} - \theta)$$

$$\times \prod_{j=2}^{n} h(y_{(1)} + z_j - \theta) dz_j \, dy_{(1)}.$$

Now, by the invariance of the regret function and (36), we can write

$$r(\theta,t) = n! \int_{-\infty}^{\infty} \int_{0}^{\infty} \dots \int_{0}^{\infty} t^2(y_{(1)}, y_{(1)} + z_2, \dots, y_{(1)} + z_n) h(y_{(1)})$$

$$\times \prod_{j=2}^{n} h(y_{(1)} + z_j) dz_j dy_{(1)} . \tag{37}$$

Further, (36) also implies that

$$t(y_{(1)}, y_{(1)} + z_2, \dots, y_{(1)} + z_n) = t(0, z_2, \dots, z_n) + y_{(1)}.$$

It is now evident that the minimum risk estimate must minimize the inner integral in (37) for all $(z_2, \dots, z_n)$. Hence

$$t(y_{(1)}, y_{(1)} + z_2, \dots, y_{(1)} + z_n) = y_{(1)} - \frac{\int v h(v) \prod_{j=2}^{n} h(v + z_j) dv}{\int h(v) \prod_{j=2}^{n} h(v + z_j) dv},$$

so that the minimum risk invariant estimate is

$$t(y) = y_{(1)} - E(Y_{(1)} | Y_{(2)} - Y_{(1)} = y_{(2)} - y_{(1)}, \dots, Y_{(n)} - Y_{(1)} = y_{(n)} - y_{(1)} \tag{38}$$

This is the Pitman estimate of location (Pitman, 1938), and corresponds to the use of a uniform improper prior on Θ over the whole real line. An equivalent form of the estimate uses $\bar{y}$ in place of $y_{(1)}$, from which we can see that the estimate is an adjustment of the sample mean based on the ancillary statistic $(z_2, \dots, z_n)$. The Pitman estimate is admissible and hence minimax.

An alternative, and in some ways more enlightening, form of (38) is obtained by using the connexion discussed in Example 7.7 between the conditional density given the ancillary statistic and the likelihood function, normalized to integrate to one. This shows that (38) is the mean of the normalized likelihood function ; in fact the estimate is sometimes called the mean likelihood estimate. A particular consequence is that if the likelihood function is approximately symmetrical about its maximum, then the Pitman estimate is likely to differ little from the maximum likelihood estimate.

The various *ad hoc* methods of deriving point estimates serve to emphasise the point made earlier in Chapter 8 that without a prior distribution the selection of a point estimate involves somewhat arbitrary elements, when the point estimate is the final goal of the analysis. However, it is rare in the general analysis of data that the objective really is just the specification of a point estimate without even a crude specification of precision. The use of point estimates in the reduction of complex data is a different matter, and, as we argued in Chapter 8, unbiased estimates of minimum variance may then have a direct justification. Of course, such applications as the control problem of Example 11.2 do specifically require point estimates.

In a similar way problems involving future observations rather than parameter values can be given a decision-theoretic statement. In the first place the utility is defined as a function of the future observations and the decision, but, as in (17), by taking expectations this is directly converted into a function of the parameter value and the decision.

11.8   A remarkable case of inadmissibility

We now discuss an example of a point estimate which, at first sight very surprisingly, is inadmissible. Suppose that $Y_1, \ldots, Y_n$ are independently normally distributed with means $(\mu_1, \ldots, \mu_n)$ and known variance σ_0^2. It is required to estimate the vector parameter $\mu = (\mu_1, \ldots, \mu_n)$ by an estimate $t = (t_1, \ldots, t_n)$, the regret being

$$(t_1 - \mu_1)^2 + \ldots + (t_n - \mu_n)^2. \tag{39}$$

It may make the problem more vivid to think of the Y_j's as means of samples, a preliminary reduction having been made on grounds of sufficiency. In fact, with suitable transformation, the sufficient statistics for the general linear model can be made independent and the discussion can be extended to cover this case, although a general quadratic regret function is then involved.

It is "natural" to think of T_j as a function of Y_j alone, and to use the "natural" estimate $T_N = (Y_1, \ldots, Y_n)$ in which every component of μ is estimated separately. The risk function for this estimate is the sum of the expected squared errors and is therefore $n\sigma_0^2$. The constancy of the risk function suggests that the estimate is minimax and this can be proved rigorously.

In Example 10.9 we derived the posterior means of the μ_j for a similar situation, the difference being that the prior mean ξ had a prior distribution with variance χ_0 and the sampling variance of Y_j was taken to be one. The results given there apply to the present case when $\chi_0 = 0$ and when ν_0/σ_0^2 is substituted for ν_0. For the regret function (39), the posterior means are the Bayes estimates, and so we deduce from the modified version of Example 10.9 that the Bayes estimate for μ can be written in vector form

$$T_{\mathbf{B}} = (1 - \alpha)\, Y + \alpha \xi_0 1\,, \qquad (40)$$

where $\xi_0 1$ denotes a vector all of whose components are ξ_0 and

$$\alpha = \frac{\sigma_0^2}{\nu_0 + \sigma_0^2}.$$

The risk of the Bayes estimate $T_{\mathbf{B}}$ is lower than $n\sigma_0^2$ if the μ_j are near ξ_0, but much greater than $n\sigma_0^2$ if the μ_j are distant from ξ_0. This was illustrated in Fig. 11.1.

If now we suppose that the variance ν is unknown, with ξ_0 still known, then we can treat the situation as a fully parameterized empirical Bayes problem, to be dealt with by the methods of Section 10.7. In fact, we described this problem in Example 10.13, but for the more general case where ξ and the variance σ^2 are unknown. In the present case we can estimate ν by the unbiased estimate

$$\Sigma(Y_j - \xi_0)^2/n - \sigma_0^2$$

and, substituting this for ν_0 in (40), we obtain the empirical Bayes estimate

$$T_{\mathbf{E}\mathbf{B}} = \left\{1 - \frac{n\sigma_0^2}{\Sigma(Y_j - \xi_0)^2}\right\} Y + \frac{n\sigma_0^2}{\Sigma(Y_j - \xi_0)^2}\, \xi_0 1. \qquad (41)$$

This would correspond exactly to (10.66) if we defined $T_{\mathbf{E}\mathbf{B}} = \xi_0 1$ when the coefficient of Y in (41) is negative, a natural consequence of the restriction $\alpha \leqslant 1$ in (40).

There is a very important difference between (40) and (41). Both involve "shrinking" the vector Y towards the origin $\xi_0 1$; in (40) the shrinking is by a fixed factor, whereas in (41) the shrinking is by a variable factor that is negligible if Y is far from $\xi_0 1$. This makes the empirical Bayes estimate in many ways very much more appealing than the proper Bayes estimate. One might still expect the risk function of (41) to be below $n\sigma_0^2$ only if μ is suitably close to $\xi_0 1$. James and Stein (1961), however, established that if (41) is modified to

$$T_{ST} = \left\{1 - \frac{(n-2)\sigma_0^2}{\Sigma(Y_j - \xi_0)^2}\right\} Y + \frac{(n-2)\sigma_0^2}{\Sigma(Y_j - \xi_0)^2} \xi_0 1, \qquad (42)$$

then for $n \geqslant 3$ the risk function is uniformly below $n\sigma_0^2$, the risk of the "natural" estimate. Thus the "natural" estimate $T_N = Y$ is inadmissible for $n \geqslant 3$; for $n \leqslant 2$, T_N is admissible. Stein (1956) had earlier proved the inadmissibility of T_N. In fact T_{ST} also is inadmissible, even when modified to be equal to $\xi_0 1$ when the coefficient of Y is negative.

We prove the superiority of T_{ST}, following James and Stein (1961), by computing explicitly the risk function of (42).

We can, without loss of generality, take $\xi_0 = 0$ and $\sigma_0^2 = 1$, and consider the slightly more general random variable

$$T'_{ST} = \left(1 - \frac{b}{D}\right) Y, \qquad (43)$$

where $D = \Sigma Y_j^2$. The risk function of T'_{ST} for arbitrary b is

$$r(\mu, \delta'_{ST}) = E\left[\sum_{j=1}^{n} \left\{\left(1 - \frac{b}{D}\right) Y_j - \mu_j\right\}^2 ; \mu\right]$$

$$= E\left\{\left(1 - \frac{b}{D}\right)^2 D ; \mu\right\} + \rho - 2\Sigma\mu_j E\left\{\left(1 - \frac{b}{D}\right) Y_j ; \mu\right\},$$
$$(44)$$

where $\rho = \Sigma\mu_j^2$. The first term in (44) can be calculated from the fact that D has a non-central chi-squared distribution, so that the outstanding part of the calculation is the last part of the final term in (44). For this, note that

$$E\left\{(Y_j - \mu_j)/D ; \mu\right\} = \frac{\partial}{\partial\mu_j} \int \ldots \int \frac{1}{D} \prod_{k=1}^{n} \phi(y_k - \mu_k) d\mu_k$$
$$(45)$$
$$= 2\mu_j \frac{\partial E_n(D^{-1} ; \rho)}{\partial\rho},$$

where $E_n(. ; \rho)$ denotes expectation with respect to the non-central chi-squared distribution of D with n degrees of freedom and non-centrality parameter ρ. To proceed further, we use the fact that in this distribution D can be written X_{n+2J}^2, where J has a Poisson distribution of mean $\frac{1}{2}\rho$ and X_m^2 has a central chi-squared distribution with m degrees of freedom. The first consequence of this is that

$$2 \frac{\partial E_n(D^{-1};\rho)}{\partial \rho} = E_{n+2}(D^{-1};\rho) - E_n(D^{-1};\rho),$$

and substitution through (45) back into (44) gives

$$r(\mu,\delta'_{ST}) = (n - 2b) + 2\rho E_{n+2}(D^{-1};\rho) + b^2 E_n(D^{-1};\rho) . \quad (46)$$

Using the above mixture representation for D again, we have

$$\tfrac{1}{2} \rho E_{n+2}(D^{-1};\rho) = E\left(\frac{J}{X^2_{n+2J}};\rho\right),$$

so that (46) becomes

$$r(\mu,\delta'_{ST}) = n - 2b + 4b\, E\left(\frac{J}{X^2_{n+2J}};\rho\right) + b^2 E\left(\frac{1}{X^2_{n+2J}};\rho\right)$$

$$= n - 2b(n - 2) E\left(\frac{1}{n - 2 + 2J};\rho\right) + b^2 E\left(\frac{1}{n - 2 + 2J};\rho\right).$$

This risk function is minimized with respect to b at $b = n - 2$, giving the minimum risk for δ_{ST}, namely

$$r(\mu,\delta_{ST}) = n - (n - 2)^2 E\left(\frac{1}{n - 2 + 2J};\rho\right), \quad (47)$$

which is never greater than n, the risk of T_N.

If $\mu = 0$, so that shrinking of Y in T_{ST} is towards the true value, then $J = 0$ in (47) and the total estimation risk is reduced from n to 2. At the other extreme, if $\rho = \Sigma\mu_j^2$ is very large, then J is very large and the risk of the James-Stein estimate approaches n.

For simplicity, we took the prior mean to be known equal to zero. Essentially no new points arise if we allow this mean to be unknown and shrink toward the sample mean $\bar{Y}_.$. The James-Stein estimator then takes the form

$$T^*_{ST} = \bar{Y}_.\,1 + \left\{1 - \frac{(n - 3)\sigma_0^2}{\Sigma(Y_j - \bar{Y}_.)^2}\right\}(Y - \bar{Y}_.\,1) , \quad (48)$$

with risk function, for $\sigma_0^2 = 1$,

$$r(\mu,\delta^*_{ST}) = n - (n - 3)^2 E\left(\frac{1}{n - 3 + 2J};\rho^*\right), \quad (49)$$

where $\rho^* = \Sigma(\mu_j - \bar{\mu}_.)^2$. This can be derived directly from the previous results by working with the residual vector $Y - \bar{Y}.1$ in the $n - 1$ dimensional space orthogonal to the vector $\bar{Y}.1$.

In a more general treatment, a linear model could be fitted to the data and a shrinking applied in the space forming the orthogonal complement to the parameter space. The results generalize in a fairly obvious way.

We now make some general comments on the interpretation of these results. The estimate T_{ST} in (42) shrinks the vector Y towards $\xi_0 1$, and there is a reduction in risk from that of the usual estimate T_N whatever value of ξ_0 is used. The reduction is very slight unless $\xi_0 1$ is close to μ. With the estimate T_{ST}^*, the risk depends only on the scatter of the means and is still as low as 3 if the means are identical. From a practical point of view, if there is to be a worthwhile gain, then it is necessary that prior information should reliably predict small differences among the means ; to be specific, one can show from (49) that an appreciable gain is only achieved if $\Sigma(\mu_j - \bar{\mu}_.)^2 < n\sigma_0^2$. Of course, if prior information on ξ is available, then it should be used, but it would have to be quite precise in order to improve on T_{ST}^*.

Next, it is crucial in the above discussion that we assess estimates by the composite regret function (39) and not by the separate individual components. The inadmissibility of the usual estimate T_N requires there to be at least three components. In particular, we must be prepared for the James-Stein estimate to be worse than the usual estimate for certain components ; this must be so because T_N is minimax. As an extreme case of this, suppose that n is large and that $\mu_1 = \rho^{\frac{1}{2}}$, $\mu_2 = \ldots = \mu_n = 0$, with ρ of order n. The James-Stein estimate of μ_1 with $\xi_0 = 0$ in (42) can then be written approximately as

$$\left(1 - \frac{n\sigma_0^2}{\rho}\right) Y_1,$$

so that its mean squared error is approximately

$$\left(\frac{n\sigma_0^2}{n\sigma_0^2 + \rho}\right)^2 \rho,$$

which is of order n, compared to the risk σ_0^2 for the usual estimate Y_1. At the same time, the mean squared errors of the other estimates are very small, the total risk being n, according to (47).

Of course, if it is known that the first component is likely to be anomalous, then the shrinking procedure would be applied to the remaining components only. If the extreme case just described above did occur, then it would be quite apparent from the data and the first component could be treated separately. More generally, we could use a compromise decision rule that would limit the amount of shrinking. This type of rule has been discussed in detail by Efron and Morris (1972a, 1973).

There are many generalizations of the above results such as to the situation where the variance or covariance matrix is unknown, when loss functions other than squared error are involved and where parameters other than those in normal-theory linear models are of interest. The essential element is that there is a composite regret function involving three or more parameters (Brown, 1966). Some corresponding results (Stein, 1962) are available on confidence interval estimation, although the position there is less clear.

To complete this discussion, there follow a few notes on the possible practical implications of these results.

(a) If the point estimates serve only as intermediate steps in the analysis of complex data, it will probably be merely a matter of convention whether shrinking is applied at the intermediate step ; the original vector y can be recovered from t_{ST}. Since it will often be convenient to keep the intermediate quantities as nearly independent as possible, it will probably usually be best not to use estimates such as T_{ST}.

(b) If there is a well-defined stable physical random mechanism generating the different parameter values, then the empirical Bayes formulation is sensible and, provided that the distributional assumptions are reasonable and consistent with the data, an estimate such as T_{ST}, and associated confidence regions, seem entirely appropriate. In practice, one would usually shrink only in the error space of a fitted linear model, as with T_{ST}^{*}. Note, however, that two or more well separated clusters of the observations would throw doubt on the normality of the μ_j's, and in such a case one would not shrink these clusters towards each other. Compromise empirical Bayes procedures that take account of clusters can, of course, be devised (Efron and Morris, 1973).

(c) Suppose that there is no clear random mechanism generating the means, but that it is reasonable to expect the means to be quite close together, and that it is required to give a set of point estimates

of the means without explicit statement of their uncertainty. Is it
sensible to use T^*_{ST} or some similar estimate? This depends on
whether one is prepared to accept the possibility of an overall gain at
the expense of a worsening of individual estimates. For example, the
means might be significance limits of a test statistic for various
situations and the observations estimates of them by simulation
(Efron and Morris, 1972 b). The composite regret (39) can be
interpreted as the average loss in repeated use of the tables, when all
entries have equal usage. As such, if a single table is to be published,
it seems quite reasonable to give one minimizing composite regret.
On the other hand, a once-and-for-all user of the table, wanting to
enter the table at a point estimate seriously affected by the shrinking,
might in a certain circumstance prefer the original estimate, if that
were available.

11.9   Sequential decision problems

Throughout the previous discussion we have supposed that there is a
single occasion on which one of a number of possible decisions has to
be taken. In the fully specified situation, the objective is to maximize
the expected utility. It often happens, however, that there are several
different decisions to be faced ; often these are to be taken in a
particular time sequence. The theory of sequential decision-making
is a major subject and here we give only an introduction to a few key
ideas. For a more extensive account, see the book by DeGroot (1970);
further references are given in the bibliographic notes.

The decisions involved at each action point in a sequential decision
problem may be

(i) terminal decisions of the type discussed in earlier sections ; or

(ii) decisions as to what kind of observations should be taken
next, e.g. as to what factor levels should be chosen for the next stage
in a factorial type of experiment ; or

(iii) decisions as to how many, if any, observations of a given kind
should be taken.

One trivial case can be dealt with immediately. Suppose that there
are k decisions, with given decision spaces $\mathfrak{D}^{(1)}, \ldots, \mathfrak{D}^{(k)}$, all
decisions to be made on the same data, and that the utility function
is the sum of contributions

$$\sum_{j=1}^{k} u_j(\theta, d^{(j)}) \qquad (50)$$

for any $d^{(j)} \in \mathfrak{D}^{(j)}$. Then the overall expected utility is clearly maximized by maximizing each component term separately. The same conclusion would apply more generally if the data available for the jth decision does not depend on the decision taken at other action points. In such cases, therefore, we analyze the separate decision problems separately and the fact that several decision problems are based on the same or related data need not concern us. However, as soon as the utility function or the data available for one decision problem depends on other decisions in the sequence, the whole problem becomes essentially more complicated. For example, taking the optimum decision in $\mathfrak{D}^{(1)}$ may leave us very poorly placed to tackle the decision in $\mathfrak{D}^{(2)}$, and so on.

It is, in principle, possible to enlarge the decision and sample spaces, so that, for example, a single decision in the new space specifies the decisions to be taken at all stages. Then in effect we explore the consequences of every possible combination of separate decisions. Except in very simple problems, this approach is too cumbersome to be practicable.

The different decisions may be interlinked in a complex way, but from now on we suppose them taken in sequence in time.

Example 11.13. Single choice of sample size. Suppose that there are two stages of decision-making. In the first, the possible decisions $d_j^{(1)}$ correspond to choices of sample size, with $d_0^{(1)}$ being the decision to take no observations. The decision $d_j^{(1)}$ involves a utility of $-c_j$, i.e. c_j is the cost of obtaining the amount of data implied by adopting $d_j^{(1)}$. Here $\mathfrak{D}^{(1)}$ is the set of possible sample sizes available.

If we choose the experiment corresponding to $d_j^{(1)}$, we then observe a vector random variable Y having p.d.f. $f_Y^{(j)}(y;\theta)$, where θ is an unknown parameter with prior p.d.f. $f_\Theta(\theta)$. At the second stage, there is a terminal decision problem of the type considered previously. If decision $d_k^{(2)}$ is taken, then the resulting utility is $u(\theta, d_k^{(2)})$. Thus the total utility achieved is

$$u(\theta, d_k^{(2)}) - c_j \;.$$

A simple special case may clarify the situation. Let the unknown parameter θ be the mean μ of a normal distribution which has the $N(\xi_0, \nu_0)$ prior distribution, and suppose that the variance per observation has the known value σ_0^2. The second stage decisions might be any of those described in earlier sections. To be explicit, suppose that they correspond to Example 11.2 with utility function (2) ; thus we are estimating μ with squared error regret.

In the first stage, let $d_j^{(1)}$ be the decision to take j observations, this having utility

$$c_j = \begin{cases} 0 & (j = 0), \\ -k_0 - k_1 j & (j \neq 0), \end{cases}$$

where $k_0, k_1 > 0$.

A crude method of solution is to find the expected utility for each combination of first and second stage decisions, and then to take the pair of decisions giving an overall maximum. This, while quite feasible in the present example, is, as noted in the general discussion, usually very cumbersome. Instead we argue backwards in time as follows. Consider the second stage decision, given that a particular first stage decision $d_r^{(1)}$ has been taken. We now have exactly the situation contemplated earlier, the mean of the posterior distribution being taken as the estimate and the conditional expected utility given $d_r^{(1)}$ is

$$a - b(1/\nu_0 + r/\sigma_0^2)^{-1},$$

as in Example 11.5.

Now consider the first stage decision. If $d_r^{(1)}$ is taken, then the second stage will be that just discussed. Thus the total expected utility to be achieved from $d_r^{(1)}$ is

$$\begin{cases} a - b\nu_0 & (r = 0), \\ a - b(1/\nu_0 + r/\sigma_0^2)^{-1} - k_0 - k_1 r & (r = 1, 2, \dots). \end{cases} \tag{51}$$

Note that for $r = 0$ the utility only involves the prior variance.

The optimum value of r, i.e. the best decision $d_r^{(1)}$, is easily found from (51). If the value is fairly large, so that it is legitimate to treat r as a continuous variable, we find on differentiation that

$$r_{\mathrm{opt}} \doteq \sigma_0 \sqrt{(b/k_1)} - \sigma_0^2/\nu_0.$$

If k_0/k_1 and σ_0^2/ν_0 are both appreciable, then it may be best to take no observations ; this can be examined quantitatively from (51).

This approach to choosing an appropriate sample size may be

compared with that based on power functions or consideration of confidence interval width. The present analysis gives an explicit answer, once the necessary information about utilities and prior distributions is introduced; in the earlier approaches essentially the same information is introduced indirectly, for example in choosing appropriate constraints on the power function.

More generally, in the problem formulated at the beginning of this example, we can first calculate, for every $d_r^{(1)}$, the optimum second stage decision and the expected utility associated with it. Then we move back to the first stage, bearing in mind that any first stage decision will be followed by the associated optimum second stage decision.

In more complicated problems, with several stages of decision-making, the same principle can be followed. It is sometimes convenient, especially in finite problems, to represent the system graphically by a decision tree, in which there are nodes of two types

(a) decision nodes, at which the different branches correspond to different decisions among which a choice has to be made;

(b) "random" nodes, at which the different branches correspond to the different observations that may be obtained with the relevant probability distributions. In the tree, the order of increasing time is left to right; the order of solution is right to left. Figure 11.3 shows the decision tree for Example 11.13. At the decision node $\mathscr{D}^{(1)}$ the branches represent the possible decisions to take $0,1,2, \ldots$ observations; with each is the associated utility representing the cost of observations. Along each such branch is a "random" node, at which the relevant observations become available.

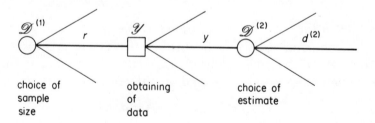

Fig. 11.3 Decision tree for simple experimental design problem of Example 11.13.

The different branches from here represent schematically the possible

observations that can result. Each leads to a second decision node
$\mathfrak{D}^{(2)}$, the branches from which are the possible terminal decisions, i.e.
point estimates of the mean given y.

   The following example, taken from El-Sayyad and Freeman (1973),
illustrates the more general sequential determination of sample size
in a simple situation.

Example 11.14. Sequential estimation for a Poisson process. Suppose
that we wish to estimate the rate ρ of a stationary Poisson process,
where the number of events in a fixed time interval $(0,t)$ has the
Poisson distribution with mean ρt. For simplicity, we take the improper
uniform prior density for $\log P$, so that the posterior density of P for
a given number n of events occurring in the interval $(0,t)$ is

$$t^n \rho^{n-1} e^{-\rho t}/\Gamma(n) . \tag{52}$$

The assumed utility function has three components, namely $-k_1 n$
for the number of observed events n, $-k_2 t$ for the length of observation time t, and $a - b (d - \rho)^2$ for the estimation of ρ by the
number d. These are additive, so that the total regret function is

$$b(d - \rho)^2 + k_1 n + k_2 t . \tag{53}$$

   In the decision tree, the last decision node corresponds to the
choice of decision rule for estimating ρ with given values of n and t.
All preceding decision nodes refer to decisions on whether or not to
continue sampling.

   For this particular problem there are two elementary sampling
procedures. In the first, we can at each stage wait for another event
to occur, and then decide whether or not to continue sampling. In
the second, we can observe for time intervals of small length Δ and
at the end of each interval decide whether or not to continue. It turns
out to be simpler to consider the second procedure, and this we shall
do. The other procedure can be shown to give very similar results in
many cases.

   Our decision tree, then, is such that at each random node we
observe the number of events occurring in a time interval of length
Δ, following which there is a decision node at which we decide
whether or not to continue sampling ; this decision depends on the
regret function (53). There are two branches from each such decision
node, one to a random node corresponding to the decision to continue
sampling and one to the terminal decision node at which we choose

the decision corresponding to a point estimate of ρ.

As explained before, we work in reverse time order in the decision analysis. Therefore, first we suppose that the decision has been reached to stop sampling, with observed number of events n after time t. With n and t fixed, the optimum strategy according to (53) is to choose d to be the posterior mean of P, which is n/t, with associated mean squared error bn/t^2. Thus, for given n and t, the minimum Bayes risk is

$$r_B(n,t) \; = \; bn/t^2 + k_1 n + k_2 t, \tag{54}$$

say. Notice that this is a unimodal function of t for any value of n.

The general argument for such problems now proceeds as follows. Suppose that we follow the optimal Bayes strategy from the point at which n and t are as observed, with associated Bayes risk $v(n,t)$. The optimal strategy corresponds either to immediate termination of sampling, or to sampling once more and following the optimal strategy from then on. In the present context, let R additional events be observed in the time interval $(t, t + \Delta]$, so that the Bayes risk of the optimal strategy from time $t + \Delta$ will be $v(n + R, t + \Delta)$. Therefore, given the situation at time t, we have the *dynamic programming equation*,

$$v(n,t) \; = \; \min\left[r_B(n,t), E\{v(n + R, t + \Delta)|n,t\}\right], \tag{55}$$

because in the optimal strategy we will choose the better of the two alternatives at time t.

If we assume that Δ is small, then by the definition of the Poisson process we have

$$\text{pr}(R = 1|P = \rho) \doteq \rho\Delta \doteq 1 - \text{pr}(R = 0|P = \rho)$$

and, conditional on (n,t), R has the density (52). Therefore

$$\text{pr}(R = 1|n,t) \doteq n\Delta/t \doteq 1 - \text{pr}(R = 0|n,t),$$

and (55) can be written

$$v(n,t) \; \doteq \; \min\left\{r_B(n,t), \frac{n\Delta}{t} v(n + 1, t + \Delta) + \left(1 - \frac{n\Delta}{t}\right) v(n, t + \Delta)\right\}. \tag{56}$$

Now suppose that it is optimal to stop sampling at the point (n,t), but not at $(n, t - \Delta)$ or before. Then, by definition, we must have the following relationships :

$v(n,t - \Delta) < r_B(n,t - \Delta), v(n,t) = r_B(n,t)$ and $v(n+1,t) \leqslant r_B(n+1,t)$.

Using these inequalities together with the representation (56) for $v(n,t - \Delta)$, we see that a lower bound for t, given n, is provided by the inequality

$$\frac{n\Delta}{t - \Delta} r_B(n + 1,t) + \left(1 - \frac{n\Delta}{t - \Delta}\right) r_B(n,t) < r_B(n,t - \Delta)$$

and substitution from (54) gives the inequality

$$k_2 + k_1 n/t - bn/t^3 > 0. \qquad (57)$$

Thus, for each value of n, values of t satisfying (57) correspond approximately to stopping points. This determines a boundary in the (n,t) plane on one side of which it is optimal to continue sampling. The error involved in using the approximation to obtain (57) can be shown to be very small.

Equation (57) can be obtained also by the following tempting argument. If

$$E\{r_B(n + R, t + \Delta)|n,t\} - r_B(n,t) > 0, \qquad (58)$$

there is an increase in expected regret by sampling for one more time interval, i.e. if (58) holds it is better to stop at (n,t) than to continue for just one more interval. Now substitution in (58) yields (57). While this argument certainly shows that sampling for just one further interval is an inferior strategy, it does not show that sampling for an appreciable further interval also is inferior. In fact, it is only in exceptional cases that the argument leading to (58) gives the right answer ; usually it will lead to premature stopping.

A dynamic programming equation similar to (55) holds for a broad class of sequential stopping problems. Normally solution by backward recursion is used ; the key is that if a sufficiently rich class of points can be found for which it is certainly best to stop, it is known that for these points $v(.)$ is equal to the Bayes risk for the optimal terminal decision. As the solution is recovered by repeated substitution in the dynamic programming equation, the continuation region is revealed by those points for which $v(.)$ is less then the terminal Bayes risk.

Bibliographical notes

Many of the basic ideas of statistical decision theory were stated by
Neyman and Pearson (1933b). A systematic theory for the situation
when there is no prior distribution was developed by Wald and set out
in a book (Wald, 1950) completed just before his death; this included
a formulation of common statistical problems in decision theory
terms. Important contributions continuing Wald's work are by
Girshick and Savage (1951) and Stein (1956 and unpublished
lecture notes). A detailed introduction is given by Ferguson (1967).
This work lead to the view, widely held for a period, that all statistical
problems should be formulated in terms of decision making.

von Neumann and Morgenstern developed an axiomatic approach
to subjective utilities in connexion with their work on game theory
(von Neumann and Morgenstern, 1953, 1st ed. 1935); see DeGroot
(1970), Fishburn (1969) and, for an elementary account, Chernoff
and Moses (1959).

In the late 1950's and early 1960's theoretical interest shifted from
the situation without prior distributions to Bayesian decision theory
in which personalistic prior distributions and utilities are central to
the argument; an authoritative account was given by Raiffa and
Schlaifer (1961). There are now a number of excellent elementary
introductions (Aitchison, 1970a; Lindley, 1971b; Raiffa, 1968).
There is also an extensive economic, psychological, sociological and
operational research literature on decision making. A review on the
practical measurement of utilities is given by Hull, Moore and Thomas
(1973).

Dynamic programming, i.e. optimal sequential decision-making is
treated in general terms by Bellman (1957). The application to
sequential sampling is given in outline by Wald (1950,p.158) and much
more explicitly by Wetherill (1961). Whittle (1964, 1965) has discussed
the finite case in elegant general terms. Lindley (1968) has applied
the decision tree to the selection of variables in multiple regression
under some rather stringent assumptions. For an introduction to the
extensive control theory literature, see Aoki (1967) and Åström
(1970).

Work on sequential stopping problems is reviewed briefly by
Lindley (1971a). Choice of sample size in one- and two-stage investigations is discussed by Yates (1952) and Grundy, Healy and Rees
(1956).

Further results and exercises

1. From data y, the posterior p.d.f. of a scalar parameter θ is found. The possible decisions d_t are indexed by a real number t. Three possible utility functions are contemplated, namely

   (i) $a - b_1(t - \theta)^2$,
   (ii) $a - b_2|t - \theta|$,
   (iii) $a - b_3(t - \theta)^4$,

where $b_k > 0$ $(k = 1,2,3)$.

Show that the Bayes decision rules for (i) and (ii) are, respectively, the mean and median of the posterior distribution, and that for (iii) it is $\mu(\theta) + \kappa(\theta)\,\sigma(\theta)$, where $\kappa^3(\theta) + 3\kappa(\theta) - \gamma_1(\theta) = 0$ and $\mu(\theta), \sigma(\theta)$ and $\gamma_1(\theta)$ are respectively the mean, standard deviation and skewness of the posterior distribution.

Explain qualitatively the reason for the difference between the answers for (i) and (ii).

[Section 11.3]

2. Suppose that Y is $N(\mu,1)$, where μ is 1 with prior probability π and -1 with prior probability $1 - \pi$. Find the Bayes rule for discriminating between the two possible values of μ when the zero-one regret function is used.

Now suppose that $Y_1, \dots, Y_n$ are conditionally independent in $N(\mu_1,1), \dots, N(\mu_n,1)$, where the μ_j are independently distributed in the above two-point prior distribution. Construct an empirical Bayes discrimination rule, estimating π from the data. Examine critically the qualitative and quantitative properties of this rule.

[Sections 11.3 and 10.7 ; Robbins, 1951]

3. To represent an interval estimation problem in decision-theoretic terms, it is assumed that for a scalar parameter θ the regret function corresponding to the interval $[t_1, t_2]$ is

$$w(\theta ; t_1, t_2) = \begin{cases} a(t_1 - \theta) + (t_2 - t_1) & (\theta < t_1), \\ (t_2 - t_1) & (t_1 \leqslant \theta \leqslant t_2), \\ b(\theta - t_2) + (t_2 - t_1) & (t_2 < \theta), \end{cases}$$

where $a, b > 0$. Taking for simplicity the situation where the posterior distribution of Θ is $N(\bar{y}, \sigma_0^2/n)$, explore the connexion between the effective confidence interval and a and b.

Generalize the discussion to the construction of an interval estimate for the mean of a normal distribution with unknown variance, using the conjugate prior distributions of Example 10.5, and a scale invariant regret function.

[Sections 11.3 and 11.7 ; Aitchison and Dunsmore, 1968]

4. There are m normal distributions of variance σ^2 whose means $\mu_1, \ldots, \mu_m$ have independently the prior distribution $N(\xi_0, \nu_0)$. A number r of observations are available independently from each distribution and the one with largest observed mean is selected as "best". The utility achieved by selecting a population of mean μ is $a\mu$. Show, for example by first finding the conditional expected utility given the largest observed mean, that the expected utility of the procedure is

$$a\xi_0 + \frac{ag_{mm}\nu_0}{(\nu_0 + \sigma_0^2/r)^{\frac{1}{2}}},$$

where g_{mm} is the expected value of the largest of m observations from the standard normal distribution.

[Section 11.3]

5. Let $S = \Sigma b(Y_j)/n$ be the sufficient statistic for the one-dimensional parameter exponential family (2.35), with $\mu(\theta) = E(S ; \theta)$. Using the Cramér-Rao lower bound of Section 8.3, show that any estimate $T = t(Y)$ which has uniformly lower risk than S for estimating $\mu(\theta)$ with regret function $\{t - \mu(\theta)\}^2$ must be unbiased. Hence deduce that S is admissible.

[Section 11.6 ; Girshick and Savage, 1951]

6. Suppose that Y is distributed in $N(\mu, \sigma_0^2)$ and that μ has the $N(0, \nu_0)$ prior distribution, σ_0^2 and ν_0 both being known. If risk is proportional to mean squared error, then the Bayes estimate $\delta_B(y)$ for μ has minimal Bayes risk, but its risk function is unbounded, whereas the m.l.e. $\delta_{ML}(Y)$ has minimax risk. A compromise estimate is

$$\delta_C(Y) = \begin{cases} \delta_{ML}(Y) + a & (Y < -k), \\ \delta_B(Y) & (-k \leqslant Y \leqslant k), \\ \delta_{ML}(Y) - a & (Y > k), \end{cases}$$

where a and k are chosen to make $\delta_C(.)$ a continuous function, but are otherwise arbitrary. Compare the Bayes risk and risk function of $\delta_C(Y)$ with those of the m.l. and Bayes estimates, and comment on the applicability of such a compromise.

Describe and investigate a corresponding compromise between the m.l.e. and the Stein estimate for a vector of normal means.

[Sections 11.6 and 11.8 ; Efron and Morris, 1971, 1972a]

7. Let $Y_1, \ldots, Y_n$ be i.i.d. in $N(\mu, \sigma^2)$ with both μ and σ^2 unknown. Location invariant estimates of σ^2 are necessarily functions of the maximal invariant $SS = \Sigma(Y_j - \bar{Y}_.)^2$. Use arguments parallel to those of Section 11.8 to prove that any such invariant estimate is inadmissible for regret function $w(\sigma^2, d) \propto (d - \sigma^2)^2$, by showing that

$$T = \min\left(\frac{SS}{n+1}, \frac{SS + n\bar{Y}_.^2}{n+2}\right)$$

has uniformly lower risk. Explain qualitatively why T might be a good estimate.

[Sections 11.6 and 11.8 ; Stein, 1964]

8. To compare two treatments, n_0 paired observations are available, leading to a normal posterior distribution for the difference in treatment means, the usual normal-theory assumptions having been made. There is the possibility of taking further observations and the number of such observations is to be determined, the cost of n_1 further observations being $k_0 + k_1 n_1 (n_1 = 1, 2, \ldots)$. Show that if the problem is one of estimation with squared error as loss, the discussion of Example 11.13 is directly applicable. Carry through the corresponding discussion for the two-decision problem of Example 11.3 in order to find the optimum n_1.

[Section 11.9 ; Grundy, Healy and Rees, 1956]

APPENDIX 1

DETERMINATION OF PROBABILITY DISTRIBUTIONS

Application of the methods of Chapters 3–7 and 9 requires the numerical determination of the probability distributions of statistics.

The best situation is, of course, when ·we can determine the exact distribution in a form allowing the use of readily available or easily computed concise tables, for example those of Pearson and Hartley (1970, 1972) and Fisher and Yates (1963). Sometimes a computer algorithm for the required probabilities will be suitable, although, even with the present wide availability of computers, it would usually be a limitation of the usefulness of a procedure for it to depend crucially on a computer for the fresh calculation of probability levels from each set of data.

With non-standard procedures, exact results are quite often so complicated that they can be used only for suggesting and checking approximations. Further, exact information may well be limited to the values of the first few moments, or to some other partial information about the distribution. Very often, however, it may be known that in the limit as some such quantity as the number n of observations tends to infinity, the distribution of interest approaches a limiting form, such as the normal or chi-squared distribution. While such limiting results are of great importance they need, if at all possible, to be supplemented by a check of their numerical adequacy for particular n.

A central problem, therefore, is to use limited information, such as the first few moments, to obtain numerical results about a distribution or to check on and improve results obtained from limit theorems. Chapter 6, in particular, contains many examples of test statistics for which the low order moments under the null hypothesis can be obtained fairly simply.

Sometimes it may be best to rely on efficient computer simulation for the determination of significance levels, using any theoretical information for controlling and checking the simulation. In particular, this may often be the quickest way of checking on isolated values arising in applications. We shall not discuss this further, but for some valuable comments, see Andrews *et al.* (1972).

Of course, the most satisfactory solution, short of an exact one, is to have quite narrow bounds for the required probabilities expressing, for example, the maximum error that can arise in using a limiting normal distribution. Unfortunately, such bounds are rarely available in useful form. Another type of inequality arises from applying Tchebychev's inequality and its generalizations (Godwin, 1964). These, while important in mathematical analysis, tend to give limits too wide to be useful in statistical applications.

The most widely-used technique in practice, although it is to some extent empirical, is the fitting of simple distributions usually by equating moments. In choosing a distribution to be fitted we are guided by information on the range of the distribution and on its limiting form. Subject to this, we choose a smooth distribution for which tables are easily available. Preliminary transformation of the statistic will often be wise.

Commonly used approximating distributions include the following:

(a) for non-negative random variables the distribution of $a\mathrm{X}_d^2$ or $a\mathrm{X}_d^{2c}$, where X_d^2 has a chi-squared distribution with d degrees of freedom;

(b) for non-negative random variables the log normal distribution with the correct first two moments;

(c) for random variables with known finite range, which can be taken as $(0,1)$, a beta distribution, which can for numerical purposes be transformed into the variance ratio distribution;

(d) for distributions symmetrical and with longer tails than a normal distribution, a multiple of t, the constant factor and the degrees of freedom being fitted to the second and fourth moments about the centre of symmetry.

These are the commoner frequency curves of the Pearson system. Other systems of distributions useful in particular applications are those of Johnson (1965) and Hastings *et al.* (1947). For approximating discrete distributions a continuous distribution with a continuity correction is usually adequate, although in some cases approximation by a Poisson or binomial distribution may be preferred.

Pearson and Hartley (1970, Table 42) reproduce a table showing for Pearson curves the standardized deviate for various probability levels as a function of the dimensionless third and fourth cumulants. This enables us to see immediately the effect on, say, the normal upper 2½% point $k_{0.025}^* = 1.96$ of specified skewness and kurtosis. This is especially useful for assessing the effect on a limiting normal approximation of modest amounts of skewness and kurtosis. Of course, the approach does assume substantial smoothness in the underlying distribution; thus an independent check is desirable when considerable accuracy is important.

A different type of approximation based on moments is particularly relevant when working with sums of i.i.d. random variables. For then an asymptotic expansion is possible for the cumulative distribution function (Kendall and Stuart, 1969, Vol. 1, Chapter 6). In outline, the argument is that the sum of n i.i.d. one-dimensional random variables, standardized to zero mean and unit variance, has cumulant generating function

$$nK(t/\sqrt{n}) = \frac{1}{2} t^2 + \frac{\gamma_1}{3!\sqrt{n}} t^3 + \frac{\gamma_2}{4!n} t^4 + \dots, \tag{1}$$

where $K(.)$ is the cumulant generating function for a single random variable scaled to zero mean and unit variance. The moment generating function can thus be expanded in powers of $1/n$ and inverted to give for the p.d.f. the expansion

$$\frac{1}{\sqrt{(2\pi)}} e^{-\frac{1}{2}x^2} \left[1 + \frac{\gamma_1}{6\sqrt{n}} H_3(x) + \frac{1}{n} \left\{ \frac{\gamma_2}{24} H_4(x) + \frac{\gamma_1^2}{72} H_6(x) \right\} + \dots \right], \tag{2}$$

for the cumulative distribution function the expansion

$$\Phi(x) - \frac{1}{\sqrt{(2\pi)}} e^{-\frac{1}{2}x^2} \left[\frac{\gamma_1}{6\sqrt{n}} H_2(x) + \frac{1}{n} \left\{ \frac{\gamma_2}{24} H_3(x) + \frac{\gamma_1^2}{72} H_5(x) \right\} + \dots \right], \tag{3}$$

and for the variate value corresponding to a cumulative probability $1 - \alpha$ the expansion

$$k_\alpha^* + \frac{\gamma_1^2}{6\sqrt{n}} (k_\alpha^{*2} - 1) + \frac{\gamma_2}{24n} (k_\alpha^{*3} - 3k_\alpha^*) - \frac{\gamma_1^2}{36n} (2k_\alpha^{*3} - 5k_\alpha^*) + \dots \tag{4}$$

The first two expansions are called the Edgeworth series and the last one the Fisher-Cornish inversion. In these formulae, $H_r(.)$ denotes the rth degree Hermite polynomial defined by $d^r(e^{-\frac{1}{2}x^2})/dx^r = (-1)^r H_r(x)e^{-\frac{1}{2}x^2}$.

The Fisher-Cornish inversion serves the same purpose as the table of Pearson and Hartley (1970) mentioned earlier in connexion with approximation by Pearson curves; it gives the modification to the normal value k_α^* on account of skewness and kurtosis.

Similar results apply to multivariate distributions and to conditional distributions.

There are the following general points to be made about this and similar asymptotic expansions.

(a) The expansions are directly relevant if we are dealing with sums of independent random variables. More refined expansions based on the method of steepest descent (Daniels, 1954) are possible.

(b) When dealing with functions other than sums and when the limiting distribution is not normal, similar expansions may be possible using the orthogonal polynomials associated with the limiting distribution.

(c) If it is required to approximate to an arbitrary standardized distribution of small skewness γ_1' and small kurtosis γ_2', the distribution not necessarily corresponding to a sum of independent random variables, then it may be sensible to take the density

$$\frac{1}{\sqrt{(2\pi)}} e^{-\frac{1}{2}x^2} \left\{ 1 + \frac{1}{6} \gamma_1' H_3(x) \right\} \tag{5}$$

or

$$\frac{1}{\sqrt{(2\pi)}} e^{-\frac{1}{2}x^2} \left\{ 1 + \frac{1}{6} \gamma_1' H_3(x) + \frac{1}{24} \gamma_2' H_4(x) + \frac{1}{72} \gamma_1'^2 H_6(x) \right\}. \tag{6}$$

(d) A serious drawback to these expansions, both for their original purpose and for the more empirical application (c), is that the density is non-negative for the important range of x only for fairly restricted values of γ_1' and γ_2' (Barton and Dennis, 1952; Draper and Tierney, 1972).

On the whole, especially on account of (d), the judicious use of Pearson or other families of curves is, for numerical work, likely to be preferable to the use of Edgeworth and similar expansions.

APPENDIX 2
ORDER STATISTICS

A2.1 General properties

Let $Y_1, \ldots, Y_n$ be i.i.d. in a continuous distribution with p.d.f. $f(y)$ and cumulative distribution function $F(y)$. The *order statistics* are these values arranged in increasing order; they are denoted by

$$Y_{(n\,1)} \leqslant \ldots \leqslant Y_{(nn)} .$$

The first suffix, denoting the number of observations, may often be omitted without danger of confusion. For the order statistics as a vector, we write $Y_{(n.)}$ or $Y_{(.)}$.

To find the p.d.f. of $Y_{(nr)}$, we note that $y_{(nr)} \leqslant z$ if and only if at least r out of $y_1, \ldots, y_n$ are less than or equal to z. Thus

$$\mathrm{pr}(Y_{(nr)} \leqslant z) = \sum_{t=r}^{n} \binom{n}{t} \{F(z)\}^t \{1 - F(z)\}^{n-t},$$

so that the p.d.f. is

$$f_{(nr)}(z) = \frac{n!}{r!(n-r-1)!} \{F(z)\}^{r-1} \{1 - F(z)\}^{n-r} f(z). \quad (1)$$

Alternatively, we can obtain (1) *via* the probability that $r - 1$ values are below z, one is in $(z, z + \Delta z)$, and $n - r$ are above z.

Similarly, the joint p.d.f. of $Y_{(nr_1)}$ and $Y_{(nr_2)}$ is, for $r_1 < r_2$ and $z_1 < z_2$,

$$f_{(nr_1r_2)}(z) = \frac{n!}{(r_1-1)!(r_2-r_1-1)!(n-r_2)!}$$

$$\{F(z)\}^{r_1-1} \{F(z_2)-F(z_1)\}^{r_2-r_1-1} \{1-F(z_2)\}^{n-r_2} f(z_1)f(z_2).$$

$$(2)$$

The joint p.d.f. of all the order statistics is

$$n! \, f(z_1) \ldots f(z_n) \quad (-\infty < z_1 < \ldots < z_n < \infty). \qquad (3)$$

An alternative specification, equivalent to the full $Y_{(.)}$, is the sample distribution function

$$\tilde{F}_n(y) = \{\text{propn of } Y_j \leqslant y\} = n^{-1} \sum_{j=1}^{n} \text{hv}(y - Y_{(j)}), \qquad (4)$$

where the unit Heaviside function $\text{hv}(.)$ is given by

$$\text{hv}(x) \; = \; \begin{cases} 1 & (x \geqslant 0), \\ 0 & (x < 0). \end{cases}$$

For the order statistics from a p.d.f. of the form $\tau^{-1} g\{(y - \mu)/\tau\}$, it is easily shown that

$$E(Y_{(nr)}) \; = \; \mu + g_{nr}\tau, \quad \text{cov}(Y_{(n.)}) = \tau^2 v_n^{(g)}, \qquad (5)$$

where g_{nr} and $v_n^{(g)}$ do not involve μ and τ and can be found from (1) and (2) by numerical integration. The special case of a normal distribution is of particular importance. Similarly for a p.d.f. $\tau^{-1} g(y/\tau)$ $(y \geqslant 0)$, $E(Y_{(nr)}) = g_{nr}\tau$. Simple approximate formulae will be given in the next section.

A.2.2   Special distributions

The transformation $V = F(Y)$ leads to random variables $V_1, \ldots, V_n$ that are i.i.d. in the uniform distribution on $(0,1)$; clearly $V_{(nr)} = F(Y_{(nr)})$. Further, the transformation $Z = -\log(1 - V)$ produces $Z_1, \ldots, Z_n$ that are i.i.d. in the exponential distribution of unit mean, and a corresponding set of order statistics. These transformations are of theoretical and practical value, because relatively simple results hold for the uniform and exponential order statistics.

First, it follows from (1) that the $V_{(nr)}$ have beta distributions, and in particular that

$$E(V_{(nr)}) \; = \; \frac{r}{n+1} \, , \, \text{cov}\, (V_{(nr_1)}, V_{(nr_2)}) = \frac{r_1 r_2}{(n+1)^2 \, (n+2)} \, . \quad (6)$$

Secondly, we have from (3) that the joint p.d.f. of $Z_{(n1)}, \ldots, Z_{(nn)}$ is

$$n! \exp\left(-\sum_{r=1}^{n} z_r\right) = n! \exp\left\{-\sum_{r=1}^{n} (n-r+1)(z_r - z_{r-1})\right\},$$

where $z_0 = 0$. Therefore the random variables

$$W_{(nr)} = (n-r+1)(Z_{(nr)} - Z_{(n,r-1)}) \qquad (7)$$

are i.i.d. in the exponential distribution of unit mean. It follows from (7) that

$$Z_{(nr)} = \frac{W_{(n1)}}{n} + \ldots + \frac{W_{(nr)}}{n-r+1}, \qquad (8)$$

and, therefore, that

$$e_{nr} = E(Z_{(nr)}) = 1/n + \ldots + 1/(n-r+1), \qquad (9)$$

$$\text{var}(Z_{(nr)}) = 1/n^2 + \ldots + 1/(n-r+1)^2. \qquad (10)$$

Note that (8) implies that the order statistics for an arbitrary continuous distribution can be written in the form

$$Y_{(nr)} = F^{-1}\left\{1 - \exp\left(-\frac{W_{(n1)}}{n} - \ldots - \frac{W_{(nr)}}{n-r+1}\right)\right\}. \qquad (11)$$

The fact that the $W_{(nr)}$ are i.i.d., combined with the special form of (11), shows that if $\{Y_{(nr)}\}$ for fixed n is considered as a stochastic process in the "time" variable r, then the Markov property holds.

A2.3   Asymptotic distributions

The asymptotic distribution theory of order statistics can be approached in various ways. The transformations of Section A2.2 show the essential equivalence of the order statistics from different distributions and that, for some purposes, it is enough to examine a particular originating distribution, e.g. the uniform or exponential. One modern approach is to start from a uniform distribution and to show that the sample distribution function (4) converges weakly to Brownian motion tied to the points (0,0) and (1,1) (Billingsley, 1968). This is a powerful way of dealing with functions of order statistics, but has the serious disadvantage of not leading readily to higher-order approximations.

A simple direct proof of the asymptotic normality of $Y_{(nr)}$ follows from (8) for $r = np$, $0 < p < 1$, $n \to \infty$. Because $Z_{(nr)}$ is a sum and the conditions for the central limit theorem for non-identically distributed random variables are satisfied, it follows that $Z_{(nr)}$ is asymptotically normal with mean and variance given respectively from (9) and (10) as

$$\frac{1}{n} + \ldots + \frac{1}{n - np + 1} \sim \frac{n}{n} \int_0^p \frac{dx}{1 - x} = -\log(1 - p) \quad (12)$$

and

$$\frac{1}{n^2} + \ldots + \frac{1}{(n - np + 1)^2} \sim \frac{1}{n} \int_0^p \frac{dx}{(1 - x)^2} = \frac{1}{n(1-p)}. \quad (13)$$

Therefore, by expansion of (11) around

$$\xi_p = F^{-1}(p), \quad (14)$$

we have, provided that $f(.)$ is continuous and non-zero at ξ_p, that $Y_{(nr)}$ is asymptotically normal with mean ξ_p and variance

$$\frac{p(1 - p)}{n\{f(\xi_p)\}^2}. \quad (15)$$

More generally, if $r_j = p_j n$ with $0 < p_j < 1$ $(j = 1, \ldots, k)$, then as $n \to \infty$ the order statistics $Y_{(nr_1)}, \ldots, Y_{(nr_k)}$ are asymptotically multivariate normal with mean and covariance matrix determined by (14), (15) and

$$\operatorname{cov}(Y_{(nr_j)}, Y_{(nr_l)}) \sim \frac{p_j(1 - p_l)}{nf(\xi_{p_j})f(\xi_{p_l})} \quad (r_j < r_l). \quad (16)$$

An alternative approach is to expand (1), or in the bivariate case (2), writing in (1) $z = \xi_p + xn^{-\frac{1}{2}}$, and holding x fixed as $n \to \infty$. In this way, we derive, not only the limiting normal form of the density, but also higher-order approximations. In the same way, approximations can be obtained for the moments of order statistics.

For the expected values, the crudest approximation is, from (14), that

$$E(Y_{(nr)}) = F^{-1}(r/n), \quad (17)$$

or, if (6) is used, $F^{-1}\{r/(n + 1)\}$. Equation (17) has the disadvantage of breaking down completely if $r = n$ and it is natural to look for approximations in which the right-hand side of (17) is replaced by

$$F^{-1}\left(\frac{r+a}{n+2a+1}\right).$$

Blom (1958) has analysed these approximations and, in particular, has shown that for the order statistics from $N(0,1)$ a good choice is $a = -\frac{3}{8}$.

The limiting bivariate distribution specified by (16) applies to order statistics that are $O(n)$ apart. Different results hold for order statistics close together. For example, it follows directly from (7) and (11) that the difference $Y_{(n,r+1)} - Y_{(nr)}$ is asymptotically exponentially distributed with mean $\{nf(\xi_p)\}^{-1}$, and is asymptotically independent of $Y_{(nr)}$.

As a particular application of these results, we consider the median defined by

$$\tilde{Y} = \begin{cases} Y_{(n,[\frac{1}{2}n+\frac{1}{2}])} & (n \text{ odd}), \\ \frac{1}{2}Y_{(n,[\frac{1}{2}n])} + \frac{1}{2}Y_{(n,[\frac{1}{2}n+1])} & (n \text{ even}), \end{cases}$$

which differs from $Y_{(n,[\frac{1}{2}n])}$ by $O_p(n^{-1})$, and is asymptotically normal with mean $\xi_{\frac{1}{2}} = F^{-1}(\frac{1}{2})$ and variance

$$[4n\{f(\xi_{\frac{1}{2}})\}^2]^{-1}.$$

To estimate the variance from data, without parametric assumptions, it is, in effect, necessary to estimate the p.d.f. $f(\xi_{\frac{1}{2}})$. More generally, a consistent estimate of $\{f(\xi_p)\}^{-1}$ is

$$\frac{Y_{(n,[pn]+l)} - Y_{(n,[pn]-l)}}{2l/n}, \tag{18}$$

where $l = O(n^d)$ for $0 < d < 1$ as $n \to \infty$. Continuity of the p.d.f. at ξ_p is required. In practice, for any particular n, a numerical choice of l has to be made in the light of the smoothness of $f(.)$. Estimates more elaborate than (18) can be constructed.

A2.4   Linear combinations of order statistics
The representation (11) for $Y_{(n.)}$ in terms of exponential order statistics is useful in studying not just single order statistics but, more generally, in dealing with linear combinations of the form

$$T_n = n^{-1}\Sigma c_{nr}k(Y_{(nr)}). \tag{19}$$

In particular, some of the robust estimates of location and scale studied in Section 9.4 are given by (19) with $k(y) = y$. To prove the asymptotic normality of (19), we use the transformations of Section A2.2 and write

$$T_n = n^{-1} \Sigma c_{nr} k^\dagger (Z_{(nr)}), \tag{20}$$

where $E(Z_{(nr)}) = e_{nr} = -\log\{1 - r/(n+1)\} + O(1/n)$ and $\mathrm{var}(Z_{(nr)}) = O(n^{-1})$. We then expand (20) in a Taylor series around e_{nr} and express the answer in terms of the i.i.d. variables $W_{(nr)}$. The result is

$$T_n = n^{-1} \Sigma a_{nr}(W_{(nr)} - 1) + n^{-1} \Sigma c_{nr} k^\dagger (e_{nr}) + o_p(n^{-\frac{1}{2}}), \tag{21}$$

provided that $k^\dagger(.)$ is continously differentiable almost everywhere. The coefficients a_{nr} are given by

$$a_{nr} = \frac{1}{n-r+1} \sum_{s=r}^{n} c_{ns} \frac{dk^\dagger (e_{ns})}{de_{ns}} . \tag{22}$$

The asymptotic normality of T_n now follows from the central limit theorem applied to the first term of (21); we assume that $\Sigma a_{nr}^2/n = O(1)$. Thus T_n is asymptotically normal with mean and variance given respectively by

$$n^{-1} \Sigma c_{nr} k^\dagger (e_{nr}) \quad \text{and} \quad n^{-2} \Sigma a_{nj}^2. \tag{23}$$

If c_{nr} has the form $c\{r/(n+1)\}$, then the sums in (23) can be approximated by integrals, the asymptotic mean and variance thereby respectively taking the forms

$$\int_0^1 c(u) k^\dagger \{-\log(1-u)\} du = \int c\{F(y)\} k(y)\, dy \tag{24}$$

and

$$n^{-1} \iint_{x \leqslant y} c\{F(x)\} c\{F(y)\} k'(x) k'(y) F(x)\{1 - F(y)\}\, dx\, dy . \tag{25}$$

A detailed account of this argument is given by Chernoff, Gastwirth and Johns (1967).

As an example, suppose that in order to estimate the scale parameter τ of the distribution $\tau^{-1} g\{(y-\mu)/\tau\}$, the statistic

$$T_n = \frac{a}{n(n-1)} \sum_{j>k} (Y_{(nj)} - Y_{(nk)})$$

is used, where a is a constant to be determined. Then asymptotically

$$E(T_n) = a\tau \int y \{2 G(y) - 1\} g(y) dy,$$

agreeing with (17), of course. In terms of the general linear combination (19), $k(x) = 2x - 1$. In the normal case, T_n is unbiased for σ if $a = \pi^{\frac{1}{2}}$ and the variance becomes

$$\frac{2\sigma^2}{n} \iint_{x \leqslant y} \{2\Phi(x) - 1\}\{2\Phi(y) - 1\}\Phi(x)\{1 - \Phi(y)\} dx dy = \frac{(\frac{1}{3}\pi + 2\sqrt{3} - 4)\sigma^2}{n}.$$

This is approximately 1.02 times the asymptotic variance of the usual estimate $MS^{\frac{1}{2}}$.

A2.5   Extreme value theory

The asymptotic results of Section A2.3 concern $Y_{(nr)}$, where as $n \to \infty$, $r = np$, $p \neq 0,1$. If we examine tail order statistics, different results emerge. Here we consider only the maximum value $Y_{(nn)}$; essentially identical results hold for $Y_{(n1)} = -\max(-Y_j)$. Fairly minor extensions cover statistics near but not at the extreme, e.g. $Y_{(n,n-k)}$ for fixed k.

There are two radically different cases depending on whether Y has an upper terminal. The simplest distribution with an upper terminal is the uniform distribution on $(0,1)$ and, because for this $Y_{(nn)} \simeq 1$, we write

$$Z_n = (1 - Y_{(nn)})/b_n,$$

where b_n is to be chosen. Then

$$pr(Z_n \leqslant z) = \begin{cases} 0 & (z < 0), \\ (1 - b_n z)^n & (b_n^{-1} > z > 0). \end{cases}$$

It is now clear that $b_n = 1/n$ is an appropriate standardization and that $n(1 - Y_{(nn)})$ has a limiting exponential distribution with unit mean.

More generally, with an upper terminal at θ, suppose that as $y \to \theta$

$$1 - F(y) \sim a(\theta - y)^c \quad (c > 0).$$

Then for the standardized maximum

$$Z_n = (\theta - Y_{(nn)})/b_n,$$

we have the cumulative distribution function

$$1 - \{F(\theta - b_n z)\}^n \sim 1 - \{1 - a(b_n z)^c\}^n$$

for $z \geqslant 0$ and $b_n z = O(1)$. We take $a b_n^c = n^{-1}$ to show that

$$(\theta - Y_{(nn)})(na)^{1/c}$$

has the limiting p.d.f.

$$c z^{c-1} \exp(-z^c). \tag{26}$$

Next, suppose that Y does not have an upper terminal and that as $y \to \infty$

$$1 - F(y) \sim a y^{-c} \quad (c > 0).$$

For the standardized variable $Y_{(nn)}/b_n = Z_n$, we have that

$$\text{pr}(Z_n \leqslant z) = [1 - \{1 - F(b_n z)\}]^n \sim \{1 - a(b_n z)^{-c}\}^n.$$

The choice $b_n = (an)^{1/c}$ then gives the limiting result

$$\text{pr}(Z_n \leqslant z) \sim \exp(-z^{-c}). \tag{27}$$

This covers as a special case the Cauchy distribution with $c = 1$.

Finally, suppose that $1 - F(y)$ tends to zero exponentially fast as $y \to \infty$. We return to the more general standardization $Z_n = (Y_{(nn)} - a_n)/b_n$, when

$$\text{pr}(Z_n \leqslant z) = (1 - \exp[\log\{1 - F(a_n + b_n z)\}])^n. \tag{28}$$

The crucial values of $Y_{(nn)}$ are those close to $F^{-1}(1 - 1/n)$, which we take to be a_n. Then, expanding (28) in a Taylor series, we obtain

$$\text{pr}(Z_n \leqslant z) \sim [1 - n^{-1} \exp\{- b_n z n f(a_n)\}]^n,$$

so that with $b_n^{-1} = n f(a_n)$, we have

$$\text{pr}(Z_n \leqslant z) \sim \exp(-e^{-z}), \tag{29}$$

the corresponding p.d.f. being

$$\exp(z - e^{-z}). \tag{30}$$

In a rigorous development it is necessary to assume that

$$\frac{d}{dy}\left\{\frac{1 - F(y)}{f(y)}\right\} \to 0.$$

Gnedenko (1943) showed that the three limiting distributions obtained in (26), (27) and (29) are the only possible ones.

Sometimes we use the limiting distributions directly as a model for the data and then they are taken in the form with unknown scale parameters and indices to be estimated from data in the case of (26) and (27), and with unknown scale and location parameter in the case of of (30) ; thus the p.d.f. corresponding to (30) will be

$$\frac{1}{\tau} \exp \left\{ \frac{y - \mu}{\tau} - \exp \left(-\frac{y - \mu}{\tau} \right) \right\}. \tag{31}$$

It is important that the three types of limiting extreme value distribution are all related by simple transformations to the exponential distribution. Thus if V has an exponential distribution of unit mean, then $\log V$ has the density (30).

Bibliographic notes

An account of order statistics with an extensive bibliography is given by David (1970) ; see also the earlier book of papers edited by Sarhan and Greenberg (1962). The definitive work on the limiting distribution of extremes is by Gnedenko (1943) ; an accessible account is given by Thompson (1969). Other work of importance on extremes is by Fisher and Tippett (1928) and by Gumbel (1935, 1958).

APPENDIX 3
SECOND – ORDER REGRESSION FOR ARBITRARY RANDOM VARIABLES

Let T and U be arbitrary random variables of finite non-zero variance. Without loss of generality, suppose that $E(U) = 0$. A linear function $\gamma + \beta U$ predicts T with mean squared error

$$E\{(T - \gamma - \beta U)^2\},$$

and a simple calculation shows that this is minimized with

$$\gamma = E(T), \quad \beta = \text{cov}(T, U)/\text{var}(U).$$

We call

$$\hat{T} = E(T) + \frac{\text{cov}(T, U)}{\text{var}(U)} U$$

the *second-order regression equation* of T on U. If we write

$$T = \hat{T} + Z, \tag{1}$$

then

(i) $E(Z) = 0$,

(ii) $\text{cov}(Z, \hat{T}) = \text{cov}(Z, U) = 0$,

(iii) $\text{var}(Z) = \text{var}(T) - \dfrac{\{\text{cov}(T, U)\}^2}{\text{var}(U)}$.

According to (ii) we can, for arbitrary random variables, write T as a linear function of U plus an *uncorrelated* residual term. This is to be sharply contrasted with the assumption of a full linear regression relation, in which T is a linear function of U plus an *independent* residual term. This is a very strong assumption about the joint distribution.

It follows from (iii) that a necessary and sufficient condition that T is a linear function of U is that

$$\text{var}(T) = \frac{\{\text{cov}(T,U)\}^2}{\text{var}(U)} \tag{2}$$

and that otherwise we have the Cauchy-Schwarz inequality

$$\text{var}(T) > \frac{\{\text{cov}(T,U)\}^2}{\text{var}(U)}. \tag{3}$$

More generally, let T and $U_1, \dots, U_q$ be arbitrary random variables of finite non-zero variance, taking, without loss of generality, $E(U_r) = 0$ $(r = 1, \dots, q)$. Consider the linear function $\gamma + \beta_1 U_1 + \dots + \beta_q U_q$ minimizing

$$E\{(T - \gamma - \beta_1 U_1 - \dots - \beta_q U_q)^2\}.$$

A simple calculation shows that $\gamma = E(T)$ and that

$$\mathbf{i}\beta = c,$$

where $i_{rs} = \text{cov}(U_r, U_s)$, $c_r = \text{cov}(T, U_r)$ and $\mathbf{i}$ is assumed non-singular. If we write

$$\hat{T} = E(T) + \beta_1 U_1 + \dots + \beta_q U_q,$$

$$T = \hat{T} + Z,$$

then

(i) $E(Z) = 0$,

(ii) $\text{cov}(Z, U_r) = 0$,

(iii) $\text{var}(Z) = \text{var}(T) - \beta^{\text{T}} \mathbf{i} \beta$

$$= \text{var}(T) - c^{\text{T}} \mathbf{i}^{-1} c.$$

It follows from (iii) that a necessary and sufficient condition that T is a linear function of $U_1, \dots, U_q$ is that

$$\text{var}(T) = c^{\text{T}} \mathbf{i}^{-1} c \tag{4}$$

and that otherwise

$$\text{var}(T) > c^{\text{T}} \mathbf{i}^{-1} c. \tag{5}$$

In the context of the Cramér-Rao bound of Section 8.3, we have the special case $c^{\text{T}} = (1, 0, \dots, 0)$, when it follows from (4) and (5) that

$$\text{var}(T) \geqslant i^{11},$$

with equality if and only if T is a linear function of $U_1, \dots, U_q$.

In this book, we use these results only in connexion with the Cramér-Rao inequality of Section 8.3. The results are, however, important in other contexts, for example in connexion with representations of time series (Anderson, 1971).

REFERENCES

Aitchison, J. (1970a). *Choice against chance: an introduction to statistical decision theory*. Reading, Mass, Addison-Wesley.

Aitchison, J. (1970b). 'Statistical problems of treatment allocation' (with discussion). *J.R. Statist. Soc.*, A, **133**, 206–238.

Aitchison, J. and Dunsmore, I.R. (1968). 'Linear-loss interval estimation of location and scale parameters'. *Biometrika*, **55**, 141–148.

Aitchison, J. and Sculthorpe, D. (1965). 'Some problems of statistical prediction'. *Biometrika*, **52**, 469–483.

Aitchison, J. and Silvey, S.D. (1958). 'Maximum-likelihood estimation of parameters subject to restraints'. *Ann. Math. Statist.*, **29**, 813–828.

Aitchison, J. and Silvey, S.D. (1960). 'Maximum-likelihood estimation procedures and associated tests of significance'. *J.R. Statist. Soc.*, B, **22**, 154–171.

Aitken, A.C. and Silverstone, H. (1942). 'On the estimation of statistical parameters'. *Proc. R. Soc. Edinb.*, A, **61**, 186–194.

Altham, P.M.E. (1969). 'Exact Bayesian analysis of a 2 × 2 contingency table, and Fisher's "exact" significance test'. *J.R. Statist. Soc.*, B, **31**, 261–269.

Andersen, E.B. (1970). 'Asymptotic properties of conditional maximum-likelihood estimators'. *J.R. Statist. Soc.*, B, **32**, 283–301.

Andersen, E.B. (1973). *Conditional inference and models for measuring*. Copenhagen, Mentalhygiejnisk Forlag.

Anderson, T.W. (1958). *An introduction to multivariate statistical analysis*. New York, Wiley.

Anderson, T.W. (1971). *The statistical analysis of time series*. New York, Wiley.

Anderson, T.W. and Goodman, L.A. (1957). 'Statistical inference about Markov chains'. *Ann. Math. Statist.*, **28**, 89–110.

Andrews, D.F., Bickel, P.J., Hampel, F.R., Huber, P.J., Rogers, W.H. and Tukey, J.W. (1972). *Robust estimates of location: survey and advances*. Princeton University Press.

Andrews, D.F., Gnanadesikan, R. and Warner, J.L. (1971). 'Transformations of multivariate data'. *Biometrics*, **27**, 825–840.

Anscombe, F.J. (1948). 'Validity of comparative experiments', (with discussion). *J.R. Statist. Soc.*, A, **111**, 181–211.

Anscombe, F.J. (1953). 'Sequential estimation', (with discussion). *J.R. Statist. Soc.*, B, **15**, 1–29.

Anscombe, F.J. (1956). 'Contribution to discussion of paper by F.N. David and N.L. Johnson'. *J.R. Statist. Soc.*, B, **18**, 24–27.

Aoki, M. (1967). *Optimization of stochastic systems*. New York, Academic Press.

Åström, K.J. (1970). *Introduction to stochastic control theory*. New York, Academic Press.

Atiqullah, M. (1962). 'The estimation of residual variance in quadratically balanced least-squares problems and the robustness of the *F*-test'. *Biometrika*, **49**, 83–91.

Atkinson, A.C. (1970). 'A method for discriminating between models', (with discussion). *J.R. Statist. Soc.*, B, **32**, 323–353.

Bahadur, R.R. (1954). 'Sufficiency and statistical decision functions'. *Ann. Math. Statist.*, **25**, 423–462.

Bahadur, R.R. (1971). *Some limit theorems in statistics*. Philadelphia, S.I.A.M.

Barankin, E.W. and Gurland, J. (1951). 'On asymptotically normal, efficient estimators: I.' *Univ. Calif. Publ. Statist.*, **1**, 89–129.

Barnard, G.A. (1947). 'Review of book *Sequential analysis* by Wald, A.' *J. Amer. Statist. Assoc.*, **42**, 658–664.

Barnard, G.A. (1951). 'The theory of information', (with discussion). *J.R. Statist. Soc.*, B, **13**, 46–64.

Barnard, G.A. (1962). Prepared contribution. In *The foundations of statistical inference*, pp. 39–49, eds. Savage L.J. et al. London, Methuen.

Barnard, G.A. (1963). 'The logic of least squares'. *J.R. Statist. Soc.*, B, **25**, 124–127.

Barnard, G.A. (1972). 'The unity of statistics'. *J.R. Statist. Soc.*, A, **135**, 1–12.

Barnard, G.A., Jenkins, G.M. and Winsten, C.B. (1962), 'Likelihood inference and time series', (with discussion). *J.R. Statist. Soc.*, A, **125**, 321–372.

Barnard, G.A. and Sprott, D.A. (1971), 'A note on Basu's examples of anomalous ancillary statistics', (with discussion). In *Foundations of statistical inference*, pp. 163–176, eds. Godambe, V.P. and Sprott, D.A. Toronto, Holt, Rinehart and Winston.

Barndorff-Nielsen, O. (1973a). 'Unimodality and exponential families'. *Comm. Statist.*, **1**, 189–216.

Barndorff-Nielsen, O. (1973b). 'Exponential families and conditioning'. Thesis, University of Copenhagen.

Barnett, V.D. (1966). 'Evaluation of the maximum-likelihood estimator where the likelihood equation has multiple roots'. *Biometrika*, **53**, 151–165.

Barnett, V.D. (1973). *Comparative statistical inference*. London, Wiley.

Bartlett, M.S. (1936a). 'The information available in small samples'. *Proc. Camb. Phil. Soc.*, **32**, 560–566.

Bartlett, M.S. (1936b). 'Statistical information and properties of sufficiency'. *Proc. R. Soc.*, A, **154**, 124–137.

Bartlett, M.S. (1937). 'Properties of sufficiency and statistical tests'. *Proc. R. Soc.*, A, **160**, 268–282.

Bartlett, M.S. (1938). 'Further aspects of the theory of multiple regression'. *Proc. Camb. Phil. Soc.*, **34**, 33–40.

Bartlett, M.S. (1947). 'Multivariate analysis', (with discussion). *J.R. Statist. Soc. Suppl.*, **9**, 176–197.

Bartlett, M.S. (1953a). 'Approximate confidence intervals'. *Biometrika*, **40**, 12–19.

Bartlett, M.S. (1953b). 'Approximate confidence intervals II. More than one unknown parameter'. *Biometrika*, **40**, 306–317.

Bartlett, M.S. (1955). 'Approximate confidence intervals III. A bias correction'. *Biometrika*, **42**, 201–204.

Bartlett, M.S. (1966). *An introduction to stochastic processes.* 2nd edition. Cambridge University Press.

Bartlett, M.S. (1967). 'Inference and stochastic processes'. *J.R. Statist. Soc.*, A, **130**, 457–474.

Barton, D.E. and Dennis, K.E. (1952). 'The conditions under which Gram-Charlier and Edgeworth curves are positive definite and unimodal'. *Biometrika*, **39**, 425–427.

Basu, D. (1964). 'Recovery of ancillary information', *Sankhyā*, A, **26**, 3–16.

Basu, D. (1973). 'Statistical information and likelihood'. Report, University of Sheffield.

Bayes, T. (1763). 'Essay towards solving a problem in the doctrine of chances'. Reproduced with a bibliographical note by Barnard, G.A., *Biometrika*, **45**, (1958), 293–315.

Bellman, R. (1957). *Dynamic programming.* Princeton University Press.

Bhattacharyya, A. (1950). 'Unbiased statistics with minimum variance'. *Proc. R. Soc. Edinb.*, A, **63**, 69–77.

Billingsley, P. (1961a). 'Statistical methods in Markov chains'. *Ann. Math. Statist.*, **32**, 12–40.

Billingsley, P. (1961b). *Statistical inference for Markov processes.* University of Chicago Press.

Billingsley, P. (1968). *Convergence of probability measures.* New York, Wiley.

Birnbaum, A. (1962). 'On the foundations of statistical inference' (with discussion). *J. Amer. Statist. Assoc.*, **57**, 269–326.

Birnbaum, A. (1969). 'Concepts of statistical evidence'. In *Philosophy Science and Method: Essays in Honor of E. Nagel*, pp. 112–143, eds. Morgenbesser, S., Suppes, P. and White, M., New York, St. Martin's Press.

Birnbaum, A. (1970). 'On Durbin's modified principle of conditionality'. *J. Amer. Statist. Assoc.*, **65**, 402–403.

Blackwell, D. (1947). 'Conditional expectation and unbiased sequential estimation'. *Am. Math. Statist.*, **18**, 105–110.

Blom, G. (1958). *Statistical estimates and transformed beta-variables*. New York, Wiley.

Box, G.E.P. (1949). 'A general distribution theory for a class of likelihood criteria'. *Biometrika*, **36**, 317–346.

Box, G.E.P. and Andersen, S.L. (1955). 'Permutation theory in the derivation of robust criteria and the study of departures from assumption', (with discussion). *J.R. Statist. Soc.*, B, **17**, 1–34.

Box, G.E.P. and Hill, W.J. (1967). 'Discrimination among mechanistic models'. *Technometrics*, **9**, 57–71.

Box, G.E.P. and Henson, T.L. (1970). 'Some aspects of mathematical modelling in chemical engineering'. pp. 548–570, Proc. Conference of Scientific Comp. Centre and Inst. of Statistical Studies and Research, Cairo.

Box, G.E.P. and Kanemasu, H. (1973). 'Posterior probabilities of candidate models in model discrimination'. University of Wisconsin Technical Report.

Brillinger, D.R. (1963). 'Necessary and sufficient conditions for a statistical problem to be invariant under a Lie group'. *Ann. Math. Statist.*, **34**, 492–500.

Bross, I.D.J. (1971). 'Critical levels, statistical language, and scientific inference', (with discussion). In *Foundations of statistical inference*, pp. 500–519, eds. Godambe, V.P. and Sprott, D.A. Toronto, Holt, Rinehart andl Winston.

Brown, L.D. (1966). 'On the admissibility of invariant estimators of one or more location parameters'. *Ann. Math. Statist.*, **37**, 1087–1136.

Brown, L.D. (1971). 'Admissible estimators, recurrent diffusions, and insoluble boundary value problems'. *Ann. Math. Statist.*, **42**, 855–903.

Buehler, R.J. (1959). 'Some validity criteria for statistical inferences'. *Ann. Math. Statist.*, **30**, 845–863.

Buehler, R.J. (1971). 'Measuring information and uncertainty', (with discussion). In *Foundations of statistical inference*, pp. 330–341, eds. Godambe, V.P. and Sprott, D.A. Toronto, Holt, Rinehart and Winston.

Chapman, D.G. and Robbins, H.E. (1951). 'Minimum variance estimation without regularity assumptions'. *Ann. Math. Statist.*, **22**, 581–586.

Chernoff, H. (1951). 'A property of some Type A regions'. *Ann. Math. Statist.*, **22**, 472–474.

Chernoff, H. (1954). 'On the distribution of the likelihood ratio'. *Ann. Math. Statist.*, **25**, 573–578.

Chernoff, H., Gastwirth, J.L. and Johns, M.V. (1967). 'Asymptotic distribution of linear combinations of functions of order statistics with applications to estimation'. *Ann. Math. Statist.*, **38**, 52–72.

Chernoff, H. and Lehmann, E.L. (1954). 'The use of maximum likelihood estimates in χ^2 tests for goodness of fit'. *Ann. Math. Statist.*, **25**, 579–586.

Chernoff, H. and Moses, L.E. (1959). *Elementary decision theory*. New York, Wiley.

Chernoff, H. and Savage, I.R. (1958). 'Asymptotic normality and efficiency of certain nonparametric test statistics'. *Ann. Math. Statist.*, **29**, 972–994.

Cohen, L. (1958). 'On mixed single sample experiments'. *Ann. Math. Statist.*, **29**, 947–971.

Copas, J.B. (1969). 'Compound decisions and empirical Bayes', (with discussion). *J.R. Statist. Soc.*, B, **31**, 397–425.

Copas, J.B. (1972). 'The likelihood surface in the linear functional relationship problem'. *J.R. Statist. Soc.*, B, **34**, 274–278.

Cormack, R.M. (1968). 'The statistics of capture-recapture methods'. *Oceanogr. Mar. Biol. Ann. Rev.*, **6**, 455–506.

Cornfield, J. (1969). 'The Bayesian outlook and its application', (with discussion). *Biometrics*, **25**, 617–657.

Cox, D.R. (1958). 'Some problems connected with statistical inference'. *Ann Math. Statist.*, **29**, 357–372.

Cox, D.R. (1961). 'Tests of separate families of hypotheses'. *Proc. 4th Berkeley Symp.*, **1**, 105–123.

Cox, D.R. (1962). 'Further results on tests of separate families of hypotheses'. *J.R. Statist. Soc.*, B, **24**, 406–424.

Cox, D.R. (1964). 'Some problems of statistical analysis connected with congestion', (with discussion). In *Congestion Theory*, pp. 289–316, eds. Smith, W.L. and Wilkinson, W.E. Univ. of North Carolina Press.

Cox, D.R. (1967). 'Fieller's theorem and a generalization'. *Biometrika*, **54**, 567–572.

Cox, D.R. (1970). *The analysis of binary data*. London, Methuen.

Cox, D.R. (1971). 'The choice between alternative ancillary statistics'. *J.R. Statist. Soc.*, B, **33**, 251–255.

Cox, D.R. and Herzberg, A.M. (1972). 'On a statistical problem of E.A. Milne'. *Proc. R. Soc.*, A, **331**, 273–283.

Cox, D.R. and Hinkley, D.V. (1968). 'A note on the efficiency of least-squares estimates'. *J.R. Statist. Soc.*, B, **30**, 284–289.

Cox, D.R. and Lewis, P.A.W. (1966). *The statistical analysis of series of events*. London, Methuen.

Cox, D.R. and Snell, E.J. (1968). 'A general definition of residuals', (with discussion). *J.R. Statist. Soc.*, B, **30**, 248–275.

Cox, D.R. and Snell, E.J. (1971). 'On test statistics calculated from residuals'. *Biometrika*, **58**, 589–594.

Cramér, H. (1946). *Mathematical methods of statistics*. Princeton University Press.

D'Agostino, R.B. (1970). 'Transformation to normality of the null distribution of g_1'. *Biometrika*, **57**, 679–681.

D'Agostino, R.B. and Pearson, E.S. (1973). 'Tests for departure from normality. Empirical results for the distributions of b_2 and $\sqrt{b_1}$'. *Biometrika*, **60**, 613–622.

D'Agostino, R.B. and Tietjen, G.L. (1971). 'Simulation probability points of b_2 for small samples'. *Biometrika*, **58**, 669–672.

Daniel, C. and Wood, F.S. (1971). *Fitting equations to data.* New York, Wiley.

Daniels, H.E. (1954). 'Saddlepoint approximations in statistics'. *Ann. Math. Statist.*, **25**, 631–650.

Dar, S.N. (1962). 'On the comparison of the sensitivities of experiments'. *J.R. Statist. Soc.*, B, **24**, 447–453.

Darling, D.A.S. (1957). 'The Kolmogorov-Smirnov, Cramér-von Mises tests'. *Ann. Math. Statist.*, **28**, 823–838.

Darmois, G. (1936). 'Sur les lois de probabilité à estimation exhaustive'. *C.R. Acad. Sci.*, Paris, **200**, 1265–1266.

Darroch, J.N. (1971). 'A definition of independence for bounded-sum, non-negative, integer-valued variables'. *Biometrika*, **58**, 357–368.

Darroch, J.N. and Ratcliffe, D. (1973). 'Tests of F-independence with reference to quasi-independence and Waite's fingerprint data'. *Biometrika*, **60**, 395–401.

David, H.A. (1970). *Order statistics.* New York, Wiley.

Davies, R.B. (1969). 'Beta-optimal tests and an application to the summary evaluation of experiments'. *J.R. Statist. Soc.*, B, **31**, 524–538.

Dawid, A.P. (1970). 'On the limiting normality of posterior distributions'. *Proc. Camb. Phil. Soc.*, **67**, 625–633.

Dawid, A.P. (1973). 'Posterior expectations for large observations'. *Biometrika*, **60**, 664–667.

de Finetti, B. (1937). 'La prévision: ses lois logiques, ses sources subjectives'. *Ann. Inst H. Poincaré*, **7**, 1–68.

de Finetti, B. (1972). *Probability, induction and statistics.* London, Wiley.

DeGroot, M.H. (1970). *Optimal statistical decisions.* New York, McGraw Hill.

Dickey, J. and Lientz, B.P. (1970). 'The weighted likelihood ratio, sharp hypotheses about chances, the order of a Markov chain'. *Ann. Math. Statist.*, **41**, 214–226.

Draper, N.R. and Tierney, D.E. (1972). 'Regions of positive and unimodal series expansion of the Edgeworth and Gram–Charlier approximations'. *Biometrika*, **59**, 463–465.

Durbin, J. (1960). 'Estimation of parameters in time-series regression models'. *J.R. Statist. Soc.*, B, **22**, 139–153.

Durbin, J. (1970). 'On Birnbaum's theorem on the relation between sufficiency, conditionality and likelihood'. *J. Amer. Statist. Assoc.*, **65**, 395–398.

Durbin, J. (1973). *Distribution theory for tests based on the sample distribution function.* Philadelphia, S.I.A.M.

Edwards, A.W.F. (1972). *Likelihood*. Cambridge University Press.

Efron, B. (1971). 'Does an observed sequence of numbers follow a simple rule?' (with discussion). *J. Amer. Statist. Assoc.*, **66**, 552–568.

Efron, B. (1974). 'On second-order efficiency'. To be published.

Efron, B. and Morris, C. (1971). 'Limiting the risk of Bayes and empirical Bayes estimators-Part I: the Bayes case'. *J. Amer. Statist. Assoc.*, **66**, 807–815.

Efron, B. and Morris, C. (1972a). 'Limiting the risk of Bayes and empirical Bayes estimators-Part II: the empirical Bayes case'. *J. Amer. Statist. Assoc.*, **67**, 130–139.

Efron, B. and Morris, C. (1972b). 'Empirical Bayes on vector observations: an extension of Stein's method'. *Biometrika*, **59**, 335–347.

Efron, B. and Morris, C. (1973). 'Combining the results of possibly related experiments', (with discussion). *J.R. Statist. Soc.*, B, **35**, 379–4.

El-Sayyad, G.M. and Freeman, P.R. (1973). 'Bayesian sequential estimation of a Poisson process rate'. *Biometrika*, **60**, 289–296.

Feder, P.I. (1968). 'On the distribution of the log likelihood ratio test statistic when the true parameter is 'near' the boundaries of the hypothesis regions'. *Ann. Math. Statist.*, **39**, 2044–2055.

Feller, W. (1971). *An introduction to probability theory*. Vol. 2, 2nd edition. New York, Wiley.

Ferguson, T.S. (1958). 'A method of generating best asymptotically normal estimates with application to the estimation of bacterial densities'. *Ann. Math. Statist.*, **29**, 1046–1062.

Ferguson, T.S. (1961). 'On the rejection of outliers'. *Proc. 4th Berkeley Symp.*, **1**, 253–287.

Ferguson, T.S. (1967). *Mathematical statistics: a decision theoretic approach*. New York, Academic Press.

Filippova, A.A. (1962). 'Mises' theorem on the asymptotic behavior of functionals of empirical distribution functions and its statistical applications'. *Theory Prob. and its Appl.*, **7**, 24–57.

Finney, D.J. (1964). *Statistical method in biological assay*. 2nd edition, London, Griffin.

Fishburn, P.C. (1969). 'A general theory of subjective probabilities and expected utilities'. *Ann. Math. Statist.*, **39**, 1419–1429.

*Fisher, R.A. (1922). 'On the mathematical foundations of theoretical statistics'. *Phil. Trans. R. Soc.*, A, **222**, 309–368.

* Sir Ronald Fisher's collected papers are being issued by the University of Adelaide Press. At the time of writing, two volumes are available, covering the years up to 1931. A selection of papers, with corrections and comments by Fisher, was published by Wiley (1950), with the title *Contributions to mathematical statistics*. This is, however, no longer in print. It contained, in particular, the 1922, 1925, 1934 and 1935 papers referred to here.

Fisher, R.A. (1925). 'Theory of statistical estimation'. *Proc. Camb. Phil. Soc.*, **22**, 700–725.

Fisher, R.A. (1934). 'Two new properties of mathematical likelihood'. *Proc. R. Soc.*, A, **144**, 285–307.

Fisher, R.A. (1935). 'The logic of inductive inference', (with discussion). *J.R. Statist. Soc.*, **98**, 39–82.

Fisher, R.A. (1950). 'The significance of deviations from expectation in a Poisson series'. *Biometrics*, **6**, 17–24.

Fisher, R.A. (1953). 'The expansion of statistics'. *J.R. Statist. Soc.*, A, **116**, 1–6.

Fisher, R.A. (1956). 'On a test of significance in Pearson's Biometrika tables, (no. 11)'. *J.R. Statist. Soc.*, B, **18**, 56–60.

Fisher, R.A. (1966). *Design of experiments.* 8th edition. Edinburgh, Oliver and Boyd.

Fisher, R.A. (1973). *Statistical methods and scientific inference.* 3rd edition, Edinburgh, Oliver and Boyd.

Fisher, R.A. and Tippett, L.H.C. (1928). 'Limiting forms of the frequency distribution of the largest or smallest member of a sample'. *Proc. Camb. Phil. Soc.*, **24**, 180–190.

Fisher, R.A. and Yates, F. (1963). *Statistical tables for biological, agricultural and medical research.* 6th edition. Edinburgh, Oliver and Boyd.

Fraser, D.A.S. (1961). 'The fiducial method and invariance'. *Biometrika*, **48**, 261–280.

Fraser, D.A.S. (1968). *The structure of inference.* New York, Wiley.

Fraser, D.A.S. (1971). 'Events, information processing, and the structured model', (with discussion). In *Foundations of statistical inference*, pp. 32–55, eds. Godambe, V.P. and Sprott, D.A. Toronto, Holt, Rinehart and Winston.

Gaver, D.P. and Hoel, D.G. (1970). 'Comparison of certain small-sample Poisson probability estimates'. *Technometrics*, **12**, 835–850.

Girshick, M.A. and Savage, L.J. (1951), 'Bayes and minimax estimates for quadratic loss functions'. *Proc. 2nd Berkely Symp.*, 53–73.

Gnedenko, B.V. (1943). 'Sur la distribution limite du terme maximum d'une série aléatoire'. *Ann. Math.*, **44**, 423–453.

Gnedenko, B.V. (1967). *The theory of probability.* 4th edition. New York, Chelsea.

Godambe, V.P. (1960). 'An optimum property of regular maximum likelihood estimation'. *Ann. Math. Statist.*, **31**, 1208–1211.

Godambe, V.P. and Thompson, M.E. (1971). 'The specification of prior knowledge by classes of prior distributions in survey sampling estimation', (with discussion). In *Foundations of statistical inference*, pp. 243–258, eds. Godambe, V.P. and Sprott, D.A. Toronto, Holt, Rinehart and Winston.

Godwin, H.J. (1964). *Inequalities in distribution functions.* London, Griffin.

Good, I.J. (1950). *Probability and the weighing of evidence.* London, Griffin.

Good, I.J. (1965). *The estimation of probabilities.* Cambridge, Mass, M.I.T. Press.

Goodman, L.A. (1953). 'Sequential sampling tagging for population size problems'. *Ann. Math. Statist.*, 24, 56–69.

Goodman, L.A. (1968). 'The analysis of cross-classified data: independence, quasi-independence and interactions in contingency tables with or without missing entries'. *J. Amer. Statist. Assoc.*, 63, 1091–1131.

Gray, H.L. and Schucany, W.R. (1972). *The generalized jackknife statistic.* New York, Marcel Dekker.

Grundy, P.M. and Healy, M.J.R. (1950). 'Restricted randomization and quasi-Latin squares'. *J.R. Statist. Soc.*, B, 12, 286–291.

Grundy, P.M., Healy, M.J.R. and Rees, D.H. (1956). 'Economic choice of the amount of experimentation', (with discussion). *J.R. Statist. Soc.*, B, 18, 32–55.

Gumbel, E.J. (1935). 'Les valeurs extrêmes des distributions statistiques'. *Ann. Inst. H. Poincaré*, 5, 115–158.

Gumbel, E.J. (1958). *Statistics of extremes.* New York, Columbia University Press.

Guttman, I. (1970). *Statistical tolerance regions.* London, Griffin.

Hájek, J. (1969). *Nonparametric statistics.* San Francisco, Holden Day.

Hájek, J. and Sïdák, Z. (1967). *Theory of rank tests.* New York, Academic Press.

Hall, W.J., Wijsman, R.A. and Ghosh, J.K. (1965). 'The relationship between sufficiency and invariance with applications in sequential analysis'. *Ann. Math. Statist.*, 36, 575–614.

Halmos, P.R. and Savage, L.J. (1949). 'Application of the Radon-Nikodym theorem to the theory of sufficient statistics'. *Ann. Math. Statist.*, 20, 225–241.

Halperin, M. (1970). 'On inverse estimation in linear regression'. *Technometrics*, 12, 727–736.

Hampton, J.M., Moore, P.G. and Thomas, H. (1973). 'Subjective probability and its measurement'. *J.R. Statist. Soc.*, A, 136, 21–42.

Hastings, C., Mosteller, F., Tukey, J.W. and Winsor, C.P. (1947). 'Low moments for small samples: a comparative study of order statistics'. *Ann. Math. Statist.*, 18, 413–426.

Hinkley, D.V. (1972). 'Time-ordered classification'. *Biometrika*, 59, 509–523.

Hinkley, D.V. (1973). 'Two-sample tests with unordered pairs'. *J.R. Statist. Soc.*, B, 35, 337–346.

Hodges, J.L. and Lehmann, E.L. (1963). 'Estimates of location based on rank tests'. *Ann. Math. Statist.*, 34, 598–611.

Hoeffding, W. (1948). 'A class of statistics with asymptotically normal distribution'. *Ann. Math. Statist.*, **19**, 293–325.

Hoeffding, W. (1965). 'Asymptotically optimal tests for multinomial distributions', (with discussion). *Ann. Math. Statist.*, **36**, 369–408.

Hogg, R.V. (1956). 'On the distribution of the likelihood ratio'. *Ann. Math. Statist.*, **27**, 529–532.

Hotelling, H. (1931). 'The generalization of Student's ratio'. *Ann. Math. Statist.*, **2**, 360–378.

Huber, P.J. (1964). 'Robust estimation of a location parameter'. *Ann. Math. Statist.*, **35**, 73–101.

Huber, P.J. (1972). 'Robust statistics: a review'. *Ann. Math. Statist.*, **43**, 1041–1067.

Huber, P.J. (1973). 'Robust regression: asymptotics, conjectures and Monte Carlo'. *Ann. Statist.* **1**, 799–821.

Hull, J., Moore, P.G. and Thomas, H. (1973). 'Utility and its measurement'. *J.R. Statist. Soc.*, A, **136**, 226–247.

James, G.S. (1956). 'On the accuracy of weighted means and ratios'. *Biometrika*, **43**, 304–321.

James, G.S. (1959). 'The Behrens-Fisher distribution and weighted means'. *J.R. Statist. Soc.*, B, **21**, 73–90.

James, W. and Stein, C. (1961). 'Estimation with quadratic loss'. *Proc. 4th Berkeley Symp.*, **1**, 361–379.

Jeffreys, H. (1961). *Theory of probability*. 3rd edition. Oxford, Clarendon Press.

Johns, M.V. (1974). 'Non-parametric estimation of location'. *J. Amer. Statist. Assoc.*, **69**, to appear.

Johnson, N.L. (1965). 'Tables to facilitate fitting S_U frequency curves'. *Biometrika*, **52**, 547–558.

Johnson, R.A. (1970). 'Asymptotic expansions associated with posterior distributions'. *Ann. Math. Statist.*, **41**, 851–864.

Kalbfleisch, J.D. (1974). 'Sufficiency and conditionality'. Unpublished paper.

Kalbfleisch, J.D. and Sprott, D.A. (1970). 'Application of likelihood methods to models involving large numbers of parameters', (with discussion). *J.R. Statist. Soc.*, B, **32**, 175–208.

Kempthorne, O. (1952). *The design and analysis of experiments.* New York, Wiley.

Kempthorne, O. (1966). 'Some aspects of experimental inference'. *J. Amer. Statist. Assoc.*, **61**, 11–34.

Kempthorne, O. and Doerfler, T.E. (1969). 'The behaviour of some significance tests under experimental randomization'. *Biometrika*, **56**, 231–248.

Kempthorne, O. and Folks, L. (1971). *Probability, statistics, and data analysis.* Ames, Iowa, Iowa State University Press.

Kendall, M.G. (1961). 'Natural law in the social sciences'. *J.R. Statist. Soc., A,* **124**, 1–16.

Kendall, M.G. (1962). *Rank correlation methods.* 3rd edition. London, Griffin.

Kendall, M.G. (1973). 'Entropy, probability and information'. *Rev. Int. Inst. Statist.,* **41**, 59–68.

Kendall, M.G. and Stuart, A (1967–69). *Advanced theory of statistics* vols. 1–3, (3rd 2nd and 2nd editions). London, Griffin.

Kiefer, J. (1952). 'On minimum variance estimators'. *Ann. Math. Statist.,* **23**, 627–629.

Kiefer, J. and Wolfowitz, J. (1956). 'Consistency of the maximum likelihood estimator in the presence of infinitely many incidental parameters'. *Ann. Math. Statist.,* **27**, 887–906.

Klotz, J. (1973). 'Statistical inference in Bernoulli trials with dependence'. *Ann. Statist.,* **1**, 373–379.

Koopman, B.O. (1936). 'On distribution admitting a sufficient statistic'. *Trans. Amer. Math. Soc.,* **39**, 399– 409.

Krutchkoff, R.G. (1967). 'Classical and inverse regression methods of calibration'. *Technometrics,* **9**, 425–439.

Kullback, S. (1968). *Information theory and statistics.* New York, Dover.

Kulldorff, G. (1957). 'On the conditions for consistency and asymptotic efficiency of maximum likelihood estimates'. *Skand. Akt.,* **40**, 129–144.

Lachenbruch, P.A. and Mickey. M.R. (1968). 'Estimation of error rates in discriminant analysis'. *Technometrics,* **10**, 1–11.

Lancaster, H.O. (1969). *The chi-squared distribution.* New York, Wiley.

Lauritzen, S. (1975). 'Sufficiency prediction and extreme models'. *Scandinavian J. Statist.,* to appear.

Lawley, D.N. (1956). 'A general method for approximating to the distribution of likelihood ratio criteria'. *Biometrika,* **43**, 295–303.

LeCam, L. (1953). 'On some asymptotic properties of maximum likelihood estimates and related Bayes' estimates'. *Univ. Calif. Publ. Statist.,* **1**, 277–329.

LeCam, L. (1956). 'On the asymptotic theory of estimation and testing hypotheses'. *Proc. 3rd Berkeley Symp.,* **1**, 129–156.

LeCam, L. (1970). 'On the assumptions used to prove asymptotic normality of maximum likelihood estimates'. *Ann. Math. Statist.,* **41**, 802–828.

Lehmann, E.L. (1959). *Testing statistical hypotheses.* New York, Wiley.

Lehmann, E.L. and Scheffé, H. (1950). 'Completeness, similar regions, and unbiased estimation, Part I'. *Sankhyā,* **10**, 305–340.

Lehmann, E.L. and Scheffé, H. (1955). 'Completeness, similar regions, and unbiased estimation. Part II'. *Sankhyā,* **15**, 219–236.

Lindley, D.V. (1958). 'Fiducial distributions and Bayes' theorem'. *J.R. Statist. Soc.*, B, **20**, 102–107.

Lindley, D.V. (1961). 'The use of prior probability distributions in statistical inference and decisions'. *Proc. 4th Berkeley Symp.*, **1**, 453–468.

Lindley, D.V. (1968) 'The choice of variables in multiple regression', (with discussion). *J.R. Statist. Soc.*, B, **30**, 31–66.

Lindley, D.V. (1971a). *Bayesian statistics, a review*. Philadephia, S.I.A.M.

Lindley, D.V. (1971b). *Making decisions*. London, Wiley–Interscience.

Lindley, D.V. and Smith A.F.M. (1972). 'Bayes' estimates for the linear model', (with discussion). *J.R. Statist. Soc.*, B, **34**, 1–41.

Linnik, Yu.V. (1968). *Statistical problems with nuisance parameters*. Translations of mathematical monographs, no. 20 (from the 1966 Russian edition). New York, American Mathematical Society.

Mann, H.B. and Wald, A. (1943). 'On stochastic limit and order relationships'. *Ann. Math. Statist.*, **14**, 217–226.

Mardia, K.V. (1972). *Statistics of directional data*. London, Academic Press.

Maritz, J. (1970). *Empirical Bayes' methods*. London, Methuen.

Mehta, J.S. and Srinivasan, R. (1970). 'On the Behrens–Fisher problem'. *Biometrika*, **57**, 649–655.

Menon, M.V. (1966). 'Characterization theorems for some univariate probability distributions'. *J.R. Statist. Soc.*, B, **28**, 143–145.

Miller, R.G. (1966). *Simultaneous statistical inference*. New York, McGraw Hill.

Miller, R.G. (1974). 'The jackknife: a review'. *Biometrika*, **61**, 1–15.

Mitchell, A.F.S. (1967). 'Discussion of paper by I.J. Good'. *J.R. Statist. Soc.*, B, **29**, 423–424.

Mood, A.M. (1943). 'On the dependence of sampling inspection plans upon population distributions'. *Ann. Math. Statist.*, **14**, 415–425.

Moran, P.A.P. (1970). 'On asymptotically optimal tests of composite hypotheses'. *Biometrika*, **57**, 47–55.

Moran, P.A.P. (1971). 'Maximum-likelihood estimation in non-standard conditions'. *Proc. Camb. Phil. Soc.*, **70**, 441–450.

Nelder, J.A. (1965a). 'The analysis of randomized experiments with orthogonal block structure I. Block structure and the null analysis of variance'. *Proc. R. Soc.*, A, **283**, 147–162.

Nelder, J.A. (1965b). 'The analysis of randomized experiments with orthogonal block structure II. Treatment structure and the general analysis of variance'. *Proc. R. Soc.*, A, **283**, 163–178.

Neyman, J. (1937). 'Outline of a theory of statistical estimation based on the classical theory of probability'. *Phil. Trans. R. Soc.*, A, **236**, 333–380.

Neyman, J. (1949). 'Contribution to the theory of the χ^2 test'. *Proc. Berkeley Symp.*, 239–273.

Neyman, J. (1952). *Lectures and conferences on mathematical statistics*. 2nd edition. Washington, U.S. Dept Agric. Grad. School.

Neyman, J. (1959). 'Optimal asymptotic tests of composite statistical hypotheses'. In *Probability and statistics*, pp. 213–234, ed. Grenander, U. Stockholm, Almqvist and Wiksell.

Neyman, J. (1960). 'Indeterminism in science and new demands on statisticians'. *J. Amer. Statist. Assoc.*, **55**, 625–639.

Neyman, J. (1967). *A selection of early papers of J. Neyman*. Cambridge University Press.

Neyman, J. (1969). 'Statistical problems in science. The symmetric test of a composite hypothesis'. *J. Amer. Statist. Assoc.*, **64**, 1154–1171.

Neyman, J. and Pearson, E.S. (1928). 'On the use and interpretation of certain test criteria for purposes of statistical inference'. *Biometrika*, A, **20**, 175–240 and 263–294.

Neyman, J. and Pearson, E.S. (1933a). 'On the problem of the most efficient tests of statistical hypotheses'. *Phil. Trans. R. Soc.*, A, **231**, 289–337.

Neyman, J. and Pearson, E.S. (1933b). 'The testing of statistical hypotheses in relation to probabilities *a priori*'. *Proc. Camb. Phil. Soc.*, **24**, 492–510.

Neyman, J. and Pearson, E.S. (1936). 'Contributions to the theory of testing statistical hypotheses. I. Unbiased critical regions of type A and type A_1'. *Stat. Res. Mem.*, **1**, 1–37.

Neyman, J. and Pearson, E.S. (1967). *Joint statistical papers*. Cambridge University Press.

Neyman, J. and Scott, E.L. (1948). 'Consistent estimates based on partially consistent observations'. *Econometrica*, 16, 1–32.

Norden, R.H. (1972). 'A survey of maximum likelihood estimation'. *Rev. Int. Inst. Statist.*, **40**, 329–354.

O'Neill, R. and Wetherill, G.B. (1971). 'The present state of multiple comparison methods', (with discussion). *J.R. Statist. Soc.*, B, **33**, 218–250.

Papaioannu, P.C. and Kempthorne, O. (1971). *On statistical information theory and related measures of information*. Aerospace research laboratories report, Wright–Patterson A.F.B., Ohio.

Pearson, E.S. (1956). 'Some aspects of the geometry of statistics'. *J.R. Statist. Soc.*, A, **119**, 125–146.

Pearson, E.S. (1966). *The selected papers of E.S. Pearson*. Cambridge University Press.

Pearson, E.S. and Hartley, H.O. (1970). *Biometrika tables for statisticians*. Vol. 1, 3rd edition. Cambridge University Press.

Pearson, E.S. and Hartley, H.O. (1972). *Biometrika tables for statisticians*, Vol. 2, Cambridge University Press.

Pearson, E.S. and Kendall, M.G., eds., (1970), *Studies in the history of statistics and probability.* London, Griffin.

Pearson, K. (1900). 'On the criterion that a given system of deviations from the probable in the case of a correlated system of variables is such that it can reasonably be supposed to have arisen from random sampling'. *Phil. Mag. Series 5,* **50,** 157–175.

Peers, H.W. (1971). 'Likelihood ratio and associated test criteria'. *Biometrika,* **58,** 577–587.

Pfanzagl, J. (1974). 'On the Behrens-Fisher problem'. *Biometrika,* **61,** 39–47.

Pierce, D.A. (1973). 'On some difficulties in a frequency theory of inference'. *Ann. Statist.,* **1,** 241–250.

Pillai, K.C.S. and Jayachandran, K. (1968). 'Power comparisons of tests of equality of two covariance matrices based on four criteria'. *Biometrika,* **55,** 335–342.

Pinkham, R. (1966). 'On a fiducial example of C. Stein'. *J.R. Statist. Soc.,* B, **28,** 53–54.

Pitman, E.J.G. (1936). 'Sufficient statistics and intrinsic accuracy'. *Proc. Camb. Phil. Soc.,* **32,** 567–579.

Pitman, E.J.G. (1937a). 'Significance tests which may be applied to samples from any populations', *J.R. Statist Soc.,* Suppl., **4,** 119–130.

Pitman, E.J.G. (1937b). 'Significance tests which may be applied to samples from any populations : II. The correlation coefficient test'. *J.R. Statist. Soc.,* Suppl., **4,** 225–232.

Pitman, E.J.G. (1937c). 'Significance tests which may be applied to samples from any populations : III. The analysis of variance test'. *Biometrika,* **29,** 322–335.

Pitman, E.J.G. (1938). 'The estimation of the location and scale parameters of a continuous population of any given form'. *Biometrika,* **30,** 391–421.

Pitman, E.J.G. (1939). 'A note on normal correlation'. *Biometrika,* **31,** 9–12.

Pratt, J.W. (1965). 'Bayesian interpretation of standard inference statements', (with discussion). *J.R. Statist. Soc.,* B, **27,** 169–203.

Proschan, F. and Hollander, M. (1972). 'Testing whether new is better than used'. *Ann. Math. Statist.,* **43,** 1136–1146.

Puri, M.L., ed. (1970). *Non-parametric techniques in statistical inference.* Cambridge University Press.

Puri, M.L. and Sen, P.K. (1971). *Nonparametric methods in multivariate analysis.* New York, Wiley.

Quenouille, M.H. (1949). 'Approximate tests of correlation in time-series'. *J.R. Statist. Soc.,* B, **11,** 68–84.

Quenouille, M.H. (1956). 'Notes on bias in estimation'. *Biometrika,* **43,** 353–360.

Raiffa, H. (1968). *Decision analysis.* Reading, Mass, Addison-Wesley.

Raiffa, H. and Schlaifer, R. (1961). *Applied statistical decision theory.* Boston, Harvard Business School.

Ramsey, F.P. (1931). 'Truth and probability'. In *The foundations of mathematics and other essays.* London, Kegan Paul.

Rao, C.R. (1945). 'Information and accuracy attainable in estimation of statistical parameters'. *Bull. Calcutta Math. Soc.*, **37**, 81–91.

Rao, C.R. (1961). 'Asymptotic efficiency and limiting information'. *Proc. 4th Berkeley Symp.*, **1**, 531–545.

Rao, C.R. (1962). 'Efficient estimates, and optimum inference procedures in large samples', (with discussion). *J.R. Statist. Soc.*, B, **24**, 46–72.

Rao, C.R. (1963). 'Criteria of estimation in large samples'. *Sankhyā*, A, **25**, 189–206.

Rao, C.R. (1973). *Linear statistical inference and its applications.* 2nd edition. New York, Wiley.

Rao, C.R. and Chakravarti, I.M. (1956). 'Some small sample tests of significance for a Poisson distribution'. *Biometrics*, **12**, 264–282.

Rao, C.R. and Mitra, S.K. (1971). *Generalized inverse of matrices and its applications.* New York, Wiley.

Rasch, G. (1960). *Probabilistic models for some intelligence and attainment tests.* Copenhagen, Nielson and Lydiche.

Robbins, H. (1951). 'Asymptotically subminimax solutions of compound statistical decision problems'. *Proc. 2nd Berkeley Symp.*, 131–148.

Robbins, H. (1956). 'An empirical Bayes approach to statistics'. *Proc. 3rd Berkeley Symp.*, **1**, 157–163.

Rosenblatt, M.R. (1971). 'Curve estimates'. *Ann. Math. Statist.*, **42**, 1815–1842.

Roy, S.N. (1953). 'On a heuristic method of test construction and its use in multivariate analysis'. *Ann. Math. Statist.*, **24**, 220–238.

Roy, S.N., Gnanadesikan, K. and Srivastava, J.N. (1971). *Analysis and design of certain quantitative multiresponse experiments.* Oxford, Pergamon.

Sarhan, A.E. and Greenberg, B.G., eds. (1962). *Contributions to order statistics.* New York, Wiley.

Savage, L.J. (1954). *The foundations of statistics.* New York, Wiley.

Savage, L.J. (1962). *The foundations of statistical inference.* London, Methuen.

Savage, L.J. (1970). 'Comments on a weakened principle of conditionality'. *J. Amer. Statist. Assoc.*, **65**, 399–401.

Scheffé, H. (1959). *The analysis of variance.* New York, Wiley.

Scheffé, H. (1970). 'Multiple testing versus multiple estimation. Improper confidence sets. Estimation of directions and ratios'. *Ann. Math. Statist.*, **41**, 1–29.

Seber, G.A.F. (1973). *The estimation of animal abundance and related parameters.* London, Griffin.

Shapiro, S.S., Wilk, M.B. and Chen, H.J. (1968). 'A comparative study of various tests for normality'. *J. Amer. Statist. Assoc.*, **63**, 1343–1372.

Shenton, L.R. and Bowman, K. (1963). 'Higher moments of a maximum-likelihood estimate'. *J.R. Statist. Soc.*, B, **25**, 305–317.

Sills, D.L., ed. (1968). *International encyclopedia of the social sciences*. New York, Macmillan and Free Press.

Silvey, S.D. (1959). 'The Lagrangian multiplier test'. *Ann. Math. Statist.*, **30**, 389–407.

Silvey, S.D. (1970). *Statistical inference*. Harmondsworth, Penguin.

Solari, M. (1969). 'The "maximum likelihood solution" of the problem of estimating a linear functional relationship'. *J.R. Statist. Soc.*, B, **31**, 372–375.

Stein, C. (1945). 'A two-sample test for a linear hypothesis whose power is independent of the variance'. *Ann. Math. Statist.*, **16**, 243–258.

Stein, C. (1956). 'Inadmissibility of the usual estimator for the mean of a multivariate normal distribution'. *Proc. 3rd Berkeley Symp.*, **1**, 197–206.

Stein, C. (1959). 'An example of wide discrepancy between fiducial and confidence intervals'. *Ann. Math. Statist.*, **30**, 877–880.

Stein, C. (1962). 'Confidence sets for the mean of a multivariate normal distribution', (with discussion). *J.R. Statist. Soc.*, B, **24**, 265–296.

Stein, C. (1964). 'Inadmissibility of the usual estimator for the variance of a normal distribution with unknown mean'. *Ann. Inst. Statist. Math.*, **16**, 155–160.

Stone, M. (1969). 'The role of significance testing : some data with a message'. *Biometrika*, 56, 485–493.

Stone, M. (1973). 'Discussion of paper by B. Efron and C. Morris'. *J.R. Statist. Soc.*, B, **35**,

Stone, M. and Springer, B.G.F. (1965). 'A paradox involving quasi prior distributions'. *Biometrika*, **52**, 623–627.

Stone, M. and von Randow, R. (1968). 'Statistically inspired conditions on the group structure of invariant experiments and their relationships with other conditions on locally compact topological groups'. *Zeit. Wahr. und Verw. Geb.*, **10**, 70–80.

Takeuchi, K. (1971). 'A uniformly asymptotically efficient estimator of a location parameter'. *J. Amer. Statist. Assoc.*, **66**, 292–301.

Tallis, G.M. (1969). 'Note on a calibration problem'. *Biometrika*, **56**, 505–508.

Thompson, W.A. (1969). *Applied probability*. New York, Holt, Rinehart and Winston.

Tukey, J.W. (1958). 'Bias and confidence in not quite large samples', (abstract). *Ann. Math. Statist.*, **29**, 614.

Tukey, J.W. (1960). 'A survey of sampling from contaminated distributions'. In *Contributions to probability and statistics*, pp. 448–485, eds. Olkin, I. *et al*. Stanford University Press.

Tukey, J.W. and Wilk, M.B. (1966). 'Data analysis and statistics: an expository overview'. AFIPS Conf. *Proc. Fall Joint Comp. Conf.*, **29**, 695–709.

Uthoff, V.A. (1970). 'An optimum test property of two well-known statistics'. *J. Amer. Statist. Assoc.*, **65**, 1597–1600.

von Mises, R. (1947). 'On the asymptotic distribution of differentiable statistical functions'. *Ann. Math. Statist.*, **18**, 309–348.
von Neumann, J. and Morgenstern, O. (1953). *Theory of games and economic behaviour*. 3rd edition. Princeton University Press.

Wald, A. (1941). 'Asymptotically most powerful tests of statistical hypotheses'. *Ann. Math. Statist.*, **12**, 1–19.
Wald, A. (1943). 'Tests of statistical hypotheses concerning several parameters when the number of observations is large'. *Trans. Amer. Math. Soc.*, **54**, 426–482.
Wald, A. (1947). *Sequential analysis*. New York, Wiley.
Wald, A. (1949). 'Note on the consistency of maximum likelihood estimate'. *Ann. Math. Statist.*, **20**, 595–601.
Wald, A. (1950). *Statistical decision functions*. New York, Wiley.
Walker, A.M. (1969). 'On the asymptotic behaviour of posterior distributions'. *J.R. Statist. Soc.*, B, **31**, 80–88.
Walker, A.M. (1970). 'Discussion of paper by A.C. Atkinson'. *J.R. Statist. Soc.*, B, **32**, 345–346.
Watson, G.S. (1958). 'On chi-squared goodness-of- fit tests for continuous distributions', (with discussion). *J.R. Statist. Soc.*, B, **20**, 44–72.
Welch, B.L. (1947a). 'The generalization of 'Student's' problem when several different population variances are involved'. *Biometrika*, **34**, 28–35.
Welch, B.L. (1947b). 'On the studentization of several variances'. *Ann. Math. Statist.*, **18**, 118–122.
Welch, B.L. (1951). 'On the comparison of several mean values: an alternative approach'. *Biometrika*, **38**, 330–336.
Welch, B.L. (1956). 'Note on some criticisms made by Sir Ronald Fisher'. *J.R. Statist. Soc.*, B, **18**, 297–302.
Wetherill, G.B. (1961). 'Bayesian sequential analysis'. *Biometrika*, **48**, 281–292.
White, L.V. (1973). 'An extension of the general equivalence theorem to non-linear models'. *Biometrika*, **60**, 345–348.
Whittle, P. (1964). 'Some general results in sequential analysis'. *Biometrika*, **51**, 123–141.
Whittle, P. (1965). 'Some general results in sequential design', (with discussion). *J.R. Statist. Soc.*, B, **27**, 371–394.
Wijsman, R.A. (1958). 'Incomplete sufficient statistics and similar regions'. *Ann. Math. Statist.*, **29**, 1028–1045.

Wilks, S.S. (1938). 'The large-sample distribution of the likelihood ratio for testing composite hypotheses'. *Ann. Math. Statist.*, **9**, 60–62.

Williams, E.J. (1962). 'Exact fiducial limits in nonlinear estimation'. *J.R. Statist. Soc.*, B, **24**, 125–139.

Williams, E.J. (1969a). 'Regression methods in calibration problems', (with discussion). *Bull. Inst. Internat. Statist.*, **43**, 1, 17–28.

Williams, E.J. (1969b). 'A note on regression methods in calibration'. *Technometrics*, **11**, 189–192.

Wolfowitz, J. (1947). 'The efficiency of sequential estimates and Wald's equation for sequential processes'. *Ann. Math. Statist.*, **18**, 215–230.

Yates, F. (1952). 'Principles governing the amount of experimentation in developmental work'. *Nature*, **170**, 138–140.

Yates, F. (1968). 'Theory and practice in statistics'. *J.R. Statist. Soc.*, A, **131**, 463–475.

Zacks, S. (1971). *The theory of statistical inference*. New York, Wiley.

AUTHOR INDEX

SUBJECT INDEX

Absolute test of significance 83
Acceptance sampling 366, 407, 413, 426, 436
Accuracy, intrinsic, *see* Intrinsic accuracy
Adaptive estimate, *see* Robust estimate
Adequacy of fit, *see* Goodness of fit test
Admissibility, definition of 431
Alternative hypothesis, *see* Significance test
Analysis of covariance 168
Analysis of variance
   multivariate 59, 173, 177, 178, 328, 361
   one-way
      invariant test in 168
      likelihood for 17
      maximum likelihood theory for 290, 330
      rank test for 191
   random effects model in 59, 117, 177, 403, 407
   randomization model in 197, 203, 205
   two-way 58, 205
   see also Linear model
Ancillary statistic
   Bayesian inference, in 369, 444
   choice between alternative 43, 110, 129
   definition of 32, 35
   information with 109
   invariance of 44, 440
   non-uniqueness of 33-, 44, 110, 129
   use of 33, 115, 167, 220, 221, 247, 302, 444
Approximation to probability distribution 67, 183, 185-, 462-
Association, measure of 393
   see also Dependence
Asymptotic expansion
   Behrens-Fisher problem, in 153-
   Edgeworth and Fisher-Cornish 464, 465
Asymptotic normality, *see* Maximum likelihood estimate; Maximum likelihood ratio test; Order statistics; Posterior distribution; Rank test, linear; Robust estimate; U statistic

Asymptotic relative efficiency *see* Efficiency
Autoregressive process
   estimates for 274, 301
   likelihood for 58, 144
   m.l.e. for 301
   prediction of 60
   sufficiency in 60
   tests for 144-

Bayes's theorem 53, 365-
Bayesian decision theory
   admissibility of 432
   Bayes risk in 429, 438
   Bayes rule in 418, 432, 438
   coherency principle in 46, 47, 424
   decision function in 415
   decision tree in 452, 454
   dynamic programming in 456, 458
   formulation of 415-
   hypothesis testing in 55, 413, 414, 419-, 441, 459
   improper prior, with 432, 444
   inadmissibility in 431-, 445-
   interval estimation in 442, 459
   least favourable prior in 440
   minimax rule and 434-, 439, 443, 445, 460
   point estimation in 413, 420, 442-, 452, 455, 459
   randomized decision function in 415, 418, 438-
   regret in 419, 429, 434
   risk in 429, 435-, 461
   sequential 451-, 461
   significance test in 413, 414, 419, 421, 441, 459
   strong likelihood principle satisfied 41, 418
   utility 55, 412, 417, 422-
   see also Bayesian inference; Empirical Bayes methods; Non-Bayesian decision theory

499